Thin film growth

AF739407

Related titles:

In-situ characterisation of thin film growth
(ISBN 978-1-84569-934-5)
Recent advances in techniques to characterise thin films *in-situ* during deposition could lead to an improved understanding of deposition processes and to better, faster, diagnosis of issues with the deposition process. *In-situ* characterisation of thin film growth will provide a comprehensive review of this increasingly important topic, focusing on the techniques and concepts. Part I reviews electron diffraction techniques, including the methodology for taking observations and measurements. Part II covers photoemission techniques; the principles and instrumentation. Part III contains photon techniques for real-time characterisation of the nucleation, growth, structural and electronic properties of thin films. Part IV discusses alternative *in-situ* characterisation techniques and the trend for combining different techniques.

Electromigration in thin films and electronic devices
(ISBN 978-1-84569-937-6)
Electromigration is a significant problem affecting the reliability of microelectronic devices such as integrated circuits. Recent research has focused on how electromigration affects the increasing use by the microelectronics industry of lead-free solders and copper interconnects. Part I reviews ways of modelling and testing electromigration. Part II discusses electromigration in copper interconnects, whilst Part III covers solder and wirebonds.

Advanced piezoelectric materials
(ISBN 978-1-84569-534-7)
Piezoelectric materials produce electric charges on their surfaces as a consequence of applying mechanical stress. They are used in the fabrication of a growing range of devices such as transducers, actuators, pressure sensor devices and increasingly as a way of producing energy. This book provides a comprehensive review of advanced piezoelectric materials, their properties, methods of manufacture and applications. It covers lead zirconate titanate (PZT) piezo-ceramics, relaxor ferroelectric ceramics, lead-free piezo-ceramics, quartz-based piezoelectric materials, the use of lithium niobate and lithium in piezoelectrics, single crystal piezoelectric materials, electroactive polymers (EAP) and piezoelectric composite materials.

Details of these and other Woodhead Publishing materials books can be obtained by:

- visiting our web site at www.woodheadpublishing.com
- contacting Customer Services (e-mail: sales@woodheadpublishing.com; fax: +44 (0) 1223 832819; tel.: +44 (0) 1223 499140 ext. 130; address: Woodhead Publishing Limited, 80 High Street, Sawston, Cambridge CB22 3HJ, UK)

If you would like to receive information on forthcoming titles, please send your address details to: Francis Dodds (address, tel. and fax as above; e-mail: francis.dodds@woodheadpublishing.com). Please confirm which subject areas you are interested in.

Thin film growth

Physics, materials science and applications

Edited by
Zexian Cao

WOODHEAD PUBLISHING

Oxford Cambridge Philadelphia New Delhi

Published by Woodhead Publishing Limited,
80 High Street, Sawston, Cambridge CB22 3HJ, UK
www.woodheadpublishing.com

Woodhead Publishing, 1518 Walnut Street, Suite 1100, Philadelphia, PA 19102-3406, USA

Woodhead Publishing India Private Limited, G-2, Vardaan House, 7/28 Ansari Road,
Daryaganj, New Delhi – 110002, India
www.woodheadpublishingindia.com

First published 2011, Woodhead Publishing Limited

British Library Cataloguing in Publication Data
A catalogue record for this book is available from the British Library.

Library of Congress Control Number: 2011932611

ISBN 978-0-08-101722-7 (print)
ISBN 978-0-85709-329-5 (online)

The publisher's policy is to use permanent paper from mills that operate a sustainable forestry policy, and which has been manufactured from pulp which is processed using acid-free and elemental chlorine-free practices. Furthermore, the publisher ensures that the text paper and cover board used have met acceptable environmental accreditation standards.

Typeset by Replika Press Pvt Ltd, India
Printed by Lightning Source

Contents

Contributor contact details

(* = main contact)

Editor and Chapter 8

Z. Cao
Institute of Physics
Chinese Academy of Sciences
P.O. Box 603
100190 Beijing
P. R. China

E-mail: zxcao@aphy.iphy.ac.cn

Chapter 1

Y. Homma
Department of Physics
Tokyo University of Science
1-3 Kagurazaka, Shinjuku
Tokyo 162-8601
Japan

E-mail: homma@rs.kagu.tus.ac.jp

Chapter 2

T. Miller*
c/o Synchrotron Radiation Center
3731 Schneider Drive
Stoughton, WI 53589-USA

E-mail: tamille1@wisc.edu

T.-C. Chiang
Loomis Laboratory of Physics
University of Illinois at Urbana-Champaign
1110 West Green Street
Urbana, IL 61801-3080
USA

E-mail: tcchiang@illinois.edu

Chapter 3

Professor A. Voigt
Institute for Scientific Computing
Technische Universität Dresden
01062 Dresden
Germany

E-mail: axel.voigt@tu-dresden.de

Chapter 4

Y. Kajikawa
Graduate School of Engineering
The University of Tokyo
2-11-16 Yayoi Bunkyo-ku
Tokyo 113-8656
Japan

E-mail: kaji@ipr-ctr.t.u-tokyo.ac.jp

Chapter 5

S. Kowarik*
Institut für Physik
Humboldt-Universität zu Berlin
Newtonstr. 15
D-12489 Berlin
Germany

E-mail: stefan.kowarik@physik.hu-berlin.de

A. Hinderhofer, A. Gerlach and F. Schreiber
Institut für Angewandte Physik
Universität Tübingen
Auf der Morgenstelle 10
D-72076 Tübingen
Germany

Chapter 6

D. Ye*
Department of Physics
Virginia Commonwealth University
Richmond, VA 23284
USA

E-mail: dye2@vcu.edu

T.-M. Lu
Department of Physics
Applied Physics and Astronomy
Rensselaer Polytechnic Institute
110 8th Street
Troy, NY 12180
USA

Chapter 7

F. Ramiro-Manzano, E. Bonet, I. Rodríguez and F. Meseguer*
Centro de Tecnologías Físicas
Unidad Asociada ICMM-CSIC/UPV
Universidad Politécnica de Valencia
Avda. Naranjos. Acceso J
Edificio 8B, entrada K, 1 piso
46022 – Valencia
Spain

E-mail: fmese@fis.upv.es

Chapter 9

B. H. Hong*
Department of Chemistry
Sungkyunkwan University
Suwon 440-746
S. Korea

E-mail: byunghee@skku.edu

and

SKKU Advanced Institute of Nanotechnology (SAINT) and Center for Human Interface Nanotechnology (HINT)
Sungkyunkwan University
Suwon 440-746
S. Korea

H. R. Jeon
Department of Chemistry
Sungkyunkwan University
Suwon 440-746
S. Korea

Chapter 10

Johann Coraux*
Institut Néel
CNRS-UJF
25 rue des Martyrs
38042 Grenoble
CEDEX 9
France

E-mail: johann.coraux@grenoble.cnrs.fr

A. T. N'Diaye, C. Busse and T. Michely
II. Physikalisches Institut
Universität zu Köln
Zülpicher Strasse 77
Köln
Germany

Chapter 11

N. Nilius
Fritz-Haber-Institut der MPG
D-14195 Berlin
Germany

E-mail: nilius@fhi-berlin.mpg.de

Chapter 12

X. Q. Wang
State Key Laboratory of Artificial Microstructure and Mesoscopic Physics
School of Physics
Peking University
Yiheyuan Road 5
Beijing 100871
P. R. China

A. Yoshikawa*
Graduate School of Electrical and Electronic Engineering
Chiba University
1-33 Yayoi-cho
Inage-ku
Chiba 263-8522
Japan

E-mail: yoshi@cute.te.chiba-u.ac.jp

Chapter 13

F. Foucher*
Centre de Biophysique Moléculaire
UPR CNRS 4301
rue Charles Sadron
45071 Orléans
CEDEX 2
France

E-mail: frederic.foucher@cnrs-orleans.fr

C. Coupeau, J. Colin, A. Cimetière and J. Grilhé
Institut P'
UPR CNRS 3346
Laboratoire de Physique des Matériaux
Université de Poitiers – UFR Sciences
SP2MI, Boulevard Marie et Pierre Curie
BP 30179
86 962 Futuroscope
Chasseneuil CEDEX
France

E-mail: christophe.coupeau@univ-poitiers.fr
jerome.colin@univ-poitiers.fr
alain.cimetiere@univ-poitiers.fr
jean.grilhe@univ-poitiers.fr

Chapter 14

J. Song*
Department of Mechanical and Aerospace Engineering
University of Miami
Coral Gables, FL 33146
USA

E-mail: jsong8@miami.edu

J. Wu and Y. Huang
Department of Mechanical and Department of Civil and Environmental Engineering Northwestern University
Evanston, IL 60208
USA

Chapter 15

S.-G. Lu and Q. M. Zhang*
Materials Research Institute and Department of Electric Engineering
The Pennsylvania State University
University Park, PA 16802
USA

E-mail: qxz1@psu.edu

Z. Kutnjak
Jozef Stefan Institute
1000 Ljubljana
Slovenia

Chapter 16

H. Guclu*
Department of Biostatistics
University of Pittsburgh
Pittsburgh, PA 15261
USA

E-mail: guclu@pitt.edu

T. Karabacak
Department of Applied Science
University of Arkansas at Little Rock
Little Rock, AR 72204
USA

M. Yuksel
Department of Computer Science & Engineering
University of Nevada – Reno
Reno, NV 89557
USA

Preface

It is well known to researchers who work on thin films that there are currently more than 200 books published that deal with various aspects of thin film science and technology, but this fact does not exclude the necessity for new volumes in this field. None of the existing books has tried to be exhaustive and, more importantly, the field itself keeps marching forward at an elevated pace and in more diversified directions. The latter point is quite understandable since thin films provide the most efficient, flexible and thoughtful uses of solid materials; they offer a uniquely versatile material base for the development of novel technologies. In fact, industries based on the utilization of thin-film devices constitute the strongest driving force for our economy. The attempt to meet the requirements of making more advanced thin-film devices in turn has led to the innovation or invention of deposition techniques and of tools for the characterization of growth processes, and along with this technical progress the understanding of the fundamental physics behind the seemingly endless spectrum of phenomena concerning thin-film growth is also becoming increasingly clear. The publication of a new book on thin-film growth is justified by the fast development of this enterprise alone.

With the current book we intended to summarize the most recent advancements in the field of thin-film growth, focusing on the introduction of new growth and characterization methods, the application-oriented developments and, more importantly, the understanding of the physics relevant to the growth of some particular films, which are at the forefront of thin-film science today. The chapters included in the book aim to present highlights on those topics that have attracted the special attention of the authors and the editor. We hope they can be useful to an interdisciplinary and varied audience including science and engineering graduate students working on thin films, researchers and working professionals who want to have an overview of the field in the recent years or have a strong interest in the new growth and characterization techniques.

The book contains 16 chapters grouped into two parts. Part I comprises five contributions introducing the theory of thin-film growth, including measuring nucleation and growth processes in thin films (Homma), quantum electronic stability of atomically uniform films (Miller and Chiang), phase-field modelling of thin film growth (Voigt), analysis of surface roughness evolution (Kajikawa), and modelling of deposition processes based on real-time observation (Kowarik *et al.*). The 11 chapters in Part II deal with the advancement of techniques for thin-film growth and relevant device fabrication procedures, including silicon nanostructured films grown on templated surfaces by oblique angle deposition (Ye and Lu), study of phase transitions in colloidal crystal thin films (Ramiro-Manzano *et al.*), growth for thermally unstable noble-metal nitrides by reactive magnetron sputtering (Cao), growth of graphene layers for thin films (Hong and Jeon), epitaxial growth of graphene on single-crystal metal surfaces (Coraux *et al.*), determination of electronic properties and adsorption behaviour of polar films (Nilius), polarity control and growth on polar substrates (Wang and Yoshikawa), understanding of buckling (Foucher *et al.*) and controlled buckling of thin films on compliant substrates for stretchable electronics (Song *et al.*), electrocaloric effect in ferroelectric polymer films (Lu *et al.*), and network behaviour in thin films and nanostructure growth dynamics (Guclu *et al.*). From the very beginning we cherished the hope that these chapters can serve as a guide to the advancement of thin-film science and that graduate students and specialists in both academic and industrial institutions may find the book a valuable reference. As editor I gratefully acknowledge the contributing authors of these chapters, who are all renowned experts in the corresponding subfield of thin-film science. I thank them all for having spent considerable time and effort in compiling these high-level contributions.

Lastly, I am profoundly grateful to the editorial staff of Woodhead Publishing, including Dr Cliff Elwell (commissioning editor), Ms Laura Pugh (commissioning editor) and Miss Nell Holden (project editor), for their sustained support during the two-year preparation of this volume.

Zexian Cao
Beijing

Part I

Theory of thin film growth

1
Measuring nucleation and growth processes in thin films

Y. HOMMA, Tokyo University of Science, Japan

Abstract: This chapter focuses on the observation of crystal growth processes by tracking the motion of atomic steps, which are the growth front of the crystal. After reviewing the theoretical model for atomic step motion, the chapter discusses atomic step imaging by scanning electron microscopy, and reveals basic growth modes using this technique: two-dimensional-island nucleation and step-flow modes both in growth and evaporation, and morphological instability of atomic steps between these two modes.

Key words: atomic steps, scanning electron microscopy, molecular beam epitaxy, step flow, island nucleation.

1.1 Introduction

Recent progress in surface observation techniques has enabled crystal surfaces to be observed at the atomic scale. In particular, with scanning tunnelling microscopy (STM) (Binnig and Rohrer, 1983), the motion of individual atoms on the surface can be tracked (Swartzentruber, 1996). However, at high temperatures where crystal growth takes place, atoms diffuse on the surface so fast that any existing methods cannot catch up with their movements. In vapour phase growth, those diffusing atoms, termed adatoms, finally evaporate or meet an atomic step where they are incorporated into the crystal. By the incorporation of adatoms, the motion of atomic steps takes place. The motion of atomic steps is much slower than that of adatoms. Furthermore, atomic resolution is not necessary for the observation of atomic steps which have a one-dimensional structure on the surface. Therefore, we are able to observe the crystal growth processes by tracking the motion of atomic steps which constitute the growth front of the crystal. For this purpose, electron microscopy techniques with a high sensitivity to atomic steps are useful: these include reflection electron microscopy (REM) (Osakabe *et al.*, 1980) and low energy electron microscopy (LEEM) (Bauer, 1994). These two techniques utilize electron diffraction from the surface, and thus are sensitive to the surface morphology at atomic steps.

In this chapter, however, we use a different electron microscopy, scanning electron microscopy (SEM), to observe the motion of atomic steps. As the growth method, we focus on the molecular beam epitaxy (MBE), where the crystal surface in a vacuum is exposed to a molecular flux of evaporated source material. Before explaining the principle of atomic step imaging by SEM, we review the theoretical description of atomic step motion. The theoretical model was established in 1951 by Burton, Cabrera and Frank (BCF model) (Burton *et al.*, 1951). Interestingly, this is far earlier than the establishment of MBE and surface characterization techniques.

1.2 Basic theory of epitaxial growth

We consider step-flow growth on a vicinal surface with parallel atomic steps. We take the x-axis perpendicular to the atomic steps as shown in Fig. 1.1(a). Atoms supplied from a molecular beam are adsorbed and migrate freely on the surface (adatoms). The time evolution of adatom density, $\rho(x,t)$, is described by the following diffusion equation:

$$\frac{\partial \rho(x,t)}{\partial t} = D_s \frac{\partial^2}{\partial x^2} \rho(x,t) - \frac{1}{\tau_v} \rho(x,t) + F \quad [1.1]$$

where D_s is the diffusion coefficient of adatoms, F is the flux of impinging atoms, $1/\tau_v$ is the evaporation probability of an adatom. The first term on the right-hand side in Eq. 1.1 represents the diffusion of adatoms, the second term represents the loss by evaporation, and the third term is the increment due to deposition. F is expressed by using the partial pressure, p, of the molecule composed of η atoms usable for growth:

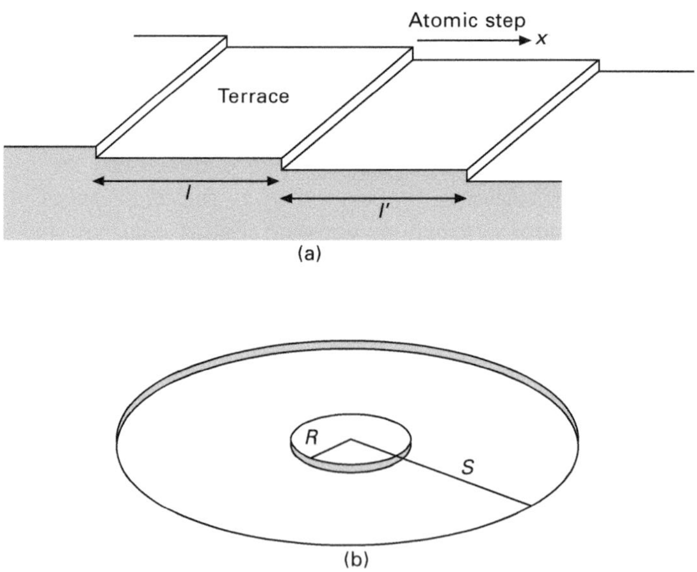

1.1 Models of crystal surfaces with (a) parallel atomic steps and (b) circular atomic steps.

$$F = \frac{\eta p}{\sqrt{2\pi m k_B T}}$$

where k_B is the Boltzmann constant, m is the mass of the molecule, and T is the absolute temperature.

To solve Eq. 1.1, we assume that the rate of attachment and detachment of adatoms at the steps is much faster than that of diffusion. That is, the adatom density at the step becomes an equilibrium value ρ_0. When the surface diffusion is much faster than the motion of steps, $\dot{\rho} = 0$. Then, for an isolated step at $x = 0$, the boundary conditions are $\rho(0) = \rho_0$, $\rho(\infty) = F\tau_v$. These give the following solution

$$\rho(x) = \rho_0 + (F\tau_s - \rho_0)\left[1 - \exp\left(-\frac{|x|}{\lambda_s}\right)\right]$$

where $\lambda_s^2 = D_s\tau_v$, and λ_s is the surface diffusion length of adatoms. The step velocity v is determined by the adatom flux at the step

$$j(0) = D_s \frac{d\rho(0)}{dx}$$

For an isolated step, assuming the fluxes from upper and lower terraces are equal (symmetric),

$$v_\infty = \frac{2D_s}{n_0}\left|\frac{d\rho}{dx}\right|_{x=0} = 2(F\tau_v - \rho_0)\frac{D_s}{n_0\lambda_s} = (F\tau_v - \rho_0)\frac{2\lambda_s}{n_0\tau_s} \quad [1.2]$$

where $n_0 = a^{-2}$ is the density of lattice (lattice constant a) on the surface. The step velocity is proportional to $2\lambda_s$. This means that only atoms impinging within λ_s from the step contribute to the step motion, and other atoms evaporate before reaching the step. This growth mode occurs when the diffusion length is much smaller than the step spacing. For the opposite case, when the diffusion length is much larger than the step spacing:

$$v = \frac{v_\infty}{2\lambda_s}\frac{l + l'}{2} \quad [1.3]$$

where l and l' are the widths of upper and lower terraces, respectively.

For circular steps, we need to solve the diffusion equation in the cylindrical coordinates. The general solution is expressed as

$$\rho(r) = F\tau_v + A(\rho_0 - F\tau_v)I_0(r/\lambda_s) + B(\rho_0 - F\tau_v)K_0(r/\lambda_s)$$

where K_0 and I_0 are the modified Bessel functions of order zero, and A and B are the proportional constants. For the circular step between an inner terrace of radius R on a larger terrace of radius S as shown in Fig. 1.1(b),

$$A \equiv A_{R,S} = \frac{K^S - K^R}{I^R K^S - I^S K^R}, \; B \equiv B_{R,S} = \frac{I^R - I^S}{I^R K^S - I^S K^R}$$

where $K^X = K_0 (X/\lambda_s)$, $I^X = I_0(X/\lambda_s)$. The step velocity is obtained as the sum of the adatom fluxes from inner and outer terraces:

$$v = \frac{D_s}{n_0 \lambda_s}(F\tau_v - \rho_0)\left[\frac{(K^R - K^s)I_1^R + (I^R - I^s)K_1^R}{I^R K^S - I^S K^R} + \frac{I_1^R}{I^R}\right] \quad [1.4]$$

where $K_1^X = K_1(X/\lambda_s)$, and $I_1^X = I_1(X/\lambda_s)$ are the modified Bessel functions of order 1 (Finnie and Homma, 2000b).

1.3 Observation method of atomic steps

For the observation of atomic steps on growing surfaces, we employed *in-situ* SEM. Atomic steps can be observed with a conventional SEM instrument, but the contrast is so faint that it is easily hidden by the contamination of the surface due to electron irradiation during SEM imaging. We used an ultrahigh vacuum SEM instrument equipped with Knudsen cells for molecular beam epitaxy (Homma *et al.*, 1994). The secondary electron detector was set to the side of the specimen in parallel to the axis of the specimen stage tilting. The primary electron beam of 25 keV was incident at a grazing incidence, 5–30° to the surface, to enhance the sensitivity to the atomic scale surface structures. In this article, the primary electron beam was incident from the bottom direction of each image. The image foreshortening due to oblique incidence is corrected in most images. The atomic steps appear bright when the primary electron beam goes down the atomic step staircase, while they appear dark when the primary electron beam goes up the staircase (Homma *et al.*, 1991). Another factor influencing the atomic step contrast is the location of the secondary electron detector (Homma *et al.*, 1993b). When the primary electron beam goes parallel to the atomic steps, the atomic steps appear bright when the steps are facing the detector. Conversely, they appear dark when they face away from the detector. Those are topographic contrasts of atomic steps, which are similar to macroscopic scale step contrasts, and can be used for imaging of steps as small as the monatomic layer of the crystal. This observation method is used in Section 1.4.

An entirely different type of contrast can be used for atomic step imaging. This is the surface phase contrast utilizing surface phase transition. In the following, we explain the 7 × 7–1 × 1 contrast on Si(111) surfaces.

A clean Si(111) surface in ultrahigh vacuum takes a long range ordered structure, the 7 × 7 reconstruction, at below the transition temperature (~860°C) (Florio and Robertson, 1970). The 7 × 7 reconstruction starts to occur at the atomic step edge during cooling from a high temperature. In SEM images, a 7×7 domain appears brighter than a 1 × 1 domain without reconstruction (Homma *et al.*, 1993a). Therefore, if 7 × 7 domains form continuously along an atomic step, the step is easily observed by SEM even with a low magnification. One such method is to keep the sample 1–2°C below the transition temperature, and observe thin 7 × 7 domains along atomic steps. Another method is to rapidly quench the sample from above the transition temperature towards room temperature, thus forming continuous 7 × 7 domains along atomic steps.

The image in Fig. 1.2 was observed using the former method. Using a Si(111) wafer with a small miscut angle, 0.01°, the sample was kept 1°C below the transition temperature. The 1 × 1–7 × 7 phase transition starts from the upper edge of an atomic step, and only a thin region of the upper terrace becomes 7×7 and appears bright. Since the width of continuous 7×7 regions is negligible to the large step spacing, 1.5 μm, the atomic steps are highlighted in the image. This method can be used for the observation of atomic step behaviour not only at around the phase transition temperature but also at higher temperatures by reducing the temperature just for observation purposes.

An example of the quenching method is shown in Fig. 1.3 (Homma and Finnie, 2002). The Si sample with a flat terrace as large as a 100 μm square was rapidly quenched from 1230°C (see Section 1.5). Concentric atomic steps with a spacing of ~20 μm can be seen. In between the concentric steps there exist smaller circular steps. Those are monolayer holes on the annular terraces. A magnified image of the small circular step is shown in Fig. 1.3(b). The lower half of the atomic step appears bright, while the

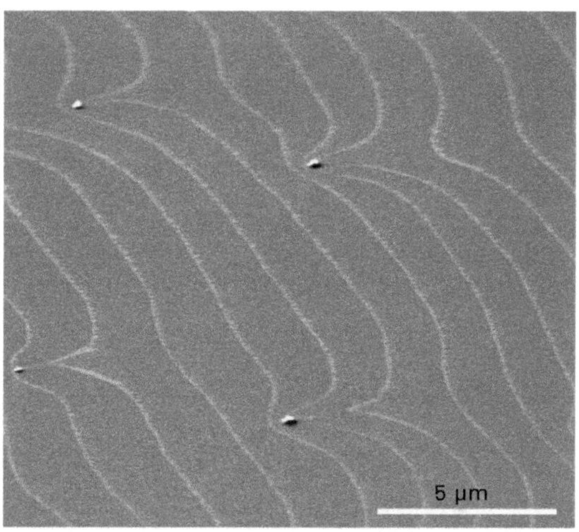

1.2 SEM image of atomic steps on Si(111) surface observed at the 1 × 1–7 × 7 phase transition temperature. Atomic steps are decorated with continuous 7 × 7 regions which appear bright in the SEM image.

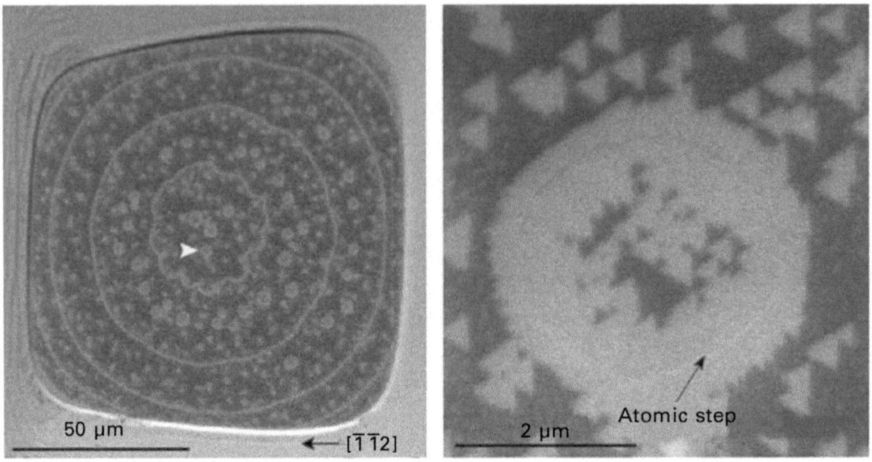

(a) (b)

1.3 SEM image of atomic steps on Si(111) surface observed by the quenching method (Homma and Finnie, 2002). Bright regions are 7 × 7 reconstructed domains. (a) Low magnification image. (b) High magnification image observed at the arrow head shown in (a).

upper half appears dark. This contrast change is due to the difference in the primary electron incidence relative to the atomic step edge, as explained above. The topographic effect alone produces the atomic step image in SEM, but the contrast is not high enough to be recognized in a low magnification image.

Both the upper and lower terraces near an atomic step turn to the 7 × 7 phase after quenching. The width of the 7 × 7 phase on the upper terrace (outside of the hole) is 0.3–0.4 μm, and that on the lower terrace (inside of the hole) is roughly twice as large. Therefore, the atomic step is observed as a ~1 μm wide line. Meanwhile, many triangular 7 × 7 domains can be seen on the terrace in the image shown in Fig. 1.3(b). Since these triangular domains distribute discretely, only atomic steps continuously decorated by 7 × 7 domains are highlighted in the low magnification image shown in Fig. 1.3(a).

In Sections 1.5 and 1.6, we discuss the atomic step behaviour on Si(111) surfaces at high temperatures based on observations using the quenching method. Although it is not a real-time observation method, the effect of quenching on the step motion is negligible because phenomena occurring on a large scale are observed on a huge terrace. It is interesting that macroscopic observations by SEM in the 10–100 μm range can reveal phenomena which relate to the attachment and detachment of atoms to/from atomic steps.

1.4 Two-dimensional-island nucleation and step-flow growth modes

In MBE growth, two basic growth modes exist depending on the ratio of the adatom diffusion length and the terrace width: the step-flow growth mode where growth proceeds through atomic step progression, and two-dimensional (2D) island nucleation growth mode where growth proceeds through island coverage increase. Here, we show those two growth modes on GaAs surfaces observed by SEM.

On a GaAs(001) surface, a monolayer step consists of Ga and As double layers with a height of 0.28 nm. In Fig. 1.4, the process of one monolayer evolution of a GaAs(001) surface is shown using the characteristic of scanning imaging (Homma *et al.*, 1995): acquisition of one image takes a certain period while growth proceeds during the period. In the present case, the image acquisition time and the monolayer growth period are 70 s and 50–55 s, respectively. Therefore, ~1.3 monolayers grow during one frame imaging, and the process can be recorded in the image. Note that the growth stage differs from the top to the bottom of the image. Growth started at the beginning of imaging (at the top of the image). Initially, atomic steps are visible. Then, small spots appear at about a quarter from the top. These are 2D islands nucleating on the surface. Then, the islands grow rapidly

1.4 SEM image of GaAs (001) surface during MBE growth (Homma *et al.*, 1995). Growth started at the top of the image and one monolayer growth is completed at about three-quarters of the image length from the top.

in a lateral direction, causing coalescence of the islands. At three-quarters from the top, holes are seen as the result of island coalescence, and finally one monolayer is completed. Atomic steps can be recognized at around the completion of one monolayer. The second layer islands appear in the bottom part of the image. In this way, an SEM image can record the growth of one monolayer in real time for the 2D island nucleation growth mode. In the initial period of one monolayer growth, islands are not observed. This is not due to the resolution of SEM, but reflects the surface reconstruction process of GaAs(001) surface. In the 2×4 structure, which is observed before MBE growth, As atoms corresponding to quarter monolayer growth are lacking, thus Ga atom rearrangement is necessary before island nucleation can occur (Osaka *et al.*, 1995).

The reason for the use of a slow growth rate is to make the islands large. The island size also relates to the surface diffusion length of adatoms. Thus, the Ga-stabilized GaAs(111) surface (so-called (111)A surface), where the surface diffusion length is much larger than that on the (001) surface, was employed. The island growth process on this surface is shown in Fig. 1.5 (Yamaguchi and Homma, 1998). In this case, a higher image acquisition time was used and the same area of the surface was repeatedly observed during growth. Owing to the large island size, nucleation, growth and coalescence processes are clearly seen in Fig. 1.5. The islands are triangular, reflecting the three-fold symmetry of the (111) surface. Furthermore, step-flow growth occurs without nucleation of island near the step edge. The step velocity is higher parallel to the original steps than that in the perpendicular direction. Precise analyses of the step velocity under different Ga supplying rate showed

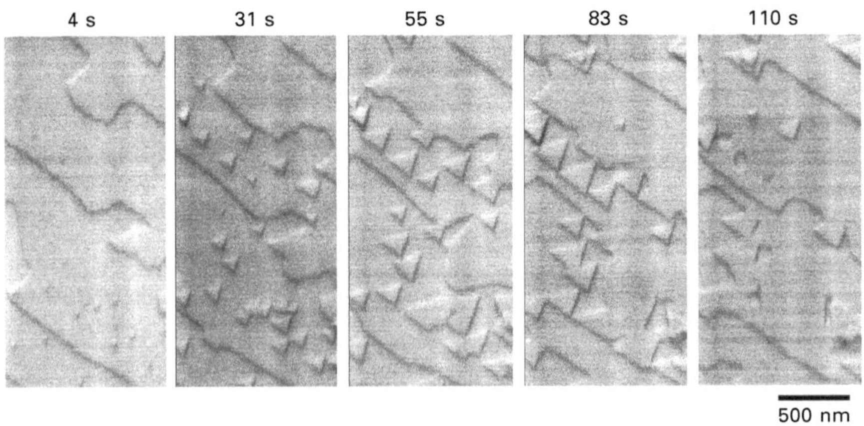

1.5 SEM image sequences showing the monolayer growth process of GaAs(111)A surface (Yamaguchi and Homma, 1998). Each image corresponding to 8 s growth starting from the time shown at the top.

that the step velocity was well described by Eq. 1.2 for an isolated step. This is because the step spacing is much larger than the Ga diffusion length on the GaAs(111)A surface. Fitting the measured step velocity dependence on the Ga flux gave the Ga diffusion length of ~100 nm. Note that the step spacing is more than 1 μm on the surface.

1.5 The motion of atomic steps on a growing and evaporating Si(111) surface

Here, we consider the atomic step motion during evaporation at high temperatures on the surface with an atomic step array with regular intervals as shown in Fig. 1.6. A mathematical expression for this is given in the Appendix. On the high temperature surface, adatoms are released from an atomic step and migrate freely onto the terrace. An atomic step acts as an emitter and a sink of adatoms simultaneously, and thus the adatom concentration at the atomic step reaches an equilibrium value. In the case of evaporation, the adatom concentration at the middle of a terrace decreases due to desorption. Adatoms are supplied from the atomic step for compensation of the decrease, causing retraction motion of the atomic step. This is the step-flow evaporation. Adatoms can also be created directly on the middle of a terrace, though the probability is small. In this case, an adatom and an advacancy are created simultaneously. An advacancy can be annihilated by recombination with an adatom. Advacancies diffuse much more slowly than adatoms, because an advacancy consists of surrounding surface atoms. When the terrace size is large enough, the advacancy concentration at the centre of the terrace increases to the extent that advacancies form 2D holes at the centre of the terrace. This is the counterpart of the 2D island nucleation in growth. Usually, only step-flow evaporation is observed on Si(111) surfaces, because the diffusion length for adatoms is much larger than the atomic step spacing on normal substrates.

In the meantime, we consider a crater (hole) on a crystalline surface as illustrated in Fig. 1.7. In the crater, there exists a slope whose inclination

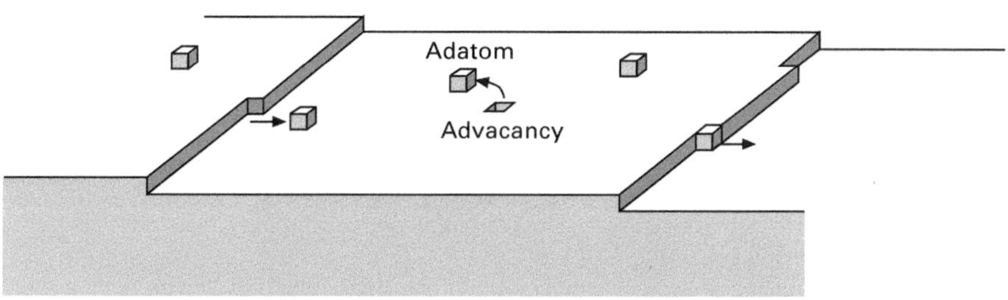

1.6 Schematic illustration of adatoms and advacancies on the high temperature surface with atomic step allay.

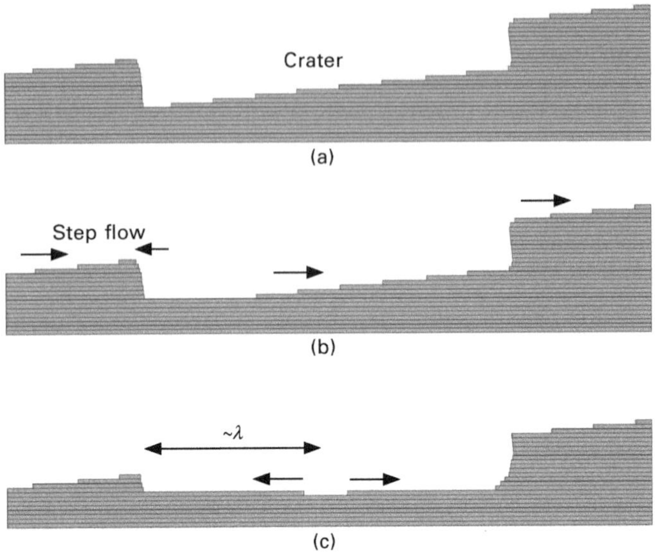

1.7 Schematic illustration of step-flow evaporation of the surface with a crater. (a) Initial surface. (b) Step-flow directions in the crater during evaporation. (c) Macro-vacancy formation after the bottom terrace size exceeds the adatom diffusion length.

direction is opposite to the average inclination of the surface, i.e., the step-flow direction is opposite to the average surface. Therefore, the very bottom terrace of the crater expands during step-flow evaporation (Homma *et al.*, 1997). When the bottom terrace becomes large enough, a new terrace appears due to 2D-macro-vacancy nucleation in the centre of the terrace. The lower terrace expands in the same way as the initial terrace, and the macro-vacancy nucleation is repeated. As a result, a set of concentric circular steps is created at the bottom of the crater as shown in Fig. 1.8. The spacing of the circular steps is on the order of adatom diffusion length (see Section 1.9.2), and thus varies with the temperature. Figure 1.9 shows the temperature dependence of the atomic step spacing observed using a huge crater created on a Si(111) surface by oxygen-ion bombardment in a secondary-ion mass spectrometer (Homma *et al.*, 1998).

The step spacing decreases with increasing temperature up to 1200°C. This relates to the decrease in the surface diffusion length caused by the increase in the desorption probability of adatoms: the surface diffusion coefficient D_s is given by

$$D_s = a^2 \upsilon_0 \exp(-W_{sd}/k_B T)$$

where a is the lattice constant, υ_0 is the frequency factor ($\sim 10^{13}$), W_{sd} is the energy barrier of surface diffusion, and T is the absolute temperature. The

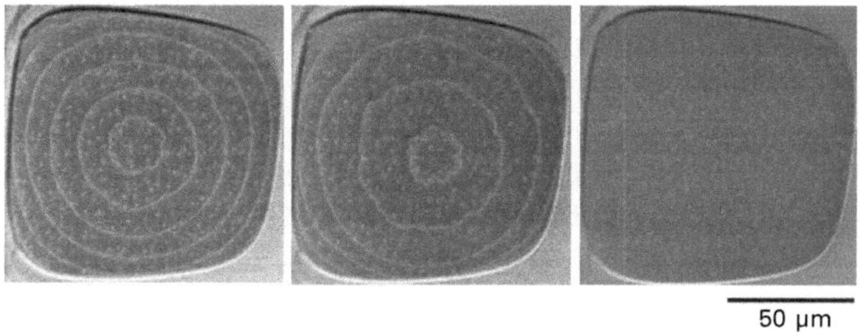

1.8 SEM images of circular atomic steps on ultra-large terrace at the bottom of a crater on Si(111) surface. The annealing temperature before quenching is (a) 1180°C, (b) 1120°C and (c) 1020°C. No macro-vacancy formation occurred in (c).

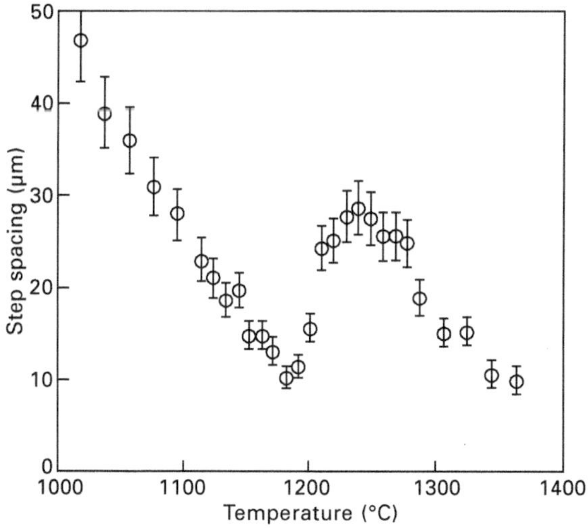

1.9 Temperature dependence of atomic step spacing observed on ultra-large terraces formed at the bottom of a crater on Si(111) surface (Homma *et al.*, 1998).

desorption probability of adatoms is expressed using the desorption barrier W_v as

$$1/\tau_v = \upsilon_0 \exp(-W_v/k_B T)$$

Using Einstein's formula $\lambda_s^2 = D_s \tau_v$, we get

$$\lambda_s = a \exp[(W_v - W_{sd})/2k_B T]$$

The slope in the range of 1000–1200°C in Fig. 1.9 gives $W_v - W_{sd} \approx 2.4$ eV.

The step spacing suddenly increases by a factor of 2.6 at around 1200°C, and decreases again for higher temperatures. This sudden increase suggests occurrence of some transition on the surface. In fact, incomplete surface melting, which is disordering of the first layer in the double layers of the Si(111) surface, has been observed by medium-energy ion scattering spectroscopy at this temperature (Hibino *et al.*, 1998). The surface structure transition may affect the surface diffusion of adatoms. Actually, the image shown in Fig. 1.3 is the surface formed by quenching from 1230°C. Small circles are seen between concentric atomic steps. These are monolayer holes located at the middle of the annular terrace between the adjacent concentric steps. When the sample passed through the transition temperature during cooling, the adatom diffusion length suddenly decreased. This caused nucleation of advacancy islands (monolayer holes) at the centre of the terrace as explained above.

By varying the annealing temperature, the spacing of the concentric atomic steps can be changed in the range 10–50 μm. If the annealing temperature is set so that the step spacing is less than half of the bottom terrace size, no nucleation of new islands occurs. Thus, an ultra-large terrace without atomic steps inside can be obtained as shown in Fig. 1.8(c) (Finnie and Homma, 2000a).

By using the ultra-large terrace created at the bottom of a crater, we can create a monolayer island or hole by growth or evaporation. With the combination of island or hole, and growth or evaporation, we can observe four cases as shown in Fig. 1.10: island shrinkage and hole expansion during evaporation; island expansion and hole shrinkage during growth. As an example, analysis of evolution of circular terraces during growth and evaporation at 940°C is shown in Fig. 1.11 (Finnie and Homma, 2000b; Homma and Finnie, 2002). The time evolution of the circular island and hole can be described by that of their radii. Experimentally obtained step velocities were well fitted to the

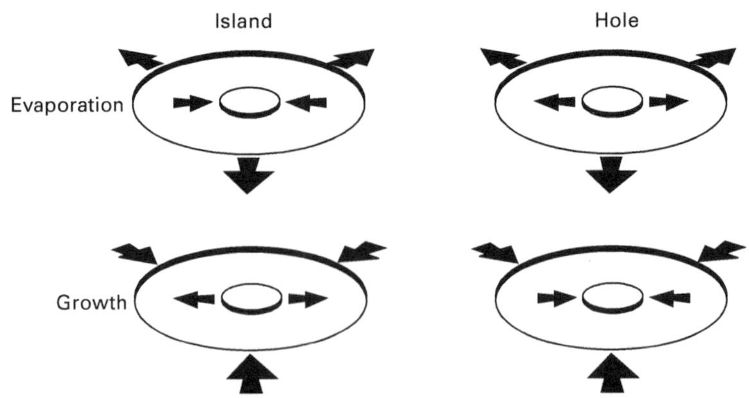

1.10 Schematic illustration of step-flow growth and evaporation of monolayer island and hole on a circular terrace.

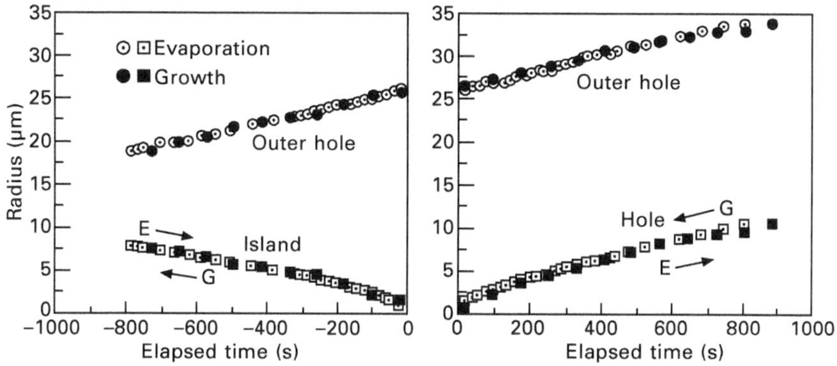

1.11 The radii of circular terraces as a function of annealing or growth time observed at 940°C on the ultra large Si(111) terrace (Homma and Finnie, 2002). The data for evaporation are overlaid with those for growth by expanding the timescale of growth by a factor of two. The time axis of growth is reversed to facilitate comparison. Arrows indicate the direction of radius change during evaporation (E) and growth (G).

solution of the BCF equation on the cylindrical coordinate (Eq. 1.4). The surface diffusion length of adatoms at 880°C estimated from the fitting was ~50 μm, which is in good agreement with the step spacing in Fig. 1.9.

As seen in Fig. 1.11, the step progression and retraction are symmetric. Also, the behaviours of hole and island are symmetric. These indicate that the contributions of upper and lower terraces are symmetric when adatoms are incorporated into or released from atomic steps. In general, adatoms from the upper terrace need to cross a higher energy barrier to be incorporated into atomic steps than those from the lower terrace. This is called the Ehrlich–Schwoebel barrier (Ehrlich and Hudda, 1966, Schwoebel and Shipsey, 1966). The present results indicate that the Ehrlich–Schwoebel barrier is not significant for the Si(111) surface at high temperatures.

1.6 Morphological instability of atomic steps

The Ehrlich–Schwoebel barrier is asymmetry in incorporating adatoms into a step between those from the upper and lower terraces. Such asymmetry can cause instability in step flow. Bales and Zangwill (1990) treated theoretically the wandering of steps in step-flow growth and showed that when the flux of adatoms from the lower terrace exceeds that from the upper terrace, the fluctuation in the atomic step is amplified, resulting in macro-scale step wandering. This unstable growth regime is located between stable step flow and 2D island nucleation in the phase diagram of growth mode as functions of impinging flux and step spacing.

As discussed above, the Ehrlich–Schwoebel barrier on the Si(111) surface at high temperatures is small and this kind of step wandering is not observed. Here, we show step instability in a different way. We utilized the fact that the adatom flux to atomic step is an increasing function of the terrace width, and distorted the symmetry between the upper and lower terraces by changing the ratio of their sizes (Finnie and Homma, 2000c). This is possible only by using an ultra-large terrace such as shown in Fig. 1.8(c).

Figure 1.12 shows the time evolution of atomic step shape on a 100 μm wide terrace (Homma *et al.*, 2001). The observation was carried out using a low Si molecular flux for nucleating only an island at the centre of the terrace at 880°C. Each frame was observed by the quenching method after growth duration starting from the flat surface. This was to avoid any extra nucleation due to repeated quenching. A small circular island is seen at the centre of the terrace in image (a). This changes to the six-fold symmetry island extending in <110> directions as seen in images (b) and (c). In this stage, the peripheral atomic step progressing from the edge of the terrace is smooth. After further growth, the periphery step becomes wandering as seen in image (d). The wandering amplitude maximizes at 0.5 monolayers, and then decreases for further growth as in image (e). The decay of wandering is due to the stabilization of step flow when the upper terrace becomes large.

While the shape of the centre island reflected the symmetry of the Si(111) surface, the periphery step was initially smooth and became wandering afterwards. This is exactly the result of step fluctuation enhancement by the large lower terrace.

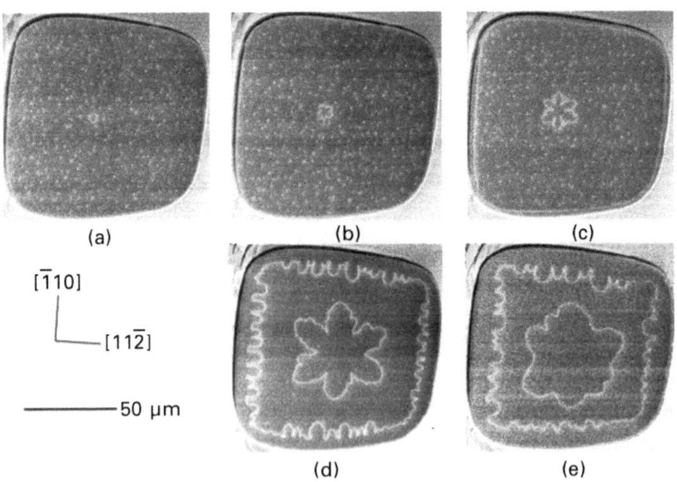

1.12 SEM image sequences showing the evolution of atomic step instability during growth at 880°C on the ultra large Si(111) terrace (Homma *et al.*, 2001). The images were obtained after growth of (a) 10 s, (b) 20 s, (c) 1 min, (d) 5 min and (e) 10 min.

1.7 Conclusion and future trends

This chapter has shown that crystal growth of Si and GaAs in molecular beam epitaxy is well described by the BCF model. That is, the surface diffusion of adatoms is the dominant process of crystal growth and the incorporation of adatoms into atomic steps is fast enough. On the evaporating surface, the behaviour of advacancies is also well described by the BCF model. The key to observe the atomic step motion of growing and evaporating Si surface at high temperatures is the use of an ultra-large terrace created at the bottom of a crater. Since the surface diffusion length of adatoms on a Si(111) surface at high temperature becomes as large as 10–50 μm, a terrace width comparable to the surface diffusion length is a prerequisite. SEM imaging with the quenching method is useful for such a large-scale observation. In this method, atomic steps were decorated with the 7×7 phase domains. In order to obtain an enhanced 7×7 domain contrast, an oblique incidence of a 25 keV primary electron beam was used. Recent progress in electron microscopy has made it easier to observe atomic steps. A low energy SEM instrument with a primary electron beam of 0.1–1 keV is commercially available these days. Surface sensitivity is largely enhanced with the low energy beam, especially when it is used in an ultrahigh vacuum. Low energy electron microscopy (LEEM), which uses projection optics of reflected electrons for imaging, has realized real-time imaging of atomic steps even at elevated temperatures. LEEM is now widely used for observation of growing surfaces of various materials (Bauer, 1994; Meyer zu Heringdorf, 2008).

1.8 References

Bales G S and Zangwill A (1990) 'Morphological instability of a terrace edge during step-flow growth', *Phys. Rev. B* 41, 5500–5508.
Bauer E (1994) 'Low energy electron microscopy', *Rep. Prog. Phys.* 57, 895–938.
Binnig G and Rohrer H (1983) '7×7 reconstruction on Si(111) resolved in real space', *Phys. Rev. Lett.* 50, 120–123.
Burton W K, Cabrera N and Frank F C (1951) 'The growth of crystals and the equilibrium structure of their surface', *Phil. Trans. Roy. Soc.* 243, 299–358.
Ehrlich G and Hudda F G (1966) 'Atomic view of surface self-diffusion: tungsten on tungsten', *J. Chem. Phys.* 44, 1039–1049.
Finnie P and Homma Y (2000a) 'Motion of atomic steps on ultraflat Si(111): constructive collisions', *J. Vac. Sci. Technol.* A18, 1941–1945.
Finnie P and Homma Y (2000b) 'Nucleation and step flow on ultraflat silicon', *Phys. Rev. B* 62, 8313–8317.
Finnie P and Homma Y (2000c) 'Stability–instability transitions in silicon crystal growth', *Phys. Rev. Lett.* 85, 3237–3240.
Florio J V and Robertson W D (1970) 'Phase transformations of the Si(111) surface', *Surf. Sci.* 22, 459–464.
Hibino H, Sumitomo K, Fukuda T, Homma Y and Ogino T (1998) 'Disordering of Si(111) at high temperatures', *Phys. Rev. B* 58, 12587–12589.

Homma Y and Finnie P (2002) 'Step dynamics on growing silicon surfaces observed by ultrahigh vacuum scanning electron microscopy', *J. Cryst. Growth* 237–239, 28–34.
Homma Y, Tomita M and Hayashi T (1991) 'Secondary electron imaging of monolayer steps on a clean Si(111) surface', *Surf. Sci.* 258, 147–152.
Homma Y, Suzuki M and Tomita M (1993a) 'Atomic configuration dependent secondary electron emission from reconstructed silicon surfaces', *Appl. Phys. Lett.* 62, 3276–3278.
Homma Y, Tomita M and Hayashi T (1993b) 'Atomic step imaging on silicon surfaces by scanning electron microscopy', *Ultramicroscopy* 52, 187–192.
Homma Y, Osaka J and Inoue N (1994) '*In-situ* observation of monolayer steps during molecular beam epitaxy of gallium arsenide by scanning electron microscopy', *Jpn. J. Appl. Phys.* 33, L563–L566.
Homma Y, Osaka J and Inoue N (1995) '*In situ* observation of surface morphology evolution corresponding to reflection high energy electron diffraction', *Jpn. J. Appl. Phys.* 34, L1187–L1190.
Homma Y, Hibino H, Ogino T and Aizawa N (1997) 'Sublimation of Si(111) surface in ultrahigh vacuum', *Phys. Rev. B* 55, R10237–R10240.
Homma Y, Hibino H, Ogino T and Aizawa N (1998) 'Sublimation of a heavily boron-doped Si(111) surface', *Phys. Rev. B* 58, 13146–13150.
Homma Y, Finnie P and Uwaha M (2001) 'Morphological instability of atomic steps observed on Si(111) surfaces', *Surf. Sci.* 492, 125–136.
Meyer zu Heringdorf F-J (2008) 'The application of low energy electron microscopy and photoemission electron microscopy to organic thin films', *J. Phys.: Condens. Matter* 20, 184007-1–184007-12.
Osaka J, Inoue N and Homma Y (1995) 'Delayed and continuous nucleation of islands in GaAs molecular beam epitaxy revealed by *in situ* scanning electron microscopy', *Appl. Phys. Lett.* 66, 2110–2112.
Osakabe N, Tanishiro Y, Yagi K and Honjo G (1980) 'Reflection electron microscopy of clean and gold deposited (111) silicon surfaces', *Surf. Sci.* 97, 393–408.
Pimpinelli A and Villain J (1994) 'What does an evaporating surface look like?', *Physica A* 204, 521–524.
Schwoebel R L and Shipsey E J (1966) 'Step motion on crystal surfaces', *J. Appl. Phys.* 37, 3682–3686.
Swartzentruber B S (1996) 'Direct measurement of surface diffusion using atom-tracking scanning tunneling microscopy', *Phys. Rev. Lett.* 76, 459–462.
Yamaguchi H and Homma Y (1998) 'Imaging of layer by layer growth processes during molecular beam epitaxy of GaAs on (111)A substrates by scanning electron microscopy', *Appl. Phys. Lett.* 73, 3079–3081.

1.9 Appendix

1.9.1 Extension of the Burton, Cabrera & Frank (BCF) model for evaporating surface

This appendix discusses the behaviour of adatoms and advacancies on an evaporating surface following the work by Pimpinelli and Villain (1994).

At high temperatures not far from the melting point, advacancies as well as adatoms exist on the surface. We need to take advacancies into account when we discuss the problems relating to evaporation on the surface. The concentration of advacancy $\sigma(x)$ on the terrace has a minimum at the centre of the terrace opposite to the concentration distribution of adatoms. When the concentration of advacancy exceeds a certain value, a macro-vacancy, one monolayer deep hole, is created.

To treat both advacancies and adatoms, we need to consider their pair creation and pair annihilation. The annihilation rate of an adatom-advacancy pair is proportional to the product of their concentrations, $\rho\sigma$. The pair creation rate at equilibrium is proportional to their equilibrium concentration, $\rho_0\sigma_0$. These values satisfy the coupled equations

$$\dot{\rho} = D\rho'' - \rho/\tau_v + \Delta\rho_0\sigma_0 - \Delta\rho\sigma + F$$

$$\dot{\sigma} = \Lambda\sigma'' + \Delta\rho_0\sigma_0 - \Delta\rho\sigma$$

where D and Λ are the diffusion coefficients of adatoms and advacancies, respectively, and Δ is the proportional constant. To linearize the equations, we introduce $\delta\rho = \rho - \rho_0$ and $\delta\sigma = \sigma - \sigma_0$, and suppose these are small (near equilibrium). Also, we neglect the external flux F. Then, we get

$$\delta\dot{\rho} \approx D\delta\rho'' - \delta\rho/\tau_\upsilon - \rho_0/\tau_\upsilon - \Delta\rho_0\delta\sigma - \Delta\,\delta\rho\sigma_0$$

$$\delta\dot{\sigma} \approx \Lambda\delta\sigma'' - \Delta\rho_0\delta\sigma - \Delta\,\delta\rho\sigma_0$$

Since the step motion is much slower than the diffusion of adatoms and advacancies, $\delta\dot{\rho} = 0$ and $\delta\dot{\sigma} = 0$. Under the boundary conditions, $\rho(\pm l/2) = \rho_0$, $\sigma(\pm l/2) = \sigma_0$, we get

$$\delta\rho(x) = \rho_0\left(\frac{\cosh(\kappa x)}{\cosh(\kappa l/2)} - 1\right) \qquad [A1.1]$$

$$\delta\sigma(x) = \sigma_0\left(1 - \frac{\cosh(\kappa x)}{\cosh(\kappa l/2)}\right)$$

where $\kappa^2 = \frac{1}{\tau_v}\frac{\rho_0}{\Lambda\sigma_0 + D\rho_0}$

In general, since adatoms diffuse much faster than advacancies, $D\rho_0 >> \Lambda\sigma_0$, then, $1/\kappa^2 \approx D\tau_v = \lambda^2$.

The step velocity for equidistant parallel steps is given by

$$v = \frac{2D}{n_0}\frac{d\rho(l/2)}{dx} - \frac{2\Lambda}{n_0}\frac{d\sigma(l/2)}{dx} = \frac{(D\rho_0 + \Lambda\sigma_0)}{n_0}\kappa\tanh(\kappa l/2)$$

1.9.2 Macro-vacancy formation

At high temperatures, the density of advacancies increases, resulting in the formation of macro-scale vacancies of monolayer deep. The edge of the macro-vacancy is an additional step. The formation of macro-vacancies is described by nucleation theory. The free energy of a macro-vacancy with radius R is expressed as

$$F(R) = 2\pi R\gamma - \pi R^2\,\delta\mu$$

where γ is the step stiffness, and $\delta\mu$ is the chemical potential. The first term is the energy increase due to step formation and the second term is the chemical potential decrease due to formation of advacancy cluster. Since the density of macro-vacancy is proportional to the Boltzmann factor, $\exp[-F(R)/k_BT]$, the clusters larger than the critical value R, which gives maximum of $F(R)$, expand, while those smaller than the critical value shrink. The maximum of $F(R)$ occurs at $R_c = \gamma/\delta\mu$,

$$F(R_c) = \pi\gamma^2/\delta\mu$$

The condition for the formation of at least one macro-vacancy on the terrace within the area of $l \times l$ is

$$l^2 \exp[-F(R_c)/k_BT] = l^2 \exp[-\pi\gamma^2/\delta\mu k_BT] \geq 1 \qquad \text{[A1.2]}$$

Since $\rho \sim \rho_0$,

$$\rho(x) = \rho_0 \exp(-\delta\mu/k_BT) \qquad \text{[A1.3]}$$

On the other hand, from Eq. A1.1,

$$\rho(x) = \rho_0 \frac{\cosh(\kappa x)}{\cosh(\kappa l/2)} \qquad \text{[A1.4]}$$

By comparing these two equations, A1.3 and A1.4, we get

$$\delta\mu(x)/k_BT = -\ln\frac{\cosh(\kappa x)}{\cosh(\kappa l/2)} \cong 1 - \frac{\cosh(\kappa x)}{\cosh(\kappa l/2)}$$

$$\text{At } x = 0,\ \delta\mu(0)/k_BT \approx \kappa^2 l^2/8 = \frac{\rho_0/\tau_v}{(D\rho_0 + \Lambda\sigma_0)} l^2/8$$

Then, inserting this result into Eq. A1.2, we get

$$l^2 \exp\left[\frac{-8\pi\gamma^2}{l^2\kappa^2(k_BT)^2}\right] \geq 1$$

or

$$\frac{8\pi\gamma^2}{l^2\kappa^2(k_BT)^2} \leq 2\ln l$$

Using typical step spacing $l \sim 10^3 - 10^4 a$,

$$\gamma / k_B T \leq l\kappa.$$

Since γ is the order of $k_B T$, we obtain the condition for macro-vacancy nucleation,

$$l \geq 1/\kappa.$$

Using $1/\kappa^2 \approx D\tau_v = \lambda^2$,

$$l \geq \lambda.$$

That is, a macro-vacancy is formed when the terrace size is comparable or larger than the adatom diffusion length.

2
Quantum electronic stability of atomically uniform films

T. MILLER and T.-C. CHIANG, University of Illinois at Urbana-Champaign, USA

Abstract: The valence electronic states in thin films are quantized by the films' boundaries. This represents a substantial modification of the band structure of the system which influences all physical properties. Dramatic variations in thermal stability are observed in thin metallic films on an atomic layer-by-layer basis. This chapter describes measurements of morphology, stability, and electronic structure of thin metallic films using angle-resolved photoemission spectroscopy, and discusses theoretical models that can predict stability based on the quantization of states. Further measurements using X-ray diffraction extend these results to structures more complex than uniform films.

Key words: quantum electronic stability, metallic quantum wells, thermal stability, atomically uniform films, angle-resolved photoemission.

2.1 Introduction

The valence electronic states in a solid-state system of nanoscale dimensions are modified relative to their bulk counterparts by the presence of the system's boundaries. These modifications, referred to generally as 'quantum size effects' (QSEs), influence all physical properties of the system. Of particular interest here is the thermal stability of a thin film. The first section of this chapter deals with the related idea of 'electronic growth' – the concept that the growth mode of a film can be governed to a large degree by the quantum effects on the valence bands. Here we are considering metallic films in which the electronic states are most amenable to a relatively simple treatment and competing effects are minimized. Next, angle-resolved photoemission (ARPES) will be described as it is applied to thin film studies. This is an important tool for studies of these phenomena because, as applied to atomically uniform films, it can reveal the electronic structure of the system while at the same time serving to precisely measure a film's thickness to an exact number of monolayers. The growth and characterization of atomically uniform films by ARPES follows. Such films may be of value for technological applications because their unusual uniformity supports quantum states with a high degree of coherence, and certainly they are important for scientific studies of quantum-size effects because of their relative simplicity and ease of characterization.

Next, measurements of quantum stability of some thin film systems will be presented. This will lead into a discussion of the general principles of thin film stability and their application to nanostructure development. Mostly this is based on a simple picture of electronic quantization akin to the 'particle-in-a-box' problem of elementary quantum mechanics. However, a section is devoted to systems where the basic quantization conditions are modified by, for example, specific substrate physical and electronic structures. Finally, we speculate on what directions future research might be expected to take.

2.2 Electronic growth

Valence electrons are responsible for the bonding of the atoms or molecules that make up a crystalline solid. In the case of a covalently bonded solid, it is intuitive that the nature of the valence states would exert a dominant role in the growth of a thin film just as they do in the formation of a molecule, via directional bonds and the filling of chemical orbitals, starting right from bonding with the substrate. For a metallic overlayer, however, of all the myriad factors influencing film growth, the valence electronic states of the overlayer would perhaps seem to be at most a minor factor. Considering the delocalized states of the metallic bond gives rise to a picture of film growth where kinetics, such as the film/substrate lattice match and efficiency of packing, and bulk thermodynamic properties, such as surface tension and melting point, are the driving forces shaping film morphology. Electronic effects due to the quantization of states by the boundaries of a smooth film have been discussed for some time, but their impact on film properties, if any, would be mitigated by a variety of effects, including loss of coherence due to lattice mismatch and defects, and in any event would eventually become irrelevant for films thicker than a few monolayers (Feibelman 1983, Feibelman and Hamann 1984). Experiment has shown, however, that this is not the case for well-ordered thin films of nanoscale thickness, where electronic effects can have a dramatic impact on thin film properties including surface reactivity (Danese, Curti *et al.* 2004, Zhang, Zhang *et al.* 2008), work function (Paggel, Wei *et al.* 2002), superconductivity (Guo, Zhang *et al.* 2004), and surface energy and thermal stability (Czoschke, Hong *et al.* 2005). Coherent electronic behavior has also been observed in films thicker than 100 monolayers and even across a mismatched substrate/film boundary (Paggel, Miller *et al.* 1999b, Speer, Tang *et al.* 2006). It is so that quantum-electronic influences on metallic film growth would be expected to diminish with increasing thickness, but they are of direct importance for practical applications as thicknesses of interest continue to shrink, and indirectly for all deposited films inasmuch as they start off as nanofilms where quantum effects should be considered in the establishment of the initial growth modes.

Of course, these quantum-mechanical effects are in competition with kinetic constraints and, depending on one's viewpoint, there are also different aspects of electronic structure, such as charge spillage, Friedel oscillations from either the substrate or the vacuum boundary, energy associated with confinement, etc., all of which may be in competition with each other in determining the morphology of a growing film (Zhang, Niu *et al.* 1998). Also, the quantum effects being considered here arise from the overall geometry of the system, and the growing film may never have the opportunity to take a suitable form for them to develop if kinetic constraints prevent it from doing so from the initial stages of growth. The term 'electronic growth' has been applied to deposition and annealing methods that permit quantum effects to be expressed.

Early experimental work on electronic growth showed that atomically flat films of Ag could be obtained on GaAs substrates by first depositing at low temperature and subsequently annealing to room temperature (RT) (Smith, Chao *et al.* 1996). By starting at low temperature, kinetic effects could be suppressed until enough material was deposited so that quantum electronic effects could stabilize the formation of a flat film upon annealing (Miyazaki and Hirayama 2008). More studies of low-temperature deposition using reflection high-energy electron diffraction with spot profile analysis (RHEED-SPA) on Pb films on Si(111) substrates revealed a preference of island step heights of seven monolayers (Budde, Abram *et al.* 2000). RHEED-SPA data contain information from a large area of the sample in which there may be many small islands of different dimensions; later work by the same group using scanning tunneling microscopy (STM) confirmed these results and found also a preference for bilayer steps, and the authors suggested a connection to quantum electronic effects (Hupalo, Kremmer *et al.* 2001). Generally, these Pb films consisted of islands with a distribution of sizes sitting on top of a continuous wetting layer. With lower growth temperatures, atomically uniform films are possible in this material system (Upton, Miller *et al.* 2004). For STM studies, a distribution of island sizes may be a desirable feature, as the spectrum of preferred heights can be directly measured on a single sample prepared at one time. However, in islands, lateral confinement occurs and the full three-dimensional structure should be considered for theoretical treatment.

For fundamental studies of quantum electronic effects, a uniform thin film gives a simple interpretation in terms of one-dimensional quantum-well states, leading to a more intuitive picture of the physics. Angle-resolved photoemission is well suited for studies of uniform films, and at the present time a wealth of information is available from many photoemission studies on quantum-well states in a variety of thin film systems (Chiang 2000, Aballe, Rogero *et al.* 2001, 2002, Bian, Miller *et al.* 2009, Evans, Alonso *et al.* 1993, Liu, Paggel *et al.* 2008, Patthey and Schneider 1994). Even for these systems of relatively simple geometry, interactions between the film and substrate can give rise to quantized electronic structures of substantial complexity (Liu, Speer *et al.* 2008).

2.3 Angle-resolved photoemission spectroscopy

ARPES is an experimental technique well suited for the measurement of the electronic structure of solids. A monochromatic photon beam of sufficient energy to overcome the sample's work function is used to eject electrons from the sample. The light source can be a laboratory one such as a resonance lamp or a laser, or synchrotron radiation using a vacuum monochromator. Synchrotron radiation is most versatile as the photon energy can be changed over a continuous range. The electrons are collected by an analyser which measures their angle of emission relative to the sample surface and their kinetic energies. Since the photon is annihilated in the photoexcitation event, energy conservation gives

$$E_{kin} = h\nu - E_b - W$$

where E_{kin} is the measured kinetic energy, $h\nu$ is the impinging photon energy, E_b is the binding energy of the electron inside the solid relative to the Fermi level (E_F) before excitation, and W is the work function. This makes it possible to relate the measured energy spectra to the spectrum of initial states inside the crystal (Hüfner 1995).

The energy states form a set of energy bands, each characterized by a dispersion relation $E_i(\boldsymbol{k})$, where E is the energy, i is a band index, and $\boldsymbol{k}$ is a wave vector referred to (somewhat loosely) as the 'crystal momentum'. Crystal momentum is analogous to momentum in free space, but takes into account the fact that electrons in a crystal are diffracted by the fixed lattice and so are in states of mixed momentum. It is customary to label states by the 'reduced crystal momentum' $\boldsymbol{k}$ which is the lowest value that could be obtained by a hypothetical individual measurement of the electron's momentum, other values being possible from the same state by additions from the discrete set of reciprocal lattice vectors. The set of all such $\boldsymbol{k}$'s fills the first Brillouin zone of k-space (Ashcroft and Mermin 1976). An example of crystal electronic structure is shown in Fig. 2.1, which shows calculated bands of Ag along various directions within the first Brillouin zone (Moruzzi, Janak *et al.* 1978, Papaconstantopoulos 1986, Smith and Mattheiss 1974).

Knowledge of the emission angle and energy equates to knowing the electron momentum after it has been emitted from the sample. Passing through the surface, the translational symmetry of the lattice is preserved in the surface plane, but not perpendicular to it, and likewise with the crystal momentum. So, photoemission can characterize electronic states in a crystal almost completely – the energy and the two in-plane components

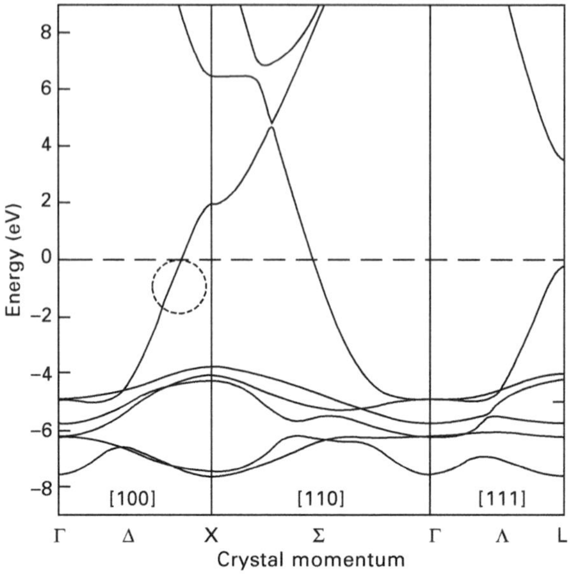

2.1 Energy band structure of silver along three high-symmetry directions. The dashed circle marks the area of the *sp* bands that contribute to the quantum-well states in Fig. 2.2.

of momentum – but the third component k_z along the surface normal is not directly determined. Applying the technique to three-dimensional systems then requires some input assumptions or special methods. Nonetheless, it has been used with great utility for measuring the band structures of many materials. ARPES gives the complete $E(\boldsymbol{k})$ relation for two-dimensional systems such as surface states, and can be applied readily also to systems that are quasi-two-dimensional, such as supported graphene (Ohta, Bostwick *et al.* 2006, 2007), certain crystal planes in high-T_c materials (Kondo, Khasanov *et al.* 2007), or layered materials such as charge-density wave compounds (Wilson and Yoffe 1969, Kidd, Miller *et al.* 2002), etc. The 'k_z problem' can also be bypassed in the application of ARPES to thin film systems that support quantum-well states (Loly and Pendry 1983).

An example of ARPES data from thin films is shown in Fig. 2.2 (Luh, Miller *et al.* 2002). The system is Ag on Fe(100), and the films were grown to promote smooth layer-by-layer growth by depositing at low temperature and subsequently annealing (Paggel, Miller *et al.* 2000). The Ag films are oriented so the surface is (100). Each curve is a photoemission energy distribution curve (EDC), created by scanning the electron analyser energy setting while keeping the emission angle and the exciting photon energy constant. The energy scale has been shifted so that zero is at the Fermi level. The different curves are from films of different thicknesses in monolayers (ML) as shown,

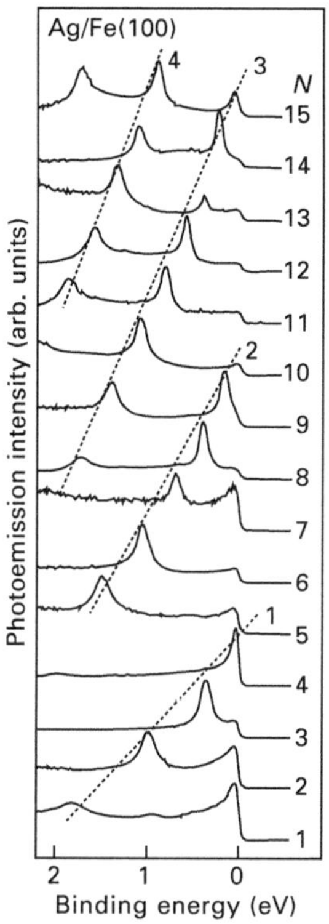

2.2 Normal emission photoemission spectra from different thicknesses of Ag on Fe(100). The thickness for each spectrum is shown as the number of monolayers *N*. The energy scale is referenced to the Fermi level. Peaks corresponding to four different quantum numbers are marked with dashed lines.

and are displaced vertically for clarity. The emission direction was normal to the sample surface. Each spectrum shows a set of discrete peaks which crowd together and become more numerous as the film thickness is increased. The spectra can be understood as follows: because the spectra were taken in normal emission, the parallel components of momenta are zero and the relevant part of the band structure is along the [100] direction. In the bulk, we see a continuous band of states in the energy range represented by these spectra (see Fig. 2.1). That band is derived from the *sp* states of Ag; it is highly dispersive and can be regarded as nearly-free-electron-like. However, in a thin film, the surface and substrate interface of the Ag film behave as reflecting barriers to electron motion, and as in the simple 'particle-in-a-box' problem of elementary quantum mechanics, the states which are continuous in the bulk become quantized. The allowed states are given by the Bohr–Sommerfeld condition:

$$2n\pi = 2k(E)d + \phi(E) \qquad [2.1]$$

where $k(E)$ is the electron wavevector component perpendicular to the surface, d is the thickness of the film, $\phi(E)$ is the total phase shift of the wavefunction on reflection from both the substrate interface and the surface, and n, an integer, is the quantum number. The term $k(E)$ is the inverse of the dispersion relation such as those represented in Fig. 2.1. With reference to the band structure, roughly (ignoring the phase shift), the allowed states can be obtained by dividing up the interval in k space between Γ and X by the number of layers in the film and reading off the corresponding energies. As in the spectra, the allowed states become more numerous and approach a continuum as the film thickness increases.

2.4 Atomically uniform films

The curves in Fig. 2.2 are labeled by integer thicknesses. Generally, films deposited using a quartz microbalance to gauge film thickness will not be controlled to such precision. However, using the two-step deposition and annealing process, these films are of atomically uniform thickness, which makes absolute thickness measurement by photoemission possible. This is illustrated by the sequence of deposition and measurement represented in Fig. 2.3 (Paggel, Miller *et al.* 1998). Data from three sequential depositions are shown, going up in the figure. The bottom and top curves are from atomically-uniform films 38 and 39 ML thick, respectively. They both show a set of individual peaks roughly evenly spaced in energy. The middle curve is from a film of intermediate thickness and, in contrast, shows pairs of peaks. By inspection, each peak in a pair corresponds to a peak in either the top or bottom spectrum. We can conclude that the middle spectrum is composed of emission from distinct areas of the film that differ in thickness by 1 ML, whereas the other two show evidence of only one thickness each.

This ability to grow atomically uniform films is extremely important for fundamental studies. Generally, a film will have a distribution of thicknesses, and photoemission spectra will be composed of many, possibly unresolved, peaks. This makes assignment of energies uncertain. Further, the peaks may appear to smoothly evolve with deposition, making an absolute determination of thickness impossible. In contrast, for the growth represented in Fig. 2.3, one knows that exactly one monolayer has been added in going from the bottom to the top spectrum, so that the microbalance can be precisely calibrated and atomically uniform films of any desired thickness can be grown. The individual sharp peaks afforded by uniformity allow photoemission measurements of quantum behavior even from rather thick films. Figure 2.4 shows spectra from films up to 119 ML (Paggel, Miller *et al.* 1999b, 2000). The presence of quantum-well peaks to this thickness demonstrates

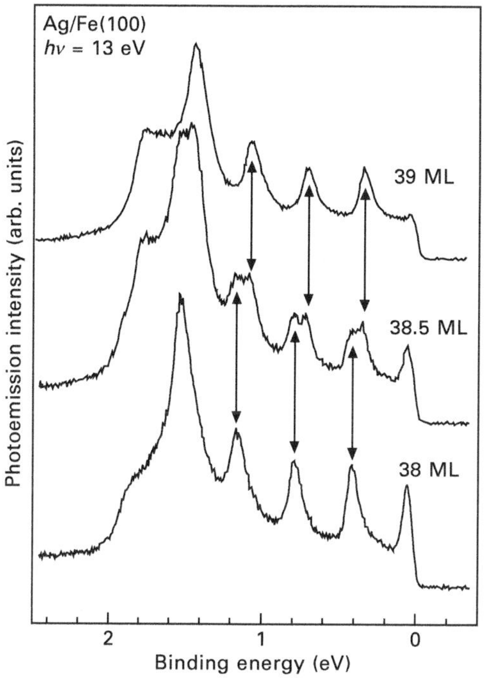

2.3 Normal emission photoemission spectra from Ag films on Fe(100) with thicknesses as shown. The top and bottom spectra are from atomically uniform films, whereas the middle spectrum is from a film of intermediate thickness. Arrows mark the positions of corresponding peaks.

a long coherence length for the electronic states in the film. Once a set of spectra like those in Figs 2.3 and 2.4 have been collected, photoemission becomes a tool for measuring the physical structure of the film on an atomic scale. Atomically uniform films have been obtained in a variety of systems, including Ag on semiconductor substrates Si(111) and Ge(111), and Pb on Si. Using these uniform films, a variety of measurements have been made of quantum behavior in thin films, including band structure (Paggel, Miller *et al.* 1999b), electron–phonon coupling (Paggel, Miller *et al.* 1999c, Paggel, Luh *et al.* 2004, Luh, Miller *et al.* 2002), variations in work function with thickness (Paggel, Wei *et al.* 2002), and thermal stability (Luh, Miller *et al.* 2001, Upton, Wei *et al.* 2004).

2.5 Quantum thermal stability of thin films

The ability of photoemission to measure the thickness of an atomically uniform film makes it an effective tool to investigate thermal stability. An example is shown in Fig. 2.5 (Luh, Miller *et al.* 2001). The two panels show the

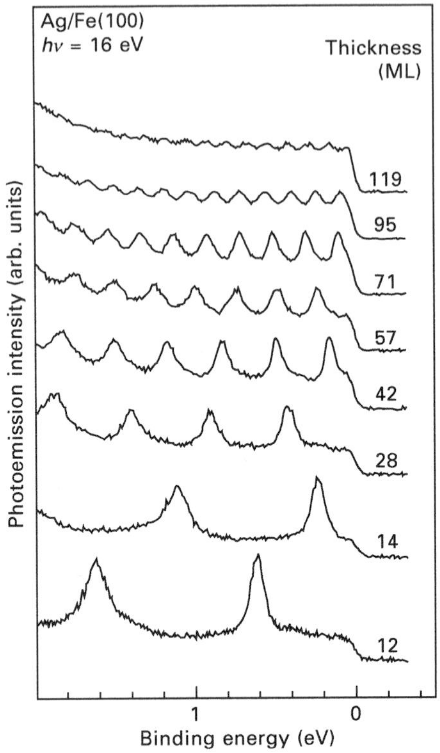

2.4 Normal emission photoemission spectra from atomically uniform Ag films of different thicknesses as shown.

evolution of spectra as two Ag films of different thicknesses (6 and 3 ML), each initially of uniform thickness, change under annealing to successively higher temperatures. The uniform films were prepared on a Fe(100) substrate by depositing at low temperature (~110K) and then annealing to produce atomically uniform films, then the sample was cooled back down before starting the measurements. The top curve in each panel was measured at this low temperature as indicated and shows a single quantum-well peak indicative of the single thickness of the film. Subsequent spectra, going down in the figure, were taken at successively higher temperatures as indicated. As the temperature is increased, some broadening of this initial peak can be seen. This is expected due to the increasing thermal agitation of the system. At some point, the spectra change qualitatively. Two other peaks on either side of the original one appear, and the original one eventually fades out. The bottom curves were taken last in the sequence, after cooling the sample back down to the base temperature. Cooling reduces the thermal broadening, sharpening the peaks, and now it can be clearly seen that the original peak has been replaced by the two new ones. By comparison to spectra taken from

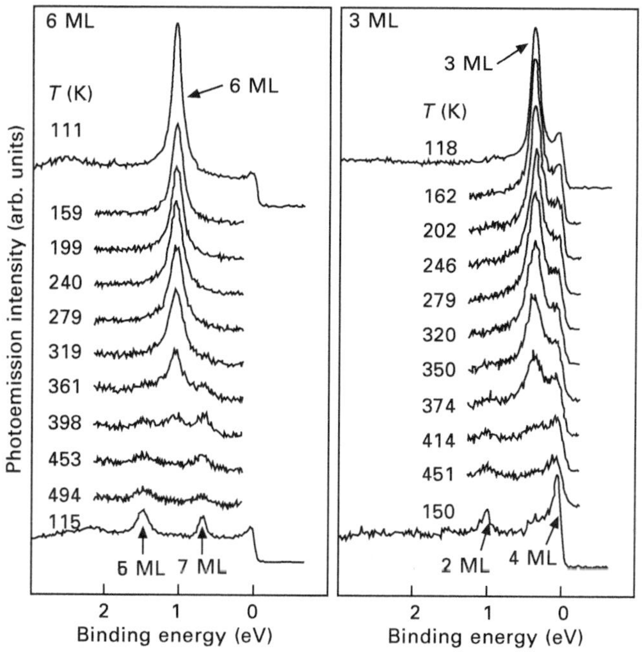

2.5 Evolution of spectra from initially uniform films under annealing. The top spectrum in each stack is an atomically uniform film of 6 ML (left) and 3 ML (right) thickness. For both cases, each spectrum below the top one was taken at successively higher temperatures as indicated, except the bottom one which was taken after the samples were allowed to cool back down to near the starting temperature. Each stack shows a single peak indicative of a uniform film which splits into two peaks as annealing proceeds. The end result is a bifurcated film.

samples of different uniform film thicknesses, the new peaks can be identified as shown in the figure: they belong to films with thicknesses differing by one monolayer from the original film. In other words, heating these films has caused them to bifurcate from the initial uniform N monolayers to a system with $N + 1$ and $N - 1$ layers.

Figure 2.6 shows the intensity of the original peak as a function of annealing temperature for a few films of different starting thicknesses (Luh, Miller *et al.* 2001). Except for the plot for the 5 ML starting thickness, each curve exhibits a well-defined breakpoint which can be assigned to a 'stability temperature' for that thickness. The 5 ML film survived the highest annealing temperature achievable in the experiment (900K), so only a lower limit could be established for its stability temperature. Stability temperatures for a range of coverages are shown in Fig. 2.7(a) (Luh, Miller *et al.* 2001). The data show dramatic differences in thermal stability between films that differ in thickness by even just one atomic layer. As mentioned above, 5 ML is

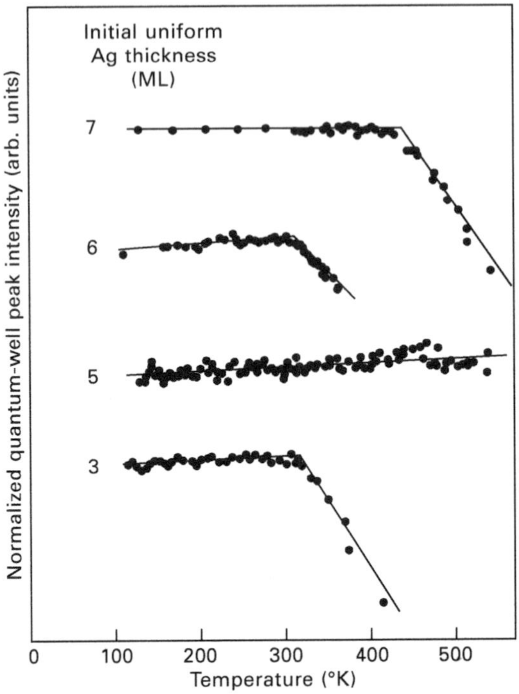

2.6 Breakdown of initially uniform films under annealing. Each curve shows the intensity of a photoemission quantum-well peak characteristic of the thickness of an initially atomically uniform film as the film was heated to successively higher temperatures. Except for the 5 ML film, all showed a distinct temperature where this peak intensity started to decrease. Dots are data points and the lines are guides to the eye.

particularly stable, while 6 ML requires care to prepare. Most films broke down at temperatures not far above RT (<400K or so), but films of thickness 1, 2, and 5 ML all survived to at least 800K. In addition, the variation in stability is not monotonic. These features point to quantum effects as the source of the variability. A clue to the exceptional stability of a 5 ML film can be seen in the spectra of quantum-well state energies for different thicknesses (Fig. 2.2). It can be seen that the highest energy state in the 5 ML spectrum lies below that of both the 4 ML and 6 ML spectra. This suggests that, in order for a 5 ML film to bifurcate, energy would have to be added to the system, making that thickness energetically stable.

A measure of stability against bifurcation can be written as:

$$\Delta(N) = \frac{1}{2}[A(N+1) + A(N-1)] - A(N)$$

where $A(N)$ is the total electronic energy of the system. $\Delta(N)$ is the discrete

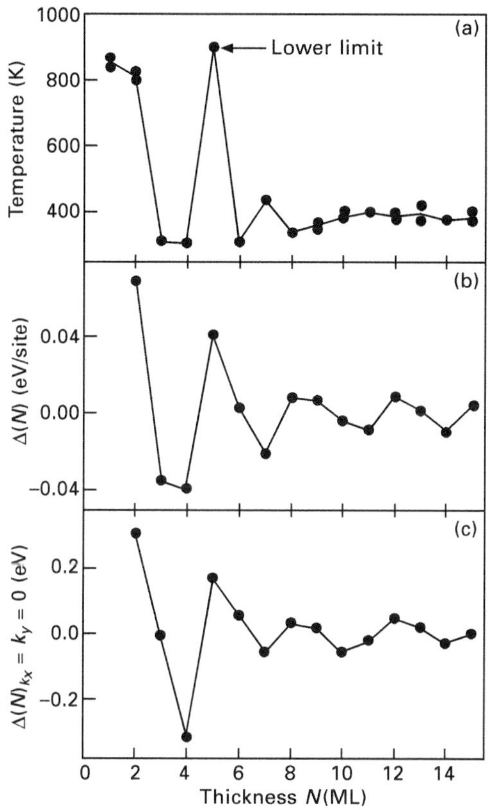

2.7 (a) Experimentally measured bifurcation temperatures for films of different starting thicknesses. The point for 5 ML is a lower limit; this was the highest temperature achieved in the experiment but the film did not break down. (b) Calculated change in energy per surface site due to bifurcation. (c) Same as (b) except only considering states visible in normal emission.

second derivative of the total electronic energy, and represents the energy needed to transform the system from an original flat film of thickness N to a bifurcated one consisting of equal areas of $N + 1$ and $N - 1$ thicknesses. So a large positive $\Delta(N)$ would characterize a stable film thickness. Here we are considering only the energy of the 'bulk' states of the film; in general other contributions, such as edge terms, are possible. To calculate $\Delta(N)$, we start with Eq. 2.1 which gives the discrete allowed values of k_z for each thickness provided the phase shift function is known. The phase shift can be experimentally determined by an analysis of the quantum well as a Fabry–Pérot interferometer with the input of quantum-well peak positions from a large number of thicknesses, as in Fig. 2.4 (Paggel, Miller *et al.* 1999b). This analysis is involved but the basic idea can be illustrated by considering a hypothetical pair of spectra from films of two different thicknesses that happen to show a quantum-well peak at the same binding energy. In this case, $k_z(E)$ and the phase shift are the same for the peaks in both spectra, but they will have different quantum numbers n. The thickness of the film is also quantized to an integer number of layers N. By writing Eq. 2.1 once for each of the two films, the system can be solved for $\phi(E)$ and $k_z(E)$ for the energy of the peak. In fact, this is a way of determining the band structure as well as the phase shift as a function of energy (Mueller, Miller *et al.* 1990, Mueller, Samsavar *et al.* 1989). $A(N)$ involves all of the filled states, not only those as seen in normal emission, but also those with non-zero parallel components of momenta as well. That is, there is a set of discrete subbands each with continuous k_x and k_y. Generally, a consideration of the entire three-dimensional band structure is needed. Figure 2.7(b) shows calculated results for $\Delta(N)$, using a realistic band structure (Smith and Mattheiss 1974). Only the nearly-free-electron-like *sp* bands are included – Ag also has a set of more deeply bound *d* states as part of its valence band structure, but these do not cross the Fermi level and are completely filled. The results can be compared to the measured stability temperatures. Two features stand out: the calculation, like the experimental results, shows significant and non-monotonic variations as a function of film thickness, and also reproduces the relatively high stability of films with $N = 2$ and $N = 5$. Experimentally, a single monolayer is also very stable. The calculation does not extend to this thickness, however, because $A(0)$ is undefined. Physically, for a single monolayer to bifurcate it must leave half of the substrate area bare. Bonding of Ag to the Fe surface might be expected to be stronger than Ag to itself, based on the fact that it does tend to wet the Fe surface. This would tend to stabilize the single monolayer without respect to quantum-well effects.

These results clearly show an important role for the valence electrons in the stability of films, even for metallic bonding where the states are distributed throughout. The extreme stability of particular thicknesses (such as the 5 ML one here) suggests that, for applications, these effects need to be considered and may be useful. Atomically uniform films were created in these studies by a fairly cumbersome procedure; however, it was noted in experiments that a 5 ML film could readily be created by desorbing a thicker film simply by heating. For these fundamental studies, the uniform films allow a simple theoretical treatment. The experimental fact that the films break down first by bifurcation also helps – the final system can be regarded as essentially two samples with different thicknesses but still each with atomic uniformity. In general, life may be more complicated, with the formation of many different thicknesses, or clusters where lateral QSE and surface energies would have to be considered along with the detailed geometries of the constituent parts.

The *sp* band of Ag is generally nearly-free-electron-like near the Fermi level, but the Fermi surface becomes non-spherical as one goes out along the (111) directions. In fact the band does not cross E_F along that direction, so the Fermi surface develops 'necks' at the zone boundaries. This potentially complicates the effect of quantization on the thermal stability, as this area would be included in the sum of energies over all states. To gain physical insight, a simpler 'stability function' can be calculated without summing over k_x and k_y, using only states with zero parallel momenta. The result is shown in Fig. 2.7(c). Comparing the two calculations we see that the simpler one actually preserves the essential features of the system's behavior. In other words, a reasonable picture of the electronic stability of these films comes from looking only at the states at the zone center, namely, the peak positions that can be seen in Figs 2.2 and 2.4. As the film thickness is increased one layer at a time, the electronic states shift towards higher binding energies, increasing the stability of the system. However, at intervals, a new state appears near the Fermi level, which increases the electronic energy in a discontinuous fashion. The physics is reminiscent of the shell model of the stability of atoms as one runs through the elements of the periodic table.

With this picture in mind, the importance of the phase shift can be readily appreciated. The entire spectrum of allowed energies of a film would shift with changing this quantity. This directly affects the total electronic energy, and also changes the points at which a new state would appear from the top of the spectrum as the thickness is increased. It is then reasonable to expect the energy landscape and the electronic stability to depend sensitively on this phase shift. In fact that is the case, as will be discussed below.

2.6 General principles of film stability and nanostructure development

Atomically uniform films of Ag on Fe(100) can be prepared over a very large range of thicknesses and their photoemission spectra display a number of peaks over a relatively wide spectrum of binding energies. These features make this system ideal for illustrating the basics of quantum electronic stability. The model presented above gives an intuitive picture, but it relies on a band structure as input (the bulk band structure was assumed) and all of the details of the surface and interface are rolled up into the phase shift parameter. In this section, the system Pb on Si(111) will be examined to extend the theoretical picture and explore some general principles of quantum electronic stability.

Atomically uniform films of Pb can be grown by depositing with the substrate at liquid nitrogen temperatures. The substrate presents a barrier to electron transmission for energy levels within the Si band gap. Outside the gap, partial transmission occurs and the quantum-well peaks become broader resonances that are more difficult to see in photoemission spectra.

This limits the number of peaks that can be seen in any given spectrum. Figure 2.8 shows normal emission photoemission intensity as a function of binding energy and film thickness for a series of depositions ranging from

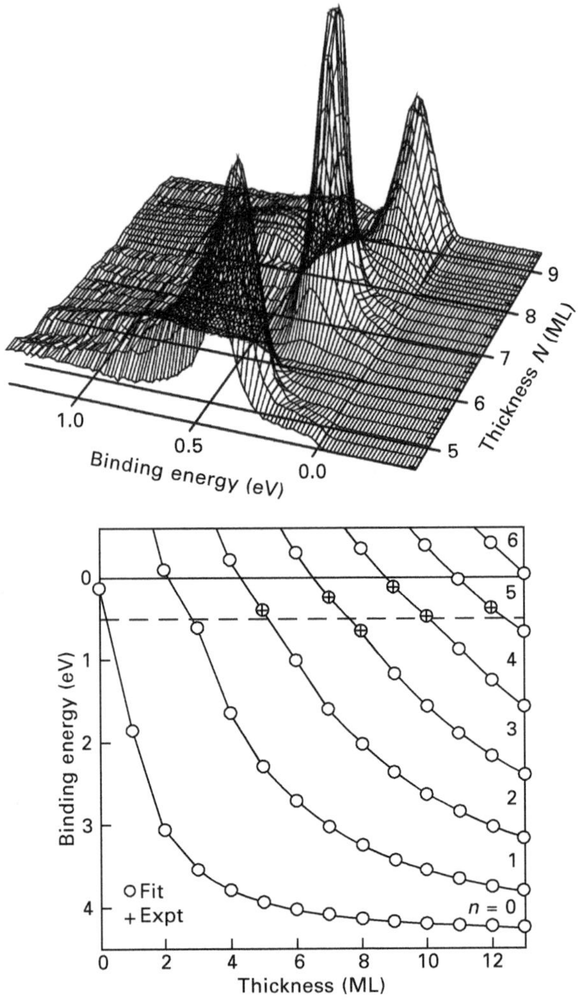

2.8 Top: 3-D mesh plot of normal emission photoemission intensity as a function of binding energy and film thickness for Pb films on Si(111). Bottom: modeled (circles) and measured (crosses) quantum-well peak binding energies for Pb on Si(111). The model is fitted to the measured peak positions and includes the band structure of Pb and the phase shift. Quantum-well peak positions corresponding to particular quantum numbers *n* are joined by lines. The dashed line represents the position of the Si band edge: quantum confinement in the film occurs for energies between the band edge and the Fermi level at zero energy.

about 5 ML to 9 ML (Upton, Wei *et al.* 2004). Pairs of peaks like those in Fig. 2.3 are not seen. However, for 6 ML no peak at all appears. The pattern can be understood by reference to the plot of the binding energies of the allowed states as a function of thickness shown in the bottom part of the figure (Luh, Miller *et al.* 2002, Smith 1985, Upton, Wei *et al.* 2004, Wei and Chou 2002). Over the thickness range from 5 to 9 ML there will be an allowed quantum-well state between the Fermi level and the Si band edge only for the odd thicknesses of 5, 7, and 9 ML. Starting at 5 ML and going up in thickness a monolayer at a time, we see the state in the 'confinement gap' for 5 ML being pulled down below the Si band edge at 6 ML, while another allowed state is still above the Fermi level and so is unoccupied, leaving no states to be seen with photoemission. Continuing up in thickness, at 7 ML this unoccupied state becomes occupied and falls within the gap region, and appears in the ARPES data. At 8 ML this state is just below the gap. It actually does appear in the photoemission data as a weak, broad feature – it is a quantum-well resonance that is not quite bound in the well. This pattern prevents the observation of pairs of peaks characteristic of adjacent monolayer thicknesses. On the other hand, the complete disappearance of peaks when no states are within the confinement gap indicates that those films are in fact atomically uniform.

The alternating-layer pattern is reflected in the growth of Pb on Si and diverse physical properties of films, including their thermal stability and even superconducting transition temperatures (Guo, Zhang *et al.* 2004). As previously mentioned, preference for bilayer height steps in the population of islands grown at different temperatures has been observed in STM measurements (Hupalo, Kremmer *et al.* 2001, Hupalo and Tringides 2002). The thermal stability of uniform Pb films has been measured by photoemission; results are shown in Fig. 2.9(a), where oscillations with a bilayer period are seen (Upton, Wei *et al.* 2004). The physical basis for the bilayer pattern is in the valence band structure. Along the (111) growth direction, the Pb *sp* band crosses the Fermi level about halfway through the Brillouin zone. As discussed above, sharp variations in physical properties are expected to be related to quantum-well states crossing the Fermi level as the film grows. As a simple illustration, consider a case where the relevant band crosses the Fermi level exactly in the middle of the zone. Ignoring the phase shift, the allowed quantum-well states are determined by dividing up the zone by the number of layers. It can easily be seen that as each layer is deposited, the spectrum of allowed energies alternates between having a state right at the Fermi level, or not. In fact, based on this picture, one can define a periodicity for quantum property variations derived from the Fermi wave vector k_F along the growth direction:

$$\Delta h = \frac{\pi}{k_F}$$

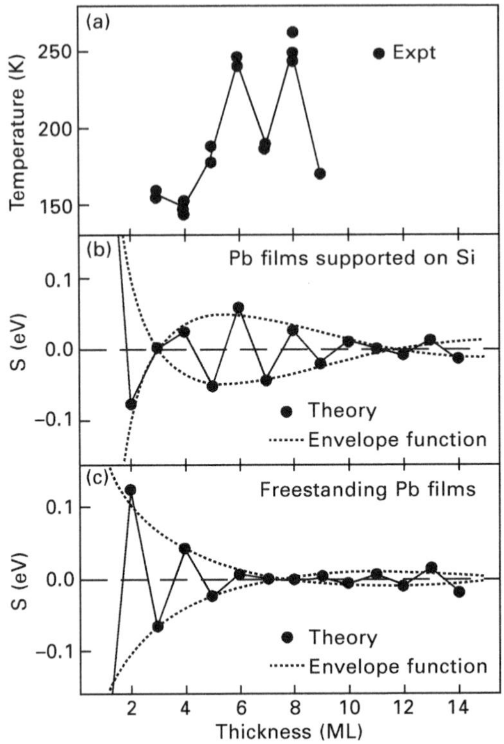

2.9 (a) Measured stability temperatures for Pb films on Si(111) as a function of thickness. Multiple measurements were made for most thicknesses: the lines are guides to the eye connecting average temperatures measured at each thickness. (b) Density-functional calculation of the surface energy per site *S* for Pb films supported on a Si substrate. Lines connect values calculated at integer monolayer thicknesses. The dotted curves show an envelope function based on a Friedel model; this illustrates the beating effect of sampling at discrete thicknesses (Upton, Wei *et al.* 2004). (c) Similar to (b) except for unsupported Pb films.

where Δh is the periodicity in thickness at which Fermi level crossings occur and is equal to half of the Fermi wavelength. Here, h is defined as a continuous variable, but a uniform film is constrained to quantized values that differ by integer numbers of monolayers. This discrete sampling is important because it can lead to aliasing which produces very different observed periods of oscillation in properties (Miller, Chou *et al.* 2009). A related complication is the definition of the Fermi wave vector. For example, using a free-electron value of $k_F = 1.58$ Å^{-1} for Pb films gives $\Delta h = 0.7$ ML, which is obviously not observed. However, taking into account the finite sampling by monolayer increments, the aliased period is 2.2 ML, a reasonable value in line with observation. Accounting for the sampling in this case is equivalent to measuring the Fermi wave vector from the Brillouin zone boundary instead of from the zone center (Wei and Chou 2002).

While 2.2 ML is not exactly an integer number of layers either, it is close enough so that over a limited range of thicknesses, bilayer oscillations in properties appear. However, over a great enough range the remainder becomes significant, leading to a beating effect and a phase reversal. Figure 2.9(b) is a plot of the second derivative of the calculated surface energies per site from first-principles slab calculations of the electronic structure (Wei and Chou 2002). Analogous with the analysis of the previous section, this quantity is a measure of the stability of a given thickness against bifurcation. The calculated results reflect the bilayer oscillations in the measured stability temperatures. The theoretical values extend to greater thicknesses than the experiment, and the beating effect leads to a diminishing of the quantum oscillations and an even–odd phase reversal around 12 ML.

Previously, the importance of the phase shift was mentioned. The first-principles calculation presented in Fig. 2.9(b) includes implicitly the nature of the interface between the film and the substrate as well as the barrier at the surface of the film. This is built-in by having substrate layers and vacuum layers as well as the film material as part of the slab model used. We note here that inclusion of the substrate is problematic – the interface is mismatched and may involve reconstructions or special adlayers which are only approximately handled in the calculation – and this could be the source of some of the discrepancies in detail with the experimental result. In any case, here the effects of the phase shift are implicitly included, and in fact an effective phase shift can be extracted from the calculation if desired. One can explore the effect of substrate interactions, for example, by removing the substrate layers from the slab model. Results for such a calculation are shown in Fig 2.9(c), which represents the stability function for a freestanding Pb film (Wei and Chou 2002). The important point to note here is that the calculated result is very sensitive to changing the substrate interface. We see cases of reversal in stability on a monolayer-by-monolayer basis, as well as a large shift in the beating pattern. Again, one can look at this as an indication of the sensitivity of quantum physical properties on the phase shift.

The foregoing examples establish atomically uniform films as an instructive 'laboratory' for the study of quantum-electronic stability. Spectra measured with ARPES from such samples give both morphological and electronic information at the same time. With the Pb on Si(111), a variety of growth modes are possible, depending on the deposition temperature and annealing recipes. How far can one go in applying the concepts learned from simple flat films to more complicated morphologies? For the determination of morphology, X-ray diffraction can be used and in some ways is more convenient than photoemission. Thickness is determined by the appearance of multilayer interference peaks between the main Bragg peaks in the diffraction pattern. Like photoemission, the measurement typically is macroscopic, the probe beam typically being of millimeter to sub-millimeter size, so measurements taken from a non-uniform sample will necessarily give superpositions of patterns from different areas of the sample (Hong, Wu *et al.* 2000). However, photoemission peak intensities can be difficult to interpret. For example, in a spectrum from a sample composed of equal areas of 5 and 6 ML thicknesses, the peak intensities associated with the two areas will in general not be equal. Even worse, the relative intensities can vary markedly with changing the exciting photon energy or the quantum numbers n of the states involved (Paggel, Miller *et al.* 1999a). This is because photoexcitation involves a matrix element that is sensitive to the relative phase of the initial and final state wavefunctions. In contrast, the analysis of X-ray diffraction data is a refined art that is readily applied to thin films.

Figure 2.10 shows X-ray diffraction data from Pb deposited on Si(111) at 180K (Hong, Wei *et al.* 2003). The stacked curves show the intensity along the Pb(10L) truncation rod at different times during a continuous deposition at a constant rate. The two main peaks are bulk Bragg peaks as noted. Between these appear smaller satellite peaks from multilayer diffraction – these peaks carry information about the thicknesses present on the sample. The behavior of the diffraction can be summarized as follows: first, there is a period of time where Pb diffraction peaks do not appear. At the same time, Si diffraction peaks (not shown) increase in intensity. This is consistent with growth limited to an adlayer, commensurate with the Si(111) surface. Then there is a period corresponding to about 1.5 ML of total deposition where nothing really happens to the intensities, suggesting Pb is present as a disordered gas on top of the adlayer. Beyond this point Pb features abruptly appear. These features include both the bulk diffraction peaks and satellite peaks, all of which increase in intensity as deposition proceeds. The line shapes, however, remain the same during this time. The pattern is characteristic of 5 ML high structures above the wetting layer, or 6 ML altogether. Deposition beyond 6 ML causes the pattern to change apparently in accord with layer-by-layer growth. The growth is summarized schematically in Fig. 2.11 (Chiang 1997). First, Pb wets the surface. Subsequent deposition leads to island growth, where all the islands are 6 ML high (from the starting surface). Eventually a completely filled 6 ML film is obtained, followed by layer-by-layer growth.

This behavior can be understood by looking at a plot of surface energies vs. film thickness (Fig. 2.12) (Boettger, Smith *et al.* 1998, Chiang 2000, Hong, Wei *et al.* 2003, Kresse and Furthmüller 1996, Materzanini, Saalfrank *et al.* 2001, Wei and Chou 2002). This plot differs from the earlier one showing the second derivative of the total energies because here we are not dealing with bifurcation of a flat layer, but rather a more general behavior, so there is

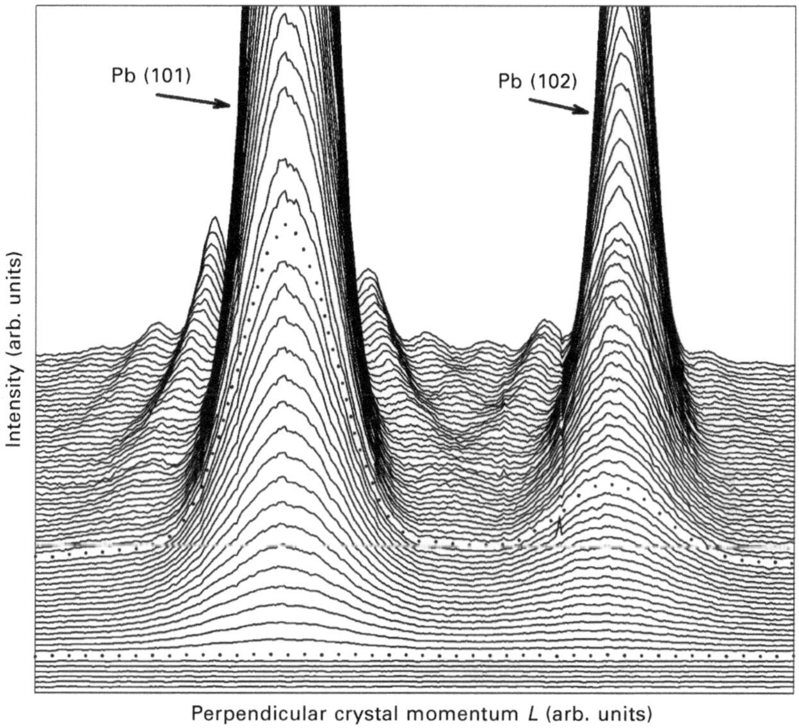

2.10 X-ray diffraction intensities along the Pb(10L) rod during deposition of Pb on Si(111). Time and the deposition amount increase uniformly for each successive scan going up in the figure; scans are displaced vertically for clarity. Two large peaks characteristic of a 6 ML thick film are marked, smaller peaks between these are multilayer diffraction for larger thicknesses. The dotted curves indicate scans at 1.7 ML (lower dotted curve) and 5.7 ML (upper dotted curve) thicknesses.

need to examine the global energy landscape. The surface energy per surface atom shows layer-by-layer variations but the important features are the deep minima at 1 and 6 ML thicknesses. The energy of any configuration of only these two thicknesses, considered as non-interacting regions, would lie on the straight line segment shown between 1 and 6 ML. It is clear since this line lies below all of the points for intermediate thicknesses that the system will prefer to phase-separate into the wetting layer plus 6 ML high islands for all intermediate amounts deposited.

So the physics of the flat film in this case predicts the behavior of a slightly more complicated system of a wetting layer plus preferred-height island growth. This study does not give information about the lateral size distribution, but apparently this is not essential to the understanding of the basics of growth. The growth temperature is important. For example, in Fig.

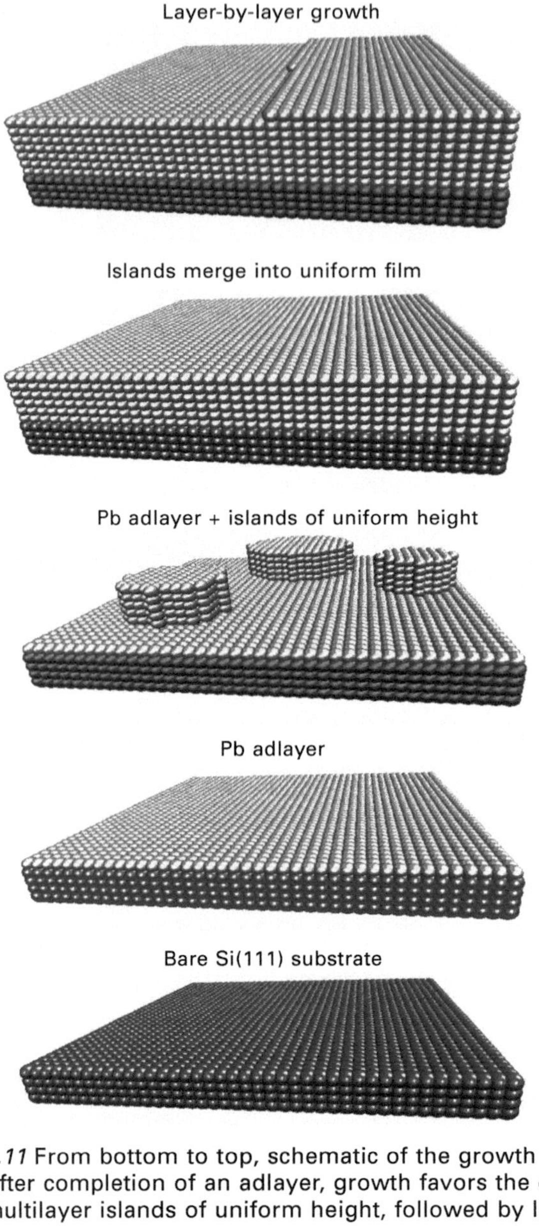

2.11 From bottom to top, schematic of the growth of Pb on Si(111). After completion of an adlayer, growth favors the development of multilayer islands of uniform height, followed by layer-by-layer growth.

2.12 we see also a significant minimum in the energy at 4 ML. A system prepared at low temperature could stay at that thickness rather than phase separate. As a general yardstick for thermal energies, $k_B T$, where k_B is Boltzmann's constant and T is the temperature, can be used, and for 180K

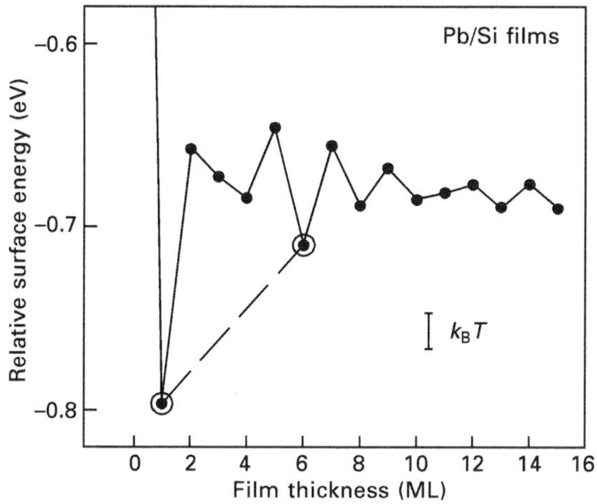

2.12 Relative surface energy derived from a density-functional calculation for different thicknesses of Pb on Si(111). Dots are calculated values and lines are guides to the eye. The dashed line joining the points at 1 and 6 ML thicknesses represents the energy of a system consisting of the adlayer plus 5 ML high islands with average thickness as plotted. The 'error bar' shows the scale of thermal energy at the growth temperature.

is shown by a bar in Fig. 2.12. Evidently this is large enough compared to the layer-by-layer variations between 1 and 6 ML so the system does not get 'stuck' at 4 ML at this temperature. It is reasonable that if the temperature is chosen suitably the bilayer oscillations in favored thicknesses would be seen instead of this phase separation.

Another X-ray diffraction study was reported on this system that explicitly explored the role of the temperature on film morphology. Figure 2.13 (Czoschke, Hong *et al.* 2005) shows the fraction of a surface covered by Pb islands of different heights, starting with an 11 ML thick nearly uniform film and subsequently annealing to the temperatures indicated. 11 ML is an unstable thickness, and with heating it starts to bifurcate and appears to be fully bifurcated into regions of 10 and 12 ML at 200K. Annealing further, we see the development of a single peak at 12 ML. From the surface energy, 10 ML should be more stable than 12 ML; however, the system started with 11 ML and the total amount of material must be conserved. To accommodate this, the system can have 12 ML islands separated by some amount of the very stable wetting layer. At higher temperatures, the film becomes disordered, and at 280K it has reached a state of local thermodynamic equilibrium with many thicknesses present. Inspection of the top curve shows that the distribution of island heights is not random, but follows what is by now a familiar pattern – a jagged series of bilayer oscillations but with the appearance of occasional phase reversals (beating effect) indicated by the

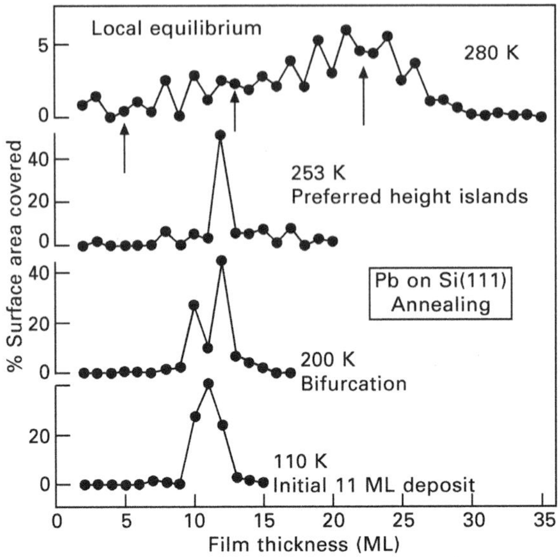

2.13 Distribution of thicknesses of Pb on Si(111) as annealling proceeded following an initial 11 ML deposition, derived from X-ray diffraction measurements. The arrows mark points where the alternating layer pattern of stability switches from even to odd numbers of monolayers and vice versa.

arrows in Fig. 2.13. Indeed, with suitable thermodynamic analysis, data such as these can be related to the calculated surface energies with good agreement. It is perhaps remarkable that even in this 'more realistic' system, which should involve many complicating factors such as lateral confinement, shape effects, registry with the substrate, etc., the simple picture afforded by atomically uniform films still gives essential insight as to its quantum electronic stability.

The importance of the phase shift has been demonstrated and emphasized in the foregoing. A natural question is, can one influence growth by adjustment of the phase shift by modification of the surface or substrate interface? The encouragement of layer-by-layer growth modes through the use of surfactants has been studied and reported on in the literature. As the name implies, these are expected to operate by reducing the surface energy. Typically, a foreign species is introduced at the start of growth that tends to 'float' on the surface, stabilizing it. In some sense this is a 'catalytic' process in that the surfactant is not consumed but acts continuously as growth proceeds (Rioux and Höchst 1992). However, typically these are viewed as acting locally, for example by saturating the surface dangling bonds, and do not involve states in the bulk of the film. For the purpose of exploiting quantum electronic effects to influence stability, intuitively it would seem to be much easier to modify the phase shift at the substrate/film interface than at the surface. However, for fundamental studies it is important that the interface remain well ordered. For Pb on Si(111), one can make several well-ordered reconstructed surfaces using 'foreign' atoms and still be able to grow Pb films on top. It turns out that by doing so, it is indeed possible to tailor the landscape of thermal stability in a dramatic fashion. Figure 2.14 shows results of experiments using Au and In as 'interfactants', compared to a straight Pb system (Ricci, Miller *et al.* 2005). These systems were prepared by first depositing the interfactant atoms and annealing to produce the following ordered surfaces: In $(\sqrt{3} \times \sqrt{3})$, Au (6×6), and Pb $(\sqrt{3} \times \sqrt{3})$. The amounts of the interfactant materials to make these reconstructions are 0.333, 0.96, and 0.33 ML, respectively. Here a ML is defined to be with respect to the Si(111) surface layer. Then, Pb layers were grown on top at 60K. The thermal stability was measured in the same way that has been outlined earlier, using photoemission to monitor the film structure. The interfactant atoms at the substrate surface change the interface

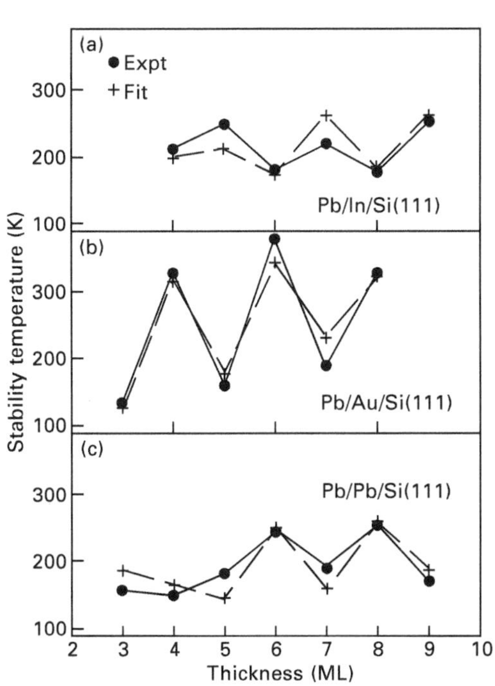

2.14 Stability temperatures for Pb films on Si(111) as a function of film thickness using three different interfactants, (a) In, (b) Au, and (c) Pb. The dots joined by solid lines are experimentally measured points, the crosses joined by dashed lines are a fit to a model from which the phase shift can be extracted (Ricci, Miller *et al.* 2005, Czoschke, Hong, *et al.* 2005). The same scales are used in all three panels so that the effect of the interfactant on thermal stability is directly visible.

barrier and hence the phase shift (Ricci, Liu *et al.* 2009). The sensitivity of the thermal stability to these changes is readily appreciated in Fig. 2.14: with In as an interfactant, films with an odd number of atomic layers are more stable than ones with even numbers of layers, whereas for the other two cases this trend is reversed. It is interesting to note, especially with a view towards applications, that all of the measured stability temperatures for films with In and Pb interfactants are below RT, whereas some of the films with Au as the interfactant are stable to temperatures above RT. The films with Au at the interface show much larger oscillations in stability temperatures, and the films with even numbers of layers are stable above 300K, while the odd-numbered films are not. This study shows that interfacial engineering can be done that has a profound effect on nanostructure growth.

2.7 Beyond the particle-in-a-box

So far we have treated quantum electronic effects within a simple one-dimensional particle-in-a-box picture. For the systems presented above, this picture is most applicable to the case of atomically uniform films of Ag on Fe(100). Here, the interface is lattice-matched, so the one-dimensional character is not disturbed, and the uniform films obviate the need to consider lateral confinement effects. However, in going to Pb on Si, lattice mismatch is introduced, so strictly speaking we no longer have one-dimensional physics. In interpreting the experimental results, it is tacitly assumed that an incommensurate interface partially contributes to random scattering which will not affect the systematics. Yet in some systems random scattering seems to play a dominant role (Miller, Samsavar *et al.* 1994). In the first-principles calculations, one or the other of the lattices is compressed or expanded to artificially produce a matched system so that the calculations are tractable. Some work has been done exploring the ramifications of these approximations. Furthermore, once structures such as islands are considered, lateral effects should come into play. Nonetheless in the cases presented above, useful insight has been obtained considering only the one-dimensional model. Obviously, the validity of this depends on the lateral sizes of the structures involved. It has been observed that interactions with the substrate introduce more complicated features into the electronic structure than have been considered so far. Here these will be only briefly mentioned because, to date, their effect on thermal stability has not been systematically explored.

Firstly, there is the nature of the confinement barrier presented by the substrate. Even in the case of Ag on Fe(100), it must be mentioned that the barrier is spin-dependent due to the ferromagnetic band structure of the substrate. Basically, the quantized states are minority spin only, and the majority spin electrons may be free to couple with the substrate states (Carbone, Vescovo *et al.* 1994). This may not have much effect on thermal stability, however, because to the extent that some states are not confined they become broader resonances and so their effect on quantum behavior would be expected to be less discernible. Like scattering at an incommensurate interface, it seems reasonable to assume this would present a minor perturbation, and the general agreement between experiment and theory bears this out. Similarly, for Pb on Si the interface barrier really only exists for binding energies within the Si band gap, below which one observes broader and weaker resonances. Some detailed studies have been done on the interaction of states from a metallic overlayer with semiconductor substrate states. Observed effects include modification of the in-plane dispersions of the metallic overlayer states (Upton, Miller *et al.* 2005), distortion of the quantum-well spectrum by the substrate band edge in Ag films on Ge(111) (Tang, Basile *et al.* 2004), coherent coupling of Ag film states with states in the band-bending region of a Si substrate (despite the mismatched lattices) (Speer, Tang *et al.* 2006), and complex electronic structures caused by umklapp scattering from an incommensurate Ag/Ge(111) interface (Tang, Lee *et al.* 2006). It remains to be seen what impact, if any, these interesting quantum electronic effects have on thermal stability of films.

2.8 Future trends

For applications, nanostructures with small lateral dimensions are likely to be of interest. In the extreme, with sizes in the order of a nanometer or so, lateral confinement and registry with the substrate would be expected to be important for the determination of structure and stability, and theory would look more like molecular or cluster calculations than something derived from the physics of the infinite crystal. Still, as has been demonstrated, for nanoscale metal structures a lot can be learned from an approximate nearly-free-electron model and it seems reasonable to extend it from uniform films to islands, stripes, or other shapes of technological interest, if only for its value in providing intuitive insight. Already mentioned above are some areas which might benefit from more experimentation and theoretical development even for flat films. On the experimental side, photoemission and X-ray diffraction are both limited for the study of structures of small lateral size by their macroscopic probe beams. Recent developments in microprobes for both techniques promise to improve this situation, but nanoscale resolution is difficult to achieve. For fundamental understanding, measuring both the physical and electronic structure of a nanostructure is needed. STM has been used with great success in measuring island physical structures. With care, layer counting can be done with STM as well. Scanning tunneling spectroscopy can give some information as to the electronic structure of single nanostructures; it lacks the capability to do the complete band mapping of ARPES but recently some very interesting results regarding quantum-well states and the electron–phonon interaction have been obtained (Wang, Zhang *et al.* 2009). Results such as these will undoubtedly spur further research in those directions.

2.9 Acknowledgments

Most of the materials chosen for inclusion in this review are based on work supported by the US Department of Energy (grant DE-FG02-07ER46383). The photoemission experiments discussed herein were performed at the Synchrotron Radiation Center (SRC), University of Wisconsin-Madison, which is supported by the US National Science Foundation (grant DMR-09-06444). We acknowledge the ACS Petroleum Research Fund and the US National Science Foundation (grant DMR-05-37588) for partial support of the beamline operations at SRC.

2.10 References

Aballe, L., Rogero, C., Gokhale, S., Kulkarni, S. and Horn, K., 2001. Quantum-well states in ultrathin aluminium films on Si(111). *Surface Science,* **482–485**(Part 1), 488.

Aballe, L., Rogero, C. and Horn, K., 2002. Quantum-size effects in ultrathin Mg films: electronic structure and collective excitations. *Surface Science*, **518**(1–2), 141.

Ashcroft, N.W. and Mermin, N.D., 1976. *Solid State Physics*. Philadelphia, PA: Saunders.

Bian, G., Miller, T. and Chiang, T.-C., 2009. Electronic structure and surface-mediated metastability of Bi films on Si(111)-7 × 7 studied by angle-resolved photoemission spectroscopy. *Physical Review B (Condensed Matter and Materials Physics)*, **80**(24), 245407.

Boettger, J.C., Smith, J.R., Birkenheuer, U., Rösch, N., Trickey, S.B., Sabin, J.R. and Apell, S.P., 1998. Extracting convergent surface formation energies from slab calculations. *Journal of Physics: Condensed Matter*, **10**, 893.

Budde, K., Abram, E., Yeh, V. and Tringides, M.C., 2000. Uniform, self-organized, seven-step height Pb/Si(111)-(7 × 7) islands at low temperatures. *Physical Review B (Condensed Matter and Materials Physics)*, **61**(16), 10602.

Carbone, C., Vescovo, E., Kläsges, R., Eberhardt, W., Rader, O. and Gudat, W., 1994. Spin-polarized photoemission from quantum well and interface states (invited). *Journal of Applied Physics*, **76**(10), 6966–6971.

Chiang, T.-C., 1997. Surface and bulk photoemission: some old problems and new findings. *Chinese Journal of Physics*, **35**(4), 496–508.

Chiang, T.-C., 2000. Photoemission studies of quantum well states in thin films. *Surface Science Reports*, **39**(7–8), 181.

Czoschke, P., Hong, H., Basile, L. and Chiang, T.-C., 2005. Quantum size effects in the surface energy of Pb/Si(111) film nanostructures studied by surface x-ray diffraction and model calculations. *Physical Review B (Condensed Matter and Materials Physics)*, **72**, 075402.

Danese, A.G., Curti, F.G. and Bartynski, R.A., 2004. Quantum size effect induced modification of the chemisorption properties of thin metal films. *Physical Review B (Condensed Matter and Materials Physics)*, **70**(16), 165420.

Evans, D.A., Alonso, M., Cimino, R. and Horn, K., 1993. Observation of quantum size effects in photoemission from Ag islands on GaAs(110). *Physical Review Letters*, **70**(22), 3483–3486.

Feibelman, P.J., 1983. Static quantum-size effects in thin crystalline, simple–metal films. *Physical Review B (Condensed Matter and Materials Physics)*, **27**(4), 1991–1996.

Feibelman, P.J. and Hamann, D.R., 1984. Quantum-size effects in work functions of free-standing and adsorbed thin metal films. *Physical Review B (Condensed Matter and Materials Physics)*, **29**(12), 6463–6467.

Guo, Y., Zhang, Y., Bao, X., Han, T., Tang, Z., Zhang, L., Zhu, W., Wang, E.G., Niu, Q., Qiu, Z.Q., Jia, J., Zhao, Z. and Xue, Q., 2004. Superconductivity modulated by quantum size effects. *Science*, **306**(5703), 1915–1917.

Hong, H., Wei, C.-M., Chou, M.Y., Wu, Z., Basile, L., Chen, H., Holt, M. and Chiang, T.-C., 2003. Alternating layer and island growth of Pb on Si by spontaneous quantum phase separation. *Physical Review Letters*, **90**(7), 076104.

Hong, H., Wu, Z., Chiang, T.-C., Zschack, P., Jemian, P., Chen, H. and Aburano, R.D., 2000. Reflection surface x-ray diffraction patterns: k-space images. *Review of Scientific Instruments*, **71**(8), 3132–3137.

Hüfner, S., 1995. *Photoemission Spectroscopy – Principles and Applications.* Berlin: Springer.

Hupalo, M., Kremmer, S., Yeh, V., Berbil-Bautista, L., Abram, E. and Tringides, M.C., 2001. Uniform island height selection in the low temperature growth of Pb/Si(111)-(7 × 7), *Surface Science* **493**(1–3), 526.

Hupalo, M. and Tringides, M.C., 2002. Correlation between height selection and electronic structure of the uniform height Pb/Si(111) islands. *Physical Review B (Condensed Matter and Materials Physics)*, **65**(11), 115406.

Kidd, T.E., Miller, T., Chou, M.Y. and Chiang, T.-C., 2002. Electron-hole coupling and the charge density wave transition in $TiSe_2$. *Physical Review Letters*, **88**(22), 226402.

Kondo, T., Khasanov, R., Karpinski, J., Kazakov, S.M., Zhigadlo, N.D., Ohta, T., Fretwell, H.M., Palczewski, A.D., Koll, J.D., Mesot, J., Rotenberg, E., Keller, H. and Kaminski, A., 2007. Dual character of the electronic structure of $YBa_2Cu_4O_8$: the conduction bands of CuO_2 planes and CuO chains. *Physical Review Letters*, **98**(15), 157002.

Kresse, G. and Furthmüller, J., 1996. Efficient iterative schemes for *ab initio* total-energy calculations using a plane-wave basis set. *Physical Review B*, **54**(16), 11169–11186.

Liu, Y., Paggel, J.J., Upton, M.H., Miller, T. and Chiang, T.-C., 2008. Quantized electronic structure and growth of Pb films on highly oriented pyrolytic graphite. *Physical Review B (Condensed Matter and Materials Physics)*, **78**(23), 235437.

Liu, Y., Speer, N.J., Tang, S.-J., Miller, T. and Chiang, T.-C., 2008. Interface-induced complex electronic interference structures in Ag films on Ge(111). *Physical Review B (Condensed Matter and Materials Physics)*, **78**(3), 035443.

Loly, P.D. and Pendry, J.B., 1983. Removing the limits to accurate band–structure determination by photoemission. *Journal of Physics C: Solid State Physics*, **16**(2), 423–431.

Luh, D.-A., Miller, T., Paggel, J.J., Chou, M.Y. and Chiang, T.-C., 2001. Quantum electronic stability of atomically uniform films. *Science*, **292**(5519), 1131–1133.

Luh, D.-A., Miller, T., Paggel, J.J. and Chiang, T.-C., 2002. Large electron-phonon coupling at an interface. *Physical Review Letters*, **88**(25), 256802.

Materzanini, G., Saalfrank, P. and Lindan, P.J.D., 2001. Quantum size effects in metal films: energies and charge densities of Pb(111) grown on Cu(111). *Physical Review B*, **63**(23), 235405.

Miller, T., Chou, M.Y. and Chiang, T.-C., 2009. Phase relations associated with one-dimensional shell effects in thin metal films. *Physical Review Letters*, **102**(23), 236803.

Miller, T., Samsavar, A. and Chiang, T.-C., 1994. Photoexcitation of resonances in Ag films on Ni(111). *Physical Review B (Condensed Matter)*, **50**(23), 17686–17689.

Miyazaki, M. and Hirayama, H., 2008. Thickness- and deposition temperature-dependent morphological change in electronic growth of ultra-thin Ag films on Si(111) substrates. *Surface Science*, **602**(1), 276.

Moruzzi, V.L., Janak, J.F. and Williams, A.R., 1978. *Calculated Electronic Properties of Metals*. New York: Pergamon Press.

Mueller, M.A., Miller, T. and Chiang, T.-C., 1990. Determination of the bulk band structure of Ag in Ag/Cu(111) quantum-well systems. *Physical Review B (Condensed Matter)*, **41**(8), 5214–5220.

Mueller, M.A., Samsavar, A., Miller, T. and Chiang, T.-C., 1989. Probing interfacial properties with Bloch electrons: Ag on Cu(111). *Physical Review B (Condensed Matter)*, **40**(8), 5845–5848.

Ohta, T., Bostwick, A., Seyller, T., Horn, K. and Rotenberg, E., 2006. Controlling the electronic structure of bilayer graphene. *Science*, **313**(5789), 951–954.

Ohta, T., Bostwick, A., McChesney, J.L., Seyller, T., Horn, K. and Rotenberg, E., 2007. Interlayer interaction and electronic screening in multilayer graphene investigated with angle-resolved photoemission spectroscopy. *Physical Review Letters*, **98**(20), 206802.

Paggel, J.J., Luh, D.-A., Miller, T. and Chiang, T.-C., 2004. Electronic-structure dependence of the electron–phonon interaction in Ag. *Physical Review Letters*, **92**(18), 186803.

Paggel, J.J., Miller, T. and Chiang, T.-C., 1998. Quasiparticle lifetime in macroscopically uniform Ag/Fe(100) quantum wells. *Physical Review Letters*, **81**(25), 5632–5635.

Paggel, J.J., Miller, T. and Chiang, T.-C., 1999a. Angle-resolved photoemission from atomic–layer–resolved quantum well states in Ag/Fe(100). *Journal of Electron Spectroscopy and Related Phenomena*, **101–103**, 271.

Paggel, J.J., Miller, T. and Chiang, T.-C., 1999b. Quantum-well states as Fabry-Pérot modes in a thin-film electron interferometer. *Science*, **283**(5408), 1709–1711.

Paggel, J.J., Miller, T. and Chiang, T.-C., 1999c. Temperature dependent complex band structure and electron–phonon coupling in Ag. *Physical Review Letters*, **83**(7), 1415–1418.

Paggel, J.J., Miller, T., Luh, D.-A. and Chiang, T.-C., 2000. Quantum well photoemission from atomically uniform Ag films: determination of electronic band structure and quasi-particle lifetime in Ag(100). *Applied Surface Science*, **162–163**, 78–85.

Paggel, J.J., Wei, C.M., Chou, M.Y., Luh, D.-A., Miller, T. and Chiang, T.-C., 2002. Atomic-layer-resolved quantum oscillations in the work function: theory and experiment for Ag/Fe(100). *Physical Review B (Condensed Matter and Materials Physics)*, **66**(23), 233403.

Papaconstantopoulos, D.A., 1986. *Handbook of the Band Structure of Elemental Solids*. New York: Plenum Press.

Patthey, F. and Schneider,W.-D., 1994. Layer-by-layer-resolved quantum-well states in ultrathin silver islands on graphite: a photoemission study. *Physical Review B (Condensed Matter and Materials Physics)*, **50**(23), 17560–17563.

Ricci, D.A., Liu, Y., Miller, T. and Chiang, T.-C., 2009. Analyticity of the phase shift and reflectivity of electrons at a metal–semiconductor interface. *Physical Review B (Condensed Matter and Materials Physics)*, **79**(19), 195433.

Ricci, D.A., Miller, T. and Chiang, T.-C., 2005. Controlling the thermal stability of thin films by interfacial engineering. *Physical Review Letters*, **95**(26), 266101.

Rioux, D. and Höchst, H., 1992. Sb/InP(100) interface: a precursor to surfactant-mediated Ge epitaxy. *Physical Review B*, **46**(11), 6857–6863.

Smith, A.R., Chao, K., Niu, Q. and Shih, C., 1996. Formation of atomically flat silver films on GaAs with a 'silver mean' quasi periodicity. *Science*, **273**(5272), 226–228.

Smith, N.V., 1985. Phase analysis of image states and surface states associated with nearly-free-electron band gaps. *Physical Review B (Condensed Matter and Materials Physics)*, **32**(6), 3549–3555.

Smith, N.V. and Mattheiss, L.F., 1974. Photoemission spectra and band structures of *d*-band metals. I. Practical aspects of fcc interpolation schemes. *Physical Review B (Condensed Matter and Materials Physics)*, **9**(4), 1341–1352.

Speer, N.J., Tang, S.-J., Miller, T. and Chiang, T.-C., 2006. Coherent electronic fringe structure in incommensurate silver-silicon quantum wells. *Science*, **314**(5800), 804–806.

Tang, S.-J., Basile, L., Miller, T. and Chiang, T.-C., 2004. Breakup of quasiparticles in thin-film quantum wells. *Physical Review Letters*, **93**(21), 216804.

Tang, S.-J., Lee, Y.-L., Chang, S.-L., Miller, T. and Chiang, T.-C., 2006. Umklapp-mediated quantization of electronic states in Ag films on Ge(111). *Physical Review Letters*, **96**(21), 216803.

Upton, M.H., Miller, T. and Chiang, T.-C., 2004. Absolute determination of film thickness from photoemission: application to atomically uniform films of Pb on Si. *Applied Physics Letters*, **85**(7), 1235–1237.

Upton, M.H., Miller, T. and Chiang, T.-C., 2005. Unusual band dispersion in Pb films on Si(111). *Physical Review B (Condensed Matter and Materials Physics)*, **71**(3), 33403.

Upton, M.H., Wei, C.M., Chou, M.Y., Miller, T. and Chiang, T.-C., 2004. Thermal stability and electronic structure of atomically uniform Pb films on Si(111). *Physical Review Letters*, **93**(2), 026802.

Wang, K., Zhang, X., Loy, M.M., Chiang, T.-C. and Xiao, X., 2009. Pseudogap mediated by quantum-size effects in lead islands. *Physical Review Letters*, **102**(7), 076801.

Wei, C.M. and Chou, M.Y., 2002. Theory of quantum size effects in thin Pb(111) films. *Physical Review B (Condensed Matter and Materials Physics)*, **66**(23), 233408.

Wilson, J.A. and Yoffe, A.D., 1969. The transition metal dichalcogenides: discussion and interpretation of the observed optical, electrical and structural properties. *Advances in Physics*, **18**(73), 193–335.

Zhang, Z., Niu, Q. and Shih, C., 1998. 'Electronic growth' of metallic overlayers on semiconductor substrates. *Physical Review Letters*, **80**(24), 5381–5384.

Zhang, Z., Zhang, Y., Fu, Q., Zhang, H., Yao, Y., Ma, T., Tan, D., Xue, Q. and Bao, X., 2008. Modulation of surface reactivity via electron confinement in metal quantum well films: O_2 adsorption on Pb/Si(111). *The Journal of Chemical Physics*, **129**(1), 014704.

3
Phase-field modeling of thin film growth

A. VOIGT, Technische Universität Dresden, Germany

Abstract: Step-flow models are an efficient multiscale approach to model film growth, as they allow kinetic effects due to atomic interactions to be incorporated in continuum models. We consider phase-field approximations of such models and demonstrate their applicability through various examples, including trench formation on W(110), spiral and mound growth on Pt(111) and anomalous spiral growth on Si(001).

Key words: phase-field modeling, Ehrlich–Schwoebel barrier, degenerate mobility, spiral growth.

3.1 Introduction

We think of the crystal film as an atomic landscape of terraces, separated by steps of one atomic height and build our modeling description of crystal thin film growth on it. Microscopic processes in crystal thin film growth include the deposition of atoms or molecules on the terraces of the film, adatom (adsorbed atoms) desorption from the terraces, adatom diffusion on terraces, and the attachment and detachment of adatoms to and from the steps, respectively. Through the net attachment, the position of the steps changes, which means that the upper terrace expands by overgrowing the terrace below. This induces vertical growth of the film. If a finite number of adatoms meet, they coalesce and form a new terrace, which induces vertical growth by nucleation. When a screw dislocation intersects a crystal surface, a step emerges from the intersection point. This step cannot be removed by growth or evaporation. The attachment of adatoms to such steps results in spiral growth with the step turning around the intersection point, which also leads to vertical growth.

We use a phase-field approximation of the moving boundary value problem first formulated by Burton, Cabrera, and Frank (1951) and today known as BCF-type models. Key parameters are the surface diffusion coefficient D for adatom diffusion on terraces and the deposition flux F, as well as the kinetic coefficients k governing the attachment of adatoms to a step from the lower and upper terrace, respectively. Step thermodynamics enter through the step stiffness and the equilibrium adatom concentration. In the phase-field approximation the crystal surface is represented by a continuous field which couples to the adatom concentration and is subjected to a multiwell potential that pins the phase-field variable to the prefered integer value representing the height of the discrete terraces. From a numerical point of view, it is desirable to have such a diffuse interface approximation. It smears out the discrete height function (which counts the atomic monolayers) on a length scale ε. The smeared-out height function (phase-field variable) ϕ thus (approximately) describes the position of the steps. The advantage is that both ϕ and the adatom density ρ can be discretized with respect to the same grid. Moreover, the vanishing of islands and the coalescence of steps are automatically handled. Diffuse interface approximations for BCF-type models have been introduced in Liu and Metiu (1997) and Karma and Plapp (1998). In Liu and Metiu (1997) the motion of a one-dimensional step train is analysed, whereas in Karma and Plapp (1999) the growth of a spiral is simulated. In Yeon *et al.* (2003) the effect of a lattice mismatch between substrate and film is incorporated into the phase-field approximation. The Ehrlich–Schwoebel barrier, an additional barrier an adatom at a 'step down' has to overcome in order to attach at the step, has been incorporated into these phase-field approximations by Otto *et al.* (2004). With these modifications it was possible to analyse growth instabilities using phase-field approximations. Among them are step bunching instabilities as a result of elastic interactions (Yeon *et al.* 2007), step meandering as a result of an Ehrlich–Schwoebel barrier (Hausser and Voigt 2008, Hu *et al.* 2011), mounding resulting from an Ehrlich–Schwoebel barrier and strong anisotropic step edge barriers (Yu *et al.* 2008) as well as spiral growth (Redinger *et al.* 2008, Yu *et al.* 2009).

We will review these phase-field approximations and apply them to simulate trench formation on W(110), a competition between spiral and mound growth on Pt(111) as well as anomalous spiral growth on Si(001).

3.2 Modeling

We recall the classical BCF model. The adatom density on the terraces obeys the following diffusion equation

$$\rho_t + \nabla \cdot \mathbf{j} = F, \qquad \mathbf{j} = -D\nabla\rho \tag{3.1}$$

where D is the diffusion constant and F is the deposition flux rate. Desorption of adatoms has been neglected. The fluxes of the adatoms to a step are given by

$$j_{\pm} = \pm\, \mathbf{j}_{\pm} \cdot \mathbf{n} - \rho_{\pm} v \tag{3.2}$$

where subscripts +, – denote quantities at the upper and lower terrace respectively, n denotes the normal pointing from upper to lower terrace and v is the normal velocity of the step. Assuming first-order kinetics for the attachment and detachment of adatoms at the steps, the adatom densities satisfy the following kinetic boundary conditions at the step:

$$j_{\pm} = k_{\pm}(\rho_{\pm} - \rho_{eq}) \qquad [3.3]$$

With this notation, asymmetric attachment rates $0 < k_+ < k_-$ model the Ehrlich–Schwoebel effect. The equilibrium density ρ_{eq} is given by the linearized Gibbs–Thomson relation

$$\rho_{eq} = \rho^*(1 + \Gamma k), \quad \Gamma = \Omega(\gamma + \gamma'')/(k_B T) \qquad [3.4]$$

where $\Omega = a^*a$ is the atomic area, 'a' being the lattice spacing, 'γ' the step free energy density, and $\gamma + \gamma''$ the step stiffness. Finally, the normal velocity of a step is given by

$$v = \Omega(j_+ + j_-) + a\nabla_s q, \quad q = -D_{step}\nabla_s(\Gamma k) \qquad [3.5]$$

where ∇_s denotes the tangential derivative along the step and D_{step} is the diffusion constant at atoms along the step. The second term in the velocity law 3.5 represents edge diffusion of edge atoms along the step, whereas the first term ensures mass conservation. Other effects such as step permeability or downward funneling will be neglected.

A phase-field formulation for Eqs 3.1–3.5 has been given and justified by matched asymptotic expansions by Hu *et al.* (2011). We will here consider a simpler form in which the edge diffusion term is neglected and a quasi-stationary approximation is assumed. The phase-field model then reads in the isotropic case in non-dimensional form:

$$\varphi_t + \nabla \cdot j = 1 \qquad [3.6]$$

$$(1 + \varepsilon^{-1}\zeta_2(\varphi))\, j = -\nabla\omega \qquad [3.7]$$

$$\varepsilon\zeta_1(\varphi)\varphi_t - \varepsilon\nabla^2\varphi + \varepsilon^{-1}G'(\varphi) = \omega \qquad [3.8]$$

with an appropriate normalized multiwell potential $G(\varphi)$, being a periodic function of φ with period 1, which vanishes for $\varphi = \ldots, -1, 0, 1, \ldots$, appropriate coefficients $\zeta_1(\varphi)$ and $\zeta_2(\varphi)$, which are again periodic functions of φ with period 1. They can be interpreted as friction coefficients for step and adatom movement, respectively. An appropriate choice of $\zeta_1(\varphi)$ and $\zeta_2(\varphi)$ allows the approximation of any kinetic coefficients k_+ and k_-, which can be shown using matched asymptotic analysis. The model can be extended to include anisotropic effects in the same way as done in classical phase-field models (see, e.g., Torabi *et al.* 2009).

3.3 Numerical results

We will demonstrate various simulation results of practical interest, which are validated by experiments on mound and spiral growth. We thereby highlight the influence of the Ehrlich–Schwoebel barrier, anisotropic kinetics and stress effects.

3.3.1 Trench formation

Using the phase-field model we simulate formation and growth of stripe arrays, which result from anisotropic growth of islands under the Ehrlich–Schwoebel barrier. Due to the Ehrlich–Schwoebel barrier, mounds form and coarsen during growth. Due to the anisotropy, the coarsening is slow in one direction and fast in the other direction, which causes formation of trench arrays. The anisotropy is incorporated in terms of attachment kinetics and is locally limited at the island edge aligned to the <100> orientation and instantaneous at the island edge aligned to the <110> orientation. This anisotropy suppresses coarsening in the <110> direction in the early stages. After a period of time the stripe arrays develop a quasi-periodic uniaxial structure with a selected transversal slope and a decreasing longitudinal slope as a consequence of the competition between coarsening and rougening. A detailed discussion can be found in Yu *et al.* (2008). Figure 3.1 shows a typical configuration for W(110) which is in qualitative agreement with the experimental results in Borca *et al.* (2007).

The formed arrays have the potential to be used as templates for the growth of one-dimensional nanostructures. A key feature to produce nanowires using this approach is the perfect ordering of the trenched templates. The final ordering of the trenches depends sensitively on the nucleation of islands in the submonolayer regime. By influencing the nucleation, e.g. through a periodic strain field of a buried misfit dislocation network or a compliant

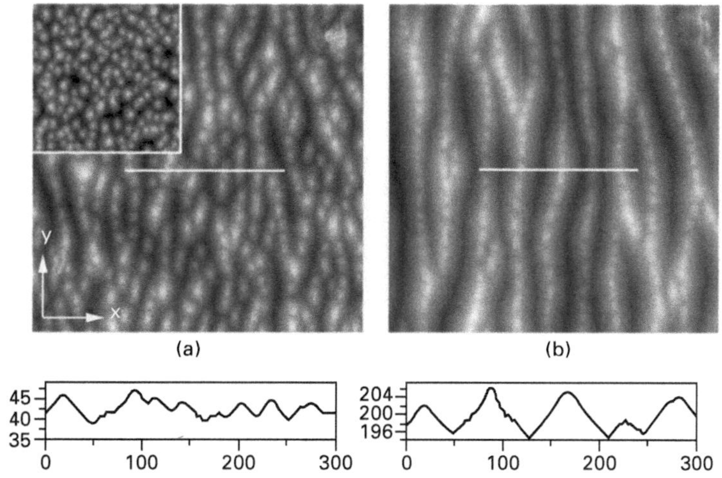

3.1 Simulated trench morphology on W(110) after deposition of (a) 40 ML and (b) 200 ML, and cross-section profile along the white line at the bottom. Inset: morphology of nucleated islands after deposition of 2 ML in cut image of the same scale as the main image. The simulations were performed by Y.-M. Yu.

substrate, ordered arrays can be formed. In Yu and Voigt (2009) such a strain field is used to nucleate islands in an ordered array which leads to perfectly ordered trenches with tunable size in the phase-field simulation.

3.3.2 Spiral and mound growth

The large-scale shape of spirals and mounds are similar. However, it is clear that the growth mechanism near the top of the structure is completely different for spiral mounds and so-called wedding cakes. The growth of wedding cakes is governed by the rate of two-dimensional nucleation of islands on the top terrace, which is enhanced by the confinement of adatoms due to the Ehrlich–Schwoebel barrier. In contrast, atoms landing near the top of the spiral mound can avoid the Ehrlich–Schwoebel barrier by moving around the core which should reduce the local adatom concentration and hence the growth rate. On the other hand the indelible presence of the step emanating from the screw dislocation obviates the need for two-dimensional nucleation, which should increase the rate of vertical growth. We use a setup in which both growth modes are possible and are interested in their competition. Two-dimensional nucleation is thereby modeled as an intrinsic feature of the phase-field approximation. If the adatom density becomes sufficiently large a new terrace nucleates as a result of the changing barrier in the multiwell potential. Figure 3.2 shows a typical surface configuration after deposition of 200 monolayers. Under these conditions in which the adatom concentration is allowed to become sufficiently large so that islands nucleate deterministically, spiral mounds were generally seen to grow higher than regular wedding cakes. The results nicely coincide with the experiments on Pt(111) surfaces in Redinger *et al.* (2008).

3.3.3 Anomalous spiral growth

The approach discussed in the previous section neglects the influence of the strain field resulting from the screw dislocation. Taking this effect into account might lead to anomalous spiral step motion as found on a Si(001) surface by Hannon *et al.* (2006) and Voorhees (2006). The anomalous behavior is attributed to interaction between the surface structure and the strain field of the screw dislocation. The screw dislocation thereby breaks two single atomic height steps. At each step, the surface phase switches between (1×2) and (2×1). The strain field of the screw dislocation causes an asymmetry between the two phases, with one phase more energetically favored than the other. The step growth is coupled with the asymmetric phase switching between (1×2) and (2×1), causing the anomalous spiral profile. The related dynamics for this growth mode is analysed by Yu *et al.* (2009). In order to account for the elastic effects we use alternating surface

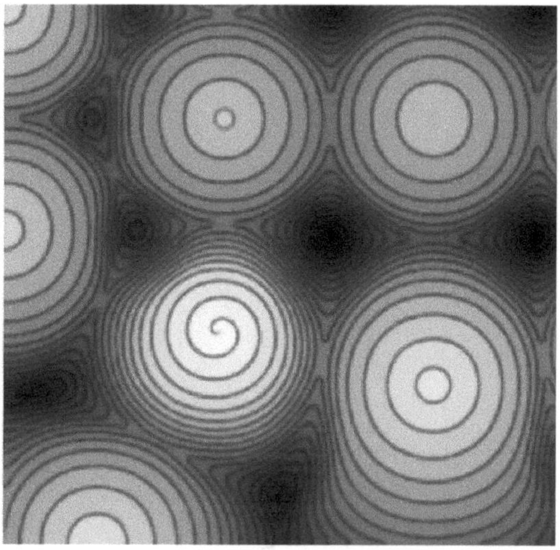

3.2 Growth spirals and mounds obtained from phase-field simulation with $D = 10$, $F = 0.2$ ML/s, $\gamma = 1$, $\rho^* = 0.1$, $k_+ = 10$ and $k_- = 1$. The surface is shown after deposition of 200 ML showing spiral mounds and wedding cakes and demonstrating the faster growth of the spiral mounds. The simulations were performed by A. Rätz.

phases of (1 × 2) and (2 × 1) with different surface energies. We define $\Delta\gamma = \gamma_{2\times1} - \gamma_{1\times2}$ corresponding to the strain field of the screw dislocation, which enters the definition of the multiwell potential. Figure 3.3 shows typical configurations under the influence of $\Delta\gamma$.

3.4 Conclusion

Phase-field models have been used to approximate BCF-type models for step flow growth. The phase-field variable thereby models the discrete height of the film. The novelty of such models is the ability to approximate not only the jump in the fluxes across the step edges but also the jump in the density field which results from the kinetic boundary conditions. This is realized through a degenerate mobility function in the phase-field equation, which can be interpreted as a degenerate Cahn–Hilliard-type equation. The phase-field approach allows the nucleation of new islands, it allows for coalescence and also for vanishing of islands, without special treatment. The derived set of equations can be solved with standard tools; however, an adaptive mesh refinement along the step edges is desirable. Using matched asymptotic analysis the connection between these phase-field models and the BCF model can be shown.

The approach is applied to understand various interesting phenomena in thin film growth. Among them are the formation of trenches as a result

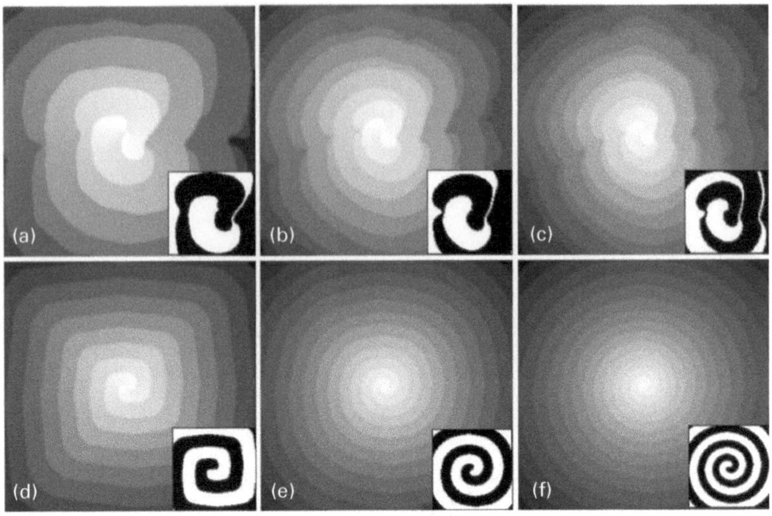

3.3 Images of the spiral growth at the steady-state regime simulated using $F = 0.005$ ML/s (left), 0.02 ML/s (middle) and 0.04 ML/s (right) for non-zero in (a)–(c) and zero in (d)–(f). The lightest gray color corresponds to the highest surface height, being around 109, 193, 231, 111, 200 and 233 atomic layers in (a)–(f), respectively. The sharp change in gray indicates the position of the step. Inset (enlarged): the corresponding simulated image of the surface termination, where the (2×1) and (1×2) phases are denoted by bright and dark, respectively, and the position of the step is indicated by switching between bright and dark. The simulations were performed by Y.-M. Yu.

of the Ehrlich–Schwoebel barrier and anisotropic kinetic coefficients, the competition between mound and spiral growth, which for typical material parameters will always be dominated by spirals, and an anomalous spiral growth mode, which results from the interaction of the strain field associated with the screw dislocation and the surface structure.

3.5 References

Borca B, Fruchart O, David P, Rousseau A, Meyer C (2007) 'Kinetic self-organization of trenched templates for the fabrication of versatile ferromagnetic nanowires', *Appl. Phys. Lett.* 90 142507.

Burton W K, Cabrera N, Frank F C (1951) 'The growth of crystals and the equilibrium of their surfaces', *Phil. Trans. R Soc. A* 243 299–358.

Hannon J B, Shenoy V B, Schwarz K W (2006) 'Anomalous spiral motion of steps near dislocations on silicon surfaces', *Science* 313 1266–1269.

Hausser F, Voigt A (2008) 'Step meandering in epitaxial growth', *J. Cryst. Growth* 303, 80–84.

Hu Z Z, Lowengrub J S, Wise S M, Voigt A (2011) 'Phase-field modeling of epitaxial growth: applications to step trains and island dynamics', *Physica D* (to appear)

Karma A, Plapp M (1998) 'Spiral surface growth without desorption', *Phys. Rev. Lett.* 81 4444–4447.

Liu F, Metiu H (1997) 'Stability and kinetics of step motion on crystal surfaces', *Phys. Rev. E* 49 2601–2616.

Otto F, Penzler P, Rätz A, Rump T, Voigt A (2004) 'A diffuse interface approximation for step flow in epitaxial growth', *Nonlin.* 17 477–491.

Redinger A, Ricken O, Kuhn P, Rätz A, Voigt A, Krug J, Michely T (2008) 'Spiral growth and step edge barriers', *Phys. Rev. Lett.* 100 035506.

Torabi S, Lowengrub J, Voigt A, Wise S (2009) 'A new phase-field model for strongly anisotropic systems', *Proc. Roy. Soc. A* 465 1337–1359.

Voorhees P W (2006) 'Materials science – step dances on surfaces', *Science* 313 1247–1248.

Yeon D H, Cha P R, Chung S I, Yonn J K (2003) 'Phase-field model for the dynamics of steps and islands on crystal surfaces', *Metals Mater. Int.* 9 67–76.

Yeon D H, Cha P R, Lowengrub J S, Voigt A, Thornton K (2007) 'Linear stability analysis for step meandering instabilities with elastic interactions and Ehrlich–Schwoebel barriers', *Phys. Rev. E* 76 011601.

Yu Y-M, Voigt A (2009) 'Directed self-organization of trenched templates for nanowire growth', *Appl. Phys. Lett.* 94 043108.

Yu Y-M, Backofen R, Voigt A (2008) 'Phase-field simulation of stripe arrays on metal bcc(110) surfaces', *Phys. Rev. E* 77 051605.

Yu Y-M, Liu B G, Voigt A (2009) 'Phase-field modeling of anomalous spiral step growth on Si(001) surfaces', *Phys. Rev. B* 79 235317.

4 Analysing surface roughness evolution in thin films

Y. KAJIKAWA, The University of Tokyo, Japan

Abstract: Surface roughness is a key factor in determining film quality and the physical properties of materials deposited by chemical vapor deposition (CVD). However, there have been few attempts to comprehensively overview the mechanism underlying the evolution of film roughness during deposition, from atomic to macroscopic scale. In this chapter, the mechanisms causing roughness evolution during CVD are surveyed from a phenomenological viewpoint. Firstly, roughness formation processes at atomic scale are reviewed, focusing on homo-epitaxial growth. Next, roughness generation during hetero- or non-epitaxial growth is discussed, as well as the process of roughness amplification. In the growth stage of a film, roughness evolution can be classified into two types: positive feedback growth and non-positive feedback growth, and these are discussed in the final part of the chapter.

Key words: surface roughness, nucleation and growth, mechanism, modeling.

4.1 Introduction

Chemical vapor deposition (CVD) involves many chemical and physical phenomena: in the gas phase with a flow of a mixture of reactant and carrier gas, in the solid phase of deposited materials, and at the surface where a variety of elementary processes occur to determine the structure and physical properties of deposited materials. By controlling operating conditions and reactor geometry, highly uniform coatings can be obtained over a broad range of scale levels, from atomic to reactor. However, CVD often produces a variety of surface roughnesses which in most cases is not suitable for application. Because currently the roughness is suppressed by trial-and-error adjustments of operating conditions, at factory level, many wafers are wasted to find a good recipe for the process. Although a better understanding of the underlying phenomena enables better design of CVD equipment and efficient adjustment of operating conditions, there are few comprehensive surveys on the underlying mechanism of surface roughness evolution.

To date, a number of reviews have been published on roughness evolution during vapor deposition, owing to its importance in application and also to scientific research interests (Meakin, 1993; Krug, 1997; Politi *et al.*, 2000).

These reviews usually focus only on the specific phenomena causing roughness evolution, such as anisotropic diffusion and shadowing instability, and its strict mathematical expressions. They can provide scholars with guidelines for simulating roughness evolution under virtual conditions. However, from a pragmatic view, such approaches do not meet the common demand to restrain the roughness evolution. The aim of this chapter is to review roughness evolution processes during CVD from a phenomenological viewpoint, rather than focusing on mathematical details.

4.2 Roughness during homo-epitaxial growth

4.2.1 Thermodynamics

The surface of a film can become rough during homo-epitaxial growth through both thermodynamic and kinetic factors. The surface of a single crystal under thermodynamic equilibrium has an atomically smooth structure at zero temperature. However, at high temperatures, the surface of a single crystal is atomically rough even under thermodynamic equilibrium. The transition temperature at which the surface of a single crystal in thermodynamic equilibrium becomes rough is called the roughening temperature, and the transition is called the roughening transition (Burton *et al.*, 1951). This can be understood by the increase in entropy with increasing temperature. In this case, the surface with atomic-scale roughness is the most stable. At temperatures higher than the roughening temperature, facets disappear and isolated crystals have a curved morphology which reflects a rough surface on the facets (Heyraud and Métois, 1987).

4.2.2 Kinetics

Below the roughening temperature, surface roughness can appear due to kinetic factors when there is not sufficient relaxation to reach thermodynamic equilibrium. This process is called kinetic roughening. In kinetic roughening, adatoms do not have sufficient time to migrate on the surface to a stable position such as kinks and steps. One of the factors influencing roughness evolution in kinetic roughening is the nucleation and growth of islands. Another is surface diffusion governed by the Ehrlich–Schwoebel barrier (Ehrlich and Hudda, 1966; Schwoebel and Shipsey, 1966).

Figure 4.1 schematically illustrates elementary processes during homo-epitaxial growth. Nucleation on the two-dimensional layer involves the arrival and adsorption of reactants on the surface (deposition), dissociation of coordinated molecules to become adatoms, their migration along the surface (diffusion), and aggregation of adatoms to create islands. These processes (adsorption, diffusion, and aggregation) are the causes of roughness evolution.

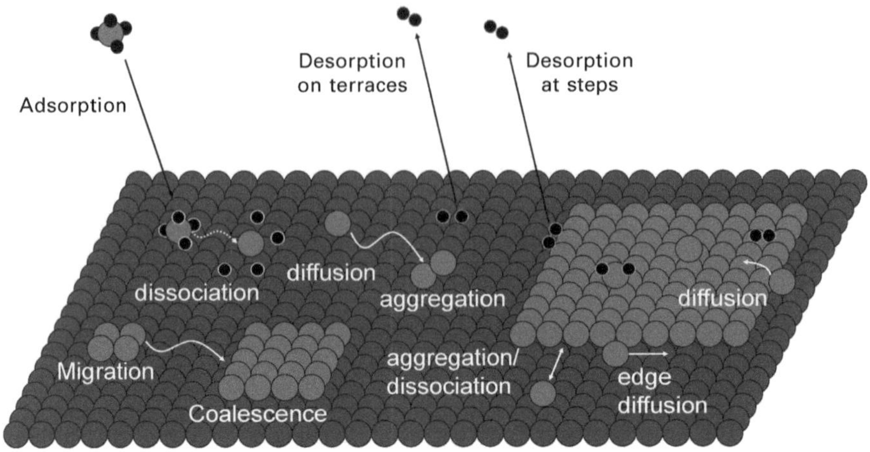

4.1 Elementary processes on the growing surface.

During epitaxial growth, adatoms arriving on the substrate surface migrate to nucleate 2D islands, and these islands expand laterally by capturing other migrating adatoms. When the nucleation rate on each layer is slow, islands grow laterally by step-flow growth, and as a result the surface becomes smooth because adatoms bond preferentially to steps and kinks in the existing islands. When islands are created at a fast nucleation rate, the surface formed is rough. Therefore, controlling nucleation rate is crucial to suppress surface roughness, which is in turn governed by a variety of processes illustrated later.

Another factor we must consider is surface diffusion between different layers. Atoms arriving on the islands also contribute to this lateral layer-by-layer growth when the diffusion barrier at the edge of the island is small. This barrier is called the Ehrlich–Schwoebel barrier. When the barrier becomes too large to allow interlayer transportation, 3D island growth occurs. This is called the Schwoebel effect. At sufficiently high temperatures, this effect is negligible. An off-oriented substrate is also used to enhance the step-flow growth and realize the smooth surface, minimizing the chance of 3D growth. However, in this vicinal surface, existing steps are sometimes observed to form bunches, giving a rough surface (Matsunami and Kimoto, 1997). This step bunching is caused by the larger deposition flux on larger terraces, by the step edge barrier and by different levels of diffusion on the different terraces (Saito and Uwaha, 1994).

Predicting island density and size distribution from an atomistic view is the primary task of nucleation theories. Figure 4.1 shows the elementary processes taking place on the surface that contribute to surface roughness at the atomic level. Because it is difficult to simultaneously account for all of the elementary processes in theoretical models, such as kinetic Monte Carlo simulations (KMC), usually only a subset of the complete set of processes are accounted for in a given model. There are a number of papers on nucleation models simulating these processes. One goal of nucleation models is to provide information on the maximum island density, N_{max}, such as $N_{max} \sim (F/D)^{\chi}$, where F is the flux of the growth species, D is the surface diffusivity of adatoms, and χ is a scaling exponent.

Determining χ for various processes is one goal of most nucleation studies. Control of nucleation density is an important key factor in restraining roughness during homo-epitaxial growth. In order to maintain step-flow growth, the island density must be kept as small as possible, which requires low F and large D depending on χ. However, when 3D growth of islands becomes dominant, such a low nucleation density becomes the cause of rough surfaces. This is because small island density allows large spaces where islands can grow and results in a rough surface caused by agglomeration of large islands. We can use nucleation theory to deduce not only island density but also the dominant processes (Kajikawa *et al.*, 2004a).

The simplest nucleation theory is the deposition-diffusion-aggregation (DDA) model, which, as its name suggests, includes the processes of deposition, diffusion, and aggregation. Because review papers describing nucleation theory have already been published (Venables *et al.*, 1984; Stoyanov and Kaschiev, 1981; Ratsch and Venables, 2003), in this chapter we review theoretical efforts to model deposition, diffusion, aggregation and other complex elementary processes such as island migration and defect trapping. Although understanding of such elementary processes can be used as a basis for understanding the root cause of roughness evolution during homo-epitaxial growth, it is not possible with current nucleation models to simultaneously include all the physical processes that can take place. Therefore, we review indices for determining which processes are active in a given system and therefore must be included in a corresponding model. The following models focus mainly on homo-epitaxial growth. However, it is possible to obtain a fundamental understanding of the elementary processes relating to other types of film growth (e.g. hetero-epitaxial and non-epitaxial growth) which determine the morphology of deposited films.

Deposition-diffusion-aggregation (DDA) model

The DDA model assumes that growing species adsorb onto a surface, diffuse, and then collide with other adatoms to create stable islands. A phenomenological version of the DDA model has already been described in the literature (Villain *et al.*, 1992; Jensen *et al.*, 1997, 1998), and is summarized in the following.

The adatom density, ρ, is determined by a balance between deposition and capture by islands, so that

$$\rho \sim F/DN \quad [4.1]$$

where F is the deposition rate per surface site per unit of time, D is the surface diffusion rate of adatoms, and N is the island density. If the contribution of direct impingement to nucleation is negligible, the nucleation rate of islands can be estimated as

$$j \sim D\rho^2 \sim N_{max}/t_c \quad [4.2]$$

where N_{max} is the maximum island density obtained and t_c is the time when islands come into contact. N_{max} is obtained at time t_c. In Eq. 4.2, we assume the nucleation rate is constant. For islands with a 3D shape, the growth rate of islands can be expressed as

$$d(R^3)/dt \sim D\rho \quad [4.3a]$$

If the rate of edge diffusion and deformation is not sufficiently fast to maintain a 3D structure, then Eq. 4.3a is expressed as

$$d(R^{d_f})/dt \sim D\rho \quad [4.3b]$$

where d_f is the fractal dimension of the islands. At the limit of slow edge diffusion and deformation, the island shape is determined by diffusion-limited-aggregation (DLA), with $d_f = 1.7$ (Tang, 1993). For fast edge diffusion and no deformation, islands become 2D disks, and $d_f = 2$. At $t = t_c$

$$R \sim N^{-1/2} \quad [4.4]$$

so that N_{max} becomes (Villain *et al.*, 1992; Tang, 1993):

$$N_{max} \sim (F/D)^{2/4+d_f} \quad [4.5]$$

Therefore, $N_{max} \sim (F/D)^{2/7}$ for 3D islands, and $N_{max} \sim (F/D)^{1/3}$ for 2D islands, which agrees well with Kinetic Monte Calo (KMC) simulations (Villain *et al.*, 1992).

There are some variations on the DDA model. In the DDA model, atoms are assumed to deposit at random positions on a surface at flux F. In some cases temporal fluctuations of the deposition flux, like the situation in modulated CVD, are treated by both the rate equation (RE) and KMC models (Jensen and Niemeyer, 1997; Combe and Jensen, 1998). Another model includes anisotropic diffusion. In most deposition processes, surface diffusion is expected to occur as 2D Brownian diffusion. Other modes of diffusion such as anisotropic and long-range diffusion can modify χ from 1/3 to 1/4 for 2D islands (Villain *et al.*, 1992; Amar *et al.*, 1998).

Evaporation

Isolated adatoms can evaporate from the surface, and models including evaporation, deposition, diffusion, and aggregation have been developed.

Using the phenomenological approach, Jensen *et al.* (1997, 1998) derived N_{max} as follows. If evaporation exists, the adatom density quickly approaches a stable value after deposition starts, i.e.,

$$\rho \sim F\tau_e \quad [4.6]$$

where τ_e is the mean time of evaporation for an adatom. When direct impingement exists, this increases the number of islands linearly according to

$$j \sim (F + D\rho)\rho \quad [4.7]$$

For 3D islands, island growth is governed by

$$d(R^3)/dt \sim F(X_s + R)^2 \quad [4.8]$$

where X_s is the mean diffusion length of adatoms before desorption, expressed as $X_s = (D\tau_e)^{1/2}$. Combining these equations, the maximum island density can be expressed as

$$N_{max} \sim (F\tau_e)^{2/3}(1 + X_s^2)^{2/3} \quad [4.9]$$

From the RE approach, Jensen *et al.* (1997, 1998) also derived island growth rates and deposition rates when coalescence starts. The scaling exponent was checked using KMC simulations, which showed good agreement with Eq. 4.9.

Edge diffusion/deformation

A monomer reaching an existing island sticks to the edge of it. If it cannot move, diffusion limited aggregation (DLA) occurs. When islands are small, islands remain compact to minimize the free energy. The fractal dimension of islands formed during DLA is about 1.7 (Tang, 1993). Therefore, if edge diffusion and deformation are not included in the model, it is possible to overestimate the island size, which decreases the accuracy of the simulated island density and size distribution (see Eq. 4.5). Detailed models including edge diffusion and deformation have been presented by Amar and Family (1996), Jensen *et al.* (1999) and Krug (2000).

Direct impingement

Direct impingement onto existing islands can also change the scaling behavior. Most theoretical models neglect this effect, and assume that monomers deposited on top of islands has no effect on the system. This implies that either the Ehrlich–Schwoebel barrier prevents monomers from falling onto the bottom layer or that the diffusion rate onto the second layer is lower than that onto the bottom layer. Direct impingement is important when either the island coverage becomes large, or there exists a difference in sticking probability between the substrate and the islands, which is a typical situation in CVD (Lantz *et al.*, 1994; Werner *et al.*, 1992; Kajikawa *et al.*, 2004b). When a large anisotropy of the sticking probability exists between islands and substrate, $N_{max} \sim F^{2/3}$, which corresponds to the high evaporation limit in Eq. 4.9 (Jensen *et al.*, 1997).

Aggregation, aggregation/dissociation, breakup

The DDA model assumes that if two adatoms occupy neighboring sites, they stick together to form an island. This is often called irreversible aggregation. More general models include reversible aggregation, where small islands can dissociate. According to Walton's relation, assuming the existence of a critical size, s^*, below which islands tend to dissociate, the density of critical islands, N_{s*}, is given by $N_{s*} \sim \rho^{s^*}$ (Venables *et al.*, 1984; Stoyanov and Kaschiev, 1981). The nucleation rate in DDA models (Eq. 4.2) can therefore be written as:

$$j \sim D\rho N_{s*} \sim D\rho^{s^*+1} \qquad [4.10]$$

For 3D islands, using Eqs. 4.1, 4.3a, and 4.10, we can write

$$N_{max} \sim (F/D)^{2s^*/2s^*+5} \qquad [4.11]$$

Therefore, including dissociation increases the scaling exponent (Ratsch *et al.*, 1994, 1995), decreases the island density (Salik, 1985), and changes the island size distribution (Chambliss and Johnson, 1994). Adding dissociation and desorption processes to DDA models yields (Jensen *et al.*, 1998):

$$N_{max} \sim (F\tau_e)^{2s^*/3}(1 + X_s^2)^{2/3} \qquad [4.12]$$

More accurate expressions for representing aggregation/dissociation processes have been developed. Detailed analysis has been focused on aspects of these processes including Ostwald ripening (Bales and Zangwill, 1997), effect of surfactants (Markov, 1996; Kotrla *et al.*, 2000), and strain in hetero-epitaxy (Dobbs *et al.*, 1997).

Migration

In the DDA model the possibility of dimers, trimers, or larger islands diffusing on the substrate is neglected. Island diffusion has been observed in experiments (Kern *et al.*, 1979; Jensen, 1999) and in molecular dynamics simulations (Deltour *et al.*, 1997; Luedtke and Landman, 1999). Villain *et al.* (1992) included the dimer mobility and derived the scaling form, $N_{max} \sim (F/D_1D_2)^{1/5}$ for 2D islands, where D_1 and D_2 represent the diffusivity of the monomer and dimer respectively. For $D_1 = D_2$, $\chi = 2/5$, which is higher than the value of $\chi = 1/3$ for 2D islands in the DDA model. Lee and Gupta (2000) considered dissociation from dimers by adding a dimer mobility, and found that χ increased to greater than 2/5. The effect of island migration has been studied by assuming the diffusivity of islands can be expressed as a power of their size (Mulheran and Robbie, 2001),

$$D_s = D_1 s^{-\gamma} \ (0 \leq \gamma) \quad [4.13]$$

Increasing the island mobility decreases the island density and increases the scaling exponent χ. For example, Mulheran and Robbie (2001) showed that for $\gamma = 9$, $\chi = 0.35$, but that for $\gamma = 0$, $\chi = 0.45$.

Migration affects not only χ, but also the coverage at maximum island density, θ_{max}, and the island size distribution. Mulheran and Robbie (2001) also showed that θ_{max} can be scaled as $\theta_{max} \sim (F/D)^{\omega}$ and the effect of migration is that for $\gamma = 9$, $\theta_{max} = 0.05$, and for $\gamma = 0$, $\theta_{max} = 0.45$. Moreover, the scaled island size distributions are a strong function of island mobility. Kuipers and Palmer (1996) observed that with increasing island mobility, the scaled size distribution changes and becomes bimodal. The largest mode also exists without island mobility, and as the island mobility increases, it becomes higher, narrower, and shifts to larger island sizes. The second mode is created by the formation of new, small islands.

Coalescence

Island density is reduced by coalescence. Coalescence occurs when island coverage is relatively high. For immobile islands, N_{max} occurs at $\theta_{max} = 0.15$. For mobile islands, θ_{max} is decreased and depends on F and D. Most modeling studies have neglected the coalescence term by focusing on the region where θ is sufficiently small that coalescence does not take place. Modeling studies have been presented that include both nucleation and coalescence (Amar *et al.*, 1994; Bartelt and Evans, 1994; Mulheran and Robbie, 2001). Coalescence without island mobility is called static coalescence, where the loss rate of islands by static coalescence is generally expressed as (Venables *et al.*, 1984; Stoyanov and Kaschiev, 1981):

$$(d(R^2)/dt)N^2 \quad [4.14a]$$

Coalescence with island mobility is called dynamic coalescence. Templier *et al.* (2000) proposed an expression for the loss rate of islands by dynamic coalescence as:

$$\alpha(d((R + R_m)^2)/dt)N^2 \quad [4.14b]$$

where α is a constant and R_m is a coefficient representing the effective distance of island migration. By fitting R_m to experimentally observed temporal changes of island density, Templier *et al.* (2000) showed that R_m

$\propto R^{-1/2}$. Urbonavičius *et al.* (1999) and Galdikas (2002) also expressed the dynamic coalescence rate as

$$\alpha C_{\text{col}}\{N(R + R_{\text{m}})^2 - 1\} \quad [4.14c]$$

where $C_{\text{col}} = 0$ when $N(R + R_{\text{m}})^2 < 1$, and $C_{\text{col}} = 1$ when $N(R + R_{\text{m}})^2 > 1$. When islands touch each other, they coalesce and assume a thermodynamically stable form. In droplet growth models (Family and Meakin, 1989; Carrey and Maurice, 2001), coalescence is assumed to be sufficiently fast that islands instantaneously attain thermodynamic equilibrium. However, when island coalescence occurs, islands sometimes have a non-equilibrium shape due to their large size. Therefore, models should include the kinetic effect of coalescence. The rate of coalescence can be expressed by the Nichols–Mullins (1965) equation:

$$\tau_{\text{coal}} \sim \frac{kTR^4}{D_s\gamma_s\Omega^{4/3}} = \frac{R^4}{B} \quad [4.15]$$

where k is Boltzmann's constant, T is temperature, and D_s is the surface diffusivity on islands. By using the droplet growth model, Carrey and Maurice (2001) derived an analytical solution of the transition thickness, f_{trans}, when islands cannot attain their equilibrium shape as $f_{\text{trans}} \sim (B/F)^{d_f - d_s/4 - d_f + d_s}$, where d_s is the surface dimensionality and d_f is the island dimensionality. For 3D islands on 2D substrates, $f_{\text{trans}} \sim F^{-1/3}$. Monte-Carlo simulations indicate that $f_{\text{trans}} \sim F^{-0.288}$ (Carrey and Maurice, 2001). However, droplet growth models do not include adatom diffusion. The assumption of no atomic diffusion is not realistic for deposition processes. Later Carrey and Maurice (2002) made KMC simulations including adatom diffusion and static coalescence kinetics. KMC simulations including atomic diffusion results in $f_{\text{trans}} \sim F^{-\alpha}$, where α increases from 0.16 to 0.29 for B increasing from 1×10^{-42} to 1×10^{-33} [m^4/s]. They also derived an expression for the percolation thickness, f_{perc}, when the film becomes continuous. For no atomic diffusion or for atomic diffusion with small B (i.e., coalescence is neglected), $f_{\text{perc}} = 1.5 f_{\text{trans}}$. For atomic diffusion with large B (i.e., coalescence is fast), $f_{\text{perc}} = 1.9 f_{\text{trans}}$. For small B, $f_{\text{perc}} \sim F^{-0.16}$ (Carrey and Maurice, 2001), which is similar to the expression $f_{\text{sat}} \sim (F/D)^{1/7}$ in the DDA model for 3D islands (Jensen *et al.*, 1997). A model considering grain boundary switching was proposed and investigated by KMC (Ruan and Schuh, 2010).

Defects

Defects can trap monomers and islands and therefore become nucleation sites. When defects are randomly distributed on the surface with concentration c, and when the island density is determined by defects, $N_{\max}$ becomes c and is independent of F/D. Jensen *et al.* (1997) modeled island formation, and included both defects and evaporation in a DDA model. Chambliss and Johnson (1994) modeled island formation when adatoms react with the substrate to become immobile inclusions that act as reaction centers, which is similar to the effect of defects.

In the above, we overviewed elementary processes determining film morphology on growing surfaces. Many elementary processes contribute to the process of film and roughness evolution: film-growth species adsorb on the surface, which we call monomers. On the surface they diffuse and aggregate with other monomers to create islands. If the islands created by monomer–monomer aggregation are stable, the aggregation rate determines the island density. If, however, the islands must grow past a certain critical size to become stable, attachment/detachment processes must be included in nucleation models to accurately simulate the nucleation rate of such islands. The migration of islands, as *i* decreases the island density when islands collide and coalesce, is also important. Deformation of islands can modify the space between the islands, which affects the deposition behavior of monomers. On substrates with defects, defects trap the diffusing monomers, creating nucleation sites.

These elementary processes can modify the physical properties of deposited materials, such as the maximum island density, average island size, and percolation thickness, and finally surface roughness of the film. Controlling these properties is one focus of experimental research, whereas predicting their formation is of interest for theoretical research and modeling. Not much is known about some of the fundamentals of 3D island growth, including the Ehrlich–Schwoebel barrier and different sticking probabilities onto the upper layer. Further study of these processes is essential for a full understanding of roughness evolution during not only homo-epitaxial growth but also hetero- and non-epitaxial growth.

4.3 Roughness during hetero- or non-epitaxial growth

4.3.1 Roughness generation during initial deposition processes

In hetero- or non-epitaxial growth, the mechanism causing surface roughness differs between the initial deposition process on the substrate and film growth processes after the substrate is covered by the deposited materials. During the initial deposition process, the 3D islands on the substrate can be formed by both thermodynamic and kinetic factors.

During the process of deposition on substrates whose materials differ from the deposited materials, it is well known that three island growth modes exist in the initial stage: two-dimensional (2D) layer (Frank–van-der-Merve,

F-M) mode, three-dimensional (3D) island (Volmer–Weber, V-W) mode and 2D layer followed by 3D island (Stranski-Krastanov, S-K) mode (Leamy *et al.*, 1975). When thermodynamic equilibrium is attained, these growth modes determine the surface energy of materials being deposited and the underlying substrate and interfacial energy between them (Leamy *et al.*, 1975; Kern *et al.*, 1979; Hu *et al.*, 2002). In the V-W and S-K modes of growth, roughness inevitably appears in conditions of thermodynamic equilibrium. On the other hand, in the F-M mode of growth, a thermodynamically stable structure is a flat surface, and roughness is caused by kinetic factors acting later on in the growth process. In the F-M mode of growth, the processes causing surface roughness are similar to those discussed in the previous section. These growth modes are well described thermodynamically when islands are small enough to attain thermodynamic equilibrium (van den Brekel and Bloem, 1977; Chin *et al.*, 1977).

Kinetic factors can also cause 3D island formation during CVD, because of the different sticking probability of precursors onto the islands and the substrate (Puurunen, 2004; Kajikawa and Noda, 2005). In the self-limiting process, it is difficult for growth species to adsorb onto the deposits, i.e., $\eta/\eta_0 << 1$, where η is the sticking probability of precursors onto the islands and η_0 is that onto the substrate. The deposited amount saturates as deposition time t_d increases. This condition is used in atomic layer deposition (ALD) to produce flat, conformal films (Leskelä and Ritala, 2003; Kim, 2003). In random deposition processes, growth species adsorb onto the substrate and onto the deposits with equal probability, i.e. $\eta/\eta_0 \sim 1$. This condition typically occurs during physical vapor deposition (PVD). In autocatalytic processes, growth species preferentially adsorb onto the deposits, i.e., $\eta/\eta_0 >> 1$. In this condition, the surface becomes rough because of the selective growth of 3D islands. This is sometimes observed when etching of deposited material exists simultaneously with the deposition. This condition typically occurs during CVD at high temperatures. In ALD, uniform layers such as those formed in the F-M growth mode can be attained due to its self-limiting behavior. The autocatalytic processes of CVD tend to cause 3D growth, i.e., the V-W island growth mode. During CVD, island growth can be controlled by changing η_0 and η. We must also note that the difference in sticking probability of reactants onto the substrate, the first layer, and existing adsorbents and islands might become a cause of S-K growth mode (Kajikawa *et al.*, 2004c). Other processes such as contamination and particle deposition can also cause roughness generation on the substrate.

4.3.2 Roughness evolution during growth stage

In this section, we discuss roughness evolution during the growth stage after the initial deposition stage on the substrate. In this growth stage, the effect of the substrate can be ignored because the substrate is entirely covered by the film. In the V-W and S-K modes of growth, roughness necessarily forms when each island on the substrate is isolated. When the islands come into contact, coalescence occurs to reduce surface free energy. When the substrate is entirely covered by deposited materials and the coalescence of islands is complete, a flat surface can also appear in the V-W and S-K modes of growth because the film tends to reduce the surface area. However, when coalescence is not finished, which is typical of materials with high melting temperatures (Thornton, 1977; Thompson and Carel, 1996), the roughness remains on the surface. When the roughness remains, it is amplified by a variety of processes.

We can categorize the process of roughness evolution according to whether or not positive feedback exists (Kajikawa *et al.*, 2004d). Typical examples of positive feedback include shadowing and a concentration gradient in the gas phase, which will be illustrated later. An example of non-positive feedback is the growth rate difference between the crystallographic plane and composition. These can be distinguished by the morphology of the roughness. As shown in Fig. 4.2, in a case of positive feedback, the ratio of growth rates is characteristically inconstant, i.e., $R_2 > R'_2 = R_1/\cos\theta$ where R_1 is growth rate of a film, R_2 is that on the initial roughness along the direction of the long axis of the roughness and R'_2 is that along the direction of the other part of the roughness. On the other hand, in non-positive feedback, the ratio of growth rate is constant, i.e., $R_2 = R'_2 = R_1/\cos\theta$. This feature of the morphology of roughness can help researchers approximately diagnose mechanisms contributing to roughness evolution on films.

4.3.3 Roughness evolution with positive feedback

Shadowing

When the size of the roughness, x_p, is smaller than the mean free path of the precursor, λ, i.e., Knudsen number $Kn = \lambda/x_p > 1$, the roughness is enlarged

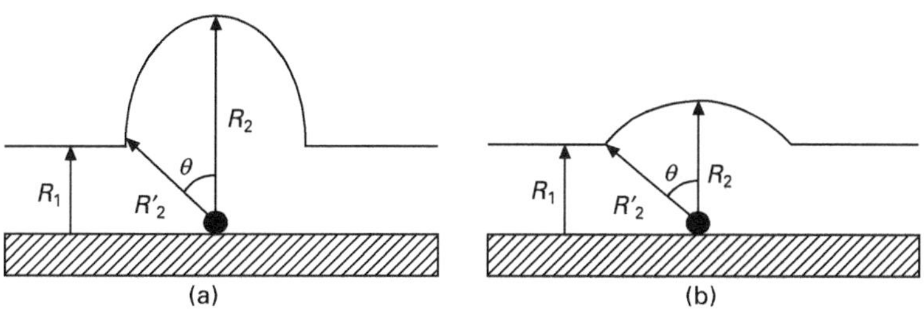

4.2 Roughness evolution during film growth (a) with and (b) without positive feedback.

by the ballistic deposition of growth species. This phenomenon is analogous to the situation where sunlight to short plants is shut off by tall plants, which is known as the 'shadowing effect' (Bales *et al.*, 1990). In shadowing growth, a film has less flux of the reactant at the bottom of the protrusion obscured by the neighboring protrusions. Researchers in CVD have used the idea of ballistic transport to study trench or via filling problems (Ikegawa and Kobayashi, 1989; Cooke and Harris, 1989; Watanabe and Komiyama, 1990; Cale and Raupp, 1990). This line-of-sight model can explain very well not only the features observed in trench evolution but also roughness evolution, although the details of the model may vary. Currently, unique structures can be fabricated by utilizing the shadowing effect such as glancing angle deposition (Robbie *et al.*, 1996; Robbie and Brett, 1997).

Roughness evolution during CVD by shadowing has been studied experimentally (Roland and Guo, 1991; Tanenbaum *et al.*, 1997; Kondo *et al.*, 1998; Flewitt *et al.*, 1999) and theoretically (Bales *et al.*, 1990; Roland and Guo, 1991; Yao and Guo, 1993; Singh and Shaqfeh, 1993; Zhao *et al.*, 2001; Drotar *et al.*, 2001). In shadowing growth, x_p grows power laws in time, $x_p \sim t_d{}^{\gamma}$, where γ is a scaling exponent. Both experimental and theoretical works focused on the scaling property of the width of the roughness, $w \sim t_d^{\beta}$, where β is called the growth exponent. If the shadowing effect only were present, one would expect $\beta = 1$ (Roland and Guo, 1991; Yao and Guo, 1993). β decreases as η becomes small or a stabilizing factor such as surface diffusion becomes dominant (Singh and Shaqfeh, 1993; Zhao *et al.*, 2001). In a-Si deposition of low pressure CVD, γ increases from 0.08 to 0.55 as temperature decreases (Zhao *et al.*, 2001). This is because, at higher temperatures, surface diffusion as a stabilizing factor becomes dominant and the sticking probability decreases. Shadowing is typical at η in the order of 0.1. When η is less than 0.01, shadowing becomes negligible, which is the same as the fact that precursors with small η can fulfill deep trenches, while when t_d increases, roughness becomes visible even at η of less than 0.01. Increasing the temperature to enhance surface diffusion and the capillary effect is one route to limiting roughness evolution by shadowing. However, high temperatures sometimes lead to an increase in the sticking probability of precursors. Therefore, the choice of precursor is important in reducing the size of roughness caused by shadowing.

Concentration gradient

When the size of the protrusion becomes larger than the mean free path of the reactants, i.e., $Kn = \lambda/x_p < 1$, the growth of the protrusion cannot be explained by ballistic deposition and is accelerated by the concentration gradients of growth species near the protrusion in the gas phase. In this regime, roughness can expand even at small η. Due to the concentration gradient in the gas phase, the top of the protrusion is exposed to growth species at higher concentrations than at the film surface, and thus higher growth rates are achieved at the top of the protrusion than at the bottom. Consequently, protrusion growth rates increase with the protrusion size, causing positive feedback between its height and its growth rate. This type of growth is observed during CVD of a variety of materials (Kajikawa *et al.*, 2004d). An example of roughness caused by the concentration gradient is shown in Fig. 4.2a.

This situation is analogous to Mullins–Sekerka instability during solidification in the liquid phase (Mullins and Sekerka, 1963, 1964). In solidification, the driving force for growth is the temperature gradient, and roughness is amplified due to temperature gradients caused by the release of latent heat at the growth front. van den Brekel and Jansen (1978) first applied their model to CVD systems by substituting the concentration gradients of growth species for temperature gradients. Since the study by van der Brekel and Jansen, theorists have extended their models to include various factors such as surface diffusion and capillarity (Palmer and Gordon, 1988, 1989; Bales *et al.*, 1989; Viljoen *et al.*, 1994; Mahalingam and Dandy, 1997, 1998; Thiart *et al.*, 2000). While many theoretical studies dealing with the diffusion-limited growth of protrusions have indicated the importance of surface diffusion and capillarity as stabilizing factors, these factors are of limited importance in simulating protrusion growth in CVD processes in the continuum regime because, for the micrometer-sized protrusions that are typical for $Kn = \lambda/x_p < 1$, these stabilizing factors are negligible (Ravi, 1992; Mahalingam and Dandy, 1997; Kajikawa *et al.*, 2004d). According to the 1D boundary layer model considering only the concentration gradient, the height of the protrusion relative to its height at $t_d = 0$ ($x_P(0)$) is expressed by (Kajikawa *et al.*, 2004d):

$$\frac{x_P(t_d)}{x_P(0)} = \exp\left(\frac{g_0 t_d}{D_g/k_s}\right) \qquad [4.16]$$

where D_g and k_s are diffusion coefficient of reactants in the gas phase and surface reaction rate, respectively. Because $g_0 t_d$ represents film thickness f, the height of the protrusions is expected to increase exponentially with decreasing D_g/k_s and with increasing f. To obtain films without abnormal protrusions, D_g/k_s must be maintained at a larger size than the desired film thickness. As shown in Eq. 4.16 in the continuum regime, the evolution of the roughness expressed by exponential growth is more rapid than that in a ballistic regime expressed by power-law, $x_p \sim t_d^{\gamma}$. The concentration gradient of reactants around the protrusion under diffusion-limited conditions is a typical cause for this type of structure.

Thermophoresis

The final cause of positive feedback growth is preferential particle deposition by thermophoresis. In long-residence-time or high-temperature growth regions, particles are often produced in the gas phase. Particle thermophoresis is sometimes utilized to enhance the film growth rate (Komiyama, 1993). However, it also becomes a cause of unusual rough structures (Bouyer *et al.*, 2001; Kajikawa *et al.*, 2002). Thermophoretic particle transport is affected by the temperature gradients near the roughness. In a cold-wall reactor, the film temperature is usually higher than the gas temperature, so that thermophoresis drives particles away from the film. However, because the roughness has a large solid angle, the temperature around it is lower than the gas temperature for radiation, which causes a thermophoretic force from the gas phase to the roughness. As a result, fibers or trumpet-like structures are produced (Goela and Taylor, 1988; Bouyer *et al.*, 2001; Kajikawa *et al.*, 2002). When particles are deposited on the roughness, the porous structure in the roughness reduces the heat transfer within the roughness, and therefore the temperature gradient around it becomes larger and more particles are deposited on it by thermophoresis. Due to this positive feedback, roughness is amplified, sometimes at an extremely high rate, in the order of mm/h (Kajikawa *et al.*, 2002). To restrain the growth, either the particle-generation rate must be reduced or the thermophoretic force towards the trumpet must be reduced. While the existence of this type of growth is shown by computational fluid dynamics (CFD) simulation (Kajikawa *et al.*, 2002), there is no analytical solution to the roughness evolution.

4.3.4 Roughness evolution with non-positive feedback

As shown above, the cause of roughness evolution is the spatial difference in growth rate. In positive feedback mechanisms, the growth rate differs between the top and base of the roughness, and the difference between them increases as the deposition proceeds. There are also cases where roughness expands without positive feedback. This non-positive feedback growth is also divided into two cases: the growth rate differs within an area of roughness, and between roughness and other parts of the film. The crystallographic effect is typical for the former and is usually observed in polycrystalline materials.

Crystallographic orientation

Each crystallographic plane usually has a different growth rate because of the difference in sticking probability and surface diffusion rate on each plane (Kajikawa *et al.*, 2003; Kajikawa, 2006). The difference in growth rates among different crystallographic planes is a typical mechanism causing roughness evolution without positive feedback (Mizushima *et al.*, 1999; Noda *et al.*, 2002; Iwamoto *et al.*, 2003). According to the evolutionary selection rule (van der Drift, 1967), in polycrystalline film depositions, grains with the fastest growth direction normal to the substrate envelop the other grains and finally dominate the film. The fastest growing grains, however, are enclosed by the slowest growing planes, and so the film surface does not become flat. Roughness evolution by evolutionary selection is modeled in a number of simulations (Wild *et al.*, 1990, 1993; Barrat *et al.*, 1996; Paritosh *et al.*, 1999). The roughness caused by the crystallographic difference in growth rates within a grain inevitably appears in polycrystalline films, and the resulting film structure is often called texture. When the difference in growth rate among crystallographic planes has substantially increased, rod-like structures appear on a film (Takeyama *et al.*, 2006). To restrain the surface roughness, modifying the process conditions during deposition to alter the fastest growth direction is effective (Wild *et al.*, 1993; Paritosh *et al.*, 1999).

Composition

This difference in growth rate is also caused by other factors such as composition (Liu *et al.*, 1996) and crystallinity (Ross *et al.*, 2000; Liu *et al.*, 2001a,b). For example, in AlN/TiN composite CVD films, Ti-rich phases grow faster than Al-rich phases. Therefore, cone structures of a Ti-rich phase appear in Al-rich flat films. Smooth films are successfully synthesized by the introduction of a TiN interlayer on the substrate prior to AlN/TiN composite deposition (Liu *et al.*, 1996). In amorphous-crystalline mixed films, the crystalline part of the film sometimes has a higher growth rate, which leads to crystalline cone structures in amorphous films.

Catalytic activity also influences the growth rates. The vapor-liquid-solid (VLS) growth mechanism produces the 1D structure by extreme anisotropic growth (Wagner and Ellis, 1964). Catalytic activity also works at the vapor–solid interfaces. For example, a submonolayer of iodine atoms adsorbed on the copper film surface catalyzes the surface reaction and increases the growth rate (Hwang and Lee, 2000). Spatial differences in catalysis concentration may cause roughness evolution but by utilizing it, a unique deposition method can be realized. In recent works, Cu superfilling with catalyst-enhanced chemical vapor deposition (CECVD) has also been proposed using iodine as a catalytic surfactant (Shim *et al.*, 2002; Josell *et al.*, 2003; Lee *et al.*, 2005). Because the iodine adsorbed on the Cu seed layer is gradually accumulated onto the bottom of the submicron features as the Cu film grows, the film growth rate of Cu is continuously accelerated on the bottom and leads to superconformal Cu bottom-up filling in CECVD of Cu. This is in good contrast with the reduction in roughness by equalizing the concentration profile of catalytic elements to cancel out the catalytic activity in AlN/TiN composite films.

4.4 Future trends

Compared to during homo-epitaxial growth, roughness evolution during hetero- or non-epitaxial growth has not been investigated in depth, despite its practical importance. This is partly because it is difficult to investigate complex processes, as scientific research tends to study focused processes under well-designed experimental conditions and theoretically under simplified conditions, in order to make calculation feasible. While a previous survey does exist (Kajikawa, 2008), there is much room to explore.

The first step to model and control such complex processes is to identify the essential processes that contribute to the roughness evolution. There are three types of model: mechanistic, phenomenological, and empirical.

A detailed mechanistic model considering all elementary processes is feasible, because the nature of the industry is rapidly changing and a detailed study of physical mechanisms is often impractical (Edgar *et al.*, 2000). Models must be developed rapidly if they are to be used in developing the process, and often this is accomplished using phenomenological empirical models.

The phenomenological model, as focused on in this chapter, is an efficient and effective one because we can model the process quickly and control it based on a fundamental understanding. Non-dimensional indicators like the scaling exponent of nucleation theory can help us to identify essential processes to be included in the model (Kajikawa *et al.*, 2004a).

The final model is empirical. Empirical models are built directly from experimental data to fit the trends in experimental measurements (Edgar, 2004). Data from designed experiments are often used to construct empirical models. A recent trend in this direction is to use *in-situ* monitoring and combinatorial chemistry for efficient control of the process. Such models work well when experimental data are readily available and the underlying detailed mechanisms are not well understood. For example, scanning tunneling microscopy (Flewitt *et al.*, 1999), X-ray reflectivity (Kellerman *et al.*, 1997), single wavelength ellipsometry (Hu *et al.*, 1996), spectroscopic ellipsometry (Fujiwara *et al.*, 2001; Volintiru *et al.*, 2008), optical reflectometry (Deatcher *et al.*, 2003; Kajikawa *et al.*, 2004c) and UV spectrophotometer (Lee *et al.*, 2010) are used for *in-situ* monitoring of deposition processes during CVD. Another recent trend is the increasing attention paid to combinatorial chemistry even in vapor deposition processes (Koinuma and Takeuchi, 2004; Noda *et al.*, 2004). The challenge in engineering is to develop an effective and efficient methodology for practical use, both theoretically and experimentally.

4.5 References

Amar J G, Family F (1996) 'Kinetics of submonolayer and multilayer epitaxial growth', *Thin Solid Films*, 272, 208–222.

Amar J G, Family F, Lam P M (1994) 'Dynamic scaling of the island-size distribution and percolation in a model of submonolayer molecular-beam epitaxy', *Phys Rev B*, 50, 8781–8797.

Amar J G, Family F, Hughes D C (1998) 'Submonolayer epitaxial growth with long-range (Levy) diffusion', *Phys Rev E*, 61, 7130–7136.

Bales G S, Zangwill A (1997) 'Self-consistent rate theory of submonolayer homoepitaxy with attachment/detachment kinetics', *Phys Rev B*, 55, R1973–R1976.

Bales G S, Redfield A C, Zangwill A (1989) 'Growth dynamics of chemical vapor-deposition', *Phys Rev Lett*, 62, 776–779.

Bales G S, Bruinsma R, Eklund E A, Karunasiri R P U, Rudnick J, Zangwill A (1990) 'Growth and erosion of thin solid films', *Science*, 249, 264–268.

Barrat S, Pigeat P, Bauer-Grosse E (1996) 'Three-dimensional simulation of CVD diamond film growth', *Diam Relat Mater*, 5, 276–280.

Bartelt M C, Evans J W (1994) 'Nucleation and growth in metal-on-metal homoepitaxy – rate-equations, simulations and experiments', *J Vac Sci Technol A*, 12, 1800–1808.

Bouyer E, Henne R H, Müller M, Schiller G (2001) 'Si_3N_4 fibers synthesized by thermal plasma CVD', *Chem Vap Deposition*, 7, 139–142.

Burton W K, Cabrera N, Frank F C (1951) 'The growth of crystals and the equilibrium structure of their surfaces', *Philos Trans Roy Soc London*, A243, 299–358.

Cale T S, Raupp G B (1990) 'Free molecular-transport and deposition in cylindrical features', *J Vac Sci Technol B*, 8, 649–655.

Carrey J, Maurice J L (2001) 'Transition from droplet growth to percolation: Monte Carlo simulations and an analytical model', *Phys Rev B*, 63, 245408.

Carrey J, Maurice J L (2002) 'Scaling laws near percolation during three-dimensional cluster growth: a Monte Carlo study', *Phys Rev B*, 65, 205401.

Chambliss D D, Johnson K E (1994) 'Nucleation with a critical cluster-size of zero – submonolayer fe inclusions in Cu(100)', *Phys Rev B*, 50, 5012–5015.

Chin J, Gantzel P K, Hudson R G (1977) 'Structure of chemical vapor-deposited silicon-carbide', *Thin Solid Films*, 40, 57–72.

Combe N, Jensen P (1998) 'Changing thin-film growth by modulating the incident flux', *Phys Rev B*, 57, 15553–15560.

Cooke M J, Harris G (1989) 'Monte-Carlo simulation of thin-film deposition in a rectangular groove', *J Vac Sci Technol A*, 7, 3217–3221.

Deatcher C J, Liu C, Pereira S, Lada M, Cullis A G, Sun Y J, Brandt O, Watson I M (2003) '*In situ* optical reflectometry applied to growth of indium gallium nitride epilayers and multi-quantum well structures', *Semicond Sci Technol*, 18, 212–218.

Deltour P, Barrat J L, Jensen P (1997) 'Fast diffusion of a Lennard–Jones cluster on a crystalline surface', *Phys Rev Lett*, 78, 4597–4600.

Dobbs H T, Vvedensky D D, Zangwill A, Johansson J, Carlsson N, Seifert W (1997) 'Mean-field theory of quantum dot formation', *Phys Rev Lett*, 79, 897–900.

Drotar J T, Zhao Y P, Lu T M, Wang G C (2001) 'Surface roughening in low-pressure chemical vapor deposition', *Phys Rev B*, 64, 125411.

Edgar T F (2004) 'Process operations: when does controllability equal profitability?', *Computer Chem Engr*, 29, 41–49.

Edgar T F, Butler S W, Campbell W J, Pfeiffer C, Bode C, Hwang S B, Balakrishnan

K S, Hahn J (2000) 'Automatic control in microelectronics manufacturing: practices, challenges, and possibilities', *Automatica*, 36, 1567–1603.

Ehrlich G, Hudda F G (1966) 'Atomic view of surface self-diffusion – tungsten on tungsten', *J Chem Phys*, 44, 1039–1049.

Family F, Meakin P (1989) 'Kinetics of droplet growth-processes – simulations, theory, and experiments', *Phys Rev A*, 40, 3836–3854.

Flewitt A J, Robertson J, Milne W I (1999) 'Growth mechanism of hydrogenated amorphous silicon studied by *in situ* scanning tunneling microscopy', *J Appl Phys*, 85, 8032–8039.

Fujiwara H, Kondo M, Matsuda A (2001) 'Real-time spectroscopic ellipsometry studies of the nucleation and grain growth processes in microcrystalline silicon thin films', *Phys Rev B* 63, 115306.

Galdikas A (2002) 'Thin film deposition onto the rough surface: phenomenological investigations', *Thin Solid Films*, 418, 112–118.

Goela J S, Taylor R L (1988) 'Monolithic material fabrication by chemical vapor-deposition', *J Mater Sci*, 23, 4331–4339.

Heyraud J C, Métois J J (1987) 'Equilibrium shape of an ionic-crystal in equilibrium with its vapor (NaCl)', *J Cryst Growth*, 84, 503–508.

Hu M, Noda S, Komiyama H (2002) 'A new insight into the growth mode of metals on TiO_2(110)', *Surf Sci*, 513, 530–538.

Hu Y Z, Diehl D J, Zhao C Y, Wang C L, Liu Q, Irene E A, Christensen K N, Venable D, Maher D M (1996) 'Real time investigation of nucleation and growth of silicon on silicon dioxide using silane and disilane in a rapid thermal processing system', *J Vac Sci Technol B*, 14, 744–750.

Hwang E S, Lee J (2000) 'Surfactant-catalyzed chemical vapor deposition of copper thin films', *Chem Mater*, 12, 2076–2081.

Ikegawa M, Kobayashi J (1989) 'Deposition profile simulation using the direct simulation Monte-Carlo method', *J Electrochem Soc*, 136, 2982–2986.

Iwamoto C, Shen X Q, Okumura H, Matsuhata H, Ikuhara Y (2003) 'Termination mechanism of inversion domains by stacking faults in GaN', *J Appl Phys*, 93, 3264–3269.

Jensen P (1999) 'Growth of nanostructures by cluster deposition: experiments and simple models', *Rev Mod Phys*, 71, 1695–1735.

Jensen P, Niemeyer B (1997) 'The effect of a modulated flux on the growth of thin films', *Surf Sci*, 384, L823–L827.

Jensen P, Larralde H, Pimpinelli A (1997) 'Effect of monomer evaporation on a simple model of submonolayer growth', *Phys Rev B*, 55, 2556–2569.

Jensen P, Larralde H, Meunier M, Pimpinelli A (1998) 'Growth of three-dimensional structures by atomic deposition on surfaces containing defects: simulations and theory', *Surf Sci*, 412/413, 458–476.

Jensen P, Combe N, Larralde H, Barrat J L, Misbah C, Pimpinelli A (1999) 'Kinetics of shape equilibration for two dimensional islands', *Eur Phys J B*, 11, 497–504.

Josell D, Kim S, Wheeler D, Moffat T P, Pyo S G (2003) 'Interconnect fabrication by superconformal iodine-catalyzed chemical vapor deposition of copper', *J Electrochem Soc*, 150, C368–C373.

Kajikawa, Y (2006) 'Texture development of non–epitaxial polycrystalline ZnO films', *J Cryst Growth*, 289, 387–394.

Kajikawa Y (2008) 'Roughness evolution during chemical vapor deposition', *Mater Chem Phys*, 112, 311–318.

Kajikawa Y, Noda S (2005) 'Growth mode during initial stage of chemical vapor deposition', *Appl Surf Sci*, 245, 281–289.

Kajikawa Y, Ono H, Noda S, Komiyama H (2002) 'Growth of trumpet-like protrusions during chemical vapor deposition of silicon carbide films', *Chem Vap Deposition*, 8, 52–55.

Kajikawa Y, Noda S, Komiyama H (2003) 'Comprehensive perspective on the mechanism of preferred orientation in reactive-sputter-deposited nitrides', *J Vac Sci Technol A*, 21, 1943–1954.

Kajikawa Y, Noda S, Komiyama H (2004a) 'Use of process indices for simplification of the description of vapor deposition systems', *Mater Sci Eng B*, 111, 156–163.

Kajikawa Y, Tsuchiya T, Noda S, Komiyama H (2004b) 'Incubation time during the CVD of Si onto SiO_2 from silane', *Chem Vap Deposition*, 10, 128–133.

Kajikawa Y, Tsumura T, Noda S, Komiyama H, Shimogaki Y (2004c) 'Nucleation of W during chemical vapor deposition from WF_6 and SiH_4', *Jpn J Apply Phys*, 43, 3945–3950.

Kajikawa Y, Noda S, Komiyama H (2004d) 'A simple index to restrain abnormal protrusions in films fabricated using CVD under diffusion-limited conditions', *Chem Vap Deposition*, 10, 221–228.

Kellerman B K, Chason E, Adams D P, Mayer T M, White J M (1997) '*In-situ* X-ray reflectivity investigation of growth and surface morphology evolution during Fe chemical vapor deposition on Si(001)', *Surf Sci*, 375, 331–339.

Kern R, Laye G L, Métois J J (1979) *Current Topics in Materials Science*, ed. by E. Kaldis, North Holland, New York.

Kim H (2003) 'Atomic layer deposition of metal and nitride thin films: current research efforts and applications for semiconductor device processing', *J Vac Sci Technol B*, 21, 2231–2261.

Koinuma H, Takeuchi I (2004) 'Combinatorial solid-state chemistry of inorganic materials', *Nature Mat*, 3, 429–438.

Komiyama H (1993) in *The Expanding World of Chemical Engineering*, eds Garside J, Furusaki S, Gordon and Breach Science Publishers, New York.

Kondo M, Ohe T, Saito K, Nishimiya T, Matsuda A (1998) 'Morphological study of kinetic roughening on amorphous and microcrystalline silicon surface', *J Non-Cryst Solids*, 227–230, 890–895.

Kotrla M, Krug J, Šmilauer P (2000) 'Submonolayer epitaxy with impurities: kinetic Monte Carlo simulations and rate–equation analysis', *Phys Rev B*, 62, 2889–2898.

Krug J (1997) 'Origins of scale invariance in growth processes', *Adv Phys*, 46, 139–282.

Krug J (2000) 'Scaling regimes for second layer nucleation', *Eur Phys J B*, 18, 713–719.

Kuipers L, Palmer R E (1996) 'Influence of island mobility on island size distributions in surface growth', *Phys Rev B*, 53, R7646–R7649.

Lantz S L, Bell A E, Ford W K, Danielson D (1994) 'W-nucleation on tin from WF_6 and SiH_4', *J Vac Sci Technol A*, 12, 1032–1038.

Leamy H J, Gilmer G H, Jackson K A (1975) in *Surface Physics of Materials*, ed. Blakely J M, Academic Press, New York.

Lee S B, Gupta B C (2000) 'Influence of small-cluster mobility on the island formation in molecular beam epitaxy', *Phys Rev B*, 62, 7545–7552.

Lee H B, Kwak D K, Kang S W (2005) 'Enhancement of iodine adsorption on ruthenium glue layer for seedless CECVD of Cu' *Electrochem Solid-State Lett*, 8, C39–C42.

Lee S R, Ahn K M, Kang S M, Ahn B T (2010) 'Surface morphology control of epitaxial silicon films grown by hot wire chemical vapor deposition using hydrogen dilution, *Sol Ener Mater Sol Cells*, 94, 606–611.

Leskelä M, Ritala M (2003) 'Atomic layer deposition chemistry: recent developments and future challenges', *Angew Chem Int Ed*, 42, 5548–5554.

Liu X D, Funakubo H, Noda S, Komiyama H (2001a) 'Internal microstructure and formation mechanism of surface protrusions in Pb-Ti-Nb-O thin films prepared by MOCVD', *Chem Vap Deposition*, 7, 253–259.

Liu X D, Funakubo H, Noda S, Komiyama H (2001b) 'Influence of deposition temperature on the microstructure of Pb-Ti-Nb-O thin films by metallorganic chemical vapor deposition', *J Electrochem Soc*, 148, C227–C230.

Liu Y J, Kim H J, Takeuchi T, Egashira Y, Kimura H, Komiyama H (1996) 'Using simultaneous deposition and rapid growth to produce nanostructured composite films of AlN/TiN', *J Am Ceram Soc*, 79, 1335–1342.

Luedtke W D, Landman U (1999) 'Slip diffusion and Levy flights of an adsorbed gold nanocluster', *Phys Rev Lett*, 82, 3835–3838.

Mahalingam P, Dandy D S (1997) 'Simulation of morphological instabilities during diamond chemical vapor deposition', *Diam Relat Mater* 6, 1759–1771.

Mahalingam P, Dandy D S (1998) 'Nonlinear stability analysis of the growth surface during diamond chemical vapor deposition', *J Appl Phys*, 83, 6061–6071.

Markov I (1996) 'Kinetics of nucleation in surfactant-mediated epitaxy', *Phys Rev B*, 53, 4148–4155.

Matsunami H, Kimoto T (1997) 'Step-controlled epitaxial growth of SiC: high quality homoepitaxy', *Mater Sci Eng R*, 20, 125–166.

Meakin P (1993) 'The growth of rough surfaces and interfaces', *Phys Rep*, 235, 189–289.

Mizushima I, Koike M, Sato S, Miyano K, Tsunashima Y (1999) 'Mechanism of defect formation during low-temperature Si epitaxy on clean Si substrate', *Jpn J Appl Phys*, 38, 2415–2418.

Mulheran P A, Robbie D A (2001) 'Island mobility and dynamic scaling during thin film deposition', *Phys Rev B*, 64.

Mullins W W, Sekerka R F (1963) 'Morphological stability of a particle growing by diffusion or heat flow', *J Appl Phys*, 34, 323–329.

Mullins W W, Sekerka R F (1964) 'Stability of planar interface during solidification of dilute binary alloy', *J Appl Phys*, 35, 444–451.

Nichols F A, Mullins W W (1965) 'Morphological changes of a surface of revolution due to capillarity-induced surface diffusion', *J Appl Phys*, 36, 1826–1835.

Noda S, Kajikawa Y, Komiyama H (2002) 'Cone structure formation by preferred growth of random nuclei in chemical vapor deposited epitaxial silicon films', *Chem Vap Deposition*, 8, 87–89.

Noda S, Kajikawa Y, Komiyama H (2004) 'Combinatorial masked deposition: simple method to control deposition flux and its spatial distribution', *Appl Surf Sci*, 225, 372–379.

Palmer B J, Gordon R G (1988) 'Local equilibrium-model of morphological instabilities in chemical vapor-deposition', *Thin Solid Films*, 158, 313–341.

Palmer B J, Gordon R G (1989) 'Kinetic-model of morphological instabilities in chemical vapor-deposition', *Thin Solid Films*, 177, 141–159.

Paritosh Srolovitz D J, Battaile C C, Li X, Butler J E (1999) 'Simulation of faceted film growth in two–dimensions: microstructure, morphology and texture', *Acta Mater*, 47, 2269–2281.

Politi P, Grenet G, Marty A, Poncher A, Villain J (2000) 'Instabilities in crystal growth by atomic or molecular beams', *Phys Rep*, 324, 271–404.

Puurunen R L (2004) 'Random deposition as a growth mode in atomic layer deposition', *Chem Vap Deposition*, 10, 159–170.

Ratsch C, Venables J A (2003) 'Nucleation theory and the early stages of thin film growth', *J Vac Sci Technol A*, 21, S96–S109.

Ratsch C, Zangwill A, Šmilauer P, Vvdensky D D (1994) 'Saturation and scaling of epitaxial island densities', *Phys Rev Lett*, 72, 3194–3197.

Ratsch C, Šmilauer P, Zangwill A, Vvdensky D D (1995) 'Submonolayer epitaxy without a critical nucleus', *Surf. Sci*, 329, L599–L604.

Ravi K V (1992) 'Morphological instabilities in the low–pressure synthesis of diamond', *J Mater Res*, 7, 384–393.

Robbie K, Brett M J (1997) 'Sculptured thin films and glancing angle deposition: growth mechanics and applications', *J Vac Sci Technol A*, 15, 1460–1465.

Robbie K, Brett M J, Lakhtakia A (1996) 'Chiral sculptured thin films', *Nature*, 384, 616.

Roland C, Guo H (1991) 'Interface growth with a shadow instability', *Phys Rev Lett*, 66, 2104–2107.

Ross C, Herion J, Wagner H (2000) 'Nucleation and growth analysis of microcrystalline silicon by scanning probe microscopy: substrate dependence, local structural and electronic properties of as-grown surfaces', *J Non-Cryst Solids*, 266–269, 69–73.

Ruan S, Schuh C A (2010) 'Kinetic Monte Carlo simulations of nanocrystalline film deposition', *J Appl Phys*, 107, 073512.

Saito Y, Uwaha M (1994) 'Fluctuation and instability of steps in a diffusion field', *Phys Rev B*, 49, 10677–10692.

Salik J (1985) 'Monte-Carlo study of reversible growth of clusters on a surface', *Phys Rev B*, 32, 1824–1826.

Schwoebel R L, Shipsey E J (1966) 'Step motion on crystal surfaces', *J Appl Phys*, 37, 3682–3686.

Shim K C, Lee H B, Kwon O K, Park H S, Koh W, Kang S W (2002) 'Bottom-up filling of submicrometer features in catalyst-enhanced chemical vapor deposition of copper', *J Electrochem Soc*, 149, G109–G113.

Singh V K, Shaqfeh E S G (1993) 'Effect of surface reemission on the surface-roughness of film growth in low-pressure chemical-vapor-deposition', *J Vac Sci Technol A*, 11, 557–568.

Stoyanov S, Kaschiev D (1981) *Current Topics in Materials Science*, ed. by Kaldis E, North Holland, New York.

Takeyama T, Kajikawa Y, Takahashi N, Nakamura T, Itoh S (2006) 'Growth mechanism of α-Bi_2O_3 rods deposited under atmospheric pressure by halide CVD', *Chem Vap Deposition*, 12, 203–206.

Tanenbaum D M, Laracuente A L, Gallagher A (1997) 'Surface roughening during plasma-enhanced chemical-vapor deposition of hydrogenated amorphous silicon on crystal silicon substrates', *Phys Rev B*, 56, 4243–4250.

Tang L H (1993) 'Island formation in submonolayer epitaxy', *J Phys I*, 3, 935–950.

Templier C, Muzard S, Galdikas A, Pranevicius L, Delafond J, Desoyer J C (2000) 'A phenomenological study of the initial stages of film growth', *Surf Coat Technol*, 125, 129–133.

Thiart J J, Hlavacek V, Viljoen H J (2000) 'Chemical vapor deposition and morphology problems', *Thin Solid Films*, 365, 275–293.

Thompson C V, Carel R (1996) 'Stress and grain growth in thin films', *J Mech Phys Solids*, 44, 657–673.

Thornton J A (1977) 'High-rate thick-film growth', *Ann Rev Mater Sci*, 7, 239–260.
Urbonavičius E, Petnyčyte E, Galdikas A (1999) 'The kinetics of thin film island growth at initial stages', *Vacuum*, 53, 377–380.
van den Brekel C H J, Bloem J (1977) 'Characterization of chemical vapor-deposition processes 2', *Philips Res Rept*, 32, 134–146.
van den Brekel C H J, Jansen A K (1978) 'Morphological stability analysis in chemical vapor-deposition processes 1', *J Cryst Growth*, 43, 364–370.
van der Drift A (1967) 'Evolutionary selection a principle governing growth orientation in vapour-deposited layers', *Phillips Res Repts* 22, 267–288.
Venables J A, Spiller G D T, Hanbücken M (1984) 'Nucleation and growth of thin-films', *Rep Prog Phys*, 47, 399–459.
Viljoen H J, Thiart J J, Hlavacek V (1994) 'Controlling the morphology of CVD films', *AIChE J*, 40, 1032–1045.
Villain A, Pimpinelli A, Tang L, Wolf D (1992) 'Terrace sizes in molecular-beam epitaxy', *J Phys I*, 2, 2107–2121.
Volintiru I, Creatore M, van de Sanden M C M (2008) '*In situ* spectroscopic ellipsometry growth studies on the Al-doped ZnO films deposited by remote plasma-enhanced metalorganic chemical vapor deposition', *J Appl Phys*, 103, 033704.
Wagner R S, Ellis W C (1964) 'Vapor-liquid-solid mechanism of single crystal growth (new method: growth "catalysis" from impurity; whiskers, epitaxial, and large crystals; Si; E)', *Appl Phys Lett*, 4, 89–90.
Watanabe K, Komiyama H (1990) 'Micro/macrocavity method applied to the study of the step coverage formation mechanism of SiO_2-films by LPCVD', *J Electrochem Soc*, 137, 1222–1227.
Werner C, Ulacia J I, Hopfmann C, Flynn P (1992) 'Equipment simulation of selective tungsten deposition', *J Electrochem Soc*, 129, 566–574.
Wild C, Herres N, Koidl P (1990) 'Texture formation in polycrystalline diamond films', *J Appl Phys*, 68, 973–978.
Wild C, Koidl P, Müller-Sebert W, Walcher H, Kohl R, Herres N, Locher R, Samlenski R, Brenn R (1993) 'Chemical-vapor-deposition and characterization of smooth (100)-faceted diamond films', *Diam Relat Mater*, 2, 158–168.
Yao J H, Guo H (1993) 'Shadowing instability in three dimensions', *Phys Rev E*, 47, 1007–1011.
Zhao Y P, Drotar J T, Wang G C, Lu T M (2001) 'Morphology transition during low-pressure chemical vapor deposition', *Phys Rev Lett*, 87, 136102.

5

Modelling thin film deposition processes based on real-time observation

S. KOWARIK, Humboldt Universität zu Berlin, Germany and
A. HINDERHOFER, A. GERLACH and
F. SCHREIBER, Universität Tübingen, Germany

Abstract: We introduce time and length scales of growth processes and then review experimental techniques for real-time and *in-situ* studies. In particular, we discuss optical monitoring techniques, time resolved microscopy, and real-time scattering techniques (X-ray, He-, and electron scattering). For scattering experiments we discuss details of the analysis, in particular anti-Bragg growth oscillations as observed in X-ray- and He-scattering as well as RHEED. We illustrate real-time observation and modelling of thin film deposition with examples from organic molecular beam deposition.

Key words: time resolved surface science, MBE growth, X-ray scattering, differential reflectance spectroscopy, He-scattering, RHEED, dewetting, scaling laws.

5.1 Introduction: time resolved surface science

As is evident from the various chapters in this book, thin film growth has many different facets to it, and growth is inherently a non-equilibrium and time-dependent phenomenon. While the investigation of the final *product*, i.e. the *grown* film, with its structure and morphology, may allow one to partially reconstruct the *growth process*, frequently the *real-time* observation of the *growing* film is required to fully characterize and understand the growth in its complexity. This is particularly evident if transient phenomena occur during the growth. In this chapter, after a brief discussion in Section 5.2 of some concepts of growth and the associated time scales, we present first an overview of various techniques (Section 5.3) for real-time studies including microscopy, optical spectroscopy, and scattering and indicate their strengths and weaknesses. Scattering techniques are discussed at some length, since they are at the center of most of the case studies in Section 5.4.

For the foundations of growth and its theoretical background we refer to other chapters in this book as well as earlier books and reviews (e.g., Venables, 2000; Venables *et al.*, 1984, Zangwill, 1988; Pimpinelli and Villain, 1998; Barabási and Stanley, 1995; Michely and Krug, 2004; Krug, 1997). For the purpose of coherence, the case studies are mostly based on our own work on systems from organic molecular beam deposition (OMBD) (Witte and Wöll, 2004; Schreiber, 2004; Forrest, 1997), but it is important to stress that the concepts outlined here can be very easily transferred to other, of course also inorganic, systems. It is clear that within the scope of this chapter the references cannot be exhaustive, but with those given it should be possible to trace a more complete list of references on a given subject.

5.2 Basics of growth and relevant length of and timescales for *in-situ* observation of film deposition

Growth phenomena are extremely rich and include a number of competing processes. Adsorption and desorption processes are followed by thermalisation and diffusion on the surface terrace. Adatoms can form a new crystal grain, attach to an existing one, or cross a step-edge onto a different terrace. Depending on the relative probabilities of these different processes, the growth mode varies between layer-by-layer growth (Frank–van-der-Merwe), layer-by-layer plus island growth (Stranski–Krastanov), or island growth (Volmer–Weber). Not only the growth mode but also the morphology of the thin film varies greatly as the island size and shape depends on the above processes, leading for example to fractal morphology for diffusion limited aggregation.

An important issue in the context of experimental real-time observations of growth and the modelling of these processes is the issue of time and length scales. Length scales in growth range from 10^{-12}–10^{-3} m: the (sub-)atomic length scales determine strain and lattice parameters, diffusion length scales often range in the μm length scale, and, depending on the growth mode, macroscopic crystallite sizes are possible. Further relevant length scales are the surface roughness σ, in-plane correlation length ξ and island sizes, which often change during growth. Time resolved techniques offer the particular advantage that the dependence of length scales on film thickness d can be followed and directly compared to, for example, theoretical scaling models, which describe the time/thickness dependence by scaling laws using the roughening exponent β and the dynamic exponent z (Krug, 2004):

$$\begin{aligned} \sigma &\sim d^{-\beta} \\ \xi &\sim d^{-1/z} \end{aligned} \qquad [5.1]$$

Similarly, growth processes span many orders of magnitude from ultra-fast to hour-long timescales. In the following sections we will discuss experimental techniques with a time resolution from μs to hours, which make it possible to study some of the most important aspects of growth. To give an overview of processes occurring during growth, Figure 5.1 shows both ultra-fast and slow timescales.

Atomistic processes that underlie and determine the growth behaviour occur on ultra-fast timescales in the picosecond range. Thermalisation of the adsorbed atoms/molecules occurs via transfer of translational and rotational energy to phonon modes of the substrate and, in the case of molecular adsorbates, also into internal molecular vibrations. Typical oscillation periods lie in the range of tens of femtoseconds (an example being the 515 cm^{-1} Raman line of silicon) up to picoseconds for large molecules. The frequencies of these (surface phonon) modes often coincide with the typical attempt frequencies assumed in Monte Carlo simulations for diffusion and hopping processes. Intuitively it seems plausible to make a connection between a vibrational frequency and the frequency of attempts Γ_0 to surmount, for example, a step-edge barrier E_b according to an Arrhenius relation for the process rate k:

$$k \sim \Gamma_0 e^{-E_b/kT} \quad [5.2]$$

Indeed, a timescale of around 1ps (i.e., 1 THz) has typically been measured in experiments and deduced in density functional theory (Ratsch and Scheffler, 1998) but it is important to note that there are also other effects such as motion of a whole cluster of adatoms/admolecules, where the attempt frequency is changed due to collective effects.

Even though the atomic/molecular movement is ultra-fast and the attempt frequency very high, the often substantial energy barriers for diffusion slow down the rates by many orders of magnitude. One mechanism of surface diffusion is hopping of atoms to the nearest neighbour lattice site, while collective movements such as dimer and cluster diffusion are often neglected. For the energy barrier E_d for hopping, values of around 100 meV (e.g., 100 meV for Ag on Ag(111) (Brune *et al.*, 1995), 20 meV for sexiphenyl (Hlawacek *et al.*, 2008) and 80 meV for PTCDA (Fendrich and Krug, 2007)) were reported. Of course, these strongly depend on the given system. The diffusion parameter D then is related to the hopping rate by

$$D = \Gamma_0 \frac{l^2}{\alpha} e^{-E_d/kT} \quad [5.3]$$

with l being the jump length and α depending on the dimensionality and symmetry of motion ($\alpha = 2$ for a square lattice). Similarly to surface diffusion, also the diffusion across a step-edge is hindered by the so-called Ehrlich–Schwöbel barrier E_b (Ehrlich and Hudda, 1966, Schwoebel and Shipsey, 1966), which is crucial in determining the growth morphology of the thin film (Markov, 2004, Trofimov and mokerov, 2002). Typical values of E_b are in the range of several 100 meV: 139 meV for Ag(111) (Haftel, 2001), 670 meV for sexiphenyl (Hlawacek *et al.*, 2008), and 750 meV for

PTCDA (Fendrich and Krug, 2007). For a sample at room temperature these step-edge barriers are large compared to $k_B T$ so that crossing a step-edge is comparatively slow. Depending on the height of the barrier, the time for an atom to move down to the layer underneath (the inverse of the rate constant for crossing a step) can then range from 10 ps to 1 s, the large spread being due to the exponential dependence on the barrier height (see Fig. 5.1). In general the diffusion and step-edge barriers depend on the orientation of the underlying crystal and, unlike homoepitaxy, for thin film growth the Ehrlich–Schwöbel barrier can also be layer dependent, e.g. due to strain (Zhang *et al.*, 2009, Krause *et al.*, 2004a).

Through rate limiting steps, such as step-edge diffusion or nucleation, which at moderate temperatures are very slow or even completely frozen out, fast processes can be slowed down to relatively long timescales (see Fig. 5.1). Apart from such material-specific, intrinsic effects, extrinsic parameters start to play a role for slower processes. The most important extrinsic parameter regarding timescales is the growth rate, which is the net flux of atoms/molecules onto a surface. The growth rate determines the time to form a monolayer and therefore is the reference timescale for real-time experiments. This monolayer formation time can range from milliseconds for technological production processes up to many seconds or even hours in research. Such slow rates are mostly employed in fundamental research, as production processes have to be optimized for high throughput, so that also fast time-resolved experimental techniques are necessary or have to be developed for production monitoring. Further, the structure and morphology of the growing thin film itself can depend strongly on the growth rate, as fast kinetic growth can result in structures far from thermodynamic equilibrium if the formation of the lowest energy structure is slow.

In conclusion, growth phenomena span an extremely wide range of timescales, so far inaccessible to any single theory or experiment. Ultra-fast studies are used to determine the basic atomistic processes and calibrate rate constants used in theory. Experiments at slower timescales are very important to identify the growth scenario and follow the structure formation, strain, and morphology. It is remarkable how experiments on the second time-scale can help to deduce information on atomistic processes on the surface. Further, real-time experiments are also important technologically, because growth

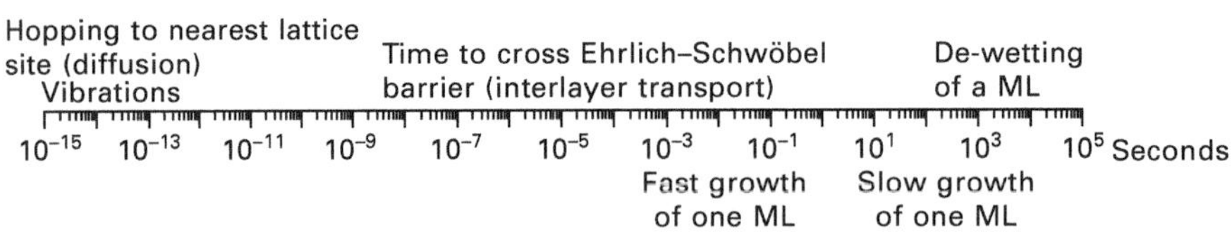

5.1 Simplified scheme of timescales for processes occurring during growth.

monitoring greatly facilitates or even enables growth of complicated nano- or hetero-structures which could not be grown without direct process control.

5.3 Experimental techniques for real-time and *in-situ* studies

Experimental techniques for real-time and *in-situ* studies face several challenges. Ideally, a wide range of timescales has to be studied, structural details have to be resolved on atomic length scales in the growth direction and on the length scale of growing islands in the in-plane direction, the observation should not interfere with growth itself, and the techniques have to be surface sensitive, that is they have to discern the small surface signal of a (monolayer) thin film from the substrate signal. Currently, no single technique is able to follow individual atomic movements on both ultra-fast and slow timescales. Correspondingly, several experimental techniques are necessary to study different time- and length scales. Without claiming to be exhaustive, Fig. 5.2 gives an overview of different techniques used for real-time measurements.

A distinction has to be made between ensemble averaging measurements and individual measurements both in space and time. In the spatial domain, ensemble measurements such as X-ray scattering provide atomic detail

<table>
<tr><th>Time / Space</th><th>Non-repeating process, continuous measurement</th><th>Repeating processes, pump-probe and spectroscopy</th></tr>
<tr><td rowspan="2">Individual details, microscopy</td><td>TEM, SEM
AFM, STM
optical microscopy</td><td>Raman spectro-microscopy</td></tr>
<tr><td colspan="2">X-ray microscopy
LEEM, PEEM</td></tr>
<tr><td rowspan="2">Space averaging, ensemble and diffraction measurements</td><td>X-ray growth oscillations
helium-atom scattering
RHEED, LEED</td><td>Pump-probe
absorption sum frequency generation</td></tr>
<tr><td colspan="2">Reflection spectroscopy</td></tr>
</table>

5.2 Simplified overview of techniques for real-time growth observation. The boundaries are, of course, not always necessarily sharp. Details of the techniques shown can be found in technique oriented chapters in this book and Lüth (2001) and Michely and Krug (2004).

averaged over a certain sample area, while microscopy techniques are able to resolve individual atomic or molecular details, albeit usually at slower time resolution. Similarly, in the time domain it is possible to distinguish between continuous real-time measurements that resolve single configurations of an on-going process on the one hand, and pump-probe or spectroscopy techniques which average over many repetitions of an ultra-fast/oscillatory process on the other (note that the data acquisition time is much longer than the process under study). Growth processes in general are non-repetitive and hard to reset or trigger by external stimuli (an exception being pulsed laser deposition; Ferguson *et al.*, 2009; Tischler *et al.*, 2006), so that pump-probe techniques do not easily lend themselves to *in-situ* growth studies. Therefore, ultra-fast time resolution is hard to achieve in growth studies, because a continuous measurement is limited by the read out speed of electronics, and, even more severely, by the time needed to acquire sufficient measurement statistics without averaging. Despite a continuous measurement on slow timescales, absorption or Raman spectroscopy also contain spectroscopic information on ultra-fast vibrational timescales of damped oscillatory processes.

In the following we will not discuss ultra-fast techniques but focus on methods that can be used for real-time measurements during growth without temporal averaging. Microscopy techniques have started to become fast enough for real-time observation and examples of LEEM (Meyer Zu Heringdorf *et al.*, 2001) or STM microscopy (Rost, 2007) show the potential of spatially resolved measurements. Nevertheless, most real-time experiments in process monitoring and basic research average over a representative sample area. Widespread techniques in both engineering and research environments are real-time reflectance/ellipsometry measurements and reflection high-energy electron diffraction (RHEED) oscillations, because of the comparative simplicity of the setup. Real-time X-ray scattering and real-time helium atom scattering are more complex experimental setups, but yield different information. The latter techniques can also be applied to different types of samples such as molecular materials, which may get damaged by high-energy electrons. In the following we will give an overview of spectroscopic and microscopic techniques for growth monitoring, and then discuss and compare helium atom, electron and X-ray scattering. We will focus on the issues conceptually relevant in the context of real-time observations and their modelling. We will not attempt to explain the implications in terms of experimental technology and rather refer to the works of Forker and Fritz (2009), Lüth (2001) and Poelsema and Comsa (1989).

5.3.1 Optical spectroscopy techniques

Optical spectroscopy techniques can detect sub-monolayers of molecular/atomic adsorbates on substrates (Forker and Fritz, 2009; Heinemeyer *et al.*, 2010).

In differential reflectance spectroscopy (DRS), spectroscopic information is gathered by acquiring the reflectivity R of the substrate before and during deposition of a thin film usually at normal incidence. The time dependent DRS signal is then calculated according to

$$DRS(t) = \frac{R(t) - R(t = 0)}{R(t = 0)} \quad [5.4]$$

where $R(t = 0)$ corresponds to the reflectivity of the bare substrate and $R(t)$ to the reflectivity at time t of the deposition when the substrate is covered by the film with thickness d. As a high photon flux measurement the detector sensitivity and dark count rate is often less crucial than high dynamic range and fast read-out speeds (typically ~10 ms). A typical time resolution is below 10 s, taking into account averaging of multiple spectra to obtain low coverage sensitivity. At normal incidence and for very thin films (i.e. for $d/\lambda << 1$) on a transparent substrate with refractive index n the DRS signal can be related directly to the imaginary part ε_2 of the dielectric function according to McIntyre and Aspnes (1971):

$$DRS(t) = \frac{8\pi d}{\lambda\cdot(1 - n^2)}\cdot\varepsilon_2 \quad [5.5]$$

This approximation simplifies the analysis greatly, although it is not possible to determine ε_1. For thicker films, the complete dielectric function (ε_1+ε_2) can be modelled by matrix method based software. The experimental set up for *in-situ* DRS is shown in Fig. 5.3.

Expanding the capabilities of DRS, reflection anisotropy spectroscopy (RAS) measures the reflectance at normal incidence for two perpendicular polarisations of light and therefore can also resolve in-plane anisotropies of the optical constants (Weightman *et al.*, 2005; Hartell *et al.*, 2001). Spectroscopic ellipsometry (Arwin and Aspnes, 1984; Aspnes *et al.*, 1990) finally expands polarisation sensitive measurements to a range of different incidence angles and therefore can also determine film thicknesses in the nm range, that is far below the wavelength of light.

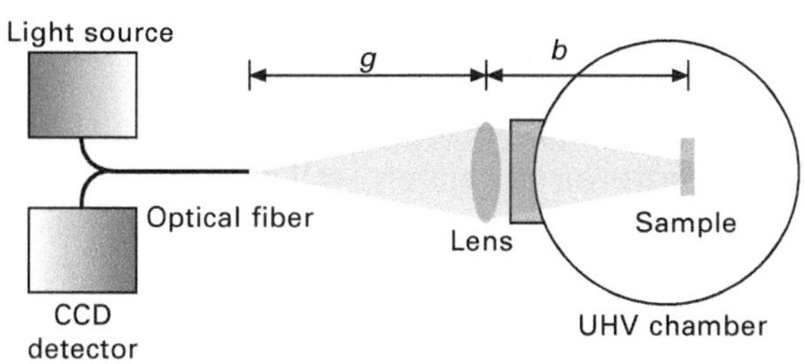

5.3 Experimental setup for *in-situ* DRS (Heinemeyer, 2009).

Typical information that can be obtained from real-time optical spectroscopy is the film thickness evolution, and spectral changes due to different structures that form during growth. The structural information is only indirectly inferred from models that describe the excited states in the thin film. Therefore structural probes such as scattering techniques are sometimes preferable as a direct measurement technique (Lazzari *et al.*, 2009; Hosokai *et al.*, 2010). Nevertheless, ellipsometry, DRS and RAS can be used as a fingerprint technique if the connection between structure and spectral features is known. Further, optical functional properties such as the optical bandgap can be determined and correlated with the changes in thin film structure during growth.

5.3.2 Microscopy

Microscopy techniques are a direct way to measure the real-space structure of a growing thin film, but at the same time most microscopy setups also severely restrict the growth environment so that they are not commonly used as standard process characterisation tools. Usually, a trade-off between temporal resolution and spatial resolution has to be found in particular for scanning microscopy techniques. Nevertheless, they combine desirable experimental properties such as observation of individual sample features in both space and time.

Scanning probe microscopy (SPM)

Scanning probe techniques have enjoyed tremendous popularity in thin film studies due to their atomic resolution capabilities and compatibility with vacuum, gas, and liquid environments. An advantage of SPM is the imaging of growth morphologies in real space, in particular island shapes, nucleation sites and nucleation densities (Rabe and Buchholz, 1991; Fischer *et al.*, 1999; Rusponi *et al.*, 1997, Kneppe *et al.*, 2000). Growth studies have been performed for solution deposition (e.g., Hyde *et al.*, 2002; Rabe and Buchholz, 1991) with a time resolution down to video rate (~30 ms). Great care has to be taken that the atomic force microscopy (AFM) tip does not disturb growth nuclei, for example by avoiding contact mode and scanning in tapping mode.

STM measurements with, in principle, atomic resolution have also been used for *in-situ* growth studies on moderately well conducting substrates. Au deposition from an evaporation source in vacuum has been studied by Rost (2007) and atomic-height steps could be imaged with a lateral resolution of 1 nm in a field of view of 500 × 500 nm. Such a real-time dataset giving microscopic insight into growth with atomic resolution makes it possible to watch growth modes such as step flow and nucleation at dislocations directly.

Great care has to be taken to avoid a disturbance of the growth, for example by shadowing the atomic flux with the tip, or coating of the tip itself with the evaporation material, which may either influence the shape of the tip or also lead to atoms diffusing from the tip onto the surface.

In conclusion, real-time STM studies are slightly challenging but gaining popularity due to technical advances. Since the focus in this chapter is on scattering techniques, for an in-depth discussion the reader is referred to Hofer *et al.* (2003) and references therein.

Electron microscopy

Electron microscopy combines good spatial resolution and a wide field of view for acquiring some surface statistics with video rate (30 ms) real-time capabilities. The best spatial resolution is obtained in transmission electron microscopy (TEM), albeit at the expense of restricting the sample to very thin substrates in the range of several 10 nm and requiring the growth apparatus to fit in the microscope. Examples of real-time TEM studies include growth of Cu on Au in a liquid cell from solution studied at 5 nm lateral resolution (Williamson *et al.*, 2003) and formation of Pt nanoparticles in solution at sub-nm resolution (Zheng *et al.*, 2009). Achieving not only structural resolution but also chemical imaging is highly desirable and X-ray fluorescence in TEM can provide additional, local chemical information but also imaging of photoelectrons gives a material-specific contrast. *In-situ* photoelectron emission microscopy (PEEM) at a time resolution of 60 s has been used to study growth of the molecular semiconductor pentacene. Using PEEM to study pentacene growth the diffusion limited dendritic growth of individual monolayers could be resolved down to 125 nm (Meyer Zu Heringdorf *et al.*, 2001).

5.3.3 Scattering methods

Elastic X-ray scattering, atom scattering and electron scattering are particularly useful for real-time observation of thin films, because they give structural information on the atomic length scale with a time resolution suited to growth dynamics. They are compatible with typical vacuum growth environments and, in particular, X-rays with their large penetrating power can be used to perform experiments in a wide range of sample environments such as vacuum, gas and liquid environments. All three scattering techniques measure an ensemble average where usually:

$$\text{beam footprint on sample} \gg \text{coherence length of structural features}$$

but the spatial resolution, which is given by the (de Broglie) wavelength, can be on an atomic length scale (for a more precise discussion of resolution, see Fenter, 2002) so that area averaged microscopic properties are sampled. Spreading the X-ray or particle flux over a macroscopic area reduces the likelihood of beam damage while at the same time providing enough flux for obtaining sufficient counting statistics with a time resolution in the sub-second range.

We will first point out common characteristics of real-time measurements with X-ray, He-atoms, and electrons before individually discussing their unique properties. In general, a part of the incident beam changes the propagation direction, and the two wave vectors k_{in} of the incident and k_{out} of the scattered beam define the scattering plane (see Fig. 5.4). The wave vector transfer $q = k_{out} - k_{in}$ points in the direction in which the periodicity d of the thin film is sampled, and strong reflections occur when the Bragg condition $q = n2\pi/d$ is fulfilled for an integer n. Both the in-plane structure ($q_{\parallel}$ to the substrate surface, Fig. 5.4b) and out-of-plane structure ($q_{\perp}$ to the sample surface along the growth direction, Fig. 5.4a) can be measured during growth. An example for following the in-plane structure is given in Section 5.4 and an observation of the in-plane reflections has been used for example in Schreiber *et al.* (1998), Schreiber (2000) and Kowarik *et al.* (2006) to follow the growth of self-assembled monolayers and other systems. There are several approaches for real-time structure analysis by scattering techniques. For instance, one can measure the (monotonous)increase or shift of a diffraction feature. Another approach is recording an oscillatory signal due to the filling of individual layers.

Oscillating signals that vary with the number of lattice planes filled during growth are of particular interest and most commonly used, because details about the growth mode can be obtained with sub-monolayer resolution. For

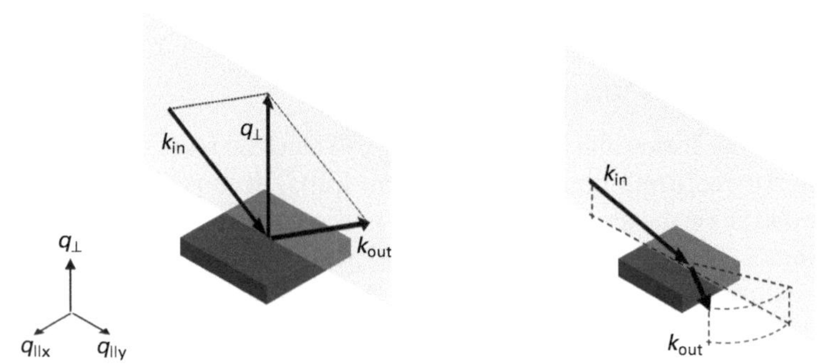

(a) (b)

5.4 (a) Specular reflectivity measurement: incoming and outgoing beams enclose the same angle with the substrate surface, and therefore the wave vector transfer q is directed along the surface normal. (b) Grazing incidence diffraction (GIXD) and non-specular scattering involves a momentum transfer parallel to the surface.

all three scattering techniques oscillations are primarily, but not exclusively, measured by observing the specular reflection, that is the scattered beam is detected for 'mirror-like' reflection ($q_\perp$ perpendicular to sample surface). For the purpose of the following discussion, here we want to distinguish two basic scattering mechanisms, which influence the specular reflection signal during growth and may lead to intensity oscillations:

(a) Out-of-phase interference between the scattering from successively growing atomic monolayers (no change in in-plane momentum $q_{||} = 0$) (see Fig. 5.5a).
(b) Loss of specular intensity into diffuse scattering 'sideways' of the specular beam due to surface corrugation that oscillates during growth $q_{||} \neq 0$) (see Fig. 5.5b).

In (a) the oscillations involve interference between atomic layers in different depths scattering out-of-phase, that is $q_\perp$ must not fulfil the Bragg condition. Then the successive filling of atomic layers will cause the specular reflectivity to oscillate according to:

$$\frac{I(t)}{I_0} = |A_{\text{substrate}}(t) \cdot e^{iq_\perp d_0} + \Sigma_n A_n(t) \cdot e^{iq_\perp nd}|^2 \qquad [5.6]$$

This equation adds the scattering amplitudes $A_{\text{substrate}}$ and A_n of both the substrate and all the n adlayers, taking into account the correct relative phase $\varphi = q_\perp d_0$ between the substrate and the adlayers due to d_0 (bonding distance between substrate and first adlayer) as well as the phase shift between adlayers separated by a distance of n times the lattice constant d. The oscillations can be most readily understood for $q_\perp = \pi/d$ where the exponential factor in the sum alternates between –1 and 1, that is adjacent layers alternate between adding to/subtracting from the scattering amplitude. Note, that this mechanism only redistributes intensity between the reflected beam and

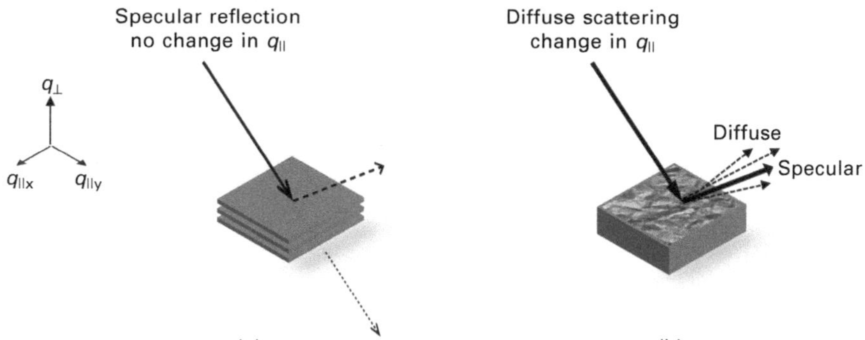

5.5 Specular scattering with a change in $q_\perp$ and diffuse scattering with a change in $q_\perp$ and $q_{||}$.

the transmitted (and eventually absorbed) beam, because the in-plane wave vector is not changed – no in-plane periodicity is probed in this oscillation mechanism. Therefore in elastic scattering where momentum is conserved the only choice for q is being normal to the surface for the reflected beam or zero for the transmitted beam.

In (b) the mechanism of intensity change during growth (monotonous or oscillatory) is distinctly different as it involves momentum transfer not only perpendicular but also parallel to the sample surface. In case of a periodicity within the surface non-specular scattering can lead, *inter alia*, to in-plane Bragg reflections, while a non-periodic corrugated surface results in broad, diffuse scattering in many directions. While the surface changes during growth also the specularly and diffusely scattered signal change in a characteristic fashion. For all three techniques of X-ray, He, and electron scattering a changing surface corrugation – for example smooth → rough (upon nucleation of islands in a new layer) → smooth (upon layer completion) – leads to an oscillation in the diffuse scattered intensity (integrated over all q) with oscillating roughness (Fuoss *et al.*, 1992; Eres *et al.*, 2002). The specular reflection intensity changes inversely to the diffuse intensity, because beam intensity which is scattered diffusely is missing in the specular reflectivity.

It is important to note that in X-ray scattering also buried layers including the substrate contribute to the scattering, that is A_n is proportional to θ_n with θ_n being the coverage of the n-th layer. In contrast, for surface sensitive helium atom scattering, only the exposed layers contribute to the scattering, that is the scattering amplitudes A_n of buried layers vanish and A_n is proportional to the uncovered fraction ($\theta_n - \theta_{n-1}$) of the n-th layer which also influences the oscillation period,

The underlying reasons (a) and (b) for growth oscillations are similar in all three techniques, but there are also important differences between the three probe types. Most importantly, the scattering cross-section and penetration depth varies greatly (see Fig. 5.6). Due to these different interaction

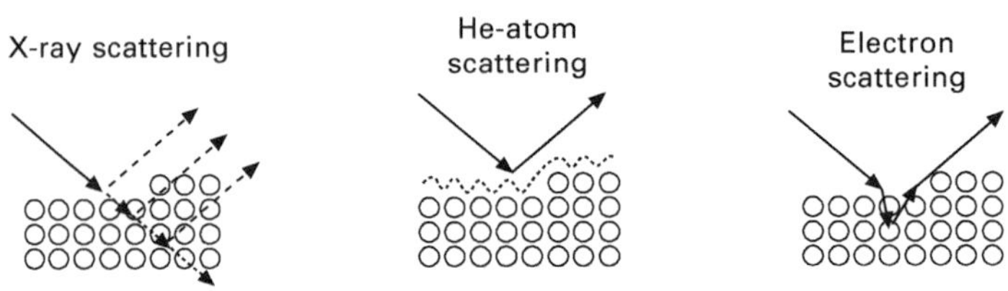

5.6 X-ray scattering penetrates the sample so that scattering occurs also from buried layers, while He-scattering is exclusively surface sensitive. Electron scattering occurs in the first few atomic layers and multiple scattering events can occur as shown.

mechanisms there are a range of differences in the real-time data which are discussed separately in the following sections.

X-ray scattering

X-ray scattering is usually performed at X-ray energies around 10 keV where the elastic scattering cross section is large and the wavelength of around 1 Å fits to studies of thin film growth on atomic length scales. It is worth noting that chemically selective scattering can be achieved by tuning the X-ray energy to specific atomic transitions where the scattering cross section is enhanced by anomalous scattering. In general, X-rays have great penetrating power so that not only the top surface is sampled, but also bulk lattice constants and buried interfaces are accessible.

Surface specificity can nevertheless be improved by using a grazing angle of incidence, so that the beam undergoes total external reflection and the X-ray intensity of the evanescent wave decays exponentially on a length scale of around 10–100 Å (depending on the film electron density) thereby reducing bulk sensitivity. This grazing incidence X-ray diffraction (GIXD) is used to determine in-plane lattice constants by scanning $q_{||}$, and it can also be used to determine the thin film unit cell if so called rod scans are performed, that is the scattering vector is chosen to have components both in $q_{||}$ and $q_{\perp}$. GIXD has been performed in real time during growth of organic molecular semiconductors (Kowarik *et al.*, 2009a) and makes it possible to follow structural changes and strain with a resolution below 0.01 $Å^{-1}$ with a time resolution of < 30 s.

Specular reflectivity measurements, for which q changes only along the surface normal, is useful to resolve changes along the surface normal, which usually coincides with the direction of growth. Experiments are typically performed at the Bragg condition to monitor the phase content, or at half the Bragg q vector which corresponds to the out-of-phase anti-Bragg condition. Both the Bragg reflection of a thin film and the reflectivity at the anti-Bragg point are usually weak enough that the kinematic approximation can be employed (the approximation breaks down close to the total reflection edge or at a strong Bragg reflection). Adapted to X-ray scattering which does not exclusively scatter at the surface but penetrates into the substrate, Eq. 5.6 can be rewritten using the form factor of the adlayer *f(q)*:

$$I_{\text{reflected}}(t) = \left| A_{\text{substrate}}(q_{\perp}) \cdot e^{\mathrm{i} q_{\perp} \mathrm{d}_0} + f(q_{\perp}) \cdot \Sigma_n \, \theta_{\mathrm{n}}(t) \cdot e^{\mathrm{i} \cdot \mathrm{n} \cdot q_{\perp} \cdot \mathrm{d}} \right|^2 \qquad [5.7]$$

Note that the substrate scattering amplitude and phase is fixed in time, and for the purpose of the time-dependent growth studies (i.e., $\theta_n(t)$ and the temporal *change* of the resulting scattering intensity) it is not important that $A_{\text{substrate}}$ may actually have to be calculated using dynamical theory.

For crystalline substrates the phase between substrate and thin film is given by $\Phi = d_0 q_\perp$ where d_0 is the distance between the substrate surface (not the physical surface, but the last lattice plane). The binding distance d_0 is an important parameter to understand interface properties such as charge carrier injection. The binding distance has been determined according to Eq. 5.7 in Alonso *et al.* (2003) which can be related to other techniques such as X-ray standing waves (Gerlach *et al.*, 2005, 2007; Koch *et al.*, 2008, Mercurio *et al.*, 2010, Yamane *et al.*, 2010).

The reflectivity for both homoepitaxy and heteroepitaxy has been calculated according to Eq. 5.7 and is shown in Fig. 5.7. In homoepitaxy the reflectivity is maximised when the surface is perfectly flat, that is the top monolayer is completely closed. When the next layer nucleates on top, the reflectivity drops and the relative change is greatest at the anti-Bragg condition, but other points in q-space oscillate with the same periodicity and may be more convenient (Krug *et al.*, 2006). The intensity of the reflectivity at this anti-Bragg point oscillates with a periodicity of one monolayer for perfect layer-by-layer growth (Stephenson *et al.*, 1999; Tischler *et al.*, 2006; Van Der Vegt *et al.*, 1995).

In heteroepitaxy the reflectivity in Fig. 5.7 is modified by the occurrence of additional Bragg reflections resulting from the overlayer and also interference fringes from reflections of the top and bottom interface of the film (Laue fringes) are visible. Further, it is important to note that the oscillation period at the anti-Bragg condition with respect to the overlayer is now changed from one monolayer to two monolayers, and the oscillation period depends on q (Weschke *et al.*, 1997, Kowarik *et al.*, 2009c). Indeed the shape of growth oscillations depends strongly on the substrate scattering amplitude and the relative phase as shown in Fig. 5.8. Compared to the case of vanishing substrate scattering, the growth oscillation amplitude can be increased if the substrate amplitude is in phase. Additional smaller scattering maxima can occur if substrate and film are out of phase. Homoepitaxy is a special case of the substrate amplitude being 180° out-of-phase and half as strong as the thin-film scattering (Dale *et al.*, 2008).

This selection of growth oscillation phenomena can be explained by the out-of-plane interference along $q_\perp$ which leaves $q_\parallel$ unchanged (model (a) above). The diffuse scattering of the surface has been neglected so far because the modulation of the out-of-plane interference is usually stronger than the diffuse scattering amplitude. For some systems the diffuse scattering, while still weaker than the specular growth oscillations, contributes to the observed oscillation amplitude (Fleet *et al.*, 2006). In conclusion, X-ray scattering offers a wide range of techniques to monitor growth experiments in real time, and due to the applicability of the kinematic scattering theory for most applications the interpretation is possible with the equations given above.

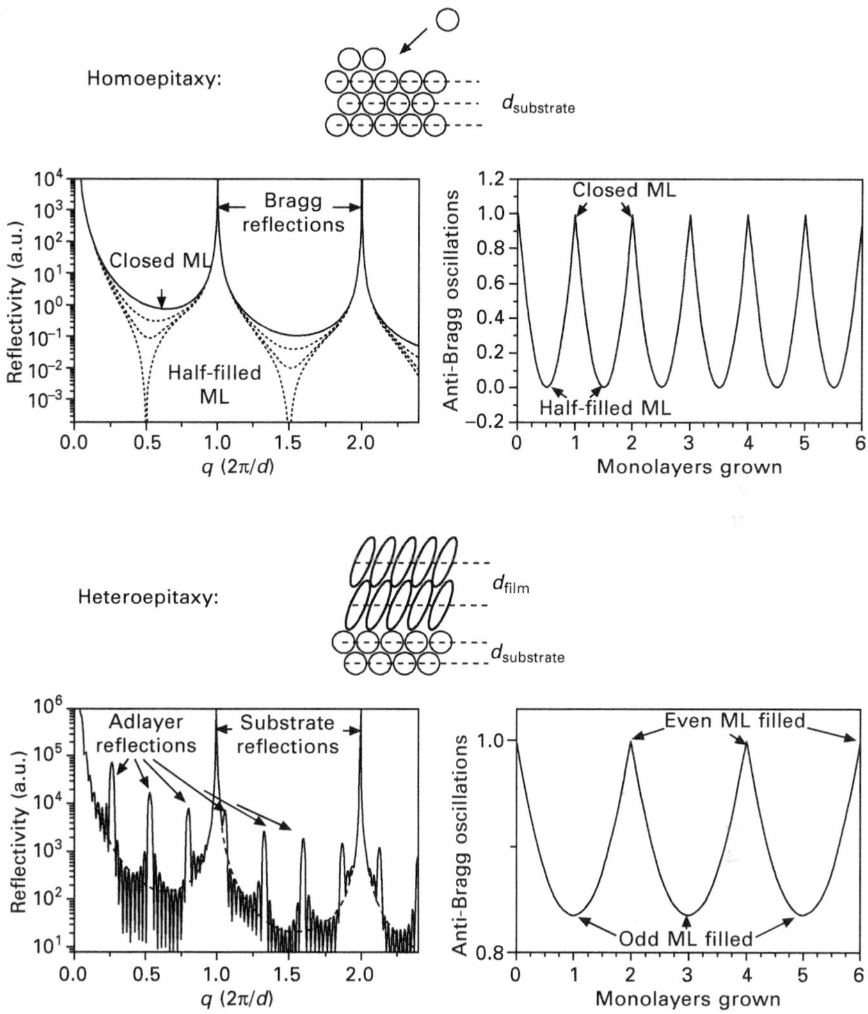

5.7 X-ray reflectivity curves (left) and growth oscillations (right) for homoepitaxy (top) and heteroepitaxy (bottom). In the simulation, perfect layer-by-layer growth is assumed. The model for heteroepitaxy consists of substrate with layer spacing $d_{Substrate}$ and nine layers with a layer spacing of d_{film}.

Helium atom scattering

In He atom scattering (HAS) the energy E_i of the impinging atoms is typically in the 10–100 meV range and therefore the de Broglie wavelength $\lambda = h/\sqrt{2mE_i}$ is in the range of 0.5–1.5 Å which is suitable for studying atomic length scales (Scoles *et al.*, 1988). This energy is several orders of magnitude lower than the energy of electrons or photons at comparable energies, and

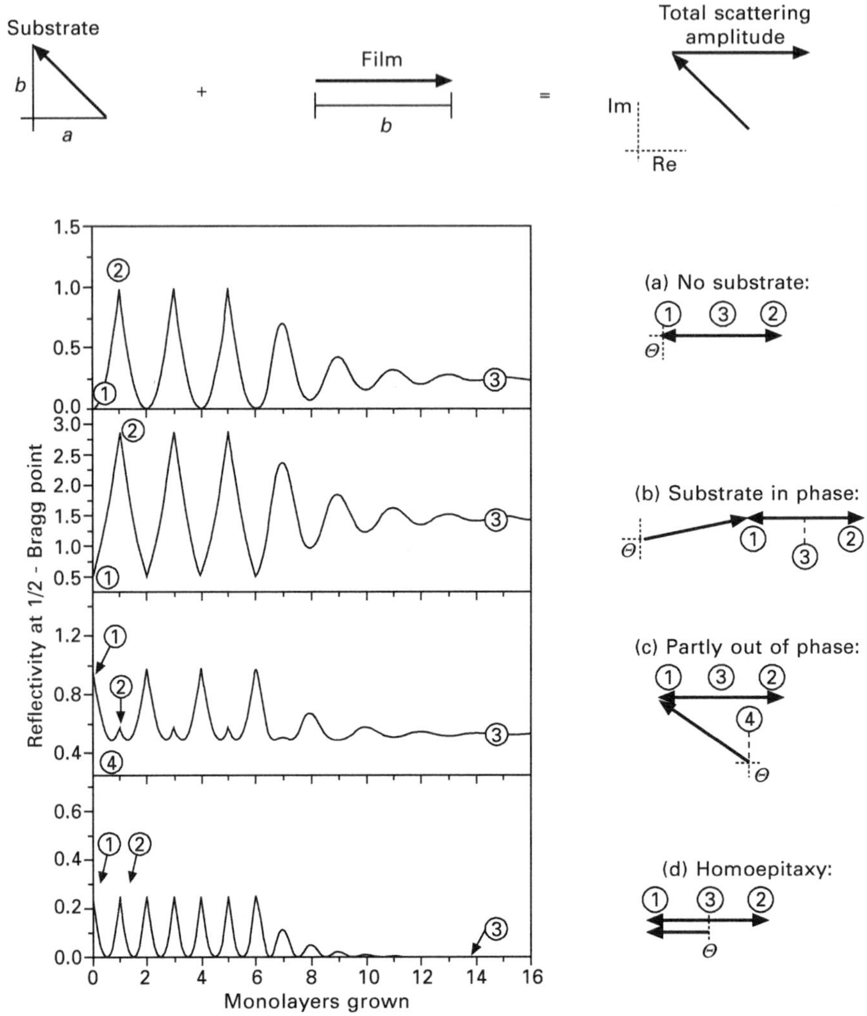

5.8 Simulation of anti-Bragg (= ½–Bragg) oscillations including substrate scattering (Kowarik *et al.*, 2009c).

therefore HAS avoids thermal or electronic excitation of the sample. The HAS energy scale of 10–100 meV is comparable to typical surface phonon energy- and timescales, and therefore inelastic scattering in HAS is an important tool to simultaneously sample dynamic surface processes as well as surface structure (Santoro *et al.*, 1987; Witte and Wöll, 1995).

Another example of such energy-resolved scattering quasi-elastic helium-atom scattering (QHAS), where the adsorbate particles moving on the surface create a moving target and therefore the helium atoms experience a small change in velocity (Frenken *et al.*, 1988). In QHAS studies the activation energy and the attempt frequency of surface diffusion have been determined and with spin echo techniques processes on timescales shorter than 0.5 μs can be detected as discussed in the review by Jardine *et al.* (2002).

Elastic HAS is also employed for structural analysis and growth oscillations, because He interacts through long range van-der-Waals forces with the sample and the cross section is large compared to X-ray and electron scattering, which makes HAS extremely surface sensitive. (Ellis *et al.*, 1995; Kern *et al.*, 1991). Therefore, the helium atoms do not penetrate the surface but get reflected 3–4 Å (depending on the kinetic energy) above the surface. Due to the large cross sections this technique is sensitive to very low coverages, such as hydrogen adatoms with a density of 1/1000 of a monolayer (Poelsema and Comsa, 1989). Elastic helium atom scattering cannot probe bulk lattice constants, as it is exclusively surface-sensitive, but it has been used to measure surface step heights (Dastoor and Allison, 2003) and is very well suited to measuring step-edge densities and faceting (Hinch *et al.*, 1990).

Real-time oscillations during growth and also the sputtering of a surface as 'reverse growth' have been studied with HAS for a wide range of systems (Farias and Rieder, 1998; Poelsema and Comsa, 1989). As the number of step-edges oscillates with the roughness development, so does the reflectivity of the surface according to mechanism (b) above, and indeed growth oscillations at the in-phase condition have been observed (Poelsema *et al.*, 1992; Farias and Rieder, 1998). Nevertheless, stronger oscillations are observed for the out-of-phase/anti-phase condition, which is mostly used in HAS growth studies and usually interpreted with the destructive interference between adjacent layers similar to mechanism (a) above. In surface scattering this leads to the oscillation period of one monolayer, because the maximum reflection of a smooth surface is diminished by the destructively interfering next monolayer until the coverage is 50%; for higher coverages the lower lying layer interferes less with the top layer until a maximum is reached again for a closed top layer. A theoretical explanation for the oscillating reflection width/intensity is given in Poelsema and Comsa (1989). For a more detailed discussion on the different information that can be learned from in-phase and anti-phase oscillations, see Poelsema *et al.* (1992) and Xu *et al.* (1991).

Electron scattering

Electron scattering is one of the most common structural probes used in UHV growth systems, because the experimental setup is simpler than X-ray or HAS real-time experiments. The wavelength is again given by the de Broglie wavelength, which for electrons can be written as $\lambda[\text{Å}] = \sqrt{150/E[eV]}$. In low energy electron diffraction (LEED) the electron energy is chosen to be in the range of 20–200 eV where the surface sensitivity is greatest due to the minimal penetration length of around 5–10 Å in this energy range (Braun,

1999, Oura *et al.*, 2003). LEED is commonly used to study the structure and possible reconstructions of surfaces in static experiments (Stadler *et al.*, 2009), but also real-time experiments during growth are performed (Floreano *et al.*, 2008) with specialised LEED setups that include an evaporation source in the LEED screen (Seidel *et al.*, 1998). The measured diffraction patterns can qualitatively be interpreted with the kinematic, single-scattering approximation, but, given the comparatively strong interaction with the substrate, multiple scattering events cannot generally be excluded. In cases where the kinematic approximation fails, dynamical scattering theory has to be used, for example by using a self-consistent multiple scattering approach (McRae, 1967). The real-time information extracted from (ultra-thin) thin film growth studied with LEED usually concerns different phases/structures that form during growth, and for determining the growth mode of thin films usually scattering of higher energy electrons is used.

The most common of all real-time growth monitoring techniques is RHEED (Braun, 1999; Cohen *et al.*, 1989; Lippmaa *et al.*, 2000), where electron energies in a typical range of 5–30 keV are used. For these higher energies relativistic corrections start to amount to a few per cent and the wavelength is calculated according to $\lambda = h/\sqrt{2mE + E^2/c^2}$. The wavelength accordingly is in the 0.01 Å range and as a consequence also the scattering angles are very small (typically in the 1° range). Because of these shallow angles the momentum component perpendicular to the surface is comparable to the values in LEED. The technique also has a similar surface sensitivity. As in X-ray and He atom scattering RHEED growth oscillations occur during sequential filling of lattice planes during growth. Modelling of growth oscillations in RHEED is more complicated than for the other two probes, as dynamic scattering theory should be used instead of kinematic theory. Nevertheless, most studies use the kinematic approximation as a qualitative description, because the rigorous dynamical treatment is computationally intensive and does not easily offer quick estimates. Due to the surface sensitivity the oscillation period in RHEED is again one monolayer as in HAS. The constructive/destructive interference leads to oscillations according to the model of Eq. 5.6 and has been applied to RHEED (Braun, 1999, Cohen *et al.*, 1989), but due to the strong electron-sample interaction and the shallow incidence angles, step-edge scattering plays an important role and therefore the diffuse scattering (mechanism (b)) also has to be included in the description:

$$\frac{I}{I_0} \sim \left| \begin{array}{c} [\mathbf{1} - \boldsymbol{\beta}_{\mathbf{d}} \boldsymbol{S}(\boldsymbol{\theta}_{\mathbf{1}})] \cdot (1 - \theta_1) \cdot e^{i\varphi} \\ + \Sigma_{n=1}^{m} \, [\mathbf{1} - \boldsymbol{\beta}_{\mathbf{d}} \boldsymbol{S}(\boldsymbol{\theta}_{\boldsymbol{n+1}}) - \boldsymbol{\beta}_{\boldsymbol{u}} \boldsymbol{S}(\boldsymbol{\theta}_{\boldsymbol{n+1}})] \cdot (\theta_n - \theta_{n+1}) \cdot e^{\mathrm{iq_\perp nd}} \\ + [[\mathbf{1} - \boldsymbol{\beta}_{\boldsymbol{u}} \boldsymbol{S}(\boldsymbol{\theta}_{\boldsymbol{m+1}})] \cdot (\theta_{m+1}) e^{\mathrm{iq_\perp (m+1)d}}] \end{array} \right|^2 \quad [5.8]$$

Following Shin and Aziz (2007), in the above equation the oscillations in the specular reflection have been modelled using Eq. 5.8 but importantly the specularly reflected intensity from each layer has been reduced by a factor $[1 - \beta_d S(\theta_{n+1}) - \beta_u S(\theta_{n+1})]$, which accounts for the step-edge scattering of the layer (θ_{n+1}) with step density $S(\theta_{n+1})$. The step reduces the specular reflection of the layer downwards of the step with an effective phenomenological constant β_d and the reflection from the upward layer by β_u, because diffuse scattering and shadowing effects occur at steps. This combination of the model (a) for interference in $q_\perp$ and model (b) for diffuse scattering with a component in $q_\parallel$ can successfully explain oscillations at the anti-phase and in-phase conditions. Importantly though there are discrepancies to fully dynamic calculations that show an increase in the specularly reflected intensity at the in-phase condition for highly stepped surfaces (Korte and Maksym, 1997), where the above equation yields a lower reflectivity. This unexpected behaviour at the in-phase conditions already points to the fact that there is no fixed relationship between the stage in the growth cycle and a feature in the growth oscillation, and maxima of the intensity do not always occur for integer layer coverages. Due to multiple scattering effects, electron probes such as RHEED, while popular as fast and relatively simple techniques for qualitative monitoring of growth, are frequently more difficult to model quantitatively (Auciello and Krauss, 2001).

We note that in addition to elastic scattering techniques of course inelastic scattering and spectroscopy techniques using electrons are also applied in surface science experiments, such as electron energy loss spectroscopy (EELS) and Auger electron spectroscopy (AES).

5.4 Experimental case studies

In the case studies we give examples for a range of typical questions that can be answered by *in-situ* and real-time growth studies: which growth mode occurs, do roughness and island size scale with film thickness, is there transient strain, are there post-growth changes, and how do structure and optical properties correlate? Here we use examples from our own work and, for the purpose of coherence of the presentation, focus on growth of molecular thin films. In particular a series of experiments on diindenoperylene (DIP) on SiO_2, will be insightful in regard to the questions above. Many concepts used here originally stem from MBE growth of atomic systems and are applicable for both molecular and atomic growth (Braun *et al.*, 2003, Fleet *et al.*, 2006, Woll *et al.*, 1999).

We note that there are of course many other systems, for which we cannot even provide an exhaustive list of references. Besides the technically important field of MBE (see examples in this book and Farrow, 1995) there are also studies of 'ablation' and dissolution, which can be seen as 'time-inverted' growth and offers some interesting ways for comparison (Murty *et al.*, 1998,

Teng *et al.*, 2001). For some of these, scattering is essentially the only way to monitor the process; in particular for dissolution/etching processes, the environment may be so harsh or hostile (extreme pH, ion sputtering and/or very high temperature) that SPM becomes almost impossible to apply since the tip would suffer in this environment.

We concentrate on examples of X-ray scattering and optical spectroscopy, but of course other real-time techniques discussed above have also been used to obtain information about growth modes and structural changes. The growth studies discussed in the following sections have been performed in a portable UHV growth chamber that is equipped with an X-ray window to allow for *in-situ* studies (Ritley *et al.*, 2001), while, simultaneously with the X-ray measurements, DRS measurements can also be performed (Hosokai *et al.*, 2010).

5.4.1 Transient strain during thin film growth

X-ray diffraction is a powerful technique to determine the crystal structure of thin films with high resolution, and for the molecule diindenoperylene (DIP) on SiO_2 the thin-film unit cell dimensions have been determined by Kowarik *et al.* (2009a). Surface-sensitive GIXD is particularly suited to study thin films, and real-time measurements during DIP growth are sensitive to sub-monolayer coverages down to 0.1 ML. Figure 5.9 (a) shows the diffraction pattern for an early growth stage in the sub-ML regime where clear in-plane reflections are visible. Once the film has grown to a thickness of 5.3 ML the reflections get stronger and due to the out-of-plane periodicity also reflections along $q_\perp$ appear. Importantly, the diffraction image shows that the (110) and (120) in-plane reflections shift significantly during growth, while the (020) reflection shifts very little. This makes it possible to determine the changes of the in-plane lattice parameter during growth as shown in Fig. 5.9(b). The unit cell can be seen to expand by 3% along the *a* direction, while the *b* direction is nearly constant. Interestingly this change from the monolayer to a multilayer structure is complete, that is the originally compressed structure of the first monolayer is converted to the multilayer structure as can be seen by the disappearance of the corresponding reflection (Kowarik *et al.*, 2006). Obviously, this transient effect would have been missed in post-growth studies.

5.4.2 Growth mode determination: X-ray anti-Bragg oscillations

During growth of DIP the X-ray reflectivity oscillates at the anti-Bragg condition as shown in Fig. 5.10. The period of oscillation between main

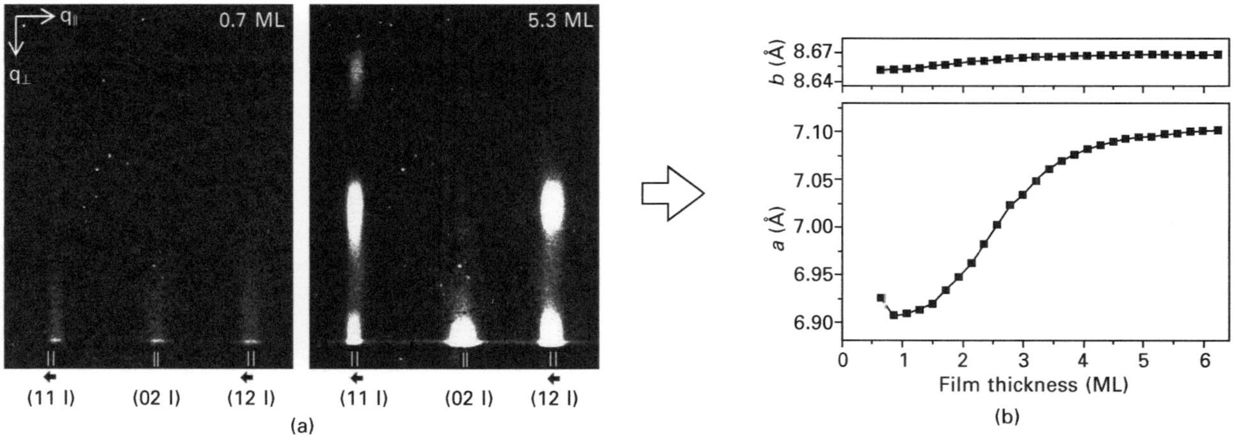

5.9 (a) Two snapshots of the GIXD pattern evolution during growth recorded with an area detector (Kowarik *et al.*, 2009a). (b) Change of the in-plane lattice parameters during growth of the first monolayers.

maxima is 2 ML, but as a result of interference between substrate and DIP scattering (see Fig. 5.8) smaller maxima appear also when odd monolayers are completed. Some observations can be directly extracted from the experimental data. In particular, the maxima indicate when a ML is completed, which makes it possible to accurately determine the growth rate. The damping of the oscillations further indicates that the film gets rougher during growth. For a more detailed understanding, quantitative growth models have to be used for fitting the data, which for organic growth was to our knowledge first done in Krause *et al.* (2004b).

Fitting experimental scattering data is typically a two-step process: (i) growth theories are used to calculate the layer coverages $\theta_{\rm n}$; (ii) using this growth scenario, the experimental observables such as growth oscillations are simulated and the underlying parameter set is varied until a fit is achieved. We restrict the discussion to an intentionally simple and transparent growth model that can reproduce the experimental data reasonably well. For a description of more advanced growth models that are used to calculate the layer coverages, the reader is referred to Cohen *et al.* (1989), Barabási and Stanley (1995), Michely and Krug (2004) and Trofimov and Mokerov (2002).

In the diffusive growth model after Cohen *et al.* (1989) the rate for a jump from layer n+1 to n is proportional to the uncovered fraction of layer n+1 and the available space in layer n:

$$\frac{d\theta_n}{d(t/\tau)} = (\theta_{\rm n-1} - \theta_n) + k_{\rm n}(\theta_{\rm n+1} - \theta_{\rm n+2})(\theta_{\rm n-1} - \theta_n) - k_{\rm n}(\theta_{\rm n-2} - \theta_{\rm n-1})(\theta_{\rm n} - \theta_{\rm n+1}) \quad [5.9]$$

Here θ_n stands for the fractional coverage of the nth layer, τ is the time to complete one monolayer, and k_n is the effective rate for interlayer transport, that is the upward and downward rates are combined. Varying k_n as a fit parameter for each layer, the anti-Bragg oscillations can be fitted from the fourth layer onwards. The first three layers have not been fitted because of the complications of transient strain, but both X-ray and AFM studies show that the first layers grow in a layer-by-layer fashion (Zhang *et al.*, 2007).

The $\theta_n(t)$ resulting from a fit using Eq. 5.9 can be calculated as shown in Fig. 5.10(b). The interface of the DIP film for a given nominal coverage (e.g. 5, 8 and 13 ML) can be calculated from these $\theta_n(t)$ evaluated at the corresponding growth times and, in general, the film roughness can be extracted as a function of time/film thickness from the layer coverages. It is obvious that after ~5 ML of deposition the roughness starts to increase, which corresponds to the strong damping of the oscillations in this growth stage. This roughness increase is due to the significant decrease in downward

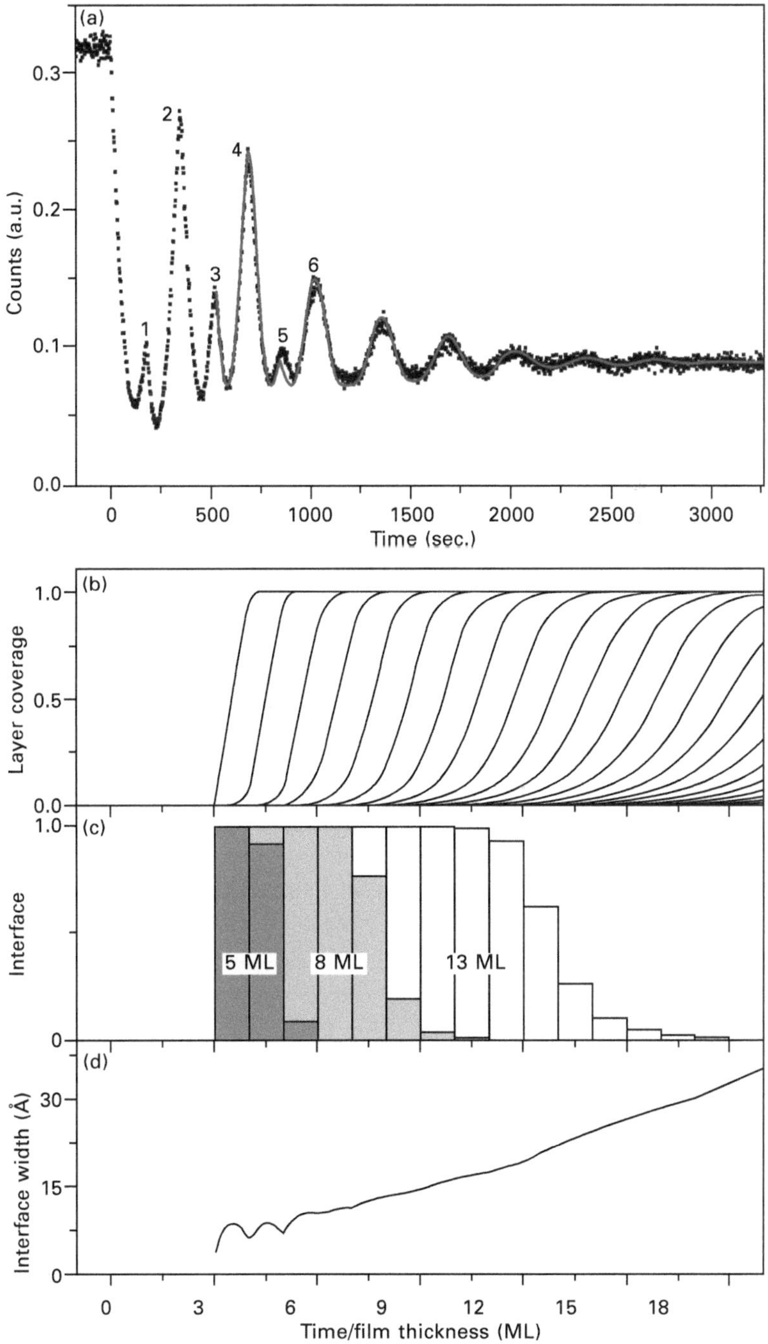

5.10 (a) X-ray anti-Bragg growth oscillation during DIP growth. (b) Individual layer coverages as function of film thickness. (c) Growth front for three thicknesses. (d) Roughness evolution during growth.

(smoothing) interlayer current. In this simple model the reduction in the interlayer transport rate cannot be directly related to an increase in the Ehrlich–Schwöbel barrier, because the diffusion probability across a step also depends on the step-edge density which is not modelled in the simple rate equation. A recent AFM study on DIP has indeed shown, though, that the Ehrlich–Schwöbel barrier is layer dependent, and confirms the trend extracted from the model used here (Zhang *et al.*, 2009).

5.4.3 Scaling laws: reflectivity and full *q*-range growth oscillations

Modern synchrotron sources make it possible to acquire not only a single point in reciprocal space but a wide q-range can be measured on a timescale of 10–100 seconds (essentially limited more by the moving speed of the X-ray diffractometer in angle dispersive experiments than by scattering intensities). Consequently, more information is acquired during the growth process, because many Fourier components of the real-space structure are sampled instead of only a single spatial frequency (Kowarik *et al.*, 2009c). Figure 5.11(a) shows a typical dataset following the evolution of the specular reflectivity from the bare substrate up to a film thickness of 12 ML. This dataset can be analysed by either simulating growth oscillations (i.e., cuts through the 3D data $f(q = \text{const.}, d)$) using the model described above, or by fitting the reflectivity curves at fixed points in time to directly extract information of the real-space structure (that is $f(q, d = \text{const.})$).

Figure 5.11(b) shows growth oscillations at q values of $\frac{1}{2}q_{\text{Bragg}}$ (the anti-Bragg condition), $\frac{2}{3}$, $\frac{3}{4}$, $\frac{4}{5}$ and $\frac{1}{1}q_{\text{Bragg}}$. The growth oscillations at q values larger than the anti-Bragg point oscillate with a slower period, and importantly also continue to oscillate after the anti-Bragg oscillations are completely damped out, that is information about later growth stages is only contained in the time evolution at larger q values in this case. The fits of growth oscillations can be performed using Eq. 5.7 which also naturally explains the increasing oscillation period on the time, i.e. thickness axis: at the anti-Bragg position complete destructive interference occurs after 2 ML are deposited resulting in a two ML oscillation period; at $q = \frac{2}{3}q_{\text{Bragg}}$ destructive inference occurs between three monolayers resulting in an oscillation period of 3 ML, and so on. Again, the rate equation model from Eq. 5.9 is combined with the kinematic scattering theory in Eq. 5.7. The result makes it possible to extract layer coverages, interlayer rate constants and evolution of surface roughness as illustrated by the example in Section 5.4.2.

As a second approach, the Parratt algorithm (Parratt, 1954) can be applied to the same 3D dataset for describing the X-ray reflectivity as a function of q for a given time, i.e. thickness d. The Parratt fit can be restricted to q

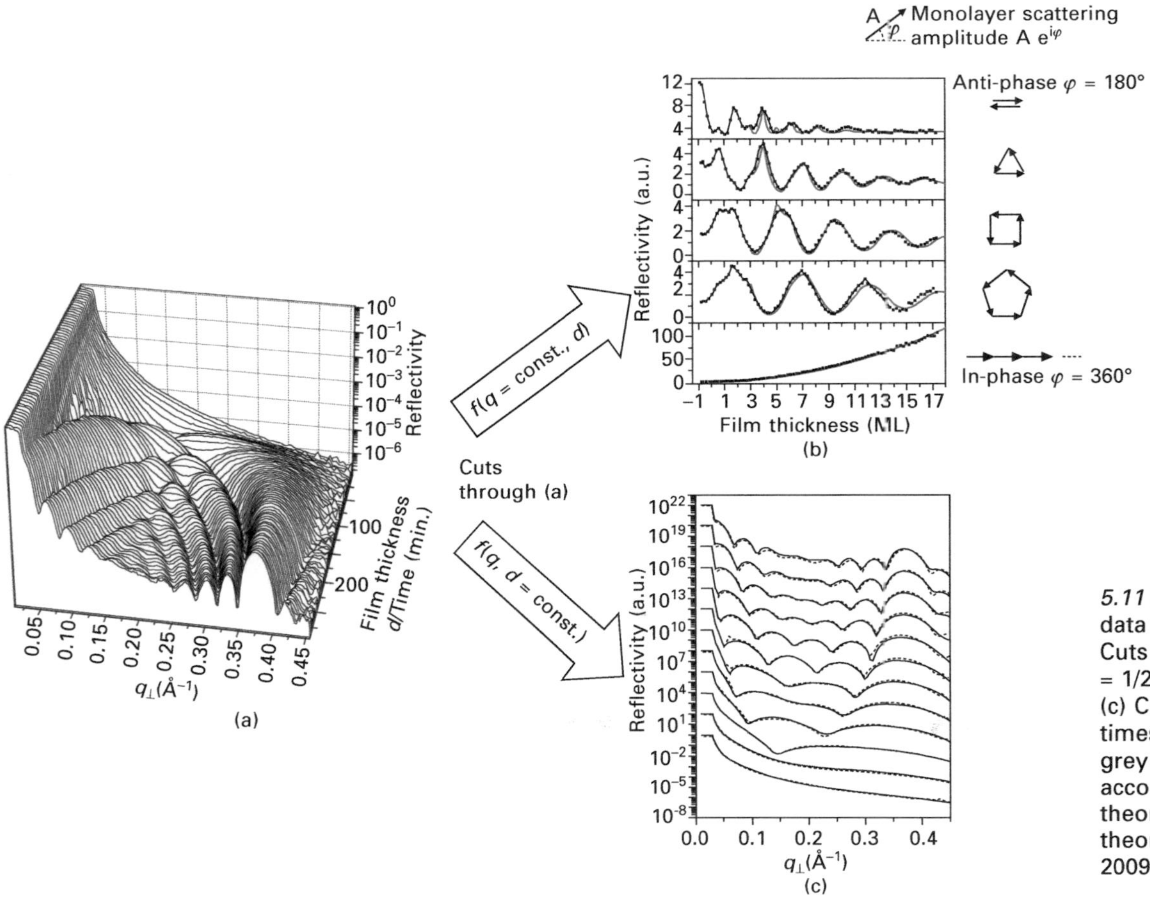

5.11 (a) Real-time reflectivity data for growth of DIP. (b) Cuts for fixed q values of q = 1/2, 2/3, 3/4, 4/5, 1/1 q_{Bragg}. (c) Cuts for fixed growth times/film thicknesses. Light grey lines are fits to the data according to the kinematic theory in (b) and Parratt theory in (c) (Kowarik *et al.*, 2009c).

$<\pi/d_{film}$ if only the film roughness is of interest (Hinderhofer *et al.*, 2010), but when including the modulation of the electron density due to the lattice periodicity, also the Bragg reflection can be fitted and the individual layer coverages can be extracted. For this approach no growth model has to be assumed and parameters such as surface roughness can be extracted directly. In Dürr *et al.* (2003) such Parratt fits have been applied to *in-situ* post-growth measurements to determine the roughening exponent β from the thickness dependency of the film roughness (Fig. 5.12). The roughening exponent β was found to be 0.75, which remarkably is higher than the roughness exponent expected for random deposition. This behaviour is similar to other molecular systems and possibly related to the additional orientational and conformational degrees of freedom in molecular growth (Yim and Jones, 2006, Amassian *et al.*, 2009, Dürr *et al.*, 2003, Kowarik *et al.*, 2007).

5.4.4 Post-deposition changes: dewetting of a molecular monolayer

Real-time techniques can not only be applied to growth itself, but also to re-arrangement and dewetting processes after the growth has stopped. This dewetting is technologically important as it can limit device stability in applications. For a monolayer of DIP the dewetting process has been followed using X-ray reflectivity as shown in Fig. 5.13. The specular reflectivity changes during evaporation of molecules onto a SiO_2 substrate at a temperature of

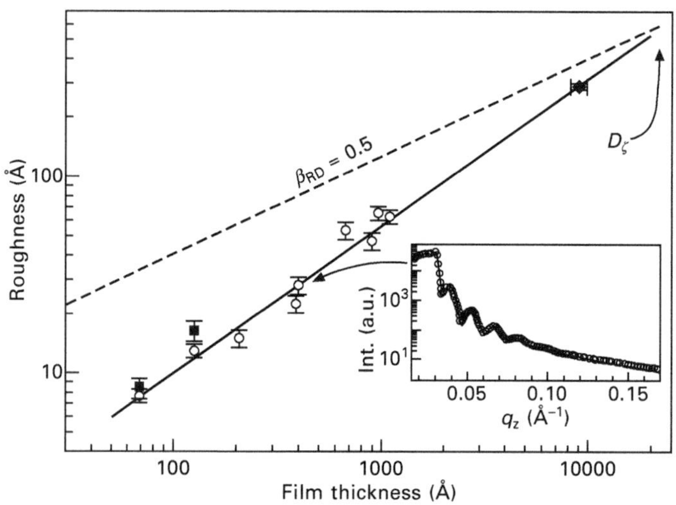

5.12 Scaling of the surface roughness with film thickness in DIP growth. Note that the data in this particular example were obtained post-growth (Dürr *et al.*, 2003).

130°C as expected for the formation of a closed monolayer.

Using the Parratt formalism (Parratt, 1954) for fitting the X-ray reflectivity, the layer coverages of the first and second monolayers have been extracted as shown in Fig. 5.13(b). As expected for layer-by-layer growth, the coverage of the first monolayer nearly completes before nucleation of the second monolayer occurs. Shortly after the second layer nucleates, the flux of molecules is stopped. Closing the shutter after completion of the first monolayer, the X-ray reflectivity continues to change indicating a post-growth re-arrangement. In particular, the coverage of the second monolayer continues to rise while the coverage of the first monolayer shrinks. The rate of change is ~ $(5.8 \pm 1.0) \times 10^{-3}$ ML/min and the fact that the loss rate in the first monolayer and the gain rate in the second monolayer are equal shows that molecules are not desorbing, but rearranging. This example again demonstrates the usefulness of real-time observation. *Ex-situ* AFM measurements confirm the conclusions regarding the dewetting (Kowarik *et al.*, 2009b), but the transient monolayer structure would be missed in post-growth measurements and of course the rate constant can only be determined in a time-resolved experiment.

5.4.5 Optical real-time studies

Following the optical properties and possible spectral changes in the growing film is also a powerful tool (Heinemeyer *et al.*, 2010, Forker and Fritz, 2009). Of course, the optical properties are related to the structure, so that it is intriguing to study these simultaneously (Hosokai *et al.*, 2010). Both optical properties and its interplay with film structure are highly relevant for organic devices, such as organic photovoltaic cells and organic light-emitting diodes.

Optical reflectance during thin film growth: DIP/SiO_2

The different molecular environment at the interface compared to the bulk of a thin film, as well as structural changes also beyond the first monolayer, can give rise to an overall shift of the optical spectra, to a redistribution of oscillator strength or to the appearance of new transitions. Spectral changes as a function of thickness d are a very general effect. In the simplest approximation these decay as $1/d$ if they are purely interface-driven. For example, we can find the appearance of a new transition in DIP growth on SiO_2 depending on film thickness (see Fig. 5.14 and Heinemeyer *et al.*, 2010).

Simultaneous optical reflectance and GIXD during thin film growth: $F_{16}CuPc/SiO_2$

Simultaneous *in-situ* real-time measurements of X-ray reflectivity (XRR),

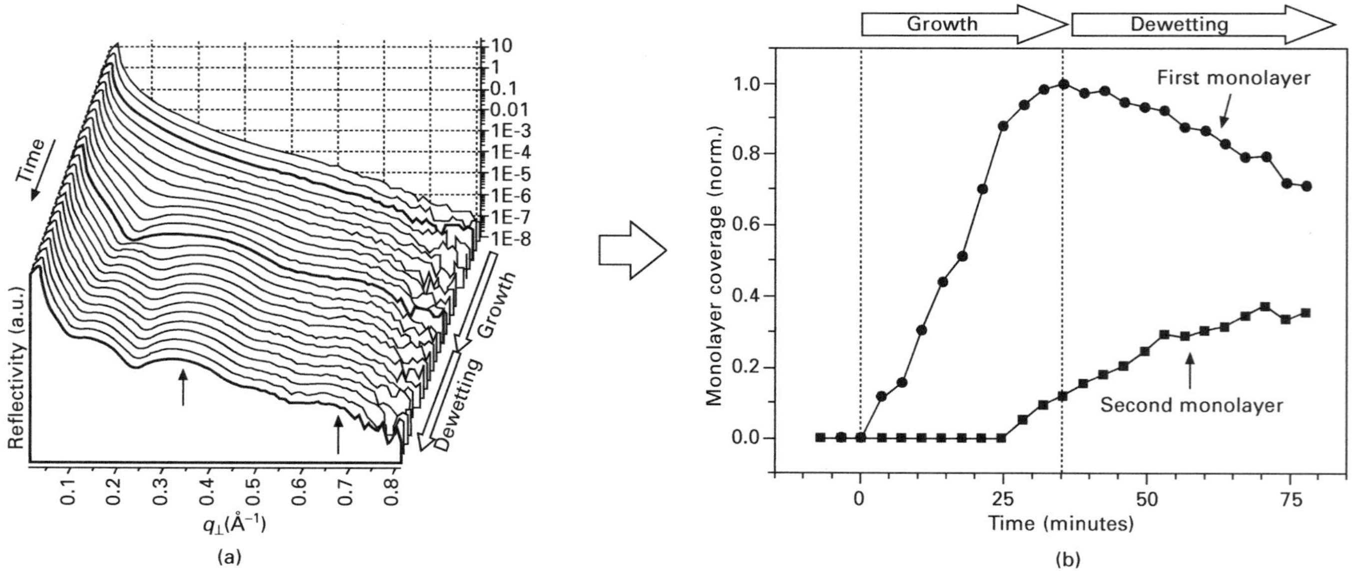

5.13 (a) X-ray reflectivity during growth and dewetting of DIP. (b) Coverages of the first and second DIP layer during growth and dewetting. The broken lines mark the opening and closing of the evaporation cell shutter (Kowarik *et al.*, 2009b).

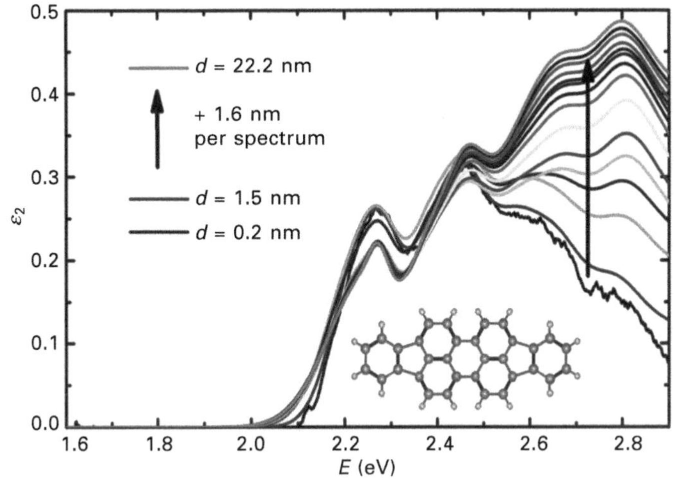

5.14 $\varepsilon_2(E)$ for a DIP film grown on glass. The thickness varies in steps of 1.6 nm from 0.2 nm to 22.2 nm. The modes at 2.25 eV (0-0), 2.48 eV (0-1) and 2.6 eV (0-2) are part of the vibronic progression, whereas the newly arising mode at 2.8 eV does not belong to it (Heinemeyer *et al.*, 2010).

grazing incidence X-ray diffraction (GIXD) and differential reflectance spectroscopy (DRS) were performed on per-fluorinated copper phthalocyanine ($F_{16}CuPc$) thin films deposited on a SiO_2/Si wafer (see Fig. 5.15(a) for the experimental setup, (b) for the real-time GIXD data taken at 303 K). Based on the X-ray data a structural transition within the first few layers of the film was established (De Oteyza *et al.*, 2006).

DRS spectra for selected film thicknesses with the corresponding numbers of molecular layers (ML) are shown in Fig. 5.16. Due to the structural transition of $F_{16}CuPc/SiO_2$ a significant variation of the spectra – especially below 2 ML – is observed. An absorption feature A and a broad shoulder B, which becomes dominant at 1 ML, are observed at 1.82 eV and 1.96 eV, respectively. In addition, a third feature C around 1.59 eV appears at 1 ML and becomes more intense with increasing film thickness.

In the literature the absorption spectra of $F_{16}CuPc$ thin films have been discussed based on molecular exciton theory (Alonso *et al.*, 2003). There, peak C is attributed to the β-phase of $F_{16}CuPc$, that is characterized by ring-over-ring (eclipsed) π-orbital stackings and an inclined alignment of the transition dipoles (see Fig. 5.16(b)). It has also been noted (Schlettwein *et al.*, 1999) that for $F_{16}ZnPc$ growth on glass substrates feature A is at similar energies as a peak found in solution spectra, suggesting that this feature could be related to an amorphous $F_{16}ZnPc$ phase in the film. With

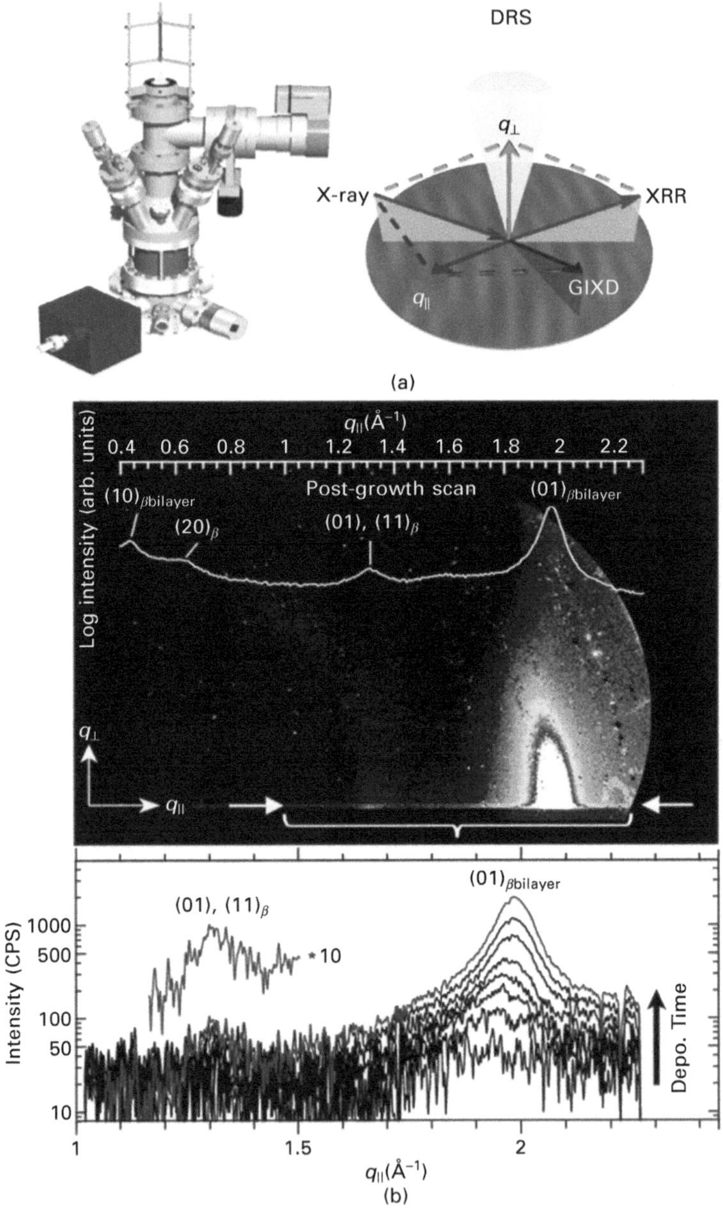

5.15 (a) Experimental setup. (b) Real-time GIXD data of $F_{16}CuPc$ deposited on SiO_2 at 303 K. From Hosokai *et al.* (2010) with permission.

that information and on the basis of the XRR data the changes in the DRS spectra can be explained as follows.

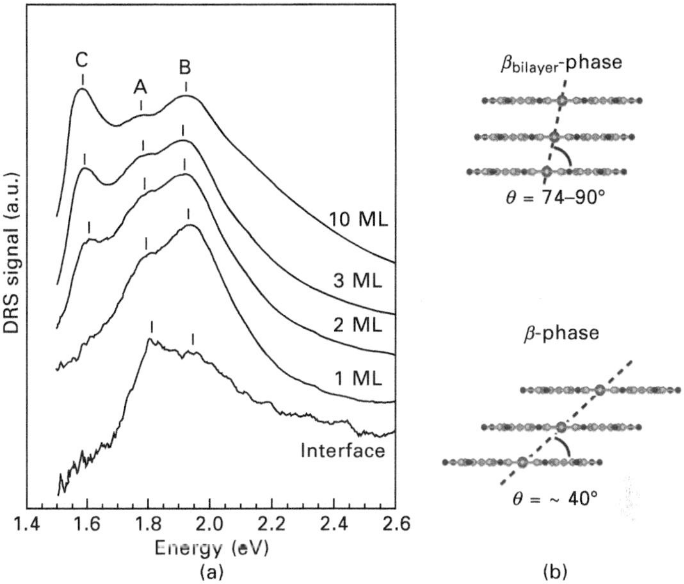

5.16 (a) Real-time DRS spectra of $F_{16}CuPc$ taken during deposition on SiO_2 at 303 K. (b) Molecular arrangement for two $F_{16}CuPc$ phases. From Hosokai *et al.* (2010) with permission.

Initially, the molecules prefer the flat-lying orientation and form an essentially amorphous film with spectra showing mainly feature A. After completion of the interface layer the molecules build up the bilayer structure reported by De Oteyza *et al.* (2006) with nearly parallel transition dipole moments, causing the observed increase of feature B. Finally, the thin film β-phase with the typical growth oscillations starts to nucleate. Above 2 ML the corresponding feature C appears in the spectrum. As is apparent from this example, a combination of several complementary real-time techniques greatly facilitates the understanding of a complicated growth mode.

5.5 Future trends

While ultra-fast experiments and atomic resolution individually can be achieved with comparative ease nowadays, the combination of fast time resolution, high spatial resolution, and surface sensitivity remains a challenge. Further, no single technique can span the wide time- and length scales that are relevant in growth processes so far.

Future progress will be driven by technological advances in monitoring techniques, allowing more and more of the above criteria to be combined. Interesting trends in X-ray scattering are photon counting pixel detectors that enable faster data acquisition, energy dispersive detectors that avoid motor scanning delays (Kowarik *et al.*, 2007), and X-ray photon correlation spectroscopy (Seydel *et al.*, 2003). New X-ray sources and in particular the X-ray laser will further enable ultra-fast X-ray scattering experiments with femtosecond time resolution. Also X-ray microscopy and tomography have been making significant progress in recent years and techniques such as interfacial X-ray microscopy (XRIM) is now established and can be applied to study growth, not only in vacuum but also in other complex environments (Fenter *et al.*, 2006).

Optical techniques continue to benefit from increased availability of (solid-state) lasers in a wide wavelength range, and in particular (attosecond-)laser science and X-ray science increasingly merge as high harmonics of lasers reach the X-ray regime. This will enable much higher temporal resolution for studies of atomic structure during thin film growth in the future.

5.6 Sources of further information and advice

Further sources of information on real-time observation of growth can be found in a range of articles (Rost, 2007; Kowarik *et al.*, 2009c; Rabe and Buchholz, 1991) and books (Auciello and Krauss, 2001; Braun, 1999). Growth modelling as the second central part of understanding the evolution of growth has been treated in a range of books on growth (Barabási and Stanley, 1995; Cohen *et al.*, 1989; Krug, 1997, 2004; Markov, 2004; Michely and Krug, 2004; Pimpinelli and Villain, 1998; Shehu *et al.*, 2010).

5.7 References

Alonso, M. I., Garriga, M., Osso, J. O., Schreiber, F., Barrena, E. & Dosch, H. (2003) Strong optical anisotropies of F_{16}CuPc thin films studied by spectroscopic ellipsometry. *Journal of Chemical Physics*, 119, 6335–6340.

Amassian, A., Desai, T. V., Kowarik, S., Hong, S., Woll, A. R., Malliaras, G. G., Schreiber, F. & Engstrom, J. R. (2009) Coverage dependent adsorption dynamics in hyperthermal organic thin film growth. *The Journal of Chemical Physics*, 130, 124701.

Arwin, H. & Aspnes, D. E. (1984) Unambiguous determination of thickness and dielectric function of thin-films by spectroscopic ellipsometry. *Thin Solid Films*, 113, 101–113.

Aspnes, D. E., Quinn, W. E. & Gregory, S. (1990) Application of ellipsometry to crystal-growth by organometallic molecular-beam epitaxy. *Applied Physics Letters*, 56, 2569–2571.

Auciello, O. & Krauss, A. R. (2001) *In situ real time characterization of thin films*. New York: Wiley.

Barabási, A. L. & Stanley, H. E. (1995) *Fractal Concepts in Surface Growth*. Cambridge: Cambridge University Press.

Braun, W. (1999) *Applied Rheed: Reflection High-energy Electron Diffraction During Crystal Growth*. Berlin: Springer.

Braun, W., Jenichen, B., Kaganer, V. W., Shtukenberg, A. G., Däweritz, L. & Ploog, K.

H. (2003) Layer-by-layer growth of GaAs(001) studied by *in situ* synchrotron X-ray diffraction. *Surface Science*, 525, 126–136.

Brune, H., Bromann, K., Röder, H., Kern, K., Jacobsen, J., Stoltze, P., Jacobsen, K. & Norskov, J. (1995) Effect of strain on surface diffusion and nucleation. *Physical Review B*, 52, R14380.

Cohen, P. I., Petrich, G. S., Pukite, P. R., Whaley, G. J. & Arrott, A. S. (1989) Birth-death models of epitaxy: I. Diffraction oscillations from low index surfaces. *Surface Science*, 216, 222–248.

Dale, D., Suzuki, Y. & Brock, J. D. (2008) *In situ* x-ray reflectivity studies of dynamics and morphology during heteroepitaxial complex oxide thin film growth. *Journal of Physics: Condensed Matter*, 20, 264008.

Dastoor, P. C. & Allison, W. (2003) Geometric contrast mechanisms in helium atom scattering: the growth of the Fe/Cu(100) system. *Physical Review B*, 67, 245403.

De Oteyza, D. G., Barrena, E., Ossó, J. O., Sellner, S. & Dosch, H. (2006) Thickness-dependent structural transitions in fluorinated copper-phthalocyanine (F16CuPc) films. *Journal of the American Chemical Society*, 128, 15052–15053.

Dürr, A. C., Schreiber, F., Ritley, K. A., Kruppa, V., Krug, J., Dosch, H. & Struth, B. (2003) Rapid roughening in thin film growth of an organic semiconductor (diindenoperylene). *Physical Review Letters*, 90, 016104.

Ehrlich, G. & Hudda, F. G. (1966) Atomic view of surface self-diffusion: tungsten on tungsten. *Journal of Chemical Physics*, 44, 1039–1049.

Ellis, J., Toennies, J. P. & Witte, G. (1995) Helium atom scattering study of the frustrated translation mode of CO adsorbed on the Cu(001) surface. *The Journal of Chemical Physics*, 102, 5059–5070.

Eres, G., Tischler, J., Yoon, M., Larson, B., Rouleau, C., Lowndes, D. & Zschack, P. (2002) Time-resolved study of $SrTiO_3$ homoepitaxial pulsed-laser deposition using surface x-ray diffraction. *Applied Physics Letters*, 80, 3379–3381.

Farias, D. & Rieder, K.-H. (1998) Atomic beam diffraction from solid surfaces. *Reports on Progress in Physics*, 61, 1575–1664.

Farrow, R. F. C. (1995) *Molecular Beam Epitaxy: Applications to Key Materials*, Park Ridge, NJ: Noyes Publications.

Fendrich, M. & Krug, J. (2007) Ehrlich–Schwoebel effect for organic molecules: direct calculation of the step-edge barrier using empirical potentials. *Physical Review B (Condensed Matter and Materials Physics)*, 76, 121302–121303.

Fenter, P. A. (2002) X-ray reflectivity as a probe of mineral-fluid interfaces: a user guide. *Reviews in Mineralogy and Geochemistry*, 49, 149–221.

Fenter, P., Park, C., Zhang, Z. & Wang, S. (2006) Observation of subnanometre-high surface topography with X-ray reflection phase-contrast microscopy. *Nat Phys*, 2, 700–704.

Ferguson, J. D., Arikan, G., Dale, D. S., Woll, A. R. & Brock, J. D. (2009) Measurements of surface diffusivity and coarsening during pulsed laser deposition. *Physical Review Letters*, 103, 256103.

Fischer, B., Brune, H., Barth, J. V., Fricke, A. & Kern, K. (1999) Nucleation kinetics on inhomogeneous substrates: Al/Au(111). *Physical Review Letters*, 82, 1732–1735.

Fleet, A., Dale, D., Woll, A. R., Suzuki, Y. & Brock, J. D. (2006) Multiple time scales in diffraction measurements of diffusive surface relaxation. *Physical Review Letters*, 96, 055508.

Floreano, L., Cossaro, A., Gotter, R., Verdini, A., Bavdek, G., Evangelista, F., Ruocco, A., Morgante, A. & Cvetko, D. (2008) Periodic arrays of Cu-phthalocyanine chains on Au(110). *The Journal of Physical Chemistry C*, 112, 10794–10802.

Forker, R. & Fritz, T. (2009) Optical differential reflectance spectroscopy of ultrathin epitaxial organic films. *Physical Chemistry Chemical Physics*, 11, 2142–2155.

Forrest, S. R. (1997) Ultrathin organic films grown by organic molecular beam deposition and related techniques. *Chemical Reviews*, 97, 1793–1896.

Frenken, J. W. M., Toennies, J. P. & Wöll, C. (1988) Self-diffusion at a melting surface observed by He scattering. *Physical Review Letters*, 60, 1727–1730.

Fuoss, P. H., Kisker, D. W., Lamelas, F. J., Stephenson, G. B., Imperatori, P. & Brennan, S. (1992) Time-resolved x-ray scattering studies of layer-by-layer epitaxial growth. *Physical Review Letters*, 69, 2791–2794.

Gerlach, A., Schreiber, F., Sellner, S., Dosch, H., Vartanyants, I. A., Cowie, B. C. C., Lee, T.-L. & Zegenhagen, J. (2005) Adsorption-induced distortion of F16CuPc on Cu(111) and Ag(111): an X-ray standing wave study. *Physical Review B*, 71, 205425.

Gerlach, A., Sellner, S., Schreiber, F., Koch, N. & Zegenhagen, J. (2007) Substrate dependent bonding distances of PTCDA – a comparative X-ray standing wave study on Cu(111) and Ag(111). *Physical Review B*, 75, 045401.

Haftel, M. I. (2001) Ehrlich–Schwoebel effect for vacancies: low-index faces of silver. *Physical Review B*, 64, 125415.

Hartell, A. D., Tok, E. S. & Zhang, J. (2001) The development of RAS and RHEED as *in situ* probes to monitor dopant segregation in GS-MBE on Si(001). *Journal of Crystal Growth*, 227, 729–734.

Heinemeyer, U. (2009) Optical properties of organic semiconductor thin films: static spectra and real-time growth studies. PhD thesis, Universität Tübingen.

Heinemeyer, U., Broch, K., Hinderhofer, A., Kytka, M., Scholz, R., Gerlach, A. & Schreiber, F. (2010) Real-time changes in the optical spectrum of organic semiconducting films and their thickness regimes during growth. *Physical Review Letters*, 104, 257401.

Hinch, B. J., Lock, A., Madden, H. H., Toennies, J. P. & Witte, G. (1990) Helium-atom scattering investigation of facetting of the Al stepped (332) surface. *Physical Review B*, 42, 1547.

Hinderhofer, A., Gerlach, A., Kowarik, S., Zontone, F., Krug, J. & Schreiber, F. (2010) Smoothing and coherent structure formation in organic-organic heterostructure growth. *EPL (Europhysics Letters)*, 91, 56002.

Hlawacek, G., Puschnig, P., Frank, P., Winkler, A., Ambrosch-Draxl, C. & Teichert, C. (2008) Characterization of step-edge barriers in organic thin-film growth. *Science*, 321, 108–111.

Hofer, W. A., Foster, A. S. & Shluger, A. L. (2003) Theories of scanning probe microscopes at the atomic scale. *Reviews of Modern Physics*, 75, 1287–1331.

Hosokai, T., Gerlach, A., Hinderhofer, A., Frank, C., Ligorio, G., Heinemeyer, U., Vorobiev, A. & Schreiber, F. (2010) Simultaneous *in situ* measurements of x-ray reflectivity and optical spectroscopy during organic semiconductor thin film growth. *Applied Physics Letters*, 97, 063301.

Hyde, M. E., Jacobs, R. & Compton, R. G. (2002) *In situ* AFM studies of metal deposition. *The Journal of Physical Chemistry B*, 106, 11075–11080.

Jardine, A. P., Ellis, J. & Allison, W. (2002) Quasi-elastic helium-atom scattering from surfaces: experiment and interpretation. *Journal of Physics: Condensed Matter*, 14, 6173.

Kern, K., Niehus, H., Schatz, A., Zeppenfeld, P., Goerge, J., & Comsa, G. (1991) Long-range spatial self-organization in the adsorbate-induced restructuring of surfaces: Cu{100}-(2 × 1)O. *Physical Review Letters*, 67, 855–858.

Kneppe, M., Dorna, V., Kohstall, P., Kot, E. & Köhler, U. (2000) A kinetic scanning tunneling microscopy study of iron silicide growth on Si(113). *Surface Science*, 454–456, 802–806.

Koch, N., Gerlach, A., Duhm, S., Glowatzki, H., Heimel, G., Vollmer, A., Sakamoto, Y., Suzuki, T., Zegenhagen, J., Rabe, J. P. & Schreiber, F. (2008) Adsorption-induced intramolecular dipole: correlating molecular conformation and interface electronic structure. *Journal of the American Chemical Society*, 130, 7300–7304.

Korte, U. & Maksym, P. A. (1997) Role of the step density in reflection high-energy electron diffraction: questioning the step density model. *Physical Review Letters*, 78, 2381.

Kowarik, S., Gerlach, A., Sellner, S., Schreiber, F., Cavalcanti, L. & Konovalov, O. (2006) Real-time observation of structural and orientational transitions during growth of organic thin films. *Physical Review Letters*, 96, 125504.

Kowarik, S., Gerlach, A., Leitenberger, W., Hu, J., Witte, G., Wöll, C., Pietsch, U. & Schreiber, F. (2007) Energy-dispersive X-ray reflectivity and GID for real-time growth studies of pentacene thin films. *Thin Solid Films*, 515, 5606–5610.

Kowarik, S., Gerlach, A., Sellner, S., Cavalcanti, L., Konovalov, O. & Schreiber, F. (2009a) Real-time X-ray diffraction measurements of structural dynamics and polymorphism in diindenoperylene growth. *Applied Physics A – Materials Science & Processing*, 95, 233–239.

Kowarik, S., Gerlach, A., Sellner, S., Cavalcanti, L. & Schreiber, F. (2009b) Dewetting of an organic semiconductor thin film observed in real-time. *Advanced Engineering Materials*, 11, 291–294.

Kowarik, S., Gerlach, A., Skoda, M. W. A., Sellner, S. & Schreiber, F. (2009c) Real-time studies of thin film growth: measurement and analysis of X-ray growth oscillations beyond the anti-Bragg point. *European Physical Journal – Special Topics*, 167, 11–18.

Krause, B., Durr, A. C., Schreiber, F., Dosch, H. & Seeck, O. H. (2004a) Late growth stages and post-growth diffusion in organic epitaxy: PTCDA on Ag(111). *Surface Science*, 572, 385–395.

Krause, B., Schreiber, F., Dosch, H., Pimpinelli, A. & Seeck, O. H. (2004b) Temperature dependence of the 2D–3D transition in the growth of PTCDA on Ag(111): a real-time X-ray and kinetic Monte Carlo study. *Europhysics Letters*, 65, 372–378.

Krug, J. (1997) Origins of scale invariance in growth processes. *Advances in Physics*, 46, 139–282.

Krug, J. (2004) Power laws in surface physics: the deep, the shallow and the useful. *Physica A*, 340, 647–655.

Krug, K., Stettner, J. & Magnussen, O. M. (2006) *In situ* surface X-ray diffraction studies of homoepitaxial electrochemical growth on Au(100). *Physical Review Letters*, 96, 246101–4.

Lazzari, R., Renaud, G., Revenant, C., Jupille, J. & Borensztein, Y. (2009) Adhesion of growing nanoparticles at a glance: surface differential reflectivity spectroscopy and grazing incidence small angle X-ray scattering. *Physical Review B*, 79, 125428.

Lippmaa, M., Nakagawa, N., Kawasaki, M., Ohashi, S. & Koinuma, H. (2000) Growth mode mapping of $SrTiO_3$ epitaxy. *Applied Physics Letters*, 76, 2439–2441.

Lüth, H. (2001) *Solid Surfaces, Interfaces and Thin Films*. Berlin: Springer-Verlag.

Markov, I. V. (2004) *Crystal Growth for Beginners, 2nd edn*. Hackensack, NJ: World Scientific Press.

McIntyre, J. D. E. & Aspnes, D. E. (1971) Differential reflection spectroscopy of very thin surface films. *Surface Science*, 24, 417–434.

McRae, E. G. (1967) Self-consistent multiple-scattering approach to the interpretation of low-energy electron diffraction. *Surface Science*, 8, 14–34.

Mercurio, G., McNellis, E. R., Martin, I., Hagen, S., Leyssner, F., Soubatch, S., Meyer, J., Wolf, M., Tegeder, P., Tautz, F. S. & Reuter, K. (2010) Structure and energetics of azobenzene on Ag(111): benchmarking semiempirical dispersion correction approaches. *Physical Review Letters*, 104, 036102.

Meyer Zu Heringdorf, F. J., Reuter, M. C. & Tromp, R. M. (2001) Growth dynamics of pentacene thin films. *Nature*, 412, 517–520.

Michely, T. & Krug, J. (2004) *Islands, Mounds and Atoms: Patterns and Processes in Crystal Growth Far From Equilibrium*. Heidelberg: Springer.

Murty, M. V. R., Curcic, T., Judy, A., Cooper, B. H., Woll, A. R., Brock, J. D., Kycia, S. & Headrick, R. L. (1998) X-ray scattering study of the surface morphology of Au(111) during Ar+ ion irradiation. *Physical Review Letters*, 80, 4713–4716.

Oura, K., Lifshits, V. G., Saranin, A. A., Zotov, A. V. & Katayama, M. (2003) *Surface Science*. Heidelberg: Springer.

Parratt, L. G. (1954) Surface studies of solids by total reflection of x-rays. *Physical Review*, 95, 359–369.

Pimpinelli, A. & Villain, J. (1998) *Physics of Crystal Growth*. Cambridge: Cambridge University Press.

Poelsema, B. & Comsa, G. (1989) *Scattering of Thermal Energy Atoms from Disordered Surfaces* Berlin/Heidelberg: Springer.

Poelsema, B., Becker, A. F., Rosenfeld, G., Kunkel, R., Nagel, N., Verheij, L. K. & Comsa, G. (1992) On the shape of the in-phase TEAS oscillations during epitaxial growth of Pt(111). *Surface Science*, 272, 269–275.

Rabe, J. P. & Buchholz, S. (1991) Direct observation of molecular-structure and dynamics at the interface between a solid wall and an organic solution by scanning tunneling microscopy. *Physical Review Letters*, 66, 2096–2099.

Ratsch, C. & Scheffler, M. (1998) Density-functional theory calculations of hopping rates of surface diffusion. *Physical Review B*, 58, 13163.

Ritley, K. A., Krause, B., Schreiber, F. & Dosch, H. (2001) A portable ultrahigh vacuum organic molecular beam deposition system for in situ X-ray diffraction measurements. *Review of Scientific Instruments*, 72, 1453–1457.

Rost, M. J. (2007) *In situ* real-time observation of thin film deposition: roughening, zeno effect, grain boundary crossing barrier, and steering. *Physical Review Letters*, 99, 266101.

Rusponi, S., Boragno, C. & Valbusa, U. (1997) Ripple structure on Ag(110) surface induced by ion sputtering. *Physical Review Letters*, 78, 2795–2798.

Santoro, G., Franchini, A., Bortolani, V., Harten, U., Toennies, J. P. & Wöll, C. (1987) Inelastic scattering of helium atoms from the phonons of the reconstructed Au(111) surface. *Surface Science*, 183, 180–188.

Schlettwein, D., Graaf, H., Meyer, J. P., Oekermann, T. & Jaeger, N. I. (1999) Molecular interactions in thin films of hexadecafluorophthalocyaninatozinc (F16PcZn) as compared to islands of N,N′-dimethylperylene-3,4,9,10-biscarboximide (MePTCDI). *The Journal of Physical Chemistry B*, 103, 3078–3086.

Schreiber, F. (2000) Structure and growth of self-assembling monolayers. *Progress in Surface Science*, 65, 151–257.

Schreiber, F. (2004) Organic molecular beam deposition: growth studies beyond the first monolayer. *Physica Status Solidi A – Applied Research*, 201, 1037–1054.

Schreiber, F., Eberhardt, A., Leung, T. Y. B., Schwartz, P., Wetterer, S. M., Lavrich, D.

J., Berman, L., Fenter, P., Eisenberger, P. & Scoles, G. (1998) Adsorption mechanisms, structures, and growth regimes of an archetypal self-assembling system: decanethiol on Au.111. *Physical Review B*, 57, 12476.

Schwoebel, R. L. & Shipsey, E. J. (1966) Step motion on crystal surfaces. *Journal of Applied Physics*, 37, 3682–3686.

Scoles, G., Bassi, D., Buck, U. & Lainé, D. (1988) *Atomic and Molecular Beam Methods*. Oxford: Oxford University Press.

Seidel, C., Poppensieker, J. & Fuchs, H. (1998) Real-time monitoring of phase transitions of vacuum deposited organic films by molecular beam deposition LEED. *Surface Science*, 408, 223–231.

Seydel, T., Madsen, A., Sprung, M., Tolan, M., Grübel, G. & Press, W. (2003) Setup for *in situ* surface investigations of the liquid/glass transition with (coherent) x rays. *Review of Scientific Instruments*, 74, 4033–4040.

Shehu, A., Quiroga, S. D., D'Angelo, P., Albonetti, C., Borgatti, F., Murgia, M., Scorzoni, A., Stoliar, P. & Biscarini, F. (2010) Layered distribution of charge carriers in organic thin film transistors. *Physical Review Letters*, 104, 246602.

Shin, B. & Aziz, M. J. (2007) Modeling RHEED intensity oscillations in multilayer epitaxy: determination of the Ehrlich–Schwoebel barrier in Ge(001) homoepitaxy. *Physical Review B*, 76, 165408.

Stadler, C., Hansen, S., Kroger, I., Kumpf, C. & Umbach, E. (2009) Tuning intermolecular interaction in long-range-ordered submonolayer organic films. *Nature Physics*, 5, 153–158.

Stephenson, G. B., Eastman, J. A., Thompson, C., Auciello, O., Thompson, L. J., Munkholm, A., Fini, P., Denbaars, S. P. & Speck, J. S. (1999) Observation of growth modes during metal-organic chemical vapor deposition of GaN. *Applied Physics Letters*, 74, 3326–3328.

Teng, H. H., Fenter, P., Cheng, L. W. & Sturchio, N. C. (2001) Resolving orthoclase dissolution processes with atomic force microscopy and X-ray reflectivity. *Geochimica et Cosmochimica Acta*, 65, 3459–3474.

Tischler, J. Z., Eres, G., Larson, B. C., Rouleau, C. M., Zschack, P. & Lowndes, D. H. (2006) Nonequilibrium interlayer transport in pulsed laser deposition. *Physical Review Letters*, 96, 226104.

Trofimov, V. I. & Mokerov, V. G. (2002) Epitaxial growth kinetics in the presence of an Ehrlich–Schwoebel barrier: comparative analysis of different models. *Materials Science and Engineering B*, 89, 420–425.

Van Der Vegt, H. A., Alvarez, J., Torrelles, X., Ferrer, S. & Vlieg, E. (1995) Indium-induced layer-by-layer growth and suppression of twin formation in the homoepitaxial growth of Cu(111). *Physical Review B*, 52, 17443–17448.

Venables, J. (2000) *Introduction to Surface and Thin Film Processes*. Cambridge: Cambridge University Press.

Venables, J. A., Spiller, G. D. T. & Hanbücken, M. (1984) Nucleation and growth of thin films. *Reports on Progress in Physics*, 47, 399–459.

Weightman, P., Martin, D. S., Cole, R. J. & Farrell, T. (2005) Reflection anisotropy spectroscopy. *Reports on Progress in Physics*, 68, 1251.

Weschke, E., Schüßler-Langeheine, C., Meier, R., Kaindl, G., Sutter, C., Abernathy, D. & Grübel, G. (1997) *q* dependence of the growth-oscillation period of x-ray reflectivity in heteroepitaxy: Ho/W(110). *Physical Review Letters*, 79, 3954–3957.

Williamson, M. J., Tromp, R. M., Vereecken, P. M., Hull, R. & Ross, F. M. (2003) Dynamic microscopy of nanoscale cluster growth at the solid-liquid interface. *Nat Mater*, 2, 532–536.

Witte, G. & Wöll, C. (1995) External vibrations of hydrocarbons on Cu(100). *The Journal of Chemical Physics*, 103, 5860–5863.

Witte, G. & Wöll, C. (2004) Growth of aromatic molecules on solid substrates for applications in organic electronics. *Journal of Materials Research*, 19, 1889–1916.

Woll, A. R., Headrick, R. L., Kycia, S. & Brock, J. D. (1999) GaN nucleation and growth on sapphire(0001): incorporation and interlayer transport. *Physical Review Letters*, 83, 4349–4352.

Xu, H., Yang, Y.-W. & Engel, T. (1991) Quantification of defects in epitaxial metal film growth: a helium diffraction investigation of the Cu/W(110) system. *Surface Science*, 255, 73–90.

Yamane, H., Gerlach, A., Duhm, S., Tanaka, Y., Hosokai, T., Mi, Y. Y., Zegenhagen, J., Koch, N., Seki, K. & Schreiber, F. (2010) Site-specific geometric and electronic relaxations at organic-metal interfaces. *Physical Review Letters*, 105, 046103.

Yim, S. & Jones, T. S. (2006) Anomalous scaling behavior and Surface roughening in molecular thin-film deposition. *Physical Review B*, 73, 161305(R).

Zangwill, A. (1988) *Physics at Surfaces*, Cambridge: Cambridge University Press.

Zhang, X. N., Barrena, E., De Oteyza, D. G. & Dosch, H. (2007) Transition from layer-by-layer to rapid roughening in the growth of DIP on SiO_2. *Surface Science*, 601, 2420–2425.

Zhang, X., Barrena, E., Goswami, D., De Oteyza, D. G., Weis, C. & Dosch, H. (2009) Evidence for a layer-dependent Ehrlich–Schwoebel barrier in organic thin film growth. *Physical Review Letters*, 103, 136101.

Zheng, H., Smith, R. K., Jun, Y.-W., Kisielowski, C., Dahmen, U. & Alivisatos, A. P. (2009) Observation of single colloidal platinum nanocrystal growth trajectories. *Science*, 324, 1309–1312.

Part II

Techniques of thin film growth

6
Silicon nanostructured films grown on templated surfaces by oblique angle deposition

D. YE, Virginia Commonwealth University, USA and
T.-M. LU, Rensselaer Polytechnic Institute, USA

Abstract: We observed that the deposition of silicon (Si) nanostructured thin films on a patterned substrate with oblique angle incidence flux gives rise to a phenomenon referred to as 'fan-out' growth, which entails the overgrowth of the structures along the direction perpendicular to the incident deposition flux. We concluded in our Monte Carlo simulations that the side-sticking of incoming particles to the previously deposited particles is the cause of 'fan-out' growth. We have designed two methods to overcome fan-out growth: 'two-phase' rotation and 'swing' rotation. By these methods, uniform slanted Si nanorods and square nanosprings were fabricated.

Key words: oblique angle deposition, fan-out growth, ballistic sticking, Monte Carlo simulation, nanostructured thin film, substrate rotation.

6.1 Introduction

The morphological evolution and pattern formation in growing surfaces have been the subject of intense study using a framework of statistical mechanics in the past decades (Barabasi and Stanley, 1995; Meakin, 1998; Marsili *et al.*, 1996). Several theoretical growth models have been proposed to describe the deposition processes and scaling behaviors of a growing surface with the atoms depositing from an orthogonal flux on a smooth substrate (Barabasi and Stanley, 1995; Kardar *et al.*, 1989; Tong and Williams, 1994). On the other hand, there has been little effort made to understand the process of atoms striking the surface from non-normal directions (Meakin and Krug, 1992; Barabasi and Stanley, 1995). This process, referred to as oblique angle deposition, can result in a porous thin film whose porosity greatly depends on the incident angle of the atoms.

In 1995, Robbie and Brett systematically studied the oblique angle deposition with substrate rotation method, which they called 'glancing angle deposition' (GLAD), at an incident angle close to 90° with respect to the substrate normal in experiments (Robbie *et al.*, 1995a,b). The substrate rotation was driven through a stepper motor and a computer programmed to control the motion of the motor. Since then, this technique has been demonstrated as a powerful tool to create complex nanostructured thin films (Robbie *et al.*, 1996; Hawkeye and Brett, 2007). Figure 6.1(a) shows a common system used for oblique angle deposition with the freedom of substrate rotation through a stepper motor. A large distance between the substrate and source is necessary to grow high quality structured thin film in oblique angle deposition. For some systems, a collimator or a nozzle can be used to minimize the angular spread of the deposition flux. The source materials can be evaporated at sufficiently low base pressures by methods such as thermal evaporation, electron-beam evaporation, pulsed laser ablation, and sputtering. Either single elements or compounds can be deposited in oblique angle deposition with a suitable evaporation method. As the incident atoms approaching the surface at an incident angle θ, as shown in Fig. 6.1(a), they will deposit on the extruding part of the surface. The lower parts of the surface are shadowed by the neighboring structures. This is called the 'global shadowing effect', as illustrated in the inset of Fig. 6.1. After landing on the surface, the atoms may diffuse across the surface to an energetically favorable site. As discussed above, the substrates can be placed in the chamber and locked in a fixed position, or can be rotated in a designed motion pattern. For both cases, a columnar thin film is the result of the competition between shadowing effect and surface diffusion. The difference is that tilted columns will grow if the substrate is fixed in one position, while vertical columns will grow if the substrate is rotated with sufficient speed, as shown in Fig. 6.1(b).

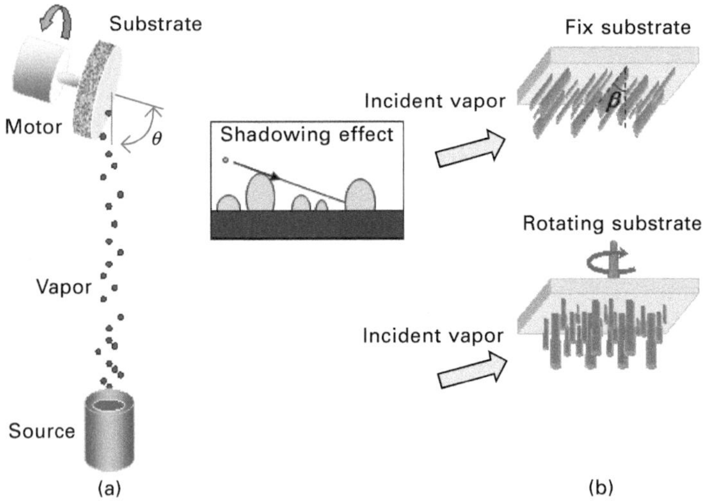

6.1 (a) Schematic of a common oblique angle deposition setup. The substrate can be rotated by a stepper motor. The shadowing effect is depicted in the inset. (b) Nanorod structures resulting from shadowing effect in oblique angle deposition. In the fixed substrate case, tilted nanorods will grow; in the rotating substrate case, vertical nanorods can be deposited.

In the case of no substrate rotation in oblique angle deposition, the columns tend to incline toward the incident flux to a tilt angle β, which is measured from the substrate normal. There is an empirical rule known as the 'tangent rule' describing the relationship of β and θ as (Nieuwenheuzen and Haanstra, 1966; Leamy *et al.*, 1980):

$$\tan\beta \cong \frac{1}{2}\tan\theta \quad [6.1]$$

The origin of the tangent rule result is generally believed to be the limited surface diffusion and the shadowing of atoms on the surface caused by the finite size of the atoms (Ablemann and Lodder, 1997). This tangent rule is not universal, however, because the tilt angles also depend on the deposition conditions and the material properties (Nakhodkin and Shaldervan, 1972; Hashimoto *et al.*, 1982; Geszti *et al.*, 1986). Even for the same material, the tangent rule will not hold for all incident angles. For larger incident angles (generally $\theta > 60°$), another empirical rule called the 'cosine rule' may give better fitting of the experimental data (Trait *et al.*, 1993; Tang *et al.*, 2003). The cosine rule can be written as:

$$\beta = \theta - \arcsin\left(\frac{1 - \cos\theta}{2}\right) \quad [6.2]$$

Experimentally, the tilt angle of cobalt columns we obtained follows the trend of the cosine rule, but has smaller values than the calculated ones using the cosine rule (Tang *et al.*, 2003).

The columnar structures can be generated in computer simulations based on the ballistic deposition model. In a (2+1)-dimensional (3D) ballistic deposition model, the simulated columnar structures have two correlation lengths ξ_x and ξ_y where the x lies in the plane of substrate normal and the incident direction of the atoms and y is perpendicular to it (Meakin and Krug, 1990, 1992). The growth of the correlation lengths as a function of time s follows a power law. The power law for ξ_x and ξ_y can be written as

$$\xi_x \sim s^{\sigma_x}, \quad \xi_y \sim s^{\sigma_y} \quad [6.3]$$

where $\sigma_x = 1/3$ and $\sigma_y = 2/3$, which are obtained from simulations by Meakin and Krug (1992). The anisotropy of the exponents in the x and y directions is reasonable in that the shadowing effect in the x direction is more severe than that in the y direction.

Unlike the case just discussed above, there has been limited research done to date that studies the growth dynamics in oblique angle deposition with substrate rotation. Smy *et al.* (2000) developed a 3D Monte Carlo simulation to study the growth mechanisms including shadowing effect, surface diffusion, and substrate rotation in oblique angle deposition with substrate rotation.

This 3D Monte Carlo simulation is based on a grid of size $N \times N \times L$ ($N = L = 500$), and a periodic boundary condition is employed (Smy *et al.*, 2000). The cubic particle approaches the system at a given oblique angle until it is deposited on the surface. After the deposition, this same particle is chosen to diffuse to a length of several lattice units (Dew *et al.*, 1992). After the sharp edges of the films are smoothed out, a variety of nanostructures including springs, vertical rods, and zigzags can be obtained and depicted very well by the simulations. Suzuki and Taga (2001) used a similar model to study the oblique angle deposition with substrate rotation. This model is also based on a simple cubic lattice and allows ballistic sticking. Ballistic sticking means a moving particle can stick to any other particles deposited on the surface if the distance between the two particles is close enough. The diffusion of the newly deposited particle is similar to a random walk in the Suzuki and Taga model.

For some applications, e.g., in the preparation of photonic crystals (Joannopoulos *et al.*, 1995), the arrangement and the position of the nanostructures must be controlled. Other important applications include micro-electro-mechanical systems, nano-magnet arrays, and high frequency resonators. Pre-patterned seeds on a substrate are necessary in order to fabricate periodic, uniform, and well-separated 3D nanostructures in oblique angle deposition. During deposition, the incoming particles only deposit on the top of the seeds exposed to the flux. One would therefore expect that well-aligned 3D nanostructures could be created that maintain the period of the initial nanoseed arrays. However, we have discovered that if the substrate is fixed in one position, a 'fan-out' structure will quickly form (Ye *et al.*, 2004; Lu *et al.*, 2005). The fan-like structure will continue to grow laterally in the direction perpendicular to the incident flux and eventually contact neighboring nanostructures. This is not desirable if one wants to create an array of equally separated nanostructures. In this chapter, the main issues to be addressed are as follows: (1) the origin of the 'fan-out' growth when periodic seed arrays are used in oblique angle deposition; (2) a method that can be used to eliminate the 'fan-out' growth. We will study the 'fan-out' growth by on-lattice Monte Carlo simulations including the mechanisms of ballistic aggregation, surface diffusion, shadowing effect, steering effect, and angular spreading of deposition flux. Experimentally, we will introduce three methods of substrate rotation that can be used to control the 'fan-out' growth and generate uniform nanostructures on the periodic seed arrays.

6.2 Preparation of templated surface for oblique angle deposition

There are many viable techniques for the preparation of templated surfaces consisting of nano-sized seeds, such as photolithography and electron beam direct writing, or e-beam lithography. In photolithography, a layer of photoresist is deposited by spin-coating. When exposed to light with a patterned mask, the exposed photoresist changes its chemical structures. After development, the exposed areas are either dissolved away or stay on the substrate depending on the property of the photoresist. The former case uses what is known as positive photoresist and requires an inverting step to have seeds protrude out of the substrate. A more advanced technique used to fabricate seeds is electron beam direct write, or e-beam lithography. In e-beam lithography, the e-beam resist is exposed to an extensive electron flux that changes its chemical structures. The advantages of e-beam lithography are: (1) the feature size of individual seeds can be < 100 nm in diameter; and (2) it is a mask-less process. The disadvantage of this technique is the time it takes to write the pattern, making large-scale production almost impossible. Typically, the whole process takes several hours in e-beam lithography. Most recently, e-beam lithography has been used to provide a mold for nanoimprint lithography. Similarly, patterned seeds with nanometer feature size are available using deep UV photolithography, but at a much added cost.

In Fig. 6.2, the templated surfaces were patterned using deep UV photolithography and e-beam lithography, respectively. The processing steps to produce this kind of seed template by deep UV photolithography are as follows. First, a layer of 1000 nm thick SiO_2 is grown on a flat Si (100) substrate. Second, arrays of holes about 120 nm in diameter are patterned in the SiO_2 layer by deep UV photolithography. The holes were arranged either in a square lattice with a ~1000 nm lattice constant or in a triangular lattice with a ~750 nm lattice constant. Third, tungsten (W) is used to fill in the holes by means of chemical vapor deposition (CVD). The thickness of the W film is about 360 nm. Finally, the surface with excess W is then removed by chemical mechanical polishing until the W in the holes is exposed. At this point, the cylindrical W seeds, also called W-pillars, are still buried inside the SiO_2 matrix. Before oblique angle deposition, the SiO_2 matrix is removed by reactive ion-etching, exposing the W seeds, as shown in Fig. 6.2(a) and (b). The height of the W seeds was measured to be 360 nm from SEM cross-sectional views. The oxide layer can also be removed using commercially available buffered oxide etch, which is a wet-chemical process. In e-beam lithography, a layer of 270 nm thick hydrogensilsesquioxane (HSQ) is used as the resist for e-beam direct writing inside a Zeiss Supra-55 scanning electron microscopy (SEM) with a Nabity pattern generator. The electron beam was accelerated by a 30 kV accelerating voltage and focused by a 7.5 μm aperture. This electron beam was used to expose a square lattice of 100 nm pillars or 50 nm pillars with a 750 nm lattice constant in the HSQ resist. After the exposure, the HSQ resist was baked at 160°C on a hot plate for 60 seconds and developed in a 25 wt% tetramethyl-ammonium-hydroxide

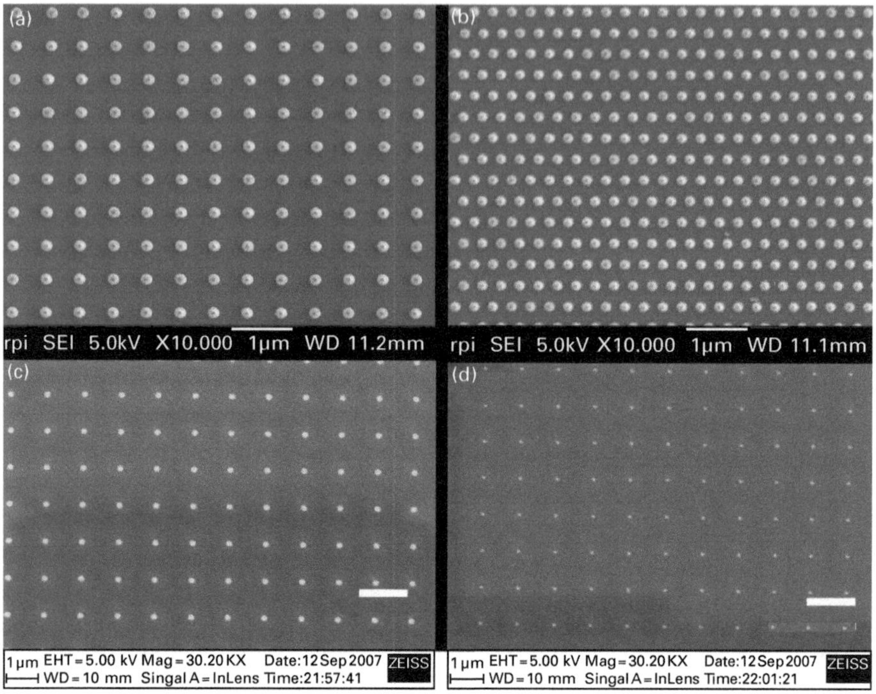

6.2 SEM top-view images of templated surfaces prepared for oblique angle deposition. (a) W pillars after removing the SiO_2 matrix. The seeds were arranged in a square lattice with a 1000 nm lattice constant. (b) W pillars were arranged in a triangular lattice with a 750 nm lattice constant. (c) Arrays of 100 nm hydrogensilsesquioxane (HSQ) pillars in a square lattice with a 750 nm lattice constant fabricated by e-beam lithography. (d) Arrays of 50 nm HSQ pillars in a square lattice by e-beam lithography.

solution for 240 seconds at room temperature. The sample was then rinsed and dried. The top-view SEM images of the patterned area with 100 nm pillars and 50 nm pillars are shown in Fig. 6.2(c) and (d), respectively. The patterned surfaces fabricated by lithography can be used as a 'hard mold' for nanoimprint lithography for massive production of cost-efficient seeds.

Malac *et al.* (1997) used e-beam lithography to prepare the patterned surfaces for oblique angle deposition. They chose poly-(methylmethacrylate) (PMMA) as the resist. After the resist is patterned, a lift-off step is followed to form pillars as the PMMA is a positive tone e-beam resist. Horn *et al.* (2004a) used a negative resist in e-beam lithography, therefore no lift-off process is involved. In addition to the seeds mentioned above, self-assembled seeds and patterned surfaces with lines or holes were used to grow nanostructures by oblique angle deposition (Zhao *et al.*, 2002a; Horn *et al.*, 2004b).

The geometry of the seeds has a significant effect on the morphology of the nanostructures grown on them in oblique angle deposition, particularly the size of individual seeds. Obviously, the size of the seeds should be sufficiently small in order to deposit a single nanostructure on each seed. We deposited Si on W seeds with a diameter of 360 nm and observed multiple nanorods grown on top of the seeds, while single nanorods can be grown on smaller W seeds with a diameter of 150 nm, all under the oblique angle deposition conditions with an 85° incident angle and uniform substrate rotation (Ye *et al.*, 2008). In oblique angle deposition on templated surfaces, the shadowing effect from the nearest seed is dominant, no matter whether or not the substrate is rotated (Ye *et al.*, 2008). We call the shadowing effect from the nearby features on the surface as 'global shadowing effect'. We can determine a cut-off angle of the incident flux θ_c in terms of the seed geometry, namely, the size of the seeds D, the height of the seeds H, and the center to center separation of the nearest seeds L, as depicted in Fig. 6.3 (a). The equation can be written as

$$\theta_c = \tan^{-1}\left(\frac{L-D}{H}\right) \qquad [6.4]$$

which is similar to the arguments by Hawkeye and Brett (2007), and Horn *et al.* (2004a,b). If the incident angle $\theta > \theta_c$, the global shadowing can affect the growth of nanostructures on each seed at the very beginning of the oblique angle deposition. In practice, the incident angle θ is much larger than θ_c. Therefore, we define another important geometric factor, 'exposure height' a, in oblique angle deposition on seeds. The exposure height determines the uniformity and morphology of the nanostructures (Ye *et al.*, 2008). It can be defined as the portion of an individual seed or nanostructure exposed to the deposition flux incorporating the shadowing from the nearest neighbors in the plane of deposition flux and the axis of seeds, as sketched in Fig. 6.3(b). Hence, the exposure height a can be represented as

$$\alpha = \frac{L - \frac{D}{2}}{\tan\theta} \qquad [6.5]$$

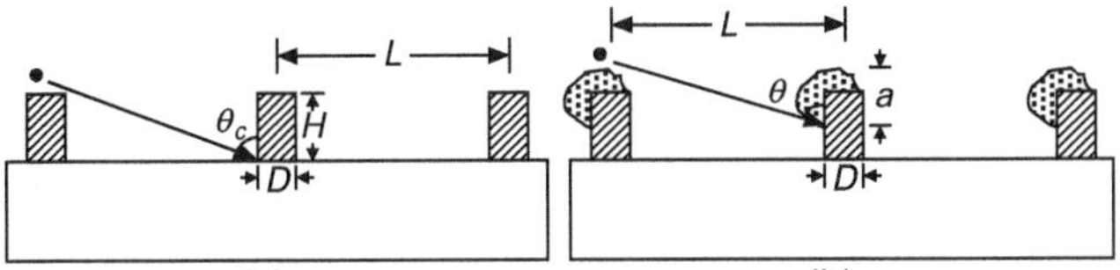

(a) (b)

6.3 Schematic of the geometry of the seeds. (a) The cut-off angle θ_c is the minimum incident angle of the flux that allows global shadowing effect. (b) The exposure height a of the growing nanostructures at a fixed incident angle θ.

From the definition, the growing portion of the nanostructure within *a* should not be affected by the global shadowing from its neighbors. We suspected that the growth of the surface in this part may be controlled by the same mechanisms as that of surface growth on a seed with normal incident flux, i.e. by surface diffusion, ballistic sticking, self-shadowing, etc.

6.3 Fan-out on templated surface with normal incident flux

6.3.1 Experimental observation of fan-out growth on templated surface

We discussed the shadowing effect and developed a geometrical model to describe the shadowing effect in oblique angle deposition in the previous section in this chapter. Now we move on to the discussion of growth mechanisms with normal incident flux in experiments and Monte Carlo simulations. In this section, we reported the creation of a fan-like structure by a simple line-of-sight evaporation technique where atoms vertically drop down onto an array of nano-sized cylindrical W seeds. Very well defined fanlike clusters are formed on these seeds. The morphology of fan-like structures will be shown and analyzed from the SEM images. A ballistic deposition model will be developed to interpret the growth mechanisms of the structures.

The templated surfaces containing W pillars in a square lattice were used for the silicon (Si) depositions with normal incident flux in our experiments. The substrates were placed at a position such that their normal vectors pointed to the source. Thus, the incident angle of the flux θ is equal to 0° measured from this surface normal vector. The angular dispersion of the flux is less than 0° due to the geometry of our deposition system, i.e. the distance between the source and the substrate is about 35 cm and the diameter of the mouth of the crucible is 1 cm. Therefore, the flux can be assumed as a parallel flux in our experiment. Before the deposition, the area containing W pillars on the substrate was centered to the incident flux using a laser beam that simulates the incoming flux.

Si (99.9999%, Alfa Aesar) was evaporated from a graphite crucible by electron-beam bombardment as described elsewhere (Wong *et al.*, 1987; Mei and Lu, 1988). The evaporation rate was measured by a quatz-crystal microbalance (QCM, Maxtek TM-350/400) and maintained at 5.0 ± 0.3 Å/s during the deposition by adjusting the power levels supplied to the crucible. Two sets of samples were prepared with normal thickness of 700 nm and 800 nm as measured by the QCM. To study the morphology of the samples, SEM images were taken using a field-emission SEM. The samples were coated by a ~10 nm thick Pt film to reduce the charging effect before loading into the SEM system. A very sharp tungsten tip was heated to release electrons in the SEM. The emission current ranges from 6.5 to 12.5 μA during operation. For the SEM cross-sectional views, the samples were cleaved in a direction along the axis of the square lattice. SEM top view images were also taken. The SEM images were used for further analysis using Photoshop (Adobe, Version 6.0), an image processing software.

We observed experimentally that the growth of Si on a templated substrate with a regular array of W pillars would give rise to a phenomenon referred to as 'fan-out' growth in normal deposition without substrate rotation (Ye and Lu, 2007a). Figures 6.4 and 6.5 demonstrate this phenomenon by the SEM top-view images and cross-sectional images of 700 nm and 800 nm thick Si films deposited on the W pillars arranged in a square lattice. Well-defined 3D fan-like structures were observed on top of each seed. If the deposition time is long enough, the nanostructures grown on the seeds will touch their neighbors, as we can see from the top-view image of those structures on the large seeds in Fig. 6.5(a). A structure with a pyramidal shape was also

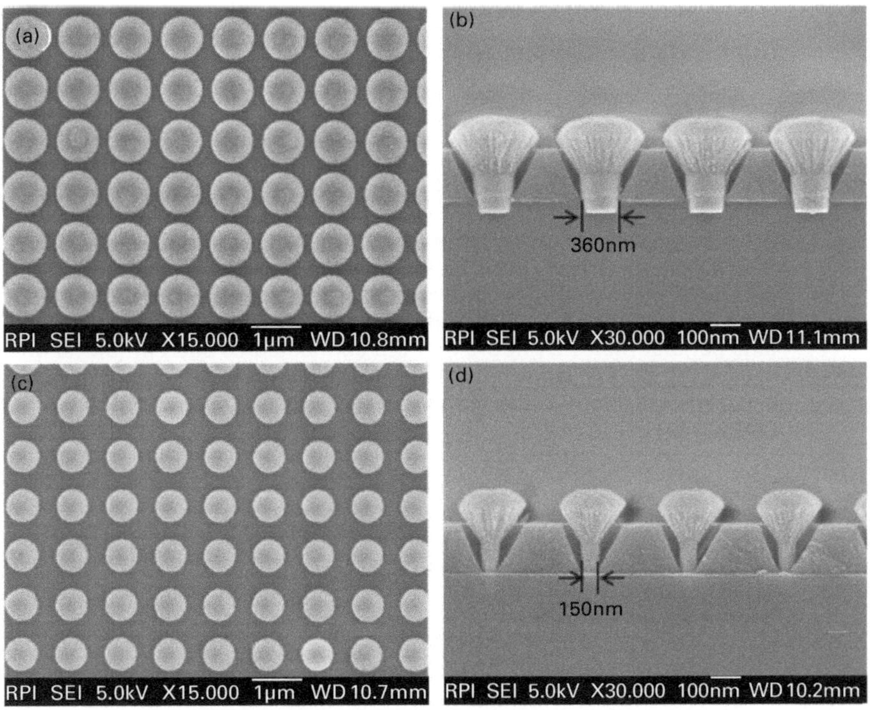

6.4 SEM images of ballistic fan-like Si structures grown on a square lattice. The thickness of the Si film is 700 nm. (a) and (b) are the top-view image and the cross-sectional image of the structures on large W pillars with a 360 nm diameter, respectively. (c) and (d) are the SEM images of similar structures on small W pillars with a 150 nm diameter.

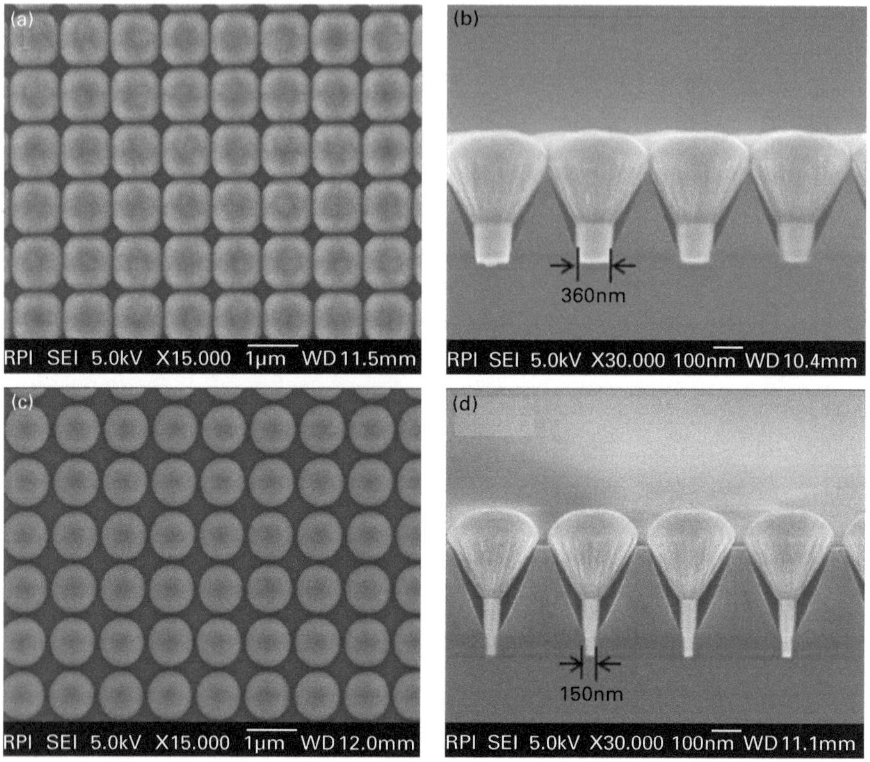

6.5 SEM images of ballistic fan-like Si structures grown on a square lattice. The thickness of the Si film is 800 nm. (a) and (b) are the top-view image and the cross-sectional image of the structures on large W pillars with a 360 nm diameter, respectively. The ballistic fans contact their neighbors and are connected as seen in (a). (c) and (d) are the SEM images of similar structures on small W pillars with a 150 nm diameter. (Adapted from Ye and Lu, 2007a, reprinted with permission.)

deposited inside the open area on the substrate between the seeds. One can see that the outlines on the side of the fans are straight and can be described geometrically by three-dimensional solid cones. Therefore, the width of the fan-like cones grew linearly with time.

6.3.2 Monte Carlo simulation of ballistic fans

There has been much study of the theory on the formation (Ramanlal and Sander, 1985; Limaye and Amritkar, 1986) of this fan-like structure as well as the scaling properties (Liang and Kadanoff, 1985; Porcu and Prodi, 1991; Krug and Meakin, 1991) of the fan based on the ballistic aggregation mechanism. Although most of the studies are in two dimensions, an extension to three-dimension simulation has also been reported (Kardar *et al.*, 1989). However, thus far very scarce experimental demonstration has been reported on the formation of this fan-like structure. In our study, a Monte Carlo (MC) simulation with nearest-neighbor ballistic sticking followed by a diffusion mechanism has been proposed to describe the fan-like structures.

Very few theoretical studies of the surface evolution concerning patterned surfaces have been carried out, mainly due to the crossover of the scales in this system. Typically, the size of individual patterns varies from tens of nanometers to several hundreds of nanometers. Each of these patterns is a result of the deposition where the atomic level surface processes and the mesoscopic processes such as mass transport cooperate to determine the growth of nanostructures on top of it. Based on this fact, the model we choose to simulate the surface growth at this scale should be able to handle the following processes (Imry, 2002):

- generating non-smooth patterned surfaces;
- multi-valued surface profiles;
- incorporation of atomic level surface processes, such as diffusion;
- complex stochastic processes in mesoscopic scales.

Ideally, multi-scale simulation techniques can provide the best solution in this situation (Castez and Salvarezza, 2006). However, in the context of limited resources, MC simulation can also address some of the issues listed above. In general, physical events including deposition of particles at random positions, surface diffusion, surface hopping, and evaporation of surface particles, can be embedded into the MC simulation.

Most recently, the importance of mass transport has been evaluated in surface growth models (Suzuki and Taga, 2001; Raible *et al.*, 2002; van Dijken *et al.*, 1999; Luedtke and Landman, 1989). This mass transport can be due to the inter-atomic interaction of the incoming particles and the surface particles (referred to as the 'steering effect'), and geometric effects such as the shadowing effect. The steering effect and other growth mechanisms will be studied by 3D MC simulations. In our MC model, a 3D lattice with dimensions $1024 \times 1024 \times 1024$ is formed by cubic lattice points. One cube represents an incident particle so that it has the dimension of a cubic lattice point. Figure 6.6 shows a schematic of the simple cubic system in this MC simulation with ballistic sticking and surface diffusion processes allowed. In the ballistic sticking model, an incident atom moving into an empty nearest-neighbor lattice point will occupy that empty position immediately. In contrast, in a solid-on-solid model, that atom will pass that empty spot unless it is the lowest position the atom can go in the system (Amar and Family, 1990). We believe that the ballistic sticking model of incident atoms comes from the inter-atomic interactions.

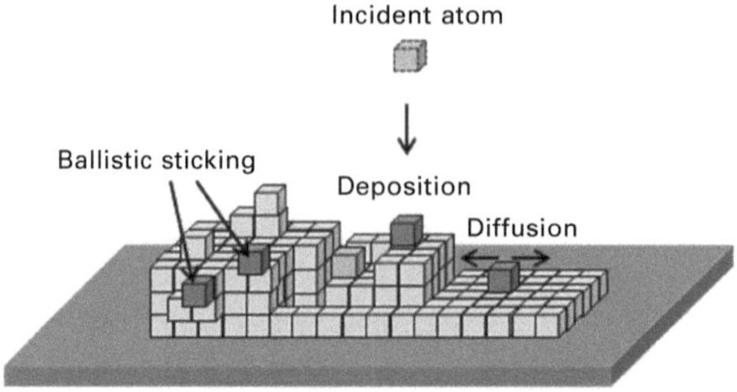

6.6 Schematic of 3D Monte Carlo simulation with ballistic sticking and surface diffusion processes on a smooth substrate. The flux of atoms approaches the surface from the top along the surface normal.

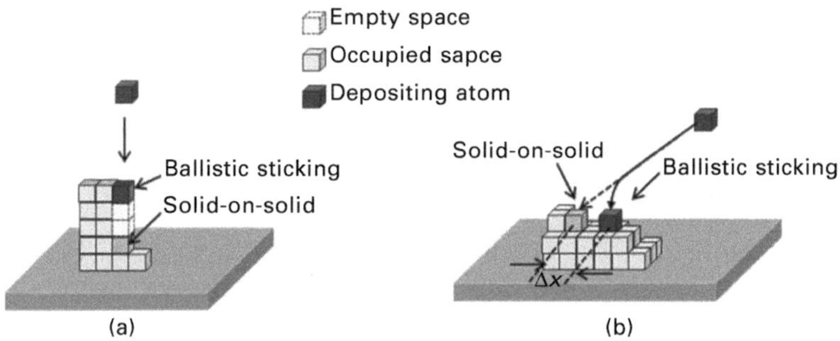

6.7 Ballistic sticking model in the MC simulations with (a) top-down and (b) oblique angle incident flux. The possible deposit sites of the incident atoms due to the ballistic sticking model and the solid-on-solid model are labeled for comparison.

The interaction of atoms is due to the Lennard–Jones (L-J) potential, which is widely used, in the form of

$$U(r) = 4\varepsilon[(\sigma/r)^{12} - (\sigma/r)^{6}], \quad [6.6]$$

where the parameters ε and σ depend on the properties of the atom. Using the equation developed by Raible *et al.* (2002), we can estimate the variation of the distance Δx from the impact point of a linear trajectory as in Fig. 6.7(b). The distance can be calculated from Raible *et al.* (2002) using the L-J potential for Si,

$$\Delta x = \int_0^\infty \mathrm{d}z\left(\tan\theta - \frac{\sin\theta}{\sqrt{\cos^2\theta - v(z)/E_0}}\right) \quad [6.7]$$

For our oblique angle deposition setup, the incident angle of the particles is close to 90°, thus we can ignore the cosine part inside the square root in

Eq. 6.7. If the kinetic energy E_0 of the incident particle is about 0.3 eV, the integral of this equation from 0.5 to 0.55 nm yields $\Delta x \approx 0.455$ nm. That means the particle was trapped by the surface already by the L-J potential in the form of Eq. 6.6. The approximated calculation shows that the particle will be pulled down from its trajectory when passing by an atom. This means that ballistic sticking is a good approximation in Si with very low kinetic energy.

3D MC simulations with and without ballistic sticking were performed in a cubic box with dimension $N \times N \times N$ ($N = 1024$). The substrate containing seeds in a square lattice is fixed. The particles were injected into the system from the top and moved downward. When the particle passes by any deposited particle in its nearest-neighbor vicinity, it will stick to that particle and become its nearest neighbor (ballistic sticking model). In order to study the mechanisms of fan-out growth, we will compare the simulation results with and without ballistic sticking mechanisms. In our deposition system, the flux is very uniform due to the large distance between the opening of the crucible and the substrate. The opening of the crucible is 1 cm, and the distance is about 35 cm. So, the angle spanned by this geometry is determined to be less than 2°. In our simulations, this angular spreading of flux is taken into account as well.

In the 3D MC simulations, arrays of lattice points are selected and occupied by particles to represent the W pillars on the templated surfaces. The cylindrical seeds are placed on a smooth surface on the bottom of the simulation box with a dimension of 1024 × 1024 units and a thickness of 8 units. The seeds have a diameter of 24 units or 36 units for the small size seeds or the large size seeds, respectively. The seeds are arranged in a square lattice with a lattice constant of 256 units. The particular reason for this arrangement is to avoid overlapping on the boundary by enforcing periodic boundary conditions. This means that if a particle moves outside of the cubic system, it will re-enter from the opposite side. The height of the seeds is 80 units. Figures 6.8 (a) and (b) show the top-view and cross-sectional images of the smaller seeds, respectively. Similar images of the large size seeds are shown in Figs 6.8(c) and (d).

First, we investigated the deposition of particles without ballistic sticking, namely, the particles approaching the surface until they are stopped by another particle deposited on the surface. When a particle is landed on the surface, a certain number K of particles, referred to as 'diffusers', within a cubic box with a side of $2a + 1$ ($a = 5$) around the deposited particle are randomly selected to diffuse on the surface (Kessler *et al.*, 1992). The number K is related to the diffusivity of material. We used the coordination numbers as the criteria of the successful diffusion. That means the diffuser will move to the new lattice point if that move increases the coordination numbers. Otherwise, it will remain at the same lattice point. The coordination

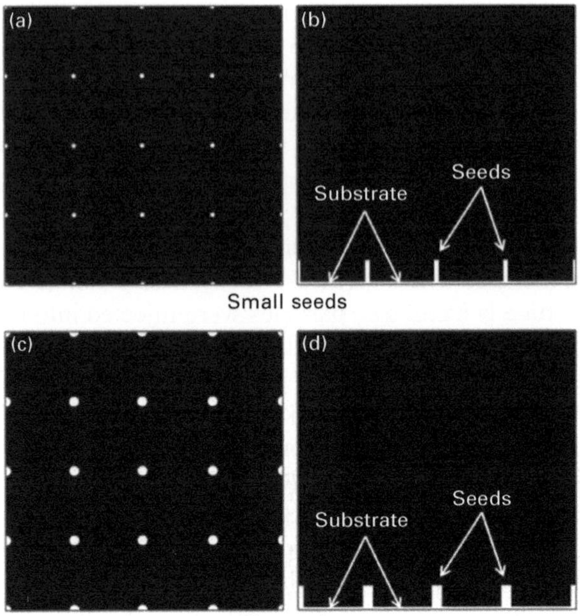

6.8 Templated surfaces containing seeds in a square lattice in the MC simulations: (a) top-view and (b) cross-sectional images of the small seeds 8 units in diameter. (c) and (d) are the top-view and the cross-sectional images of the big seeds with a diameter of 18 units. The height of the seeds is 80 units and the lattice constant of the square lattice is 256 units. The sizes of the images are 1024×1024 units.

numbers include the number of the nearest neighbors (NN) and the number of next-nearest neighbors (NNN) of the diffuser. Diffusion models that use NN and NNN numbers have also been adapted by other researchers (Suzuki and Taga, 2001; Stasevich *et al.*, 2004; Johnson *et al.*, 1994; Smilauer *et al.*, 1993). However, our diffusion model does not include the kinetic mechanism associated with the bonding energy, which is different from the models used in molecular beam epitaxial growth (Stasevich *et al.*, 2004; Johnson *et al.*, 1994; Smilauer *et al.*, 1993). The model used by Suzuki and Taga is the closest one to our model, which also deals with oblique angle deposition (Suzuki and Taga, 2001). In their approach, the site is selected from a cubic box with 3 units centered at the diffuser. The site with more coordination numbers will have a greater chance of the diffuser jumping into it according to the probability of diffusing from site A to site B:

$$P_{A\to B} = \frac{\exp(\lambda N_B)}{\sum_i \exp(\lambda N_i)} \qquad [6.8]$$

where λ is a constant and the summation is taken over all of the allowed sites

(Suzuki and Taga, 2001). After the diffuser makes a jump, it will continue the diffusion process to about 10 steps of jumping. It is not clear how the coordination numbers have been counted in the simulation model by Smy *et al.* (2000) for oblique angle deposition. From the precursor of their model, we know the diffusion model not only counts the coordination numbers but also the distance between the two sites. And the diffuser can jump over a long distance without staying on the surface (Dew *et al.*, 1992). Thus, the diffuser is more likely to jump into a saddle point on the surface in the model by Smy *et al.* (Smy *et al.*, 2000). Figure 6.9 shows the top-view and cross-sectional images of the structured films generated by the simulations on two types of seeds without ballistic sticking. The particles were deposited on the top area of individual seeds, as well as the substrate area between the seeds. A vertical rod-like structure is formed with uniform size. A continuous film was deposited in the gap between seeds as well. The number K in this simulation was 100, representing the diffusivity of the material. We tried other K numbers in simulations but we found that the results have no significant difference from those shown in Fig. 6.9. However, the simulated results as shown in Fig. 6.9 are different from the experimental results depicted in Fig. 6.4. That means the deposition model without ballistic sticking cannot be used to explain the fan-out growth phenomenon.

6.3.3 Growth exponent of ballistic fans

The 3D MC simulations with the ballistic sticking model have been carried out on the same pre-occupied seed arrays. The particles drop down randomly and

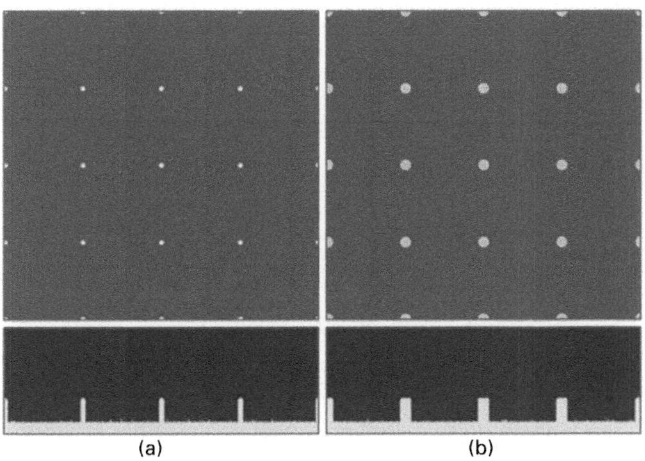

6.9 Monte Carlo simulations of particles deposited on (a) small seeds and (b) big seeds in the solid-on-solid model with diffusion. The diffusion number K is 100.

land on the surface or stick to any particle on the surface if the two particles are within the nearest-neighbor distance. It is easy to see that after deposition the differences between the local surface heights around the newly added particle should be zero or unity in magnitude. This characteristic feature is dramatically different from the non-ballistic sticking model just discussed above and is similar to the so-called 'restricted solid-on-solid growth' model (Kim and Kosterlitz, 1989). With this non-trivial model, the aggregations of particles on seed arrays form ballistic fan structures as shown in Fig. 6.10. We denoted the simulation time F as the number of total particles sent into the simulation system. The snapshots of the structures were taken at a constant

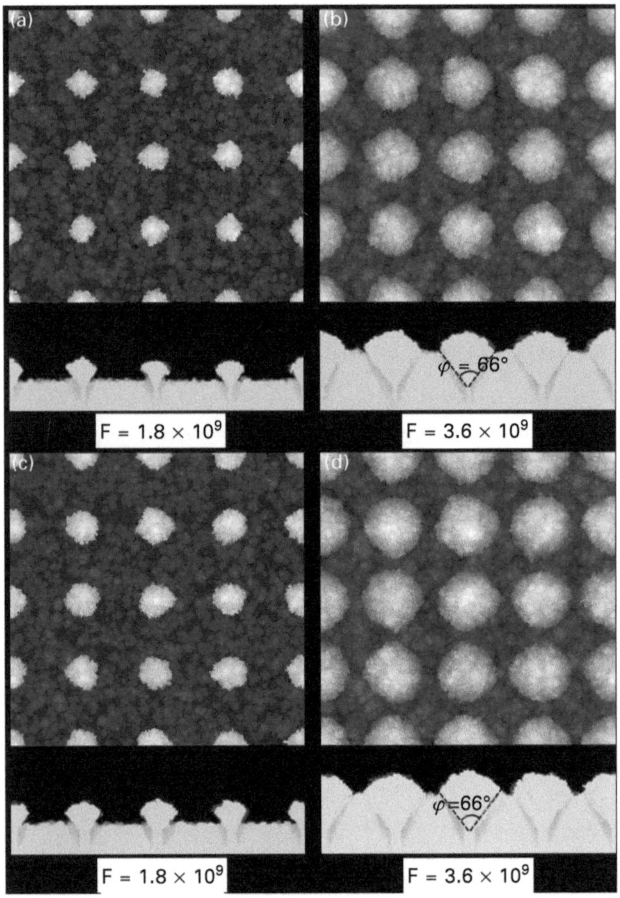

6.10 Top-view and cross-sectional snapshots of fan structures in the ballistic sticking model with normal incident flux. (a) and (b) are images of the fans on the small seeds taken at two simulation times. (c) and (d) are images of the fans on the big seeds. The diffusion parameter is set to $K = 100$.

frequency with simulation time $F = 9 \times 10^7$ particles. The simulated fan structures resemble the experimental results in that the fan-out angle φ is measured to be 66° in the simulations and the experiments. They all have a dome-shaped top with rather straight outlines on the sides.

We studied the dynamic growth of the width R of the simulated ballistic fans in terms of the power law behavior, $R = kt^p$, of the expansion of the fans in the lateral direction. The average width R of the fans was measured from the top-view images as shown in Fig. 6.10(a) and (c) plotted in Fig. 6.11. The exponent p in the power law format defines the geometrical shape of the ballistic fans. Specifically, a cylindrical shape has an exponent $p = 0$; a hyperbolic shape has an exponent $p = ½$; and a cone shape would have $p = 1$. In obtaining the p value more quantitatively from the simulations, we believe that the initial size of the cylindrical seeds has a non-trivial effect. We suppose that the size of structure is identical to the size of the seeds at the beginning of the growth. Therefore, the width R is a function of 'adjusted time' t instead of the real time t', as $R = kt^p$ and $t = t' + t_0$. Clearly, t_0 is an imaginary time that satisfies the boundary condition: $R_0 = kt_0^p$, where R_0 is the initial size of the seeds. After the adjustment of the time, it provides the possibility of comparing the simulated results with our experimental observations and with previous analytical and simulation results obtained using ballistic aggregation with a point seed. One can see from the plots in Fig. 6.11 that the growth of the ballistic fans has a power law behavior with the exponent $p \approx 1.0$. The cone-shape geometry of the fans observed

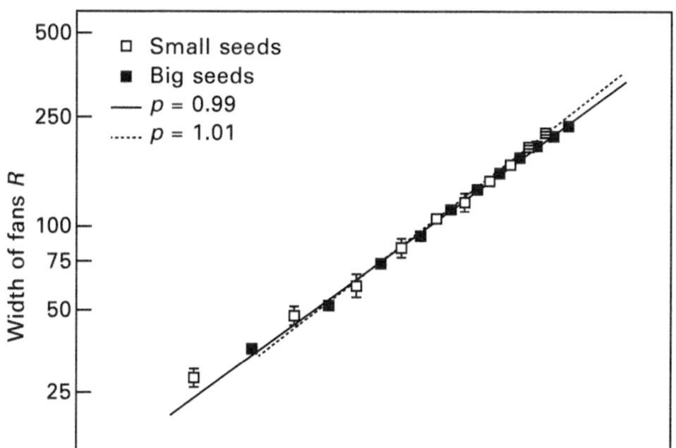

6.11 Growth of the ballistic fans on small and big seeds in Monte Carlo simulations with normal incident flux. The fitting of the curves demonstrates that the growth is linear with the exponent $p \approx 1$.

in our experiment, as shown in Fig. 6.4, is consistent with this prediction in simulations. This linear behavior is, within error, independent of the strength of diffusion, and the size of seeds. Thus, we conclude that the linear growth ($p \approx 1$) of the width of the ballistic fans may be a universal characteristic of ballistic aggregation, even when diffusion is allowed.

6.4 Fan-out growth on templated surfaces with oblique angle incident flux

We discussed the ballistic sticking model to describe the fan-out growth on templated surface with normal incident flux in the previous section. Now we move on to the oblique incident case in this section. Intuitively, the morphology of structures deposited on templated surfaces should be different for the latter case due to the global shadowing effect. Nonetheless, the ballistic sticking and surface diffusion still remains as the main mechanisms that control the growth dynamics of the nanostructured films. The question we may ask is to what extent global shadowing can affect the fan-out growth on the non-flat surface. Here, we will present our answer to this important question from the experimental and modeling results.

6.4.1 Experimental demonstration of fan-out growth with oblique angle incident flux

In oblique angle deposition of Si films, the substrates with W pillars were placed at a position such that their normal vectors were turned 5° from the horizontal plane. Thus, the incident angle of the flux θ is equal to 85° with respect to this surface normal vector. The incident flux was aligned to the axis of the square lattice and the shadowing effect only comes from the nearest neighbors. The direction of the incident flux on the plane of the substrate is assigned to be the x-axis and the perpendicular direction is assigned to be the y-axis, as shown in Fig. 6.12. We observed experimentally

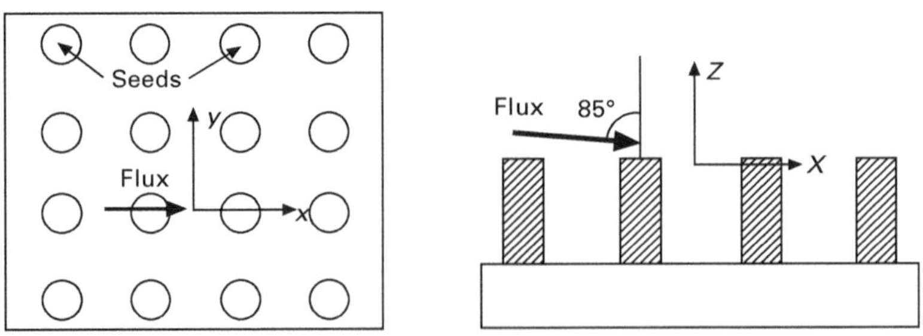

6.12 The alignment of the flux with respect to the geometry of the seed's lattice.

that the fan-out growth occurs on seeds in oblique angle deposition without substrate rotation (Ye *et al.*, 2004; Lu *et al.*, 2005). In the 'fan-out' growth, the size of the nanostructures overgrows along the direction perpendicular to the incident deposition flux which is along the y-axis (or y-direction) as labeled in Fig. 6.12. The net result is that the size of the nanostructured films cannot be controlled as they grow. If the deposition time is long enough, the nanostructured films grown on the seeds will touch and merge with their neighbors from the side along the direction of the y-axis. Thus individual nanostructures cannot be distinguished along this side. Due to the shadowing effect, the gap between the seeds along the x-direction is still maintained even with the merging occurring in the y-direction.

Figure 6.13 elucidates this phenomenon by the top-view and the cross-sectional SEM images of the Si deposition on the W-pillar seeds arranged in a square lattice. Si was deposited on the templated substrate with an 85° incidence angle. The normal thickness of the growth was 2000 nm measured by the QCM. The samples were tilted to 15° when taking the cross-sectional images. The morphology of the films grown on the small size and the large size W pillars is similar to each other. The top surface of the structure is rough and has a rectangular shape with irregular sides. The structures on

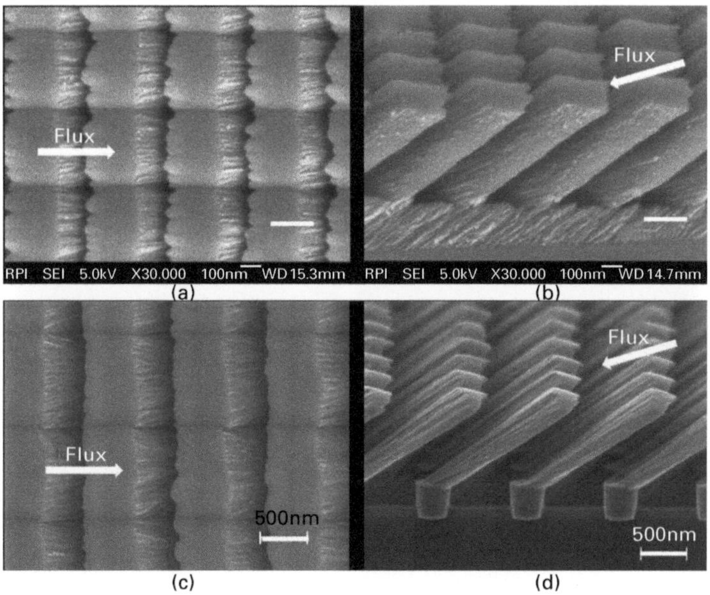

6.13 Fan-out growth of Si on W pillars with 85° incident flux. (a) Top-view and (b) tilted cross-sectional SEM images of Si film grown on small W pillars; film grown on large-sized W pillars are shown in (c) and (d) for the top-view and tilted cross-sectional images, respectively.

both of the lattices were inclined toward the source at an angle β of 53 ± 1° measured from the normal of the substrate. The size of the structures along the x-direction remained roughly constant at 340.3 ± 6.0 nm for the films grown on the square lattice seeds. From the images shown in Fig. 6.13, it is clear that the top-end surface of the structures is smoother than the surface on the sides. The side surfaces in Fig. 6.13(a) and (c) contain small fibers growing toward the source and bumpy top surfaces can be seen in Fig. 6.13(b) and (d). This suggests that the structure may be constructed by the bundling of those small Si fibers. The fan-out growth of Si films before merging is shown in Fig. 6.14(a) and (b) for the deposition on small size seeds and large size seeds, respectively. The normal thickness of the films was 800 nm as measured by the QCM. The fan-out angles φ on the seeds are measured to be 63.6 ± 0.6° and are the same for the fans on both the small and the big seeds.

6.4.2 MC simulations of fan-out growth with oblique angle incident flux

As we discussed above, a particle will likely be trapped by the already deposited particles through inter-atomic interactions or direct bonding. The particle will become part of the surface at first contact with the surface. In the oblique angle deposition with ballistic sticking mechanism, the global shadowing effect exists and comes into effect at every deposition event. The strength of diffusion is controlled by the pre-set number K. Figure 6.15 shows the top-view and the cross-sectional images of the structures deposited on the same templates as depicted in Fig. 6.8. The large square pictures are the top-view images while the small rectangular pictures are the cross-sectional images. In the top-view images, we can see the fan structures start

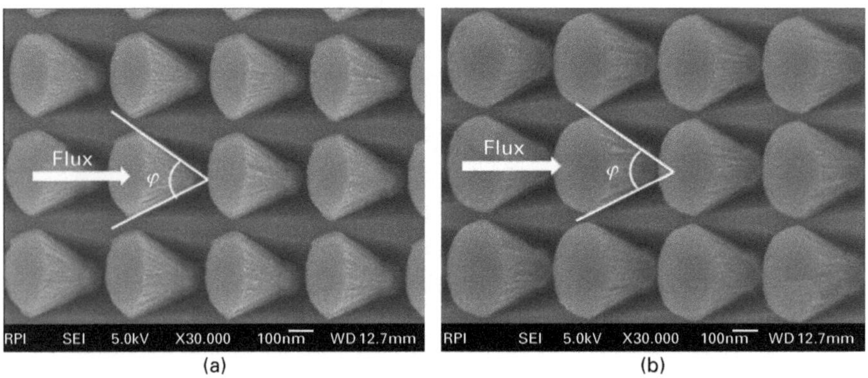

6.14 Fan-out growth of Si on (a) small-sized W pillars and (b) large-sized W pillars. (Adapted from Ye and Lu, 2007b, reprinted with permission.)

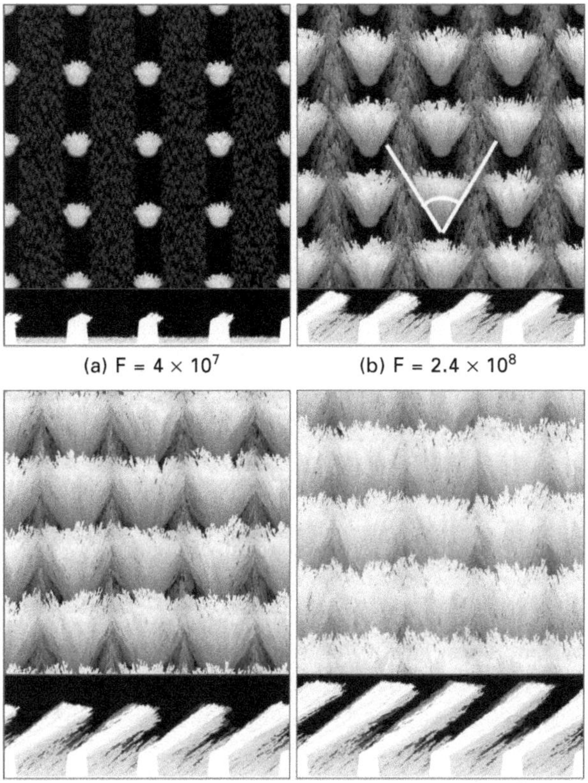

(a) $F = 4 \times 10^7$ (b) $F = 2.4 \times 10^8$

(c) $F = 4.4 \times 10^8$ (d) $F = 6.4 \times 10^8$

6.15 Top-view and cross-sectional images of fan structures in MC simulation with ballistic sticking model. The incident angle of flux is 85°. The diffusion parameter is set to $K = 100$.

to grow even in an early stage of the growth, when the number of deposited particles reaches $F = 4 \times 10^7$. The size of the simulation lattice is $1024 \times 1024 \times 1024$. The incident angle of the particles is 85° in this simulation. The diffusion parameter for this set of simulations is set to $K = 100$, which means that there are 100 attempts at diffusion within a cycle of deposition. When the deposition time is sufficiently long, the structures connect to each other from the side perpendicular to the direction of flux. From the cross-sectional image, the top end surface declined to an angle of about 8° from horizontal. This result is similar to what we observed in experiments. A series of simulations were run with different K values.

The growth of the width R of the ballistic fans was plotted as a function of the adjusted simulation time t following the method described above, as shown in Fig. 6.16. The fitting of the plots in the format of power law growth yields the growth exponent $p \approx 1$ before the fans start merging with their

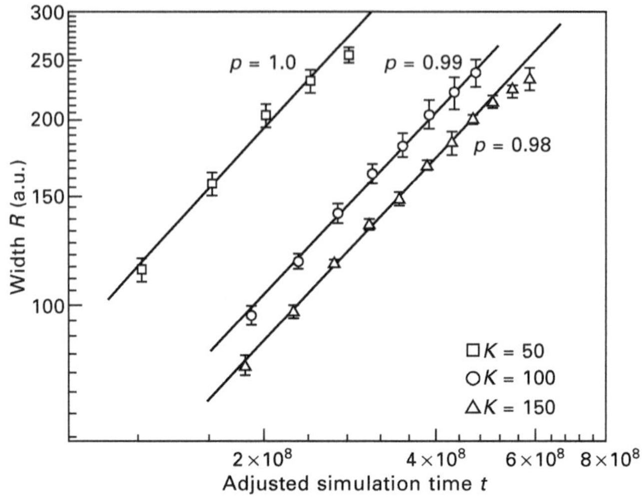

6.16 The linear growth of ballistic fans in oblique incident flux. (Source: Ye and Lu, 2007b, reprinted with permission.)

neighbors. Combined with the observations we found in the case of normal deposition, we can conclude that the fan-like structure is a result of ballistic sticking of particles. The ballistic sticking, or 'self-shadowing effect', has a characteristic growth exponent $p \approx 1$. We also found from the simulations that the fan-out angles decrease when diffusion increases. In the oblique angle deposition, the tilt angle of the structures increases as K increases. We measured the fan-out angle φ and the tilted angle β of the simulated structures. Compared to our experimental results, we can deduce that $K = 100$ fits the Si ballistic fan's fan-out angle φ and tilted angle β, which are $65.0 \pm 4.0°$ and $51.7 \pm 0.3°$ respectively, in the simulated structures.

6.5 Control of fan-out growth with substrate rotations

As previous studies revealed, it is possible to create complicated nanostructured thin films by oblique angle deposition with delicate control of substrate rotation (Robbie *et al.*, 1995a, 1996, 1998; Lakhtakia and Weiglhofer, 1995; Robbie and Brett, 1997; Kennedy *et al.*, 2002; Zhao *et al.*, 2002b). There are two types of nanostructures whose optical, mechanical, and electrical properties we are interested in: nanosprings and slanted nanorods. Isolated nanostructured thin films such as slanted nanorods and nanosprings could provide an ideal platform to study mechanical, electrical, thermal, and optical properties of materials on a nanometer scale. Therefore, the ability to uniformly deposit individual nanostructured thin films is of great practical value. These two structures are relatively difficult to fabricate by traditional lithographic methods, but using oblique angle deposition, we may achieve the required structures relatively more easily.

The growth of uniform slanted nanorod films on an existing array of seeds is challenging but achievable. As shown in the previous sections, the fan-out growth is intrinsic to the nature of thin film growth with strong self-shadowing effect. Therefore, fundamental research in controlling the fan-out growth has been the first priority in the development of this technique. One strategy to fabricate uniform nanostructured thin films is based on diminishing the fan-out growth by substrate rotation. During substrate rotation, some part of the growth front will be interrupted. The growth on that spot will start again at a later time or be terminated entirely, depending on the relative local height. Overall, each part of the growing surface will receive flux for a short period, effectively smoothing the growth front and resulting in uniform growth with no fan-out features even for a very long deposition time.

The geometric parameters of the structure, such as the tilt angle β with respective to the substrate normal, diameter R and separation D of the structures, can be tailored by changing the vapor incident angle θ, thus, changing their physical properties. In general, the column tilt angle β is less than the vapor incident angle θ, and follows the empirical tangent rule or cosine rule discussed above. However, the variation of the incident angle θ simultaneously changes the three geometric parameters (namely, β, R, and D), which causes some difficulty in controlling the geometry and the physical properties of the nanostructured films by simply changing the incident angle θ in oblique angle deposition. Therefore, it is desirable to have other methods to tailor the structures other than the variation of θ. Fortunately, there are some substrate rotation methods that can deposit nanostructured thin films with variable geometric parameters at a fixed incident flux angle θ in this research area.

In experiments, it is possible to control the rotation of the substrate at different rotation speeds, or change the direction or the rotation in every revolution of the substrate movement. Thus the substrate can be rotated in a non-uniform manner. By this means, the surface grows faster in some directions than others. Thus, the structure of the film will incline toward this direction. On the other hand, since the substrate is rotated, the growth front can still be interrupted and re-installed during growth which can potentially limit the fan-out growth. The shadowing direction is changed from time to time as well. We expect that the fan-out growth would be reduced by the rotation of substrate.

In this section, three substrate rotation methods are reviewed and discussed: 'two-phase' rotation (Ye *et al.*, 2004), 'swing' rotation (Ye *et al.*, 2005), and 'PhiSweep' technique (Jensen and Brett, 2005). In the 'two-phase' rotation method, one continuously rotates the substrate during deposition with two different speeds to finish one complete revolution. First, the substrate is rotated at a slower speed ω_1 to an angle of $\phi_1(< 2\pi)$ in the surface plane of the substrate, and then the speed will be switched immediately to a faster ω_2 in the rest of the angle $\phi_2 = 2\pi - \phi_1$ of a complete revolution (Ye *et al.*, 2004). The direction of rotation in these two parts is the same in the two-phase rotation technique. In swing rotation, the substrate is rotated to an angle $\phi(< 2\pi)$ in the plane of the substrate. The substrate is then rotated back to the initial position to cover one cycle of the motion (Ye *et al.*, 2005). A similar approach of non-uniform substrate swinging was published recently by Jensen and Brett (2005). In their method, the substrate is rotated quickly to an angle and stays at that position for a period of time, then the substrate is rotated back to the original position and stays for an equal period; i.e. the substrate is rotated to the other side with the same angle and stays for the same period before it returns. This method is given the name 'PhiSweep' in their paper (Jensen and Brett, 2005).

The fan-out growth on seeds is greatly reduced by the aforementioned substrate rotation techniques as shown in Fig. 6.17. For example, the Si nanorods can be grown to several micrometers without touching the neighbors

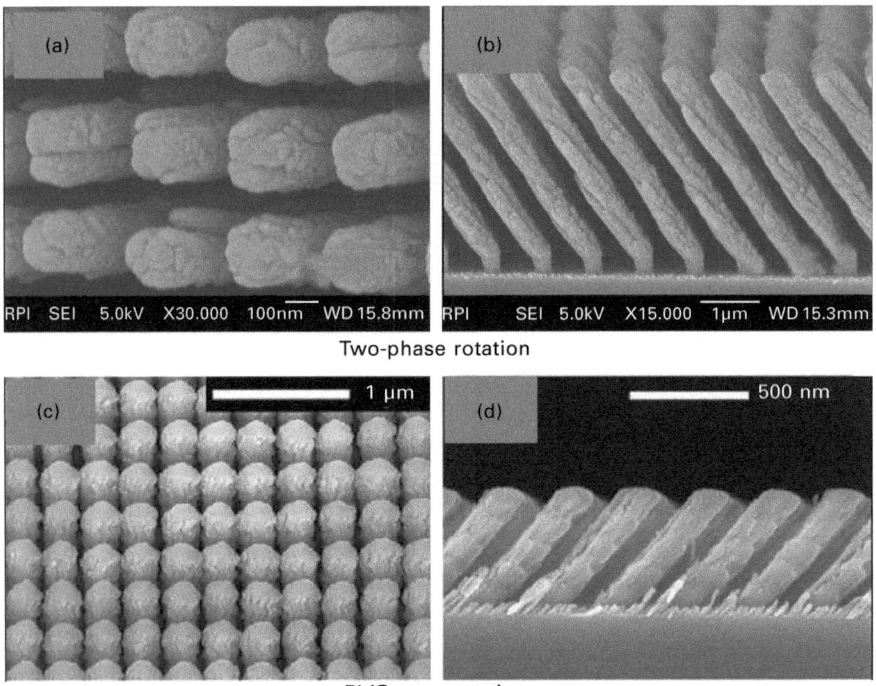

6.17 Top-view and cross-sectional SEM images of uniform Si nanostructured films grown on templated surfaces by 'two-phase' and 'PhiSweep' substrate rotation techniques. (Adapted from Gish *et al.*, 2006, reprinted with permission.)

from the side by the two-phase technique, as shown in Fig. 6.17(a) and (b). Uniform nanorods without fan-out growth were also demonstrated using the 'PhiSweep' substrate rotation method (Jensen and Brett, 2005). A set of selected SEM top-view and cross-sectional images of Si nanostructured film grown by the 'PhiSweep' technique are reproduced in Fig. 6.17(c) and (d) from the paper by Gish *et al.* (2006). Uniform nanostructured Si films on small and big seeds grown by 'swing' rotation are shown in Fig. 6.18. The swing angle was set at 90° and the incident angle of flux was 85° in this experiment.

We set up the Monte Carlo simulations to study the mechanism of reducing the fan-out growth based on the swing rotation in oblique angle deposition. Figure 6.19 shows the top-view and cross-sectional images of the Monte Carlo simulation at two different stages with swing rotation. The size of the simulation system is 1024 × 1024 × 1024. The swing angle in this simulation was fixed at $\phi = 90°$. The rotation speed in this simulation was defined by the number of particles deposited in one step of substrate rotation. In our code,

6.18 Top-view and cross-sectional SEM images of uniform Si nanostructured films grown on small and big seeds by 'swing' substrate rotation.

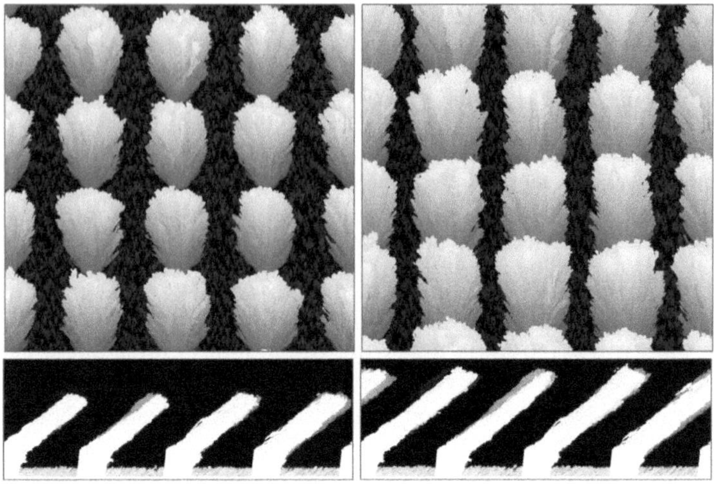

(a) $F = 4.4 \times 10^8$ (b) $F = 6.4 \times 10^8$

6.19 Monte Carlo simulation of swing rotation at $\phi = 90°$.

we change the azimuthal angles of the flux to mimic the substrate rotation because the lattice cannot be rotated. The step of increasing and decreasing this azimuthal angle is the parameter representing the substrate rotation speed. In this simulation, the rotation speed is fixed at 5000 particles per step. With swing rotation, uniform nanostructured films can be generated in our simulation of ballistic sticking of oblique incident particles on seeds (Ye and Lu, 2007b). The average tilted angle is 53.5 ± 0.3° as measured from the cross-sectional images of the structures, which is very close to the measurement in experiments. The structure grows at the beginning of the simulation and then saturates at a later simulation time to a size of 184 ± 21 lattice units when measured from the top-view images. The growth of the width R of the structures is depicted in Fig. 6.20 as a function of adjusted simulation time for the swing rotation with different swing angles. From Fig. 6.20, one can observe that the initial growth exponent is $p \approx 0.6$ for all the swing angles in the simulations, as shown in the inset of Fig. 6.20. We believe that the reduction of the exponent p from 1.0 (as in the case of no substrate rotation) to less than 0.6 is due to the substrate rotation. The global shadowing affects the saturation of the width R_{max}, where the well-separated structures stop fanning out with an exponent p approaching 0 (Ye and Lu, 2007b). With the global shadowing effect, increasing the swing angles will reduce the saturation width R_{max}, as illustrated in the inset of Fig. 6.20.

6.6 Applications and future trends

The substrate rotation techniques can produce slanted nanorods of uniform sizes, which enable the fabrication of complex nanodevices, for example,

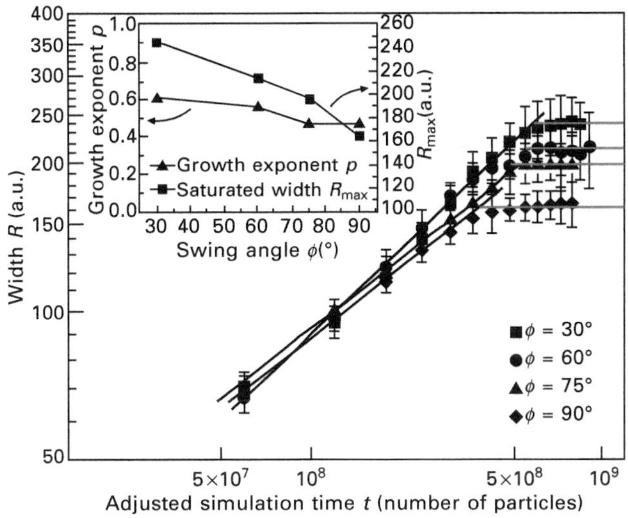

6.20 Logarithmic plot of the width R as a function of the adjusted simulation time with different swing angles. R is saturated at a maximum value R_{max} as indicated by the horizontal lines. The inset is the measurement of the exponent p and the maximum width R_{max} for different swing angles. (Source: Ye and Lu, 2007b, reprinted with permission.)

arrays of nano-cantilevers. The arrays of slanted Si nanorods deposited on templated surfaces have been used to measure the mechanical properties of these nanorods. The cantilever system of atomic force microscope was used to measure the response of the Si nanorods when forces are applied to them. The results have been reported in the literature (Gaire *et al.*, 2005). These arrays of nanorods provide an ideal system for the study of the mechanical properties of materials in nanometer size.

When a layer of slanted nanorods is deposited, the substrate can be turned 90° to deposit a second layer of slanted nanorods with the same geometry. This can be continued for many layers. Every four layers of these nanorods can form a complete turn of a square spiral as shown in Fig. 6.21. From the calculations, these spiral arrays possess a unique optical property where light with certain wavelengths cannot penetrate through the film (Toader and John, 2001). A system with this optical property is called a photonic crystal and it has rich applications in controlling of light. Recently, Si spiral photonic crystals have been deposited by the swing rotation technique (Ye *et al.*, 2007) and the 'PhiSweep' technique (Summers and Brett, 2008). The properties of the photonic crystals were tested by optical measurement as well (Ye *et al.*, 2007). However, in experiments, the Si spiral photonic crystal structures produced by oblique angle deposition cannot reach the optimized design. For optimal geometrics of Si spiral photonic crystals, the tilted angle of each arm

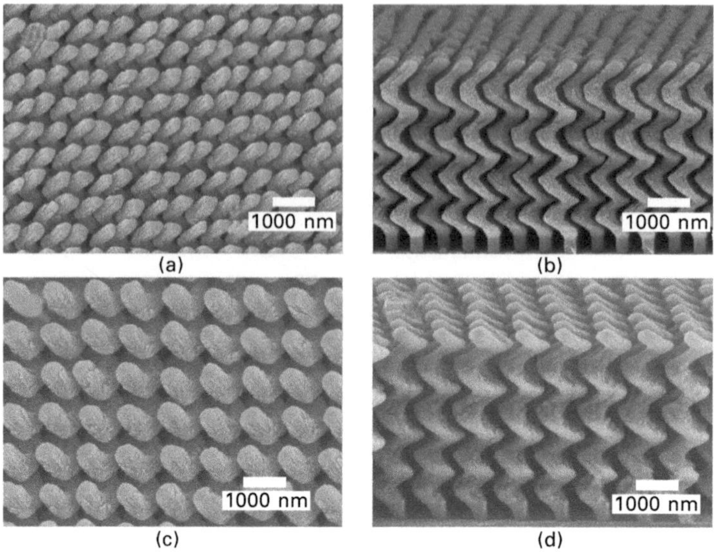

6.21 Square spirals prepared by the swing rotation on W pillars. (a) and (b) are the top-view and cross-sectional images of the spirals deposited on the triangular lattice. (c) and (d) are the top-view and cross-sectional images of the spirals deposited on the square lattice. (Source: Ye *et al.*, 2005, reprinted with permission.)

of the structure should be as large as 65° with respect to the major axis of the spirals (Toader and John, 2002). A maximum full band gap (about 16% of the gap center frequency) can be achieved with this optimal Si structure. Therefore, in the future, the growth condition of the photonic crystals should be further explored to achieve the optimized photonic crystals. Some other materials should be tested in order to fabricate the optimized structure.

On smooth surfaces, a layer of nanostructured thin film consisting of slanted nanorods has lower refractive index than a continuous film due to the porous nature of the film (Xi *et al.*, 2005). The index can be varied by changing the orientation of the nanorods. It has been tested that the stacking of multiple layers of nanostructured thin films with various refractive indices has a very low reflectance and can potentially be used as an anti-reflector in several applications (Xi *et al.*, 2007). More discussion will be presented later in this book. Although the layered films were deposited by oblique angle deposition at different incident angles, it is reasonable to extend this study using substrate rotation techniques. We would expect that the anti-reflection films are much more uniform and the refractive indices are able to be fine tuned.

In this chapter, we found that the dominant mechanism of fan-out growth is ballistic sticking that comes from the inter-atomic interactions at the atomic level. However, Monte Carlo simulation has no real dimension, so a full picture of the interactions between particles cannot be simulated. Therefore, molecular dynamics simulation might be able to handle the inter-atomic interactions. However, the system size in a typical molecular dynamics simulation is too small for the deposition of atoms on arrays of seeds. It seems that a multiscale simulation has to be developed for a better understanding of the effect of ballistic sticking in a large ensemble of atoms. In the present Monte Carlo simulation model, a simple cubic lattice is used. The extension of this simulation into other lattice structures should be interesting.

6.7 References

Ablemann L and Lodder C (1997), 'Oblique evaporation and surface diffusion', *Thin Solid Films*, 305, 1–21.

Amar J G and Family F (1990), 'Phase transition in a restricted solid-on-solid surface-growth model in 2+1 dimensions', *Phys Rev Lett*, 64, 543–546.

Barabasi A L and Stanley H E (1995), *Fractal Concepts in Surface Growth*, New York, Cambridge University Press.

Castez M F and Salvarezza R C (2006), 'Modeling growth from the vapor and thermal annealing on micro- and nanopatterned substrates', *Phys Rev E*, 73, 011607.

Dew S K, Smy T and Brett M J (1992), 'Simulation of elevated temperature aluminum metallization using SIMBAD', *IEEE Trans Commun*, 39, 1599–1606.

Gaire C, Ye D-X, Tang F, Picu R C, Wang G-C and Lu T-M (2005), 'A mechanical testing of isolated amorphous Si slanted nanorods', *J Nanosci Nanotechnol*, 5, 1893–1897.

Geszti O, Gosztola L and Seyfried E (1986), 'Cross-sectional transmission electron microscopy study of obliquely evaporated silicon oxide thin films', *Thin Solid Films*, 136, L35–L38.

Gish D A, Summers M A and Brett M J (2006), 'Morphology of periodic nanostructures for photonic crystals grown by glancing angle deposition', *Photon Nanostruct*, 4, 23–29.

Hashimoto T, Okamoto K, Hara K, Kamiya M and Fujiwara H (1982), 'Columnar structure and texture of iron films evaporated at oblique incidence', *Thin Solid Films*, 91, 145–154.

Hawkeye M M and Brett M J (2007), 'Glancing angle deposition: fabrication, properties, and applications of micro- and nanostructured thin films', *J Vac Sci Technol A*, 25, 1317–1335.

Horn M W, Pickett M D, Messier R and Lakhtakia A (2004a), 'Selective growth of sculptured nanowires on microlithographic lattices', *J Vac Sci Technol*, 22, 3426–3430.

Horn M W, Pickett M D, Messier R and Lakhtakia A (2004b), 'Blending of nanoscale and microscale in uniform large-area sculptured thin-film architectures', *Nanotechnology*, 15, 303–36.

Imry Y (2002), *Introduction to Mesoscopic Physics*, 2nd ed, New York, Oxford University Press.

Jensen M O and Brett M J (2005), 'Porosity engineering in glancing angle deposition thin films', *Appl Phys A*, 80, 763–768.

Joannopoulos J D, Meade R D and Winn J N (1995), *Photonic Crystals: Molding the Flow of Light*, Princeton, NJ, Princeton University Press.

Johnson M D, Orme C, Hunt A W, Graff D, Sudijono J, Sander L M and Orr B G

(1994), 'Stable and unstable growth in molecular beam epitaxy', *Phys Rev Lett*, 72, 116–119.

Kardar M, Parisi G and Zhang Y C (1989), 'Dynamic scaling of growing interfaces', *Phys Rev Lett*, 56, 889–892.

Kennedy S R, Brett M J, Toader O and John S (2002), 'Fabrication of tetragonal square spiral photonic crystals', *Nano Lett*, 2, 59–62.

Kessler D A, Levine H and Sander L M (1992), 'Molecular-beam epitaxial growth and surface diffusion', *Phys Rev Lett*, 69, 100–103.

Kim J M and Kosterlitz J M (1989), 'Growth in a restricted solid-on-solid model', *Phys Rev Lett*, 62, 2289–2292.

Krug J and Meakin P (1991), 'Columnar growth in oblique incidence ballistic deposition: Faceting, noise reduction, and mean-field theory', *Phys Rev A*, 43, 900–919.

Lakhtakia A and Weiglhofer W S (1995), 'On light propagation in helicoidal bianisotropic mediums', *Proc R Soc London Ser A*, 448, 419–437.

Leamy H J, Gilmer G M and Dirks A G (1980), 'The microstructure of vapor deposited thin films', *Current Topics in Mat Sci*, 6, 309–344.

Liang S and Kadanoff L P (1985), 'Scaling in a ballistic aggregation model', *Phys Rev A*, 31, 2628–2630.

Limaye A V and Amritkar R E (1986), 'Theory of growth of ballistic aggregates', *Phys Rev A*, 34, 5085–5090.

Luedtke W D and Landman U (1989), 'Molecular-dynamics studies of the growth modes and structure of amorphous silicon films via atom deposition', *Phys Rev B*, 40, 11733–11746.

Lu T-M, Ye D-X, Karabacak T and Wang G-C (2005), 'Physical self-assembly and nano-patterning', *Mat Res Soc Symp Proc*, 849, KK8.4.1–KK8.4.6.

Malac M, Egerton R F, Brett M J and Dick B (1997), 'Fabrication of submicrometer regular arrays of pillars and helices', *J Vac Sci Technol*, 17, 2671–2674.

Marsili M, Maritan A, Toigo F and Banavar J R (1996), 'Stochastic growth equations and reparametrization invariance', *Rev Mod Phys*, 68, 963–983.

Meakin P (1998), *Fractals, Scaling and Growth far from Equilibrium*, New York, Cambridge University Press.

Meakin P and Krug J (1990), 'Columnar microstructure in three-dimensional ballistic deposition', *Europhys Lett*, 11, 7–12.

Meakin P and Krug J (1992), 'Three-dimensional ballistic deposition at oblique incidence', *Phys Rev* A, 46, 3390–3399.

Mei S-N and Lu T-M (1988), 'A high ionization efficiency source for partially ionized beam deposition', *J Vac Sci Technol A*, 6, 9–11.

Nakhodkin N G and Shaldervan A I (1972), 'Effect of vapour incidence angles on profile and properties of condensed films', *Thin Solid Films*, 10, 109–122.

Nieuwenheuzen J M and Haanstra H B (1966), 'Microfractography of thin films', *Philips Tech Rev*, 27, 87–91.

Porcu F and Prodi F (1991), 'Ballistic accretion on seeds of different sizes', *Phys Rev A*, 44, 8313–8315.

Raible M, Linz S J and Hanggi P (2002), 'Growth instabilities of vapor deposited films: atomic size versus deflection effect', *Eur Phys J B*, 27, 435–442.

Ramanlal P and Sander L M (1985), 'Theory of ballistic aggregation', *Phys Rev Lett*, 54, 1828–1831.

Robbie K and Brett M J (1997), 'Sculptured thin films and glancing angle deposition: growth mechanics and applications', *J Vac Sci Technol A*, 15, 1460–1465.

Robbie K, Friedrich L J, Dew S K, Smy T and Brett M J (1995a), 'Fabrication of thin films with highly porous microstructures', *J Vac Sci Technol A*, 13, 1032–1035.

Robbie K, Brett M J and Lakhtakia A (1995b), 'First thin film realization of a helicoidal bianisotropic medium', *J Vac Sci Technol A*, 13, 2991–2993.

Robbie K, Brett M J and A. Lakhtakia A (1996), 'Chiral sculptured thin films', *Nature*, 384, 616.

Robbie K, Sit J C and Brett M J (1998), 'Advanced techniques for glancing angle deposition', *J Vac Sci Technol B*, 16, 1115–1122.

Smilauer P, Wilby M R and Vvedensky D D (1993), 'Reentrant layer-by-layer growth: a numerical study', *Phys Rev B*, 47, 4119–4122.

Smy T, Vick D, Brett M J, Dew S K, Wu A T, Sit J C and Harris K D (2000), 'Three-dimensional simulation of film microstructure produced by glancing angle deposition', *J Vac Sci Technol A*, 18, 2507–2512.

Stasevich T J, Einstein T L, Zia R K P, Giesen M, Ibach H and Szalma F (2004), 'Effects of next-nearest-neighbor interactions on the orientation dependence of step stiffness: reconciling theory with experiment for Cu(001)', *Phys Rev B*, 70, 245404.

Summers M A and Brett M J (2008), 'Optimization of periodic column growth in glancing angle deposition for photonic crystal fabrication', *Nanotechnology*, 19, 415203.

Suzuki M and Taga Y (2001), 'Numerical study of the effective surface area of obliquely deposited thin films', *J Appl Phys*, 90, 5599–5605.

Tang F, Liu D-L, Ye D-X, Zhao Y-P, Lu T-M, Wang G-C and Vijayaraghavan A (2003), 'Magnetic properties of Co nanocolumns fabricated by oblique-angle deposition', *J Appl Phys*, 93, 4194.

Toader O and John S (2001), 'Proposed square spiral microfabrication architecture for large three-dimensional photonic band gap crystals', *Science*, 292, 1133–1135.

Toader O and John S (2002), 'Square spiral photonic crystals: Robust architecture for microfabrication of materials with large three-dimensional photonic band gaps', *Phys. Rev. E*, 66, 016610.

Tong W M and Williams R S (1994), 'Kinetics of surface growth: phenomenology, scaling, and mechanisms of smoothening and roughening', *Annu Rev Phys Chem*, 45, 401–438.

Trait R N, Smy T and Brett M J (1993), 'Modelling and characterization of columnar growth in evaporated films', *Thin Solid Films*, 226, 196–201.

van Dijken S, Jorritsma L C and Poelsema B (1999), 'Steering-enhanced roughening during metal deposition at grazing incidence', *Phys Rev Lett*, 82, 4038–4041.

Wong J, Mei S-N and Lu T-M (1987), 'Al/Si(100) Schottky barrier formation using nozzle jet beam deposition', *Appl Phys Lett*, 50, 679–681.

Xi J-Q, Kim J K and Schubert E F (2005), 'Silica nanorod-array films with very low refractive indices', *Nano Lett*, 5, 1385–1387.

Xi J-Q, Schubert M F, Kim J K, Schubert E F, Chen M, Lin S-Y, Liu W and Smart J A (2007), 'Optical thin-film materials with low refractive index for broadband elimination of Fresnel reflection', *Nat. Photonics*, 1, 176–179.

Ye D-X and Lu T-M (2007a), 'Fanlike aggregations on seeds by parallel ballistic flux: experimental results and Monte Carlo simulations of the growth of three-dimensional Si structures', *Phys Rev B*, 75, 115420.

Ye D-X and Lu T-M (2007b), 'Ballistic aggregation on two-dimensional arrays of seeds with oblique incident flux: growth model for amorphous Si on Si', *Phys Rev B*, 76, 235402.

Ye D-X, Karabacak T, Lim B K, Wang G-C and Lu T-M (2004), 'Growth of uniformly aligned nanorod arrays by oblique angle deposition with two-phase substrate rotation', *Nanotechnology*, 15, 817–821.

Ye D-X, Karabacak T, Picu R C, Wang G-C and Lu T-M (2005), 'Uniform Si nanostructures grown by oblique angle deposition with substrate swing rotation', *Nanotechnology*, 16, 1717–1723.

Ye D-X, Yang Z-P, Chang A S P, Bur J, Lin S-Y, Lu T-M, Wang R Z and John S (2007), 'Experimental realization of a well-controlled 3D silicon spiral photonic crystal', *J Phys D: Appl Phys*, 40, 2624–2628.

Ye D-X, Ellison C L, Lim B-K and Lu T-M (2008), 'Shadowing growth of three-dimensional nanostructures on finite size seeds', *J Appl Phys*, 103, 103531.

Zhao Y-P, Ye D-X, Wang P-I, Wang G-C and Lu T-M (2002a), 'Fabrication of Si nanocolumns and Si square spirals on self-assembled monolayer colloid substrates', *Int J Nanosci*, 1, 87–97.

Zhao Y-P, Ye D-X, Wang G-C and Lu T-M (2002b), 'Novel nano-column and nano-flower arrays by glancing angle deposition', *Nano Lett*, 2, 351–354.

7 Phase transitions in colloidal crystal thin films

F. RAMIRO-MANZANO, E. BONET, I. RODRÍGUEZ and F. MESEGUER, Centro de Tecnologías Físicas, Unidad Asociada ICMM-CSIC/UPV, Universidad Politécnica de Valencia, Spain

Abstract: This chapter reviews the crystal order of colloidal suspensions as hard spheres when confined in wedge type cells with flat walls. This type of cell allows the study of colloidal crystal growth transitions between one and *n* monolayers.

Key words: colloidal facets, colloidal crystal, confinement, phase transitions.

7.1 Introduction

Identical objects with rounded surfaces tend to pack in an ordered manner independently of their size, being either meter size, as occurs for saw tree trunks or big pipes being transported by lorry, or millimetrical size like marbles packed in a box. In this case, marbles tend to be packed in close packed layers each ordered in a triangular symmetry. This is our everyday experience and nobody is astonished when they see such a neat periodic distribution of objects. Concerning macroscopic systems, people may guess that order is a consequence of human influence. However, this type of symmetrical distribution also appears for much smaller, submicrometric size species like colloidal particles, molecules and atoms where human influence is less believed. In the former case colloids arrange into colloidal crystals (Binks, 2006) and opals (Sanders, 1964) and in the latter they form the well-known crystalline structures (Sands, 1975) like rock salt, diamond and other fancy mineral rocks and jewels. The crystallization processes in minerals appear due mostly to electrical forces, and they have been treated in a large number of books (Kittel, 1996; Vainshtein, 1981). In the case of colloidal crystals the crystallization can appear in both charged and also in non-charged particles, dependent on both the material and the pH value of the solvent used (Pusey *et al.*, 1989; Holgado *et al.*, 1999).

In this chapter we report the case of non-charged colloids (hard spheres) where they order in a close packed manner like cannon balls in a pile. Here it is worth mentioning colloid stability is due to the double layer of ions and counterions that is governed by the DLVO theory (Derjaguin and Landau, 1941; Verwey, 1947). There are two ways non-charged identical microspheres can be arranged periodically in a box when maximizing the volume fraction occupied by the spheres. In the first, particles are ordered in a face centered cubic (FCC) symmetry, and in the second, particles pack themselves in a hexagonal close packed (HCP) structure. In both cases the filling fraction (ff) occupied by the particles represents 74% of the total volume. Although both types of particle ordering would have the same probability, Woodcock (1997) showed that the FCC arrangement would be favoured over the HCP one through the use of thermodynamic arguments. Later, Míguez *et al.* (1997) reported on colloidal suspension of submicrometric particles settled preferentially in FCC structures. There are other more efficient ways to pack particles like the Apollonius packing (Duran, 1999). However, it involves particles of different size and they lead to a quasi fractal ordering (Ramiro-Manzano *et al.*, 2006a).

Colloidal crystals have developed a great deal and at a fast pace in the last 30 years because they constitute crystallization models of much smaller species, like atoms and molecules. Also in the last ten years colloidal crystals science and technology have been of great help in developing new and cutting-edge photonic materials based on devices like photonic crystals and metamaterials (Joannopoulos *et al.*, 1995; Braun and Wiltzius, 1999; Blanco *et al.*, 2000). Photonic crystals can be defined as composite materials with a strong modulation of the refractive index in some spatial directions. The refractive index modulation produces huge diffraction effects. Light of certain frequency values is not allowed to travel through it (John, 1987; Yablonovitch, 1987; Joannopoulos *et al.*, 1997). This is the basic concept of a photonic bandgap. In this way, colloidal crystals also produce light diffraction effects, this being the origin of the pretty iridescent colours of natural opals (Sanders, 1964). Therefore, colloidal crystals can be considered as a class of photonic crystals. For many applications using colloidal crystals, just a few monolayers (ML) are enough to build up the photonic bandgap (Vlasov *et al.*, 2001). However, as colloidal crystals pack spontaneously in a FCC lattice showing its (111) facet, their application to photonic crystals is limited. In fact, the possibility of creating different lattices and facets could extend dramatically the application range of colloidal crystals (Prasad *et al.*, 2003; Ochiai and Sanchez-Dehesa, 2001; Fenollosa *et al.*, 2003).

Colloidal crystal thin films can be achieved through several methods. In most cases they are processed through self-arrangement of particles driven by convection (Míguez *et al.*, 2002, 2003; Yin *et al.*, 2001; Velikov *et al.*, 2002; Kitaev and Ozin, 2003), capillarity (Hiemenz and Rajagopalan, 1997; Knobler and Schwartz, 1999; Pan *et al.*, 2006, van Duffel *et al.*, 2001; Wolert *et al.*, 2001), gravitational (Míguez *et al.*, 1997; López *et al.*, 1997; Van Blaaderen *et al.*, 1997), evaporation (Park *et al.*, 1998; Pieranski *et al.*,

1983; Pansu *et al.*, 1984), centrifugal (Mihi *et al.*, 2006) or electrophoretic (Holgado *et al.*, 1999) forces. In the vast majority of cases it results in a thin film of FCC ordering whose surface shows a triangular symmetry facet corresponding to the (111) plane (FCC(111) for short). However, strong confinement of particle suspensions promotes new particle orderings as Pieranski *et al.* (1983) demonstrated. This confinement is generated by two solid–liquid interfaces in close proximity. However, this phenomenon is not only restricted to strong confinement. Weak confinement constituted by two interfaces, air–liquid and liquid–solid (Prevo and Velev, 2004; Mihi *et al.*, 2006), could also lead to the formation of some of these new orderings.

We consider here only the case of uncharged colloidal particles (i.e. a hard sphere system). Hard spheres represent the limit of screened colloids that could be considered unaffected by Coulomb forces. In this chapter we review how confinement in colloidal crystals can produce a manifold variety of colloidal arrangements. We also show models that give hints about the evolution dynamics of the crystallization for different confinement restrictions.

7.2 Experimental tools

In this section we explain the design of the confinement cell, the colloidal crystal growth and also their optical characterization. The design of the cell is rather similar to that used by Lu *et al.* (2001). Wedge type cells are formed by two large plates (3 cm large), the substrate plate and the covering plate. They are made of hydrophilic (treated glass) and hydrophobic (polystyrene) material, respectively. Therefore, in order to achieve large monocrystal areas of well-isolated facets, a very small wedge angle ($\approx 10^{-4}$ rad) was used. A 6 micrometers thick Mylar film, attached along only one rim of the slides, separates the plates. The wedge type cell is obtained by tightening the cell with several binder clips on the three other sides of the cell (see Fig. 7.1(a) and 7.1(d)). For thicker samples with many monolayers, we use another cell design shown in Fig. 7.1(c). In the alternative design the separator surrounds the entire cell and the plates are bent due to the action of the clips. Polystyrene particles of different sizes ranging from 245 nm up to 800 nm in diameter (Ikerlat Polymers) were employed. Particles were washed and rinsed several times with milliQ water. An aqueous suspension of particles (1 wt%) was introduced through a 2 cm high glass tube (tank) attached to the inner part of the cell through a small hole (2 mm diameter) drilled on the glass plate. Several drops were put into the small tube, which entered the cell due to capillary forces. While water leaked out and evaporated, the cell was subjected to a light ultrasonic vibration to help settle the particles. In some cases, when particles were of smaller size (245 nm) it was necessary to apply a light pressure to the feeding glass tube to compensate Brownian

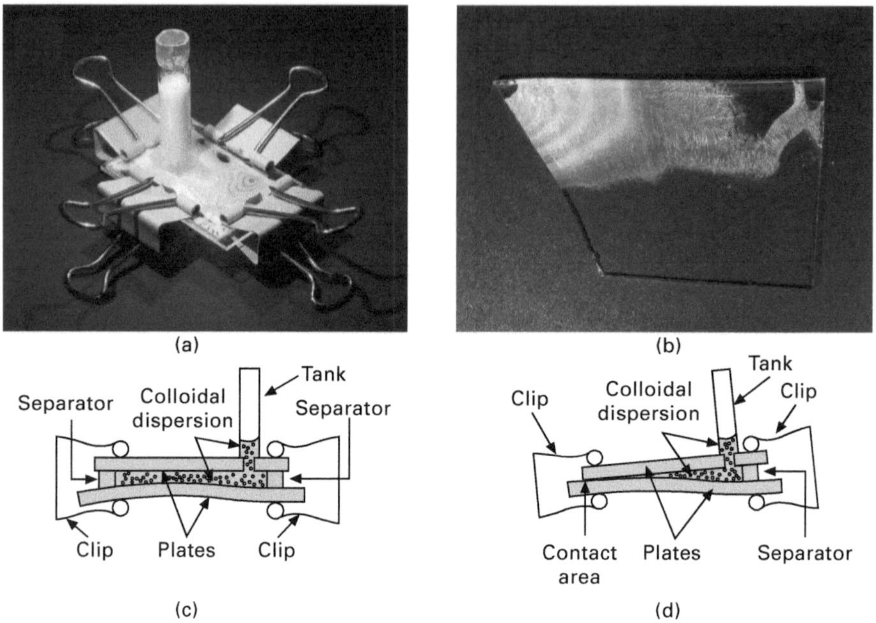

7.1 (a) Wedge cell (b) Colloidal crystal thin film sample. Descriptive scheme of the cell employed: (c) with a separator frame, and (d) with a wedge design.

forces and also speed up the water evaporation process. After about one week, the system condensed into several facets, and, finally, the sample was dried. As the polystyrene slide can be easily detached from the sample with almost no damage to the crystalline structure, very high quality thin film solid colloidal crystals are obtained. Once the sample is obtained we proceed to its characterization. We have made use of both optical reflection spectra at normal incidence and also scanning electron microscopy (SEM) for identifying the different phases appearing in the thin film. Optical reflectance spectra were recorded by using a Fourier transform infrared IFS-66 Bruker Spectrometer coupled to an optical microscope provided with a 15× Cassegrain objective lens (Ramiro-Manzano *et al.*, 2006b).

Optical reflectance of thin film colloidal crystals composed from periodical layers gives two types of features: a Bragg diffraction peak corresponding to the periodical arrangement of layers and reflectance oscillations (ripples) corresponding to Fabry–Perot interference effects of the thin film. From the position (in wavelengths) of the Bragg reflection peak and the number of Fabry–Perot oscillations, one can obtain information about the interlayer periodicity and the number of stacked layers, respectively. Figure 7.2(a) shows the reflectance spectra for the triangular facet (Δ) ordering for 1 to 5 monolayers. Ripples come from the Fabry–Perot oscillations and the central

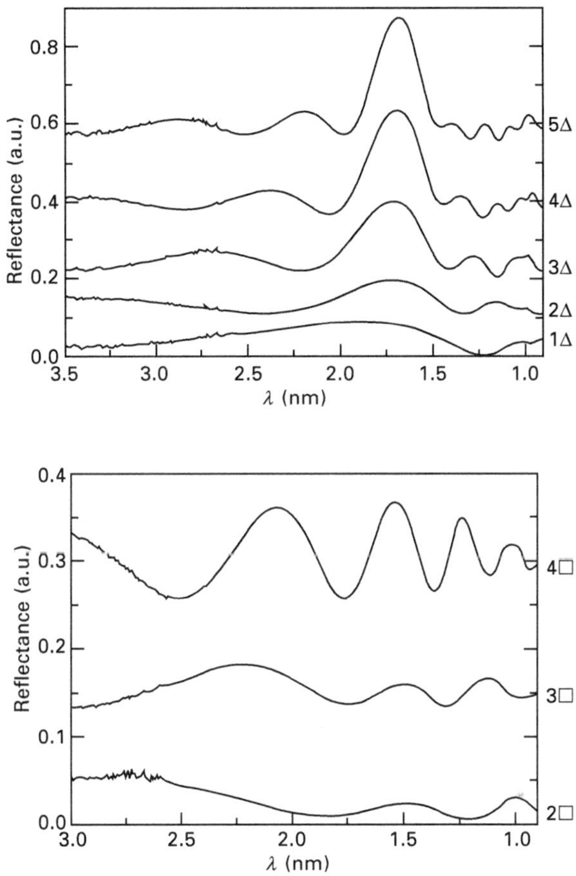

7.2 Reflectance spectra of (a) triangular facet (Δ) from 1 to 5 ML, and (b) square facet (□) from 2 to 4 ML. Spectra have been shifted for the sake of comparison.

peak corresponds to the Bragg diffraction peak along the [111] direction of the Brillouin zone of the FCC lattice. However, this type of spectra can also correspond to the HCP(001) oriented arrangement. Both phases are constituted by triangular symmetry planes. These planes could be packed either in an ABCABC or ABAB manner (see Fig. 7.3). Figure 7.3(a) forms the FCC(111) ordering and Fig. 7.3(b) the HCP(001) one. Nevertheless, SEM images of cleft edges can discriminate between both lattices. In the case of square (□) facets, corresponding to FCC(100) ordering, optical spectra do not show any Bragg peak (see Fig. 7.2(b)) and only Fabry–Perot oscillations appear. We also have performed simulated optical reflectance calculations based on scalar wave approximation (Mittleman *et al.*, 1999) that allows both the facet ordering and the number of monolayers to be obtained.

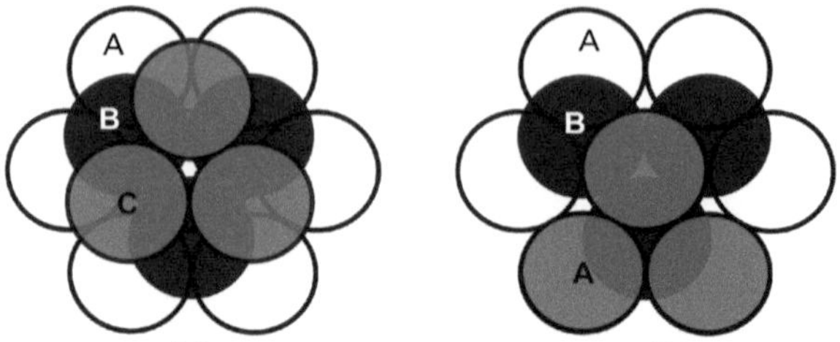

7.3 FCC and HCP models. The packed triangular plane ordering establishes the type of lattice. ABC sequence corresponds to the FCC (a) and the AB to the HCP (b).

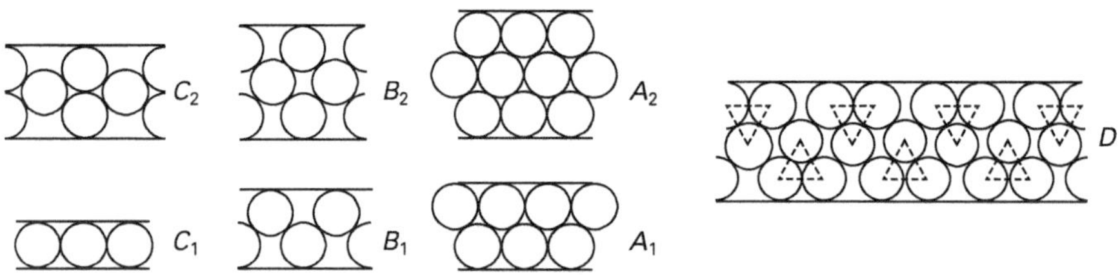

7.4 2D confinement of hard disk orderings and transitions characterized by Pieranski and Finney (1979). The arrangements B_2, B_1 and D show high resemblance to 3D orderings square, buckling and prismatic. From Pieranski and Finney (1979) with permission of the authors and the Americal Physical Society (APS).

7.3 Description of colloidal crystal phases: historical survey

7.3.1 Commensurability in two dimensions

Who of us has never played with identical coins on a table building up fancy figures? Pi. Fieranski and J. Pinney (Pieranski, 1980) turned this game into an interesting geometrical problem. A triangular close packed ordering of hard disks is the densest possible structure. However, when confined between two walls, as shown in Fig. 7.4, the influence of the limiting walls induces new geometrical distributions of disks. Pi. Pieranski and J. Finney showed how the variation of the distance between the limiting walls would generate a rich variety of disk distributions. Even more, the orderings and transitions between them (see Fig. 7.4) have a high resemblance to other 3D structures discovered many years later (Neser *et al.*, 1997).

7.3.2 Commensurability in three dimensions

As we have discussed above, hard screened colloidal suspensions (or the so -called hard sphere systems), when dried, self-arrange themselves in close packed layers. Also, spheres in each layer order in a triangular manner as in the 2D model. However, when the colloidal suspension is dried in a confined space between two plates, the new space dimension (3D) enriches much more the ordering configurations and complexity than in the case of hard disks. Pieranski *et al.*, in a seminal paper in 1983, using wedge type cells in which cell thickness increases gradually from zero to arbitrary values several times larger than the particle size, showed that only an integer number of crystalline layers may exist in equilibrium. Then, depending on the cell thickness value, different types of particle arrangements would appear. Here we review their basic characteristics so far reported.

The triangular and the square phases

Colloidal crystals in a confined geometry may be arranged not only in a triangular symmetry as would be expected for a bulk colloidal system, but they also can be ordered in a square arrangement when the cell thickness value is commensurate with the interlayer distance of the (100) facet of the FCC crystal symmetry. This was the main conclusion of the Pieranski paper (Pieranski *et al.*, 1983). They concluded that the sequence of crystallization was the following:

$$1\Delta \rightarrow 2\square \rightarrow 2\Delta \rightarrow 3\square \rightarrow \ldots \rightarrow n\Delta \rightarrow (n+1)\square \rightarrow (n+1)\Delta \qquad [7.1]$$

where Δ and $\square$ stand for triangular and square particle orderings, and 1, 2, .., n being the number of monolayers. However, there remained a number of points that required further discussion. The authors gave a theoretical background supporting their experimental results. They assumed more stable phases would be those with a maximum value of the particle density (related to ff), so they plotted the particle density as a function of the cell thickness (h) as shown in Fig. 7.5. From this graph the crystallization sequence shows abrupt changes in the particle density; i.e. only few values of h located at the maxima of the plot of Fig. 7.5 would show stable colloidal crystal orderings. These results raised a lot of interest and discussion about both identifying the intermediate colloidal crystal phases and the transition mechanisms between them.

The buckling and the rhombic phases

Later, B. Pansu, Pi. Pieranski again and his twin brother Pa. Pieranski (Pansu *et al.*, 1984) gave the first proposal for two transitions $1\Delta \rightarrow 2\square$, and $n\square \rightarrow n\Delta$. In the former case they proposed the buckling phase (see model in Fig. 7.6(a)) as a solution for the transition between a triangular monolayer and a bi-layer with a square arrangement ('escape into the third dimension'

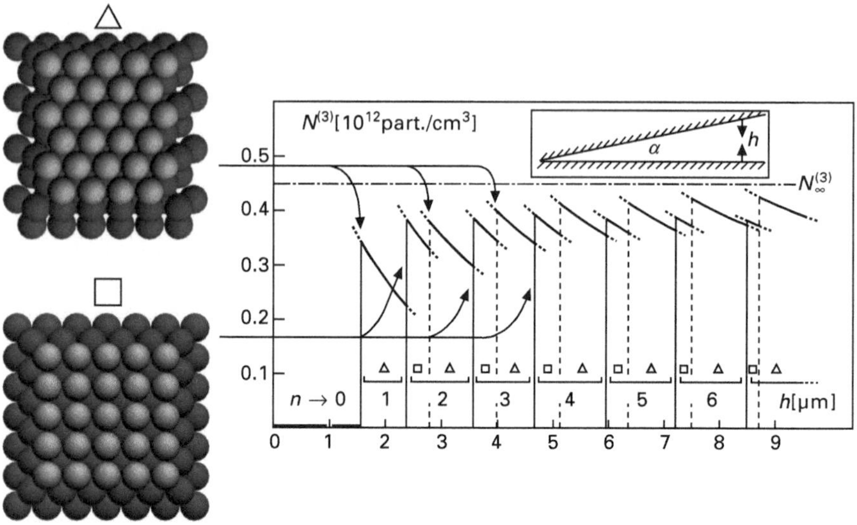

7.5 Left: 3D models of the triangular and square arrangements. Right: First sequence diagram of the particle density (related to filling fraction) versus distance between plates. The particle density plot shows abrupt changes around the square and triangular orderings. The graph is taken from Pieranski *et al.* (1983) with permission of the authors and the Americal Physical Society (APS).

they wrote) to accommodate the continuous cell thickness variation. In the second one ($n\square \rightarrow n\Delta$) they proposed the rhombic phase (see model in Fig. 7.6(b)). Figure 7.6 shows filling fraction value as a function of the reduced cell value defined as the ratio between the cell thickness value (D) and the particle diameter (Φ). Continuous lines show transition orderings that are able to link compact phases and dotted lines are unsuccessful attempts at connecting different phases. However, they could not find a generalization to explain the transition $n\Delta \rightarrow (n + 1)\square$. Therefore, the state-of-the-art for colloidal transition after the Pansu *et al.* (1984) report was the following:

$$1\Delta \rightarrow 1\mathrm{B} \rightarrow 2\square \rightarrow 2r \rightarrow 2\Delta \rightarrow 3\square \ldots\ldots\ldots\ldots\ (n-1)\ \Delta \rightarrow n\square \rightarrow nr \rightarrow n\Delta \qquad [7.2]$$

The prismatic phase

There were also contributions from other groups towards understanding and clarifying phase transition from both the experimental (Van Winkle and Murray, 1986, 1988) and theoretical (Chou and Nelson, 1993; Schmidt and Löwen, 1996, 1997) points of view. However, basically the scheme shown in Eq. 7.2 was the state-of-the-art of colloidal transitions in 1997 when Neser *et al.* (1997) introduced the prismatic phase (P). Figures 7.7 and 7.8 show

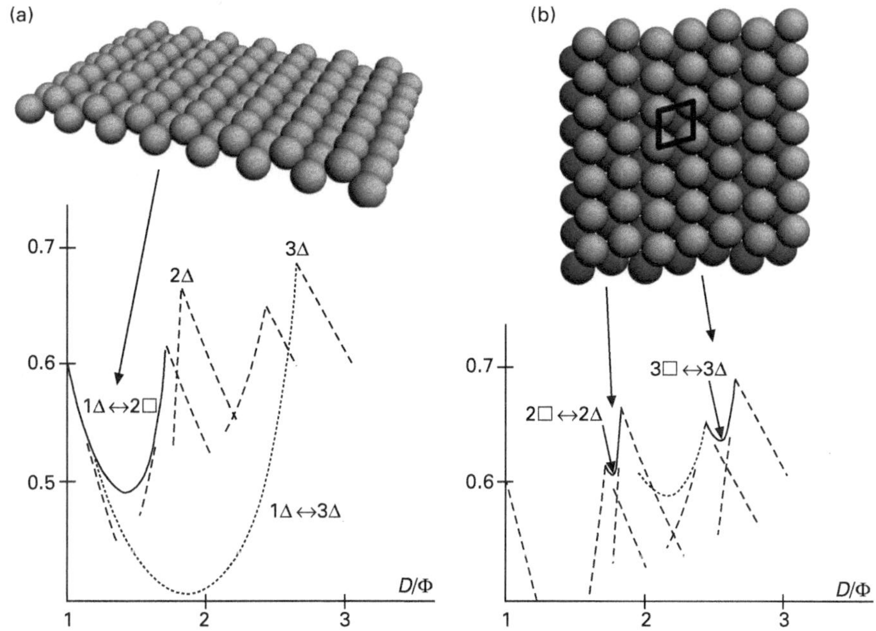

7.6 Filling fraction calculus vs normalized thickness (D/Φ) where Pieranski *et al.* made the proposal of both the buckling phase (a) and the rhombic phase (b). Graph is taken from Pansu *et al.* (1984) with permission of the authors and Édition Diffusion Presse Sciences (EDPS).

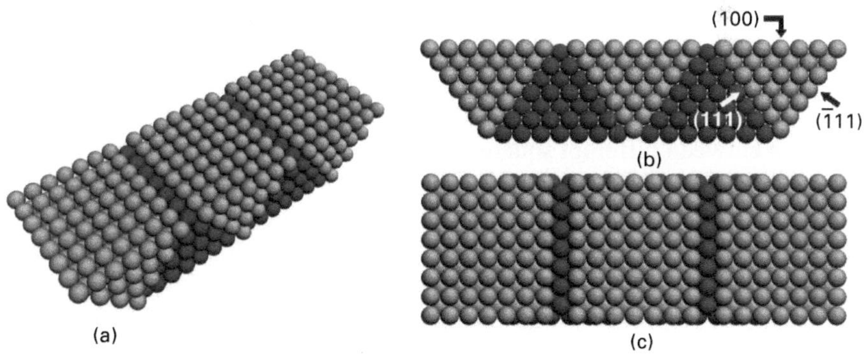

7.7 Models of the prismatic facet: (a) perspective view, (b) side view, and (c) top view. The prism side directions have been indicated, being (111) and ($\bar{1}$11) triangular, and (100) square planes.

both models of the *P* phase as well as SEM images of this new particle ordering. The *P* phase is built up by buckling the □ phase into prisms, then forming the *P* phase (see Fig. 7.7). The free surface facet (in contact with the confining walls) of the prisms is oriented along the FCC(100) direction

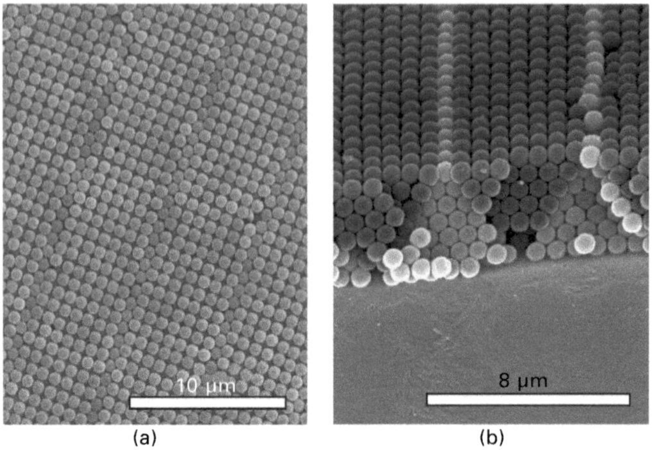

7.8 SEM images of the top (a) and cleft edges (b) of 7*P* phase. The spheres appear in shades of grey in (b) to facilitate the phase identification. From Ramiro-Manzano *et al.* (2006b) with permission of the Americal Physical Society (APS).

(square arranged particles) and the remaining prism sides correspond to the FCC(111) direction (triangular arranged particles). The prisms are surrounded by complementary inverted prisms, forming a thin film colloidal crystal with maximum filling fraction. The triangular facets of the prisms are gliding planes that allow the square phase to be transformed into the *P* phase. Figure 7.8 shows both SEM images of the top (a) and cleft edges (b) of a prismatic layering constituted of seven monolayers. It is worth stressing the top view gives information about the thickness value of the colloidal thin film, since the number of rows in the square facet (free surface facet of the prisms) is the same as the number of monolayers. This phase could explain the transition between different square facets but not between the square and the triangular particle orderings. The prismatic phase can easily be adapted to the continuous change of the gap in the wedge type cells. When cell thickness increases (this is more visible for a number of layers $n>6$), the *P* phase could be adapted itself to the cell width by introducing additional triangular planes and, simultaneously, the prisms reduce in size, as can be seen in both the model as well as the SEM images in Fig. 7.9. Finally these new triangular planes could form the HCP(001) and macled vicinal (explained in the next section). All these particle rearrangements produce a quasi-monotonous increase in the colloidal thin film thickness. We have shown the *P* phase of the colloidal system can easily be built from the FCC ordering through the introduction of triangular gliding planes. The triangular gliding plane plays a very important role in colloidal ordering as well as in solid state physics. The relative locations of the triangular gliding plane can be seen as the main

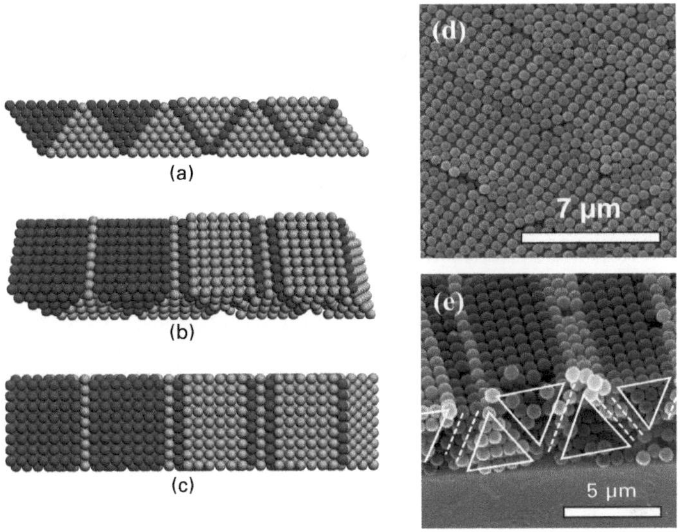

7.9 Model of how the prismatic phase adapts to the continuous increasing value of the distance between confining plates: (a) side view, (b) perspective view, and (c) top view. Different shades of grey show either prism or additional triangular planes. SEM images of adapted 7*P* phase (d) top view, and (e) cleft edge. Triangles and dashed lines highlight the prisms and the insertion of the additional triangular planes. The spheres are highlighted in (e) to facilitate comparison with the proposed models. From Ramiro-Manzano *et al.* (2006b) with permission of the Americal Physical Society (APS).

difference between the FCC and HCP symmetries. As we show later, the gliding plane transition is a general adaptation mechanism which generates rich compact orderings and enables smooth transition among them in order to adapt the ordering to the different commensurability scenarios.

As mentioned before, the bulk energy difference between the FCC and the HCP systems is very small (Woodcock, 1997). Therefore, the colloidal crystals would not pay a high price, in terms of energy, when introducing triangular gliding planes for maximizing filling fraction values in confined systems; i.e. the concept of triangular gliding planes can be extended beyond the prismatic phase described so far. In 2006 we introduced the concept of vicinal facets with triangular slip planes (Ramiro-Manzano *et al.*, 2006b), which we call here macled vicinal (MV) phase to account for the experimental results reported. We also show that under certain circumstances MV phases can lead to the HCP phase. Several authors (Fontecha *et al.*, 2005; Fortini and Dijkstra, 2006; Oguz *et al.*, 2009) have claimed a new prismatic phase P_{Δ} composed from prisms whose free surface facet is a FCC(111) plane and the other two faces (in contact with the neighbouring inverted prisms) are FCC(111) and FCC(100) facets. The P_{Δ} are more favourable when colloidal particles are charged (Oguz *et al.*, 2009). Figure 7.10 shows a model of both the triangular and the square prismatic phase.

Macled vicinal and HCP phases

The transition between different particle orderings ought to be produced with a minimum waste of energy, i.e. by introducing stacking faults in the FCC ordering. When the distance between confining plates is not commensurate with a FCC lattice, orientated along high symmetry (111) or (100) directions, particles order themselves on their vicinal structures. A FCC vicinal facet (Tanaka and Sakaki, 1989) corresponds to a crystal facet near a high symmetry direction, like (100) and (111). FCC vicinal facets consist of crystals with facets ordered in terraces of triangular or square arrangements. A vicinal facet is similar to a stair where the top surface of each step is formed by a high symmetry plane (111) or (100). In particular, the FCC(111), due to its symmetry, allows different types of vicinal orderings to be formed from the same (111) plane. Continuing with the analogy of a stair example, in this case, Fig. 7.11(a) and (b) show FCC(111) and FCC(100) ordering, highlighting with white and black lines the different vicinal planes with the same terrace size (four rows). Figures 7.12(a) and 7.13(a) show two different FCC(111) vicinal orderings and Fig. 7.14(a) shows the FCC(100) vicinal ordering.

The vicinal phase, however, does not fulfil the maximum filling fraction condition to form a stable phase in a confined geometry. In the vicinal facet we can select microcrystallites (highlighted in various shades of grey in Figs

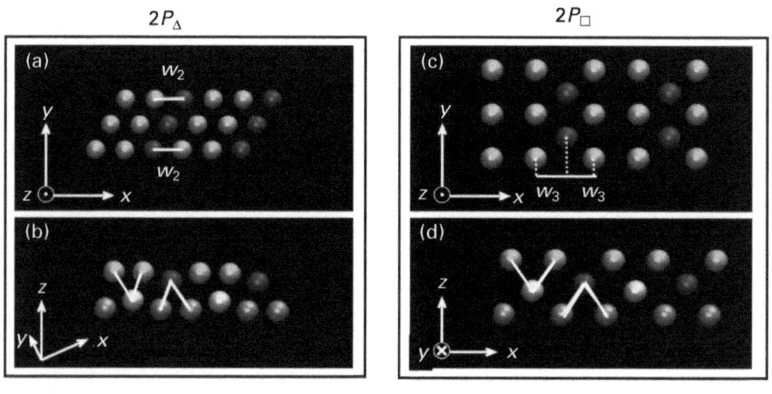

7.10 Models of prismatic phases in a charged confined system, P_Δ top view (a) and perspective view (b), and $P_\square$ top view (c) and side view (d). w_2 and w_3 are distance between prisms. From Oguz *et al.* (2009) with permission of the authors and the Institute of Physics (IOP).

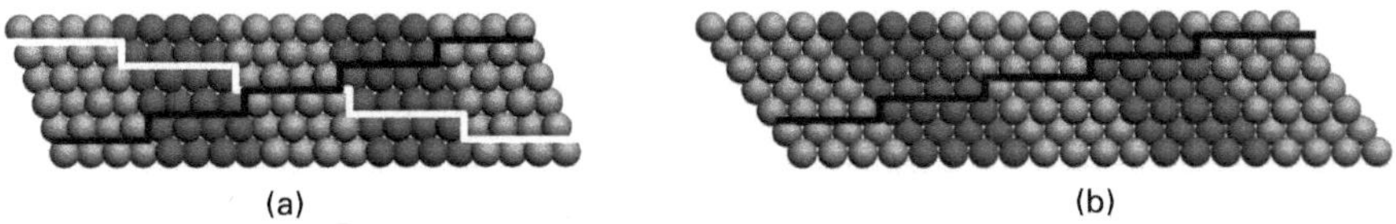

7.11 Models of the construction of vicinal arrangements: side views of (a) FCC(111) and (b) FCC(100). Black (a,b) and white (a) lines highlight two different vicinal planes of four row terraces.

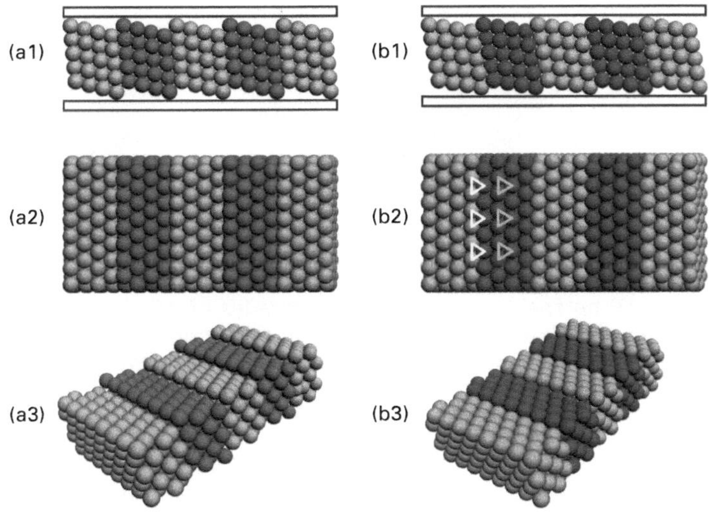

7.12 Models of (b) MVΔ_t and (a) the vicinal FCC(111) they derive from. Index numbers indicate (1) side view, (2) top view and (3) perspective view. Different shades of grey show FCC microcrystallites. Interfaces between microcrystallites in (b) indicate stacking fault plane position. Grey and white triangles highlight terrace symmetry and triangular microcrystallite connections at the terrace interface, respectively.

7.12(a)–14(a) where the interfaces between them correspond to triangular planes. The models in parts (a) of Figs 7.12–7.14 show FCC crystals oriented along vicinal directions. However, the system has the ability to produce plane dislocations in the triangular ABC stacking, i.e. by introducing stacking faults to refill the corners in the vicinal configuration (Figs 7.12–7.14 right rows). Therefore one ABC stack changes to ABA stack maximizing the particle filling fraction and forming the macled vicinal facet along the (111) direction in Figs 7.12(b) and 7.13(b) and along the (100) direction in Fig. 7.14(b). Figures 7.12(b2) and 7.13(b2) show the facets corresponding to a periodic distribution of triangular terraces with triangular connections (MVΔ_t for short) and square conection (MVΔ_s for short), respectively. In addition, Fig. 7.14(b2) shows the MV composed of square terraces with triangular connections

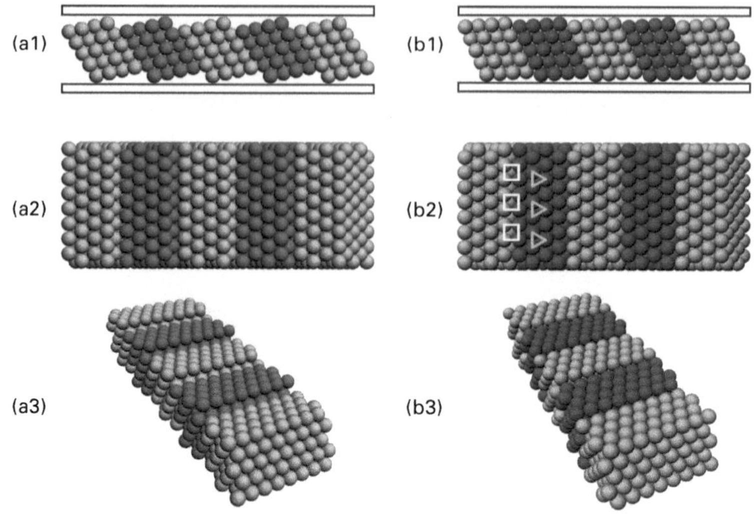

7.13 Models of (b) MVΔ_s and (a) the vicinal FCC(111) they derive from. Index numbers indicate (1) side view, (2) top view and (3) perspective view. Different shades of grey show FCC microcrystallites. Interfaces between microcrystallites in (b) indicate stacking fault plane position. Grey triangles and white squares highlight terrace symmetry and square microcrystallite connections at the terrace interface, respectively.

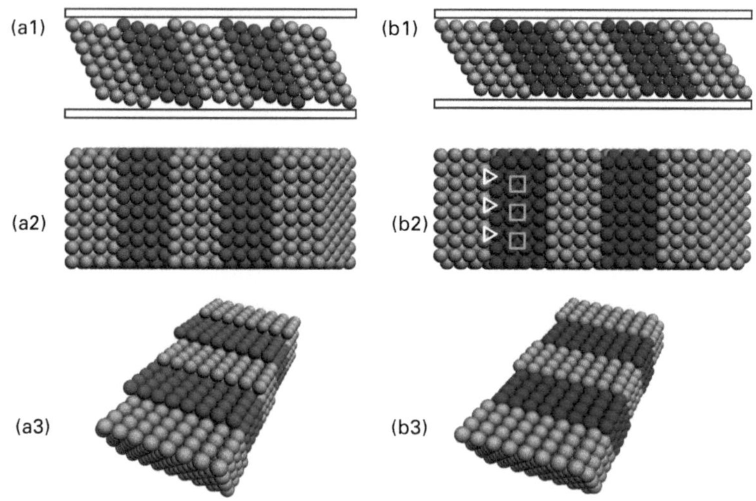

7.14 Models of (b) MV□ and (a) the vicinal FCC(100) they derive from. Index numbers indicate (1) side view, (2) top view and (3) perspective view. Different shades of grey show FCC microcrystallites. Interfaces between microcrystallites in (b) indicate stacking fault plane position. Grey squares and white triangles highlight terrace symmetry and triangular microcrystallite connections at the terrace interface, respectively.

(MV□ for short). Then, stacking faults are periodically distributed as has been highlighted by the grey to black colour change in Figs 7.12, 7.13 and 7.14. It is important to stress that (111) and (100) facets are not parallel to the confining walls, but they are slightly tilted, forming a sawtooth. Figure 7.15 shows SEM images of $MV\Delta_t$ (a-b), $MV\Delta_s$ (c) and MV□ (d) predicted by our model.

In the following we show how the $MV\Delta_t$ can connect the triangular and the HCP(100) phases (Ramiro-Manzano *et al.*, 2006b). As the distance between confining plates increases, the system should maximize the filling factor value. Then, microcrystallites get smaller because more stacking faults appear and they slightly rotate towards the upright position as shown in Fig. 7.16. Finally, when the terraces consist of two rows, the stacking fault planes are perpendicular to the confining plate and the system is formed by a periodic distribution of stacking faults every two vertical planes. It results in an ABABAB distribution of layers consistent with a HCP arrangement (see Fig. 7.16(d)). Figure 7.17 shows the SEM image of the HCP(100) arrangement predicted by our model. The free surface of this phase looks like a set of parallel strings of particles (Fig. 7.16(d)) whose distance between

7.15 SEM images of $MV\Delta_t$ (a) top view (b) cleft edge; $MV\Delta_s$ (c) top view and MV□ (d) top view. Black and white colour triangle and square symbols highlight terrace symmetry and microcrystallite connections, respectively. Image (b) is taken from Ramiro-Manzano *et al.* (2006b) with permission of the Americal Physical Society (APS).

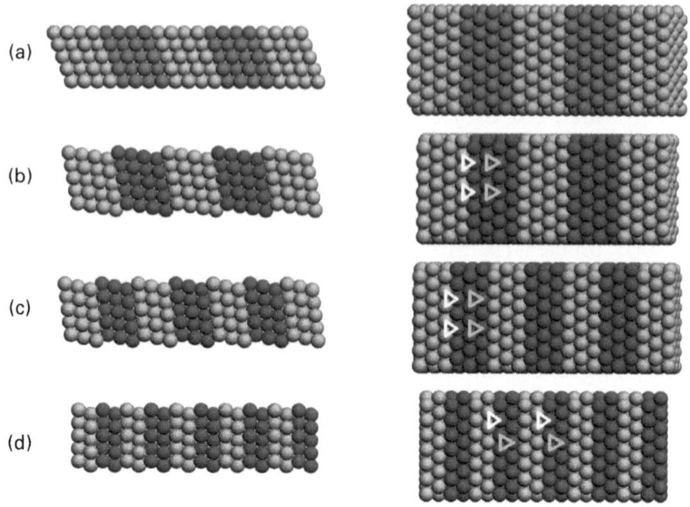

7.16 Model of the transition Δ (a) → $MV\Delta_t$ (b-c) → HCP(100) (d). White and grey colour triangles highlight terrace symmetry and microcrystallite connections, respectively.

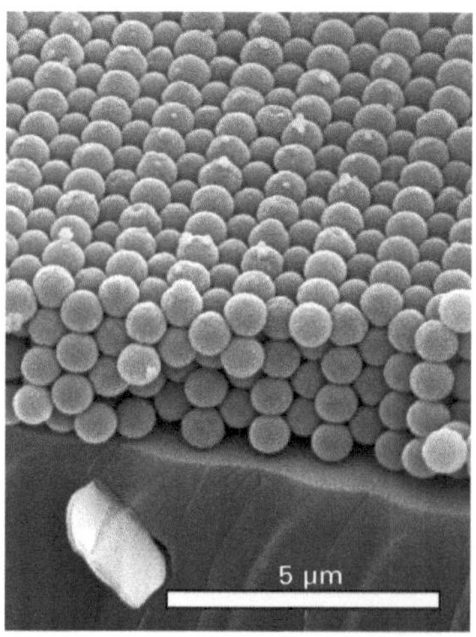

7.17 SEM image of 4HCP(100). From Ramiro-Manzano *et al.* (2007) with permission of the Americal Physical Society (APS).

them, δ, is $\delta = \sqrt{(8/3)}\Phi$. Optical spectra can give additional information about colloidal order that can be complementary to that obtained with SEM analysis. Figure 7.18 shows HCP(100) experimental spectrum (continuous line) and the theoretical calculation (dotted line) of nine monolayers with HCP symmetry oriented along the (100) direction (9 HCP(100) for short), 770

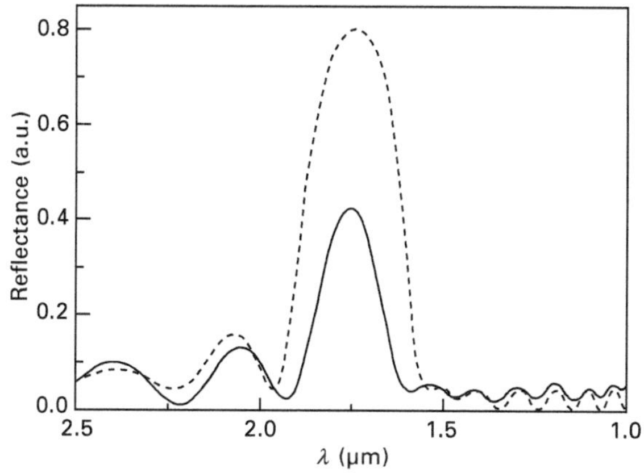

7.18 Experimental spectrum (continuous line) and theoretical calculation (dotted line) of the 9HCP(100) phase.

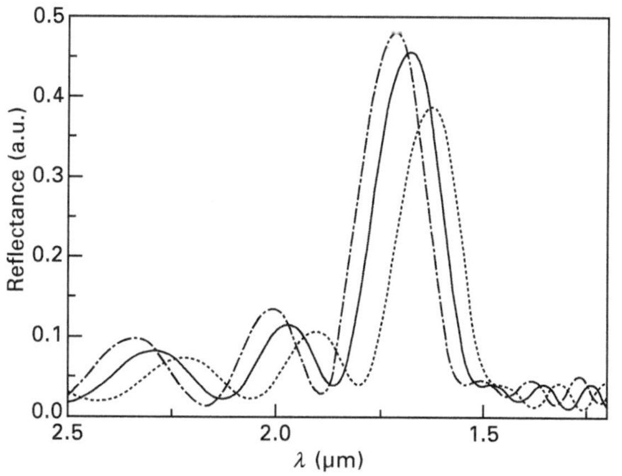

7.19 Optical spectra of 9Δ (dotted line), 9MVΔ_t (continuous line) and 9 HCP(100) (dashed line).

nm being the particle size. We have swept the light spot of the spectrometer through different regions of the colloidal crystal. Figure 7.19 shows the optical spectra of three different phases. The spectrum of nine monolayers of the MVΔ_t phase (9 MVΔ_t) shows a Bragg peak located at a wavelength value between the peak positions of the 9Δ and the 9HCP(100). As a consequence the MVΔ_t phase can be considered as an intermediate phase between the Δ and the HCP(100) particle orderings.

The MV□ and MVΔ_s arrangements can explain the transition between the FCC (100) (Fig. 7.20(a)) and the FCC(111) (Fig. 7.20(h)) through the phase HCP(011) (Fig. 7.20(d) and (e). This transition sequence is very similar to that shown in Fig. 7.16. Figure 7.20(a) shows the model of a 5□ colloidal thin

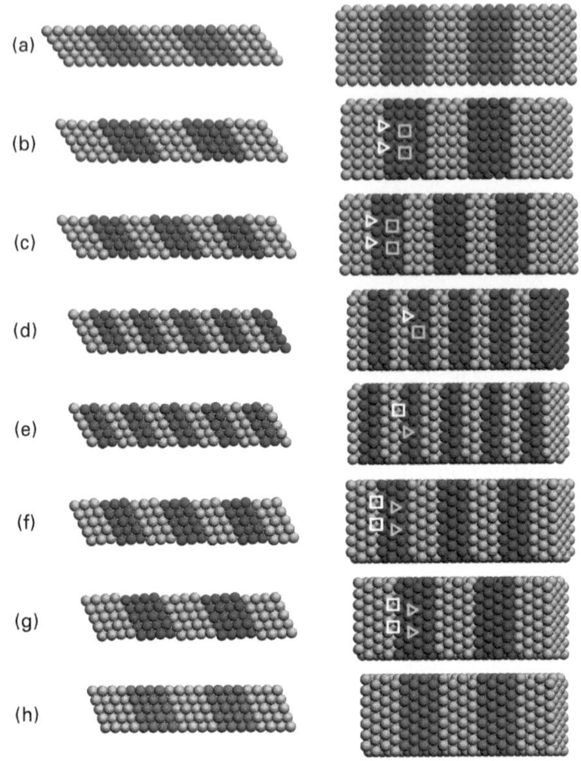

7.20 Model of the transition $n\square$ (a) → nMV$\square$ (b–c) → nHCP(011) (d–e) → nMVΔ_s (*f–g*) → Δ (h). White and grey colour symbols highlight terrace symmetry and microcrystallite connections, respectively.

film. When a periodic distribution of stacking faults is introduced it results in a colloidal arrangement like the model in Fig. 7.20(b), that corresponds to a vicinal facet MV□ with a periodic distribution of squared arranged micro-crystallites, four rows large, with triangular arranged strips in the joints between them. Figure 7.20(c) and (d) are similar to Fig. 7.20(b) but here the stacking distribution appears each three and two rows, respectively. The latter corresponds to a periodic distribution of particle strings where triangular and square orderings alternate (right panel of Fig. 7.20(d)). In fact, it corresponds to ABABAB (Fig. 7.20(d)) consistent with the HCP lattice oriented along the (011) direction (HCP(011) for short). Therefore, the HCP(011) could be considered as the stack (AB)(AB)(AB) of square facet microcrystallites (see shades of grey in Fig. 7.20(d)). This type of particle arrangement can be seen in the SEM images in Fig. 7.21.

The process of introducing periodic distribution of stacking faults can continue evolving the system from the HCP(011) phase to the Δ one. From the model of Fig. 7.20(d) one can select the appropriate slipping plane resulting

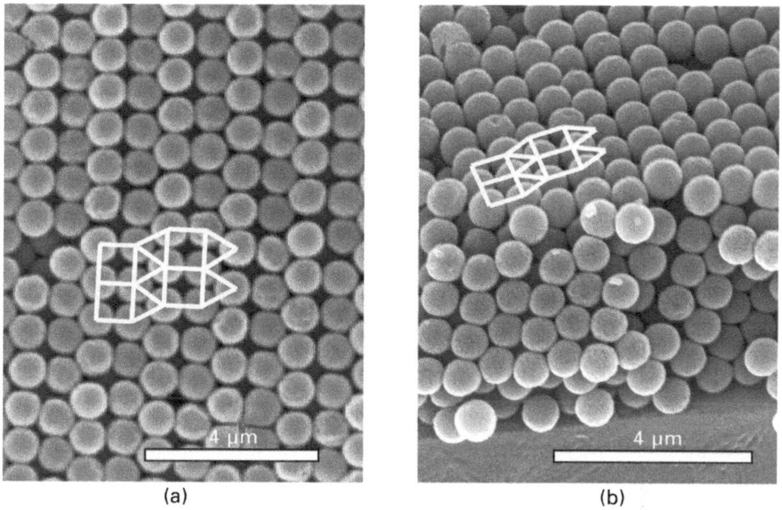

7.21 SEM images of HCP(011) (a) top view and (b) cleft edge. White colour squares and triangles highlight the surface symmetry. From Ramiro-Manzano *et al.* (2009) with permission of the Royal Society of Chemistry (RSC).

in Fig. 7.20(e) that corresponds to the MV composed by triangular terraces joined by square steps (MVΔ_s). The triangular facet growth leads the stacking faults to continue the rotation towards the upright position increasing the whole thickness (Fig. 7.20(f)–(g)). Finally, the process ends and the terrace fills all the space (the distance between stacking faults is infinite) creating the triangular facet (Fig. 7.20(h)). The MV phase can be considered as a group of particle ordering including the HCP for a two-row terrace case and the FCC facet for the case of infinite length terrace. Consequently, the MV phase opens a new phase sequence scenario connecting the following particle orderings:

$$n\square \rightarrow n\text{MV}\square \rightarrow n\text{HCP}(011) \rightarrow n\text{MV}\Delta_s \rightarrow n\Delta \rightarrow n\text{MV}\Delta_t \rightarrow n\text{HCP}(100) \qquad [7.3]$$

When this transition takes place by MV composed of only two-row terrace size (that is HCP), the slipping plane evolution is clearly observable since it covers long areas. For this reason we will include this long transition in the ordering sequence as a new phase, the HCP-like.

The HCP-like and the pre-□ phases

The HCP-like (100) phase (HCPL for short) covers large regions of the sample because it adapts itself to the progressive changes of the cell thickness by a smooth change in the interlayer spacing. Moreover, we also show the buckling phase as a particular case of the HCPL ordering. Finally, we show a very unstable colloidal arrangement, the pre-□ phase, to explain the transition $n\Delta \rightarrow (n+1)\square$.

The HCPL phase can be considered as a generalization of the HCP(100) particle arrangement where the distance between linear strings of spheres is well determined. However, this distance takes different values for the case of the HCPL phase. Figure 7.22 shows SEM images of different particle arrangements corresponding to (a) HCPL with a distance $\delta \approx 1.68\ \Phi$, (b) HCP(100) with a well-known distance $\delta = \sqrt{(8/3)}\Phi = 1.63\Phi$ and (c) HCPL with a distance $\delta \approx 1.52\Phi$. The HCPL phase has been observed in many ordering transitions $n\Delta \rightarrow (n+1)\square$. Here we discuss the transition $2\Delta \rightarrow 3\square$. This will help us to understand better the HCPL phase, and also it would allow us to introduce the pre-3□ phase. The analysis of their corresponding optical spectra, and the comparison to the SEM images (see Fig. 7.23), allows us to make an accurate assignment of the particle arrangement. Optical features in the reflectance also show gradual changes. Firstly, as the system transits between 2 and 3 monolayers a new oscillation period gradually appears. Secondly, the Bragg diffraction peak absent in the 2□ facet, clearly emerges in the 2Δ arrangement (at $\lambda = 1.7\ \mu$m). Then, for the 2HCPL phase, it shifts at longer wavelength values and smoothly fades out (at around λ = 2.2 μm) when the light spot approaches the 3□ ordering. Recently, Schöpe *et al.* (2006) have described two and three buckling phases in wet colloidal crystal thin films, that in fact correspond to HCPL phase we are describing here.

Figure 7.24 shows the calculated filling fraction in the $2\Delta \rightarrow 3\square$ transition. Here we show how the different facets evolve as the cell gap increases.

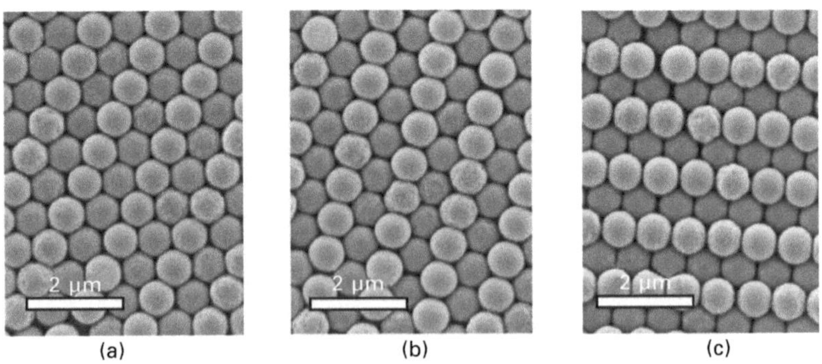

7.22 SEM images of different particle arrangements (a) HCPL $\delta \approx 1.68\Phi$, (b) HCP(100) $\delta = \sqrt{(8/3)}\ \Phi = 1.63\Phi$ and (c) HCPL $\delta \approx 1.52\Phi$, δ being the distance between linear strings of spheres and Φ the sphere diameter. From Ramiro-Manzano *et al.* (2007) with permission of the Americal Physical Society (APS).

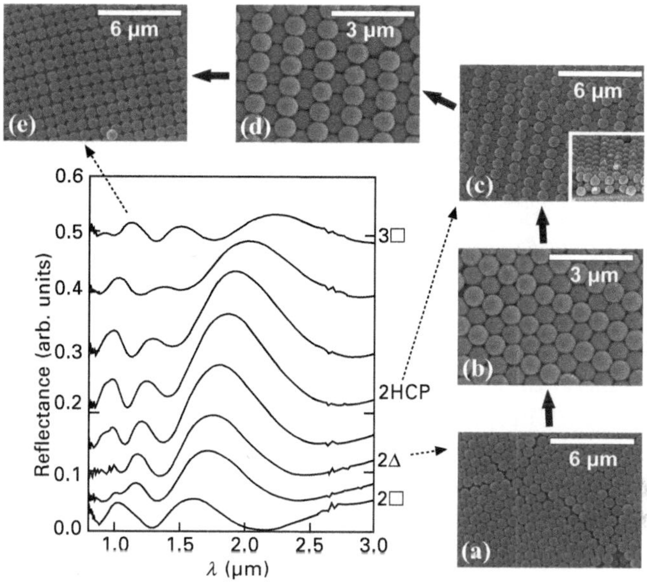

7.23 Reflectance optical spectra obtained over a path in the sample for the transition between 2Δ and 3□, being the sphere diameter 770 nm. SEM images of the sequence of the different facets: (a) 2Δ, (b,d) 2HCPL, (c) HCP(100), and (d) 3□. The inset in (c) shows a cleft edge. From Ramiro-Manzano *et al.* (2007) with permission of the Americal Physical Society (APS).

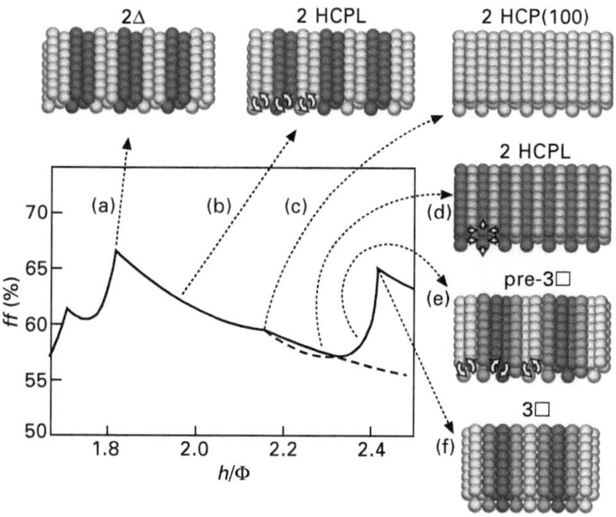

7.24 Filling fraction calculation between 2Δ and 3□ as a function of the reduced cell thickness *h*/Φ, Φ being the particle diameter. From Ramiro-Manzano *et al.* (2007) with permission of the Americal Physical Society (APS).

The local maximum at $h/\Phi = 1.82$ corresponds to the 2Δ phase. As the cell becomes thicker the system smoothly transits to the 2HCPL phase. The groups of microcrystallites artificially selected with different shades of grey can help in understanding the transitions between different particle arrangements. The 2Δ → 2HCPL(100) → 2HCP(100) transition takes place when all microcrystallites, as mention before, rotate around their longer axis (the [110] direction), thus forming, first, the 2HCPL phase (packing model (b) in Fig. 7.25) and then, the 2HCP(100) phase (packing model (c) in Fig. 7.25). As the filling fraction value also decreases monotonically, the 2HCPL is very stable, and, as expected, it can be found in large regions of the sample. An animated movie of the ordering evolution is shown in Ramiro-Manzano *et al.* (2007). The transition 2HCP(100) → 3□ continues through another 2HCPL phase (packing model (d) in Fig. 7.25). As the cell thickness increases, the upper (lowest) sublayer displaces itself vertically upwards (downwards) and the rows come closer together. Simultaneously, in order to maximize filling fraction, the rows from the second and the third

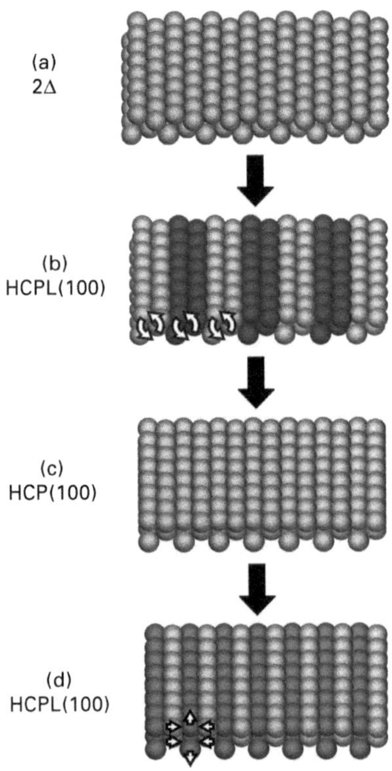

7.25 Model of the transition 2Δ (a) → 2HCPL(100) (b) → 2HCP(100) (c) → 2HCPL(100) (d). White arrows show microcrystallite rotations in (b) and sphere string movements in (d).

sublayer approach each other to form the 2HCPL phase. Therefore, packing models (b–d) in Fig. 7.25 correspond to SEM images (b–d) in Fig. 7.22.

The HCPL(100) phase has strong resemblances to the buckling arrangement (Pansu *et al.*, 1984). The buckling phase can be imagined as alternate single string rows that evolve for adapting to the gap cell of the confining plates. Figure 7.26 shows equivalent models of the evolution of the string rows adapting to the increasing cell thickness value. In Fig. 7.26(a) the groups composed of two sphere strings rotate to adapt themselves to the increasing values of the cell gap. However, in Fig. 7.26(b) the lower sublattice of parallel particle strings approach each other and the upper sublattice shifts upwards. Both models can account for the change in the particle ordering. The models of Fig. 7.26 also appear in the HCP(100) phase. In fact when the distance between the upper and lower rows of Fig. 7.26 is exactly $2\sqrt{(2/3)}\,\Phi$, the buckling phase can be considered as a monolayer of the HCP(100) ordering.

Figure 7.24 shows a new phase, the pre-3□ phase not discussed yet. This phase is a transitory ordering connecting 2HCP(100) and the 3□. Figure 7.27

7.26 The buckling phase as 1HCPL. Arrows in (a) and (b) show the HCPL transition mechanisms.

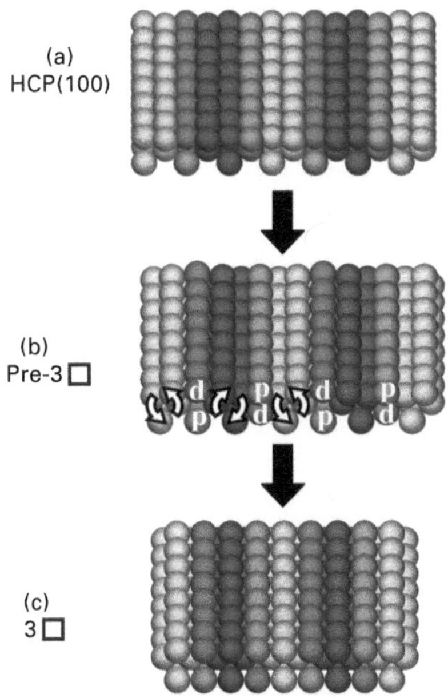

7.27 Model of the transition (a) 2HCP (100) → (b) pre-3□ → (c) 3□. White arrows show microcrystallite rotations.

shows a model of the transition 2HCP(100) → pre-3□ → 3□. The groups of microcrystallites artificially selected with different shades of grey can help in understanding the transitions between different particle arrangements. Light grey microcrystallites rotate anticlockwise, while dark grey ones rotate clockwise around the pivot row p. Simultaneously, the spheres rows d are displaced in the direction of the rotation axis, and they approach the confining plates to maximize the filling fraction value. This results in the alignment of all sphere rows in a square arrangement (Ramiro-Manzano *et al.*, 2007). Figure 7.28 shows SEM images of pre-3□ (a), pre-4□ (b–c) and pre-5□ (d) supporting this model.

In summary, we have reported on the different phases appearing in hard sphere systems. In the following we give the layering sequence in non-charged colloidal crystals confined in wedge type cells.

7.4 Phase transition sequence in colloidal crystal thin films

In the following we give our contribution to clarify the new phases as well as the transition appearing in the crystallization sequence between 1 and n ML

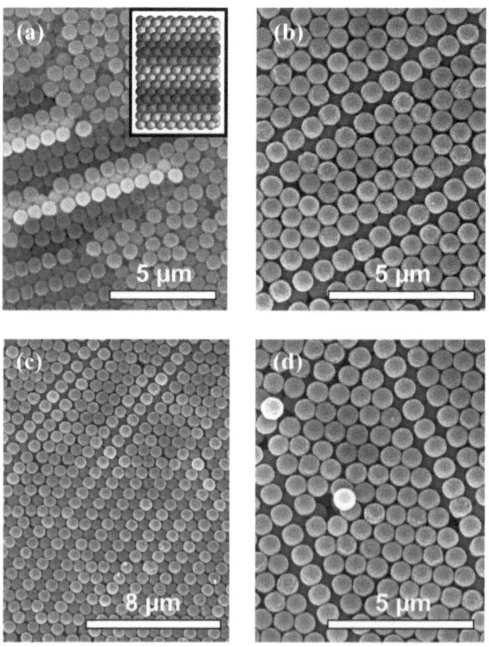

7.28 SEM images of pre-3□ (a), pre-4□ (b–c) and pre-5□ (d). Inset in (a) top view model of pre-3□. The spheres are in various shades of grey in (a) to facilitate comparison with the proposed model. Image (a) is taken from Ramiro-Manzano *et al.* (2007) with permission of the Americal Physical Society (APS).

with especial emphasis on the transition between 1 and 4 ML. We also give dynamic models that will show some of the transitions.

7.4.1 From one to four monolayers

Taking all the results reported so far by various research groups, it is possible to present an overview of the phases appearing in the transition between 1 and 4 ML. Figure 7.29 shows the calculation of the colloidal crystal filling fraction as a function of the normalized cell thickness value (h/Φ), with h and Φ being the cell thickness and the particle diameter, respectively (Ramiro-Manzano *et al.*, 2009). Figure 7.29 also shows SEM images of the different particle arrangements appearing in colloidal films between 1 and 4 ML. Continuous lines correspond to the most stable phases observed experimentally. We present SEM images of well-known phases, Δ (Pieranski *et al.*, 1983), □ (Pieranski *et al.*, 1983), *B* (Pansu *et al.*, 1984; Chou and Nelson, 1993; Schmidt and Löwen, 1996), *r* (Pansu *et al.*, 1984; Schmidt and Löwen, 1996; Messina and Löwen, 2003) and *P* (Neser *et al.*, 1997) corresponding to triangular, square, buckling, rhombic and prismatic particle arrangements, respectively. It is worth discussing the transition 4□ → 4Δ, where we show SEM images of both the 4*P* phase as well as the 4HCP(011) one. As we use dried samples, compact phases like the HCP(011) are more favoured than other colloidal arrangements. However the 4*r* phase is slightly present mixed with the HCP(011), and the HCP(011) extends over larger sample regions than the 4P arrangement. Consequently, the transition could be considered as an evolution of triangular gliding plane orderings 4□ → 4*P* → 4HCP(011) → 4Δ.

To conclude this section, the order evolution in the transition is the following:

$$1\Delta \to 1B \to 2\square \to 2r \to 2\Delta \to 2\mathrm{HCPL}(100) \to \text{pre-}3\square$$
$$\to 3\square \to 3r \to 3\Delta \to 3\mathrm{HCPL}(100) \to \text{pre-}4\square \to 4\square$$
$$\to 4P \to 4\mathrm{HCP}(011) \to 4\Delta$$

HCPL(100) includes HCP(100) because it is a particular case of HCPL, as detailed above.

7.4.2 From four to eight monolayers

The richest scenario of different colloidal arrangements corresponds to orderings between four and eight monolayers. The reason for this diversity is the inclusion of plenty of compact phases, i.e. the macled vicinals. Figure 7.15 shows SEM images of MVΔ_t (a), MV□ (b), and MVΔ_s (c and d) phases. Our experimental results confirm the existence of the pre-□ in thin colloidal

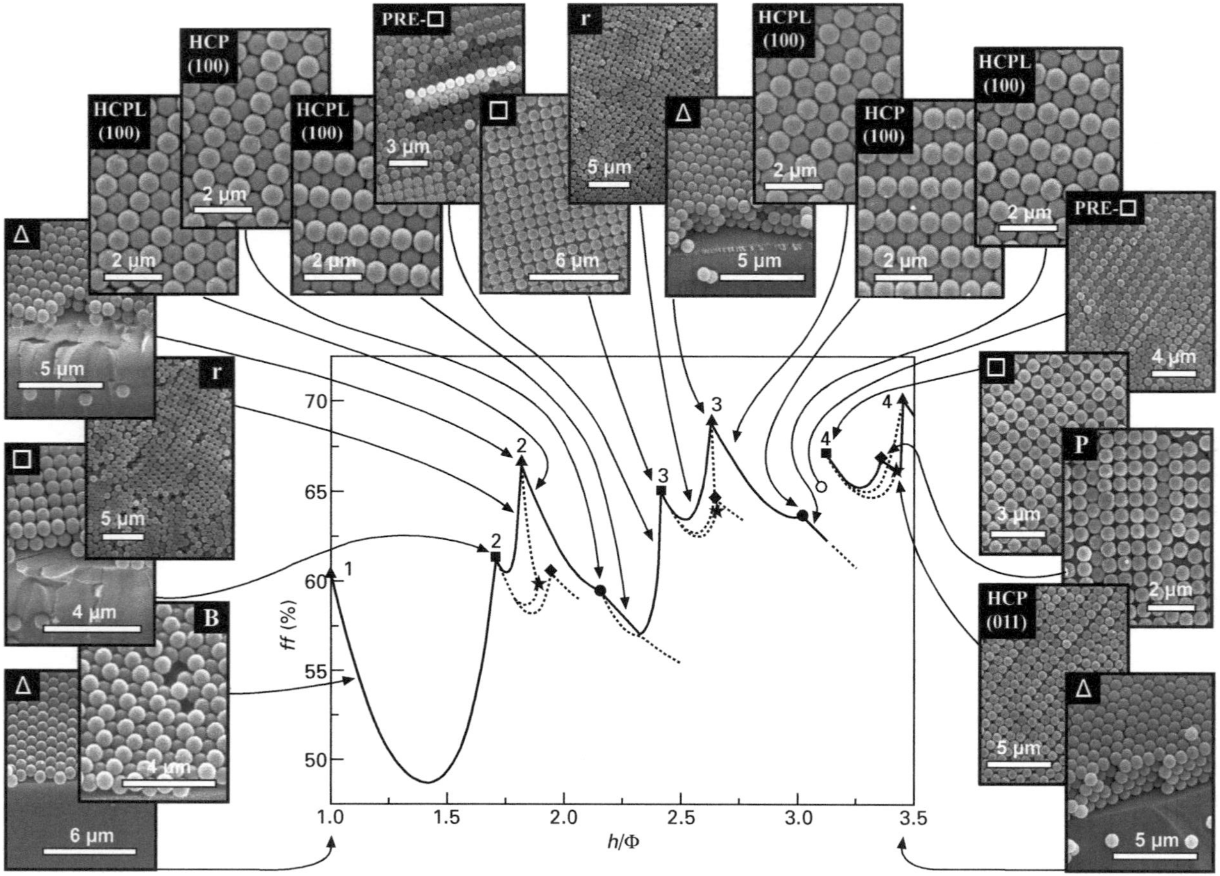

7.29 Calculation of the filling fraction between 1Δ and 4Δ as a function of the normalized cells thickness value (h/Φ), h and Φ being the cell thickness and the particle diameter, respectively. SEM images illustrate the transition sequence. From Ramiro-Manzano *et al.* (2009) with permission of the Royal Society of Chemistry (RSC).

crystal from 2 to 5 ML (i.e., pre-3□, pre-4□ and pre-5□ phases, see Fig. 7.28). However, we have not found them in thicker colloidal thin films. The experimental sequence would be the following:

$$n\Delta \rightarrow n\mathrm{MV}\Delta_t \rightarrow \text{pre-}(n+1)\square \rightarrow (n+1)\square \rightarrow (n+1)\ \mathrm{MV}\square + (n+1)\ P \rightarrow (n+1)\ \mathrm{MV}\Delta_s \rightarrow (n+1)\Delta$$

where the $\mathrm{MV}\Delta_t$ includes the HCP(100) and the $\mathrm{MV}\Delta_s$ and MV□ include the HCP (011) because the HCP is a particular case of macled vicinals.

7.4.3 Beyond eight monolayers

As the interlayer distance of the Δ facet is larger than that of the □ ordering, one can check the total thickness value of the 7Δ arrangement is roughly the same as the 8□ colloidal crystal. Therefore the Δ systems would be the most probable since they show the highest ff value. This argument holds for thicker thin films and it leads to the extinction of the □ facet and the rest of orderings connected to it (*r*, *P*, MV□). Only the MV□ can be observed with small terraces (2–3 sphere rows). Furthermore, all orderings with triangular terraces survive ($\mathrm{MV}\Delta_t$, $\mathrm{MV}\Delta_s$), especially the $\mathrm{MV}\Delta_t$ and the transition:

$$n\Delta \rightarrow n\mathrm{MV}\Delta_t \rightarrow n\mathrm{HCPL}(100) \rightarrow (n+1)\Delta$$

For large thickness, the 12Δ overpasses 12HCP (100) and therefore for thicker films Δ ordering dominates.

7.5 Conclusions and future trends

We have shown a detailed description of the crystalline phases appearing in thin film colloidal crystals when the cell thickness increases gradually. We have also shown the evolution dynamics from 1 ML to *n* ML in wedge type cells with hard sphere particles. The evolution dynamics have been exhaustively discussed and reported in the case of 1–4 ML. There are, however, many unknown aspects to be unveiled. One of them concerns the understanding of experimental findings through theoretical models (at least with hard sphere systems). Finally, the discovery of new particle orderings induced either in flat or patterned substrates could be of great interest in fundamental science and could enable new potential applications.

7.6 Acknowledgements

The authors would like to thank A. Moreno for providing wedge type cells and the Electronic Microscopy Service of the UPV for technical support. We would like to acknowledge the permission to include the figures of the following authors: Pa. Pieranski, Pi. Pieranski, P. Pansu, L. Strzelecki, E.

Oguz, R. Messina and H. Löwen. This work has been partially supported by the Spanish projects, Consolider CSD2007-046 and FIS2009-07812.

7.7 References

Binks B P (2006) *Colloidal Particles at Liquid Interfaces*, Cambridge, Cambridge University Press.

Blanco A, Chomski E, Grabtchak S, Ibisate M, John S, Leonard S W, Lopez C, Meseguer F, Miguez H, Mondia J P, Ozin G A, Toader O and Van Driel H M (2000) 'Large-scale synthesis of a silicon photonic crystal with a complete three-dimensional bandgap near 1.5 micrometres', *Nature*, 405, 437–440.

Braun P V and Wiltzius P (1999) 'Microporous materials – electrochemically grown photonic crystals', *Nature*, 402, 603–604.

Chou T and Nelson D R (1993) 'Buckling instabilities of a confined colloid crystal layer', *Phys. Rev. E*, 48, 4611–4621.

Derjaguin B V and Landau L (1941) 'Theory of the stability of strongly charged lyophobic sols and of the adhesion of strongly charged particles in solutions of electrolytes', *Acta Physicochim. URSS*, 14.

Duran J (1999) *Sands, Powders, and Grains: An Introduction to the Physics of Granular Materials*, New York, Springer.

Fenollosa R, Ibisate M, Rubio S, Lopez C, Meseguer F and Sanchez-Dehesa J (2003) 'Photonic crystal microprisms obtained by carving artificial opals', *J. Appl. Phys.*, 93, 671–674.

Fontecha A B, Schope H J, König H, Palberg T, Messina R and Löwen H (2005) 'A comparative study on the phase behaviour of highly charged colloidal spheres in a confining wedge geometry', *J. Phys.-Condes. Matter*, 17, S2779–S2786.

Fortini A and Dijkstra M (2006) 'Phase behaviour of hard spheres confined between parallel hard plates: manipulation of colloidal crystal structures by confinement', *J. Phys.-Condes. Matter*, 18, L371–L378.

Hiemenz P C and Rajagopalan R (1997) *Principles of Colloid and Surface Chemistry*, New York, Marcel Dekker.

Holgado M, Garcia-Santamaria F, Blanco A, Ibisate M, Cintas A, Miguez H, Serna C J, Molpeceres C, Requena J, Mifsud A, Meseguer F and Lopez C (1999) 'Electrophoretic deposition to control artificial opal growth', *Langmuir*, 15, 4701–4704.

Joannopoulos J D, Meade R D and Winn J N (1995) *Photonic Crystals. Molding the Flow of Light*, Princeton, NJ, Princeton University Press.

Joannopoulos J D, Villeneuve P R and Fan S H (1997) 'Photonic crystals: putting a new twist on light', *Nature*, 386, 143–149.

John S (1987) 'Strong localization of photons in certain disordered dielectric superlattices', *Phys. Rev. Lett.*, 58, 2486–2489.

Kitaev V and Ozin G A (2003) 'Self-assembled surface patterns of binary colloidal crystals', *Adv. Mater.*, 15, 75–78.

Kittel C (1996) *Introduction to Solid State Physics*, New York, John Wiley and Sons.

Knobler C M and Schwartz D K (1999) 'Langmuir and self-assembled monolayers', *Curr. Opin. Colloid Interface Sci.*, 4, 46–51.

López C, Vazquez L, Meseguer F, Mayoral R, Ocana M and Míguez H (1997) 'Photonic crystal made by close packing SiO_2 submicron spheres', *Superlattices Microstruct.*, 22, 399–404.

Lu Y, Yin Y D, Gates B and Xia Y N (2001) 'Growth of large crystals of monodispersed spherical colloids in fluidic cells fabricated using non-photolithographic methods', *Langmuir*, 17, 6344–6350.

Messina R and Löwen H (2003) 'Reentrant transitions in colloidal or dusty plasma bilayers', *Phys. Rev. Lett.*, 91, 146101.

Míguez H, Meseguer F, López C, Mifsud A, Moya J S and Vazquez L (1997) 'Evidence of FCC crystallization of SiO_2 nanospheres', *Langmuir*, 13, 6009–6011.

Míguez H, Yang S M and Ozin G A (2002) 'Colloidal photonic crystal microchannel array with periodically modulated thickness', *Appl. Phys. Lett.*, 81, 2493–2495.

Míguez H, Yang S M and Ozin G A (2003) 'Optical properties of colloidal photonic crystals confined in rectangular microchannels', *Langmuir*, 19, 3479–3485.

Mihi A, Ocana M and Míguez H (2006) 'Oriented colloidal-crystal thin films by spin-coating microspheres dispersed in volatile media', *Adv. Mater.*, 18, 2244–2249.

Mittleman D M, Bertone J F, Jiang P, Hwang K S and Colvin V L (1999) 'Optical properties of planar colloidal crystals: dynamical diffraction and the scalar wave approximation', *Journal of Chemical Physics*, 111, 345–354.

Neser S, Bechinger C, Leiderer P and Palberg T (1997) 'Finite-size effects on the closest packing of hard spheres', *Phys. Rev. Lett.*, 79, 2348–2351.

Ochiai T and Sanchez-Dehesa J (2001) 'Superprism effect in opal-based photonic crystals', *Phys. Rev. B*, 64, 245113.

Oguz E C, Messina R and Löwen H (2009) 'Crystalline multilayers of the confined yukawa system', *Epl-Europhys Lett*, 86, 28002.

Pan F, Zhang J Y, Cai C and Wang T M (2006) 'Rapid fabrication of large-area colloidal crystal monolayers by a vortical surface method', *Langmuir*, 22, 7101–7104.

Pansu B, Pieranski P and Pieranski P (1984) 'Structures of thin-layers of hard-spheres – high-pressure limit', *Journal De Physique*, 45, 331–339.

Park S H, Qin D and Xia Y (1998) 'Crystallization of mesoscale particles over large areas', *Adv. Mater.*, 10, 1028–1032.

Pieranski P (1980) 'Two-dimensional interfacial colloidal crystals', *Phys. Rev. Lett.*, 45, 569–572.

Pieranski P and Finney J (1979) 'A hard-disk system: structures of a close-packed thin layer', *Acta Crystallographica*, A35(1), 194–196.

Pieranski P, Strzelecki L and Pansu B (1983) 'Thin colloidal crystals', *Phys. Rev. Lett.*, 50, 900–903.

Prasad T, Colvin V and Mittleman D (2003) 'Superprism phenomenon in three-dimensional macroporous polymer photonic crystals', *Phys. Rev. B*, 67, 165103.

Prevo B G and Velev O D (2004) 'Controlled, rapid deposition of structured coatings from micro- and nanoparticle suspensions', *Langmuir*, 20, 2099–2107.

Pusey P N, Vanmegen W, Bartlett P, Ackerson B J, Rarity J G and Underwood S M (1989) 'Structure of crystals of hard colloidal spheres', *Phys. Rev. Lett.*, 63, 2753–2756.

Ramiro-Manzano F, Atienzar P, Rodriguez I, Meseguer F, Garcia H and Corma A (2006a) 'Apollony photonic sponge based photoelectrochemical solar cells', *Chem. Commun.*, 242–244.

Ramiro-Manzano F, Meseguer F, Bonet E and Rodriguez I (2006b) 'Faceting and commensurability in crystal structures of colloidal thin films', *Phys. Rev. Lett.*, 97, 028304.

Ramiro-Manzano F, Bonet E, Rodriguez I and Meseguer F (2007) Layering transitions between 2 triangular and 3 square. Available from: http://www.metamaterials.ctfama.org/movies/2T3S.mpg

Ramiro-Manzano F, Bonet E, Rodriguez I and Meseguer F (2009) 'Layering transitions in colloidal crystal thin films between 1 and 4 monolayers', *Soft Matter*, 5, 4279–4282.

Sanders J V (1964) 'Colour of precious opal', *Nature*, 204, 1151–1153.

Sands D (1975) *Introduction to Crystallography*, New York, Dover Publications.

Schmidt M and Löwen H (1996) 'Freezing between two and three dimensions', *Phys. Rev. Lett.*, 76, 4552–4555.

Schmidt M and Löwen H (1997) 'Phase diagram of hard spheres confined between two parallel plates', *Phys. Rev. E*, 55, 7228–7241.

Schöpe H J, Fontecha A B, König H, Hueso J M and Biehl R (2006) 'Fast microscopic method for large scale determination of structure, morphology, and quality of thin colloidal crystals', *Langmuir*, 22, 1828–1838.

Tanaka M and Sakaki H (1989) 'Formation of planar superlattice states in new grid-inserted quantum well structures', *Appl. Phys. Lett.*, 54, 1326–1328.

Vainshtein B K (1981) *Modern Crystallography*, Berlin, Springer.

Van Blaaderen A, Ruel R and Wiltzius P (1997) 'Template-directed colloidal crystallization', *Nature*, 385, 321–324.

Van Duffel B, Ras R H A, De Schryver F C and Schoonheydt R A (2001) 'Langmuir–Blodgett deposition and optical diffraction of two-dimensional opal', *J. Mater. Chem.*, 11, 3333–3336.

Van Winkle D H and Murray C A (1986) 'Layering transitions in colloidal crystals as observed by diffraction and direct-lattice imaging', *Phys. Rev. A*, 34, 562–573.

Van Winkle D H and Murray C A (1988) 'Layering in colloidal fluids near a smooth repulsive wall', *J. Chem. Phys.*, 89, 3885–3891.

Velikov K P, Christova C G, Dullens R P A and Van Blaaderen A (2002) 'Layer-by-layer growth of binary colloidal crystals', *Science*, 296, 106–109.

Verwey E J W (1947) 'Theory of the stability of lyophobic colloids', *J Phys Colloid Chem*, 51, 631–636.

Vlasov Y A, Bo X Z, Sturm J C and Norris D J (2001) 'On-chip natural assembly of silicon photonic bandgap crystals', *Nature*, 414, 289–293.

Wolert E, Setz S M, Underhill R S, Duran R S, Schappacher M, Deffieux A, Holderle M and Mulhaupt R (2001) 'Meso- and microscopic behavior of spherical polymer particles assembling at the air–water interface', *Langmuir*, 17, 5671–5677.

Woodcock L V (1997) 'Entropy difference between the face-centred cubic and hexagonal close-packed crystal structures', *Nature*, 385, 141–143.

Yablonovitch E (1987) 'Inhibited spontaneous emission in solid-state physics and electronics', *Phys. Rev. Lett.*, 58, 2059–2062.

Yin Y D, Lu Y, Gates B and Xia Y N (2001) 'Template-assisted self-assembly: a practical route to complex aggregates of monodispersed colloids with well-defined sizes, shapes, and structures', *J. Am. Chem. Soc.*, 123, 8718–8729.

8
Thin film growth for thermally unstable noble-metal nitrides by reactive magnetron sputtering

Z. CAO, Chinese Academy of Sciences, P. R. China

Abstract: Thin film growth for thermally unstable noble-metal nitrides by reactive magnetron sputtering is exemplified by the deposition of stoichiometric Cu_3N and ternary Cu_3NPd_x structures, for which a proper combination of applied voltage and working gas pressure is critical since the nitriding reaction occurs at the target surface. Nitrogen re-emission from deposits both influences the growth process and leads to protruding features via an orogenic motion mechanism of the nitride nanocrystals. Doping the cubic Cu_3N lattice gives rise to narrow band semiconductors which may exhibit various intriguing properties – in $Cu_3NPd_{0.238}$ a constant electrical resistivity has been measured in a temperature range over 200 K. The results summarized here are instructive for the deposition of other noble-metal nitrides.

Key words: noble-metal nitride, Cu_3N, reactive magnetron sputtering, stoichiometry, morphology, intercalation, narrow band semiconductors.

8.1 Introduction

8.1.1 Reactive magnetron sputtering

Magnetron sputtering is an effective physical vapor deposition method which has been widely employed in the enterprise of thin film growth. This technique is based on the sputtering of a solid target by energetic ionic species from a magnetically enhanced glow discharge. One distinct feature of the magnetron discharge is the application of a magnetic field, often with a field strength in the range of 10^2–10^3 Gauss, which serves to confine the secondary electrons in a region near the cathode, thus the closed electron drift in this small region can lead to a relatively high plasma density. Under the condition of limited electron mobility across the magnetic field, each magnetic field line is also a nearly equi-potential line of the electric field. The electric potential changes forcibly across the magnetic field, hence a strong electric field develops along the direction perpendicular to the magnetic field, accelerating the positively charged species towards the target (Baranov *et al.* 2010). The ion sputtering on the target surface plays the role of vaporizing the source materials – the sputtered atomic or molecular species generally have a kinetic energy of a few electron volts (depending on the plasma condition and the nature of the target), and they fly towards the substrate where film growth is taking place.

When a reactive gas is added into the working gas, this then turns into the so-called reactive magnetron sputtering which has been widely applied to obtain certain compound films. However, reactive magnetron sputtering as a film deposition method is a quite vague concept, since when a reactive gas is added, depending on the activity of the reactive gas and the conditions of the plasma, the reaction can occur simultaneously on the substrate, on route from the target to the substrate, and even on the target surface. This is to say that reactive magnetron sputtering is characterized by its complexity, that the sputter process in the target surface, the target process in the selvage layer, the collisional transport of the sputtered species through the gas phase and all the substrate processes are all strongly interrelated, being able to influence some specific aspects of the deposition results (Berg and Nyberg 2005, Musil *et al.* 2005).

Reactive magnetron sputtering has been very successfully applied to the growth of various oxides, with the targets being either pure metals or bulk oxides. Oxygen is a very active element that is ready to combine with most metals, vaporized or in bulk, to form oxides even in conventional atmospheric conditions. In the magnetron discharge, the oxygen can be easily ionized and then accelerated towards the target to initiate sputtering. The sputtered metal atoms and clusters have already incorporated some oxygen atoms. Although the sputtered clusters may show deficiency of oxygen with respect to the anticipated compound stoichiometry, more oxygen will be captured on the path of flight by the clusters and on the growing surface, bearing in mind that often a considerably high substrate temperature is applied, which effectively enhances the reaction of oxygen with the deposits. In fact, even a neutral atom beam of oxygen suffices in the molecular beam epitaxy or ablation growth of some oxide films. A great number of reports on the growth of thin films for various oxides are available in literature (Chambers 2010).

Unlike oxygen, use of nitrogen in reactive magnetron sputtering is less effective based on the following facts: nitrogen gas has a higher first ionization energy and a higher bonding dissociation energy than oxygen (1402.3 kJ mol^{-1} versus 1313.9 kJ mol^{-1} for the former, and 945 kJ mol^{-1} versus 495 kJ mol^{-1} for the latter), and nitrogen has a smaller dissociation energy from most metal surfaces, or a smaller bonding energy with that metal, than oxygen. Due to the large bonding dissociation energy of a nitrogen molecule, the nitrogen gas is quite 'inert' – the nitrogen molecules in the growth chamber have rarely any chance to incorporate into the deposits via reaction through collisional transport or adsorption on to the growing surface. Furthermore, this large bonding dissociation energy leads to a large fraction of molecular ions in the discharge, which is obviously an unfavorable condition for the deposition of nitride thin films. It is even a detrimental factor in the case of chemical vapor deposition of nitrides by using other kinds of plasma discharges (Cao 2001, 2002, Cao and Oechsner 2004). The smaller bonding energy with metals and the lower energy barrier for dissociation also imply the re-emission of nitrogen atoms or molecules from the deposits, a frustrating factor that is generally absent in the preparation of oxide films. In the case of preparing nitride films for noble metals such as Cu, the inertness of nitrogen ions plus the nobleness of the metals (Norskov and Hammer 1995) make the deposition of thin films of noble-metal nitrides, such as that of the cubic copper nitride, a particular challenge (Hojabri *et al.* 2010).

8.1.2 Copper nitride (Cu_3N)

Copper nitride (Cu_3N) is a binary compound in the anti-ReO_3 structure which has a cubic unit cell of a lattice constant $a = 0.382$ nm, with 12 Cu atoms occupying the middle points of the edges and eight N atoms sitting at the vertices (Fig. 8.1) (Moreno-Armenta *et al.* 2004). The first synthesis of polycrystalline powder-like Cu_3N was reported by Juza and Hahn in 1939. Cu_3N is an indirect band-gap semiconductor with a high electrical resistivity at room temperature, and it begins readily to decompose at slightly elevated temperatures, leaving behind highly conductive structures of Cu when the decomposed regions are deliberately patterned. That Cu_3N has drawn much research effort towards the growth of high-quality thin films and the investigation of its physical properties from various aspects is simply due to its decomposition without any complications or detrimental consequences,

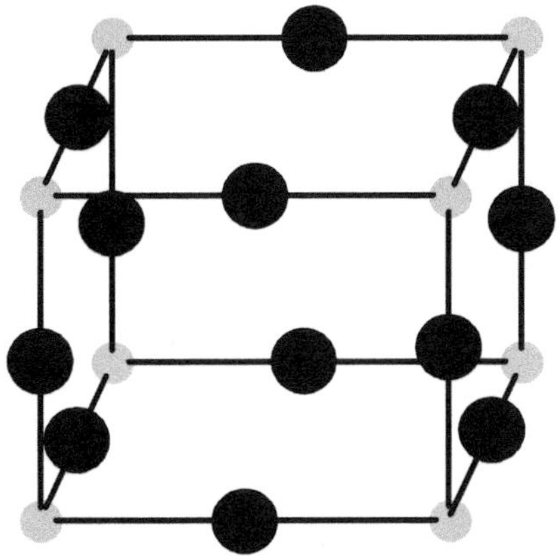

8.1 Schematic illustration of a unit cell of cubic Cu_3N: 8 Cu atoms sit at the vertices while 12 N atoms occupy the middle points of the edges.

which allows the fabrication of microscopic metallic links with maskless laser or electron beam writing (Nosaka *et al.* 2001, Maruyama and Morishita 1996). It can also be used as a sacrifice layer for the fabrication of free-standing layers of other functional materials, and in write-once optical recording. A density over Mb/cm^2 in a 1 × 1 μm dot array has been realized on Cu_3N thin film using electron beam irradiation, and the write rate can reach 3.3 Mb/s (Cremer *et al.* 2000).

The cubic Cu_3N lattice is a rather open structure, noting that the center of the unit cell is empty; hence it is expected to be a good host material. In fact, there are both theoretical and experimental investigations into the metallic Cu_4N variant where one extra Cu atom resides at the center of the unit cell (Moreno-Armenta *et al.* 2004). There is also report of the synthesis of ternary Cu_3NPd_x, where the Pd atoms are believed to occupy some of the unit cell centers (Hahn and Weber 1996). Remarkably, the interposition of another metal atom at some cell centers of the Cu_3N lattice will result in semimetals or totally metallic compounds due to the modification of the energy bands. This may provide an exemplar material system for the study of semiconducting-to-metallic transition via doping.

Due to the thermal instability of Cu_3N, the physical properties for this material reported in literature are quite inconsistent, sometimes even controversial. The temperature for the initiation of compound decomposition is believed to lie in the range from 100°C to 470°C according to different authors (Asano *et al.*, 1990, Maruyama and Morishita 1996, Wang *et al.* 1998, Nosaka *et al.* 2001). As to the band-gap of the semiconducting Cu_3N, experimental values generally fall within the range of 0.8–1.9 eV (Kim *et al.*, 2001, Borsa *et al.*, 2002), but a value as small as 0.23 eV was also declared (Hahn and Weber 1996, Ma *et al.*, 2004). The large discrepancy in the aforementioned data arises obviously from the unstable nature of copper nitride – in the growth stage the thermal instability results in Cu-rich samples which have a deteriorated structure dispersed with Cu nanoparticles, while in the characterization stage the energetic light or electron beams may destroy the samples to some extent. For both characterization and application, it is desirable to obtain some Cu_3N thin film samples that are strictly stoichiometric and of a uniform microstructure. We will see that due to the weak bonding between the Cu and N atoms, this is, however, a challenge that is not easy to meet.

8.1.3 Reactive magnetron sputtering for Cu_3N

Since nitrogen molecules do not combine with metal atoms, by reactive magnetron sputtering of Cu target with nitrogen, the incorporation of nitrogen to Cu occurs neither in the growing surface, nor on the transport path, but solely on the target surface via plasma sputtering processes. In fact, on the transport path and in the growing film what is of concern is the nitrogen re-emission from the clusters or the film. (We have no evidence for the former, but for the latter it is quite obvious from the morphology studies on the deposits. See below). Therefore, the key factor for successful deposition of stoichiometric Cu_3N thin films requires a deliberate adjustment of the working parameters, including the gas pressure, gas composition and applied power. Bearing in mind that, while a sufficiently high substrate temperature is favorable for obtaining a satisfactory crystallinity of the films, it may, however, foster the re-emission of nitrogen from the deposits, hence the substrate temperature has to be carefully chosen based on the evaluation of its influence on the film quality.

The film growth for Cu_3N by RF reactive magnetron sputtering was first tried by Terada *et al.* in 1989, where the working gases are Ar and N_2 in a ratio of 3:2 (unclearly specified). The success of the method for obtaining crystalline Cu_3N thin films was judged by the Cu_3N (001) and (002) reflection on the X-ray diffraction patterns. It was claimed that the resulting deposits are 'almost perfect insulators', but no data of electrical resistivity were specified in the literature.

By using dc magnetron sputtering, and selecting a substrate temperature from room temperature (RT) to 150°C, Reddy *et al.* obtained (111)-oriented films, where the partial pressure of nitrogen of 1×10^{-3} mbar is only one fiftieth of the total sputtering pressure (Reddy *et al.* 2007). The electrical resistivity of the films has been reduced from 8.7×10^{-1} to 1.1×10^{-3} Ωm, which we will see obviously comes from the deviation from chemical stichiometry, i.e., the deposits suffer from deficiency of N. The same research group has also investigated the influence of applied power, the partial pressure on the characters of the deposited films. They claim that single phase films of copper nitride were obtained at a sputtering power of 75 watt with an electrical resistivity of 5.8×10^{-2} Ωcm and an optical band-gap of 1.84 eV (Reddy *et al.* 2007). In a more recent publication, Dorranian *et al.* found that the N_2 partial pressure influenced the structural, electrical and optical properties of the deposited films. The X-ray diffraction measurement showed the change of the preferred orientation of the Cu_3N samples from Cu-rich (111) planes to N-rich (100) planes (Dorranian *et al.* 2010).

In the past few years it has gradually become clear that the deposition of high-quality – by quality we refer to stoichiometry, structure, and other film features such as orientation, morphology – Cu_3N films by reactive magnetron sputtering using nitrogen gas is far from trivial. Various working parameters including the power supply, total pressure of the working gas and the partial pressure of nitrogen therein, and the substrate temperature should be balanced on the basis of careful characterization of the deposits, to which a complementary set of analytical tools should be employed.

8.2 Deposition of stoichiometric Cu_3N

We adopted the radio frequency (27.12 MHz) reactive magnetron sputtering method to grow Cu_3N thin films on Si (100) wafers in a custom-designed system, with a gas mixture of argon and nitrogen. The high-purity (5N) Cu target is 60 mm in diameter. In order to suppress the nitrogen re-emission by heat, the substrate was held at 60°C just to achieve adequate adhesion for the deposits. During reactive sputtering growth, the RF power was tuned to a maximum of 150 W, and the working pressure was maintained at 9.0×10^{-1} Pa with a total gas flow rate of 4.0 sccm. The proportion of N_2 in the working gas is another variable for the optimization of film composition, and consequently the structure for this material. Prior to each deposition, the target was sputtered for 30 minutes to obtain a steady material flow to the substrate.

The low substrate temperature provokes less deteriorating effects in the deposits, and consequently samples of prescribed compositions can be well reproduced. A low substrate temperature and higher working pressure can effectively suppress the formation of pure metal phase in the deposits. In order to avoid the destructive effect of the probing particles upon the samples, repeated measurements have been conducted on the equally fresh samples for the analysis of the unstable materials considered here.

With a fixed power supply of 150 W, the copper content in the deposits declines steadily with increasing nitrogen proportion, and at a nitrogen proportion of 90% it approaches the value 76.9% (Fig. 8.2). Further increase of nitrogen in the working gas failed to obtain an expected stoichiometric

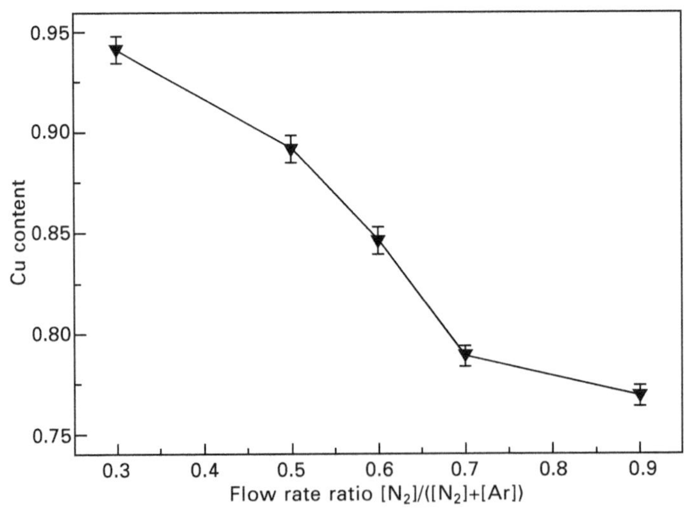

8.2 Cu content in the copper nitride thin films deposited with varied nitrogen proportion in the working gas.

Cu_3N deposit, even with pure nitrogen at this power supply. This failure in obtaining a strictly stoichiometric copper nitride deposit has nothing to do with the substrate temperature because nitrogen re-emission from compound is unlikely to occur at 60°C, but rather lies in the nature of magnetron sputtering. Since the reaction with nitrogen can only take place in the region of the plasma ring on the surface of the Cu target, it is the energetic sputtering there that prevents the formation of stoichiometric Cu_3N clusters. The unwanted presence of elemental copper precipitates in the deposits provides an explanation for the discrepancy or even controversy in many characterizations of the film properties, as will be made clear later.

X-ray diffraction confirmed the nanocrystallinity of the deposits containing Cu_3N and Cu biphases. With a Cu content below 78.8%, the reflections from the cubic Cu_3N phase become dominant, as can be seen from Fig. 8.3. The reflections assigned to the {001}- and {002}-planes of the Cu_3N phase remain unchanged, while the reflection from the {111}-plane of the Cu_3N phase shifts irregularly through the sequence (a)–(e). This is quite reasonable since the nitrogen re-emission most preferably occurs on the {111}-planes, thus their spacing is susceptible to the variation of the overall composition of the deposits. The lattice constant calculated from this reflection oscillates between 3.868 and 3.797 Å.

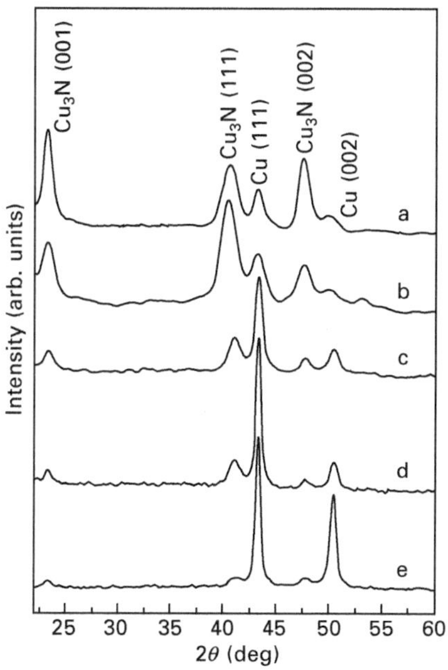

8.3 X-ray diffraction patterns for the copper nitride thin films with a Cu content of (a) 76.9%, (b) 78.8%, (c) 84.5%, (d) 89.2%, and (e) 94.0%.

For Cu_3N, the {111}-planes consist exclusively of either Cu atoms or N atoms, hence nitrogen loss most preferably occurs on these planes, which sets a limit on the size of the Cu_3N crystallites (to be discussed below). At the grain boundaries, the Cu atoms agglomerate to form nanocrystals, generally being ~5 nm in size as revealed by transmission electron microscopy. The existence of juxtaposed Cu-rich {111}-planes of Cu_3N and the Cu nanocrystals implies an uncertain factor in determining the optical and, in particlar, electrical properties. By measuring the electrical resistivity, a sudden drop was observed when the content of Cu was over 78.8% (Fig. 8.4), suggesting that a new mechanism for the electrical conductance was now switched on. Writing the composition of this sample in the form $Cu_{15.2}$–$Cu_{63.6}N_{21.2}$, and recalling that the excessive Cu atoms exist both as Cu nanocrystals and on the Cu-rich {111} planes of Cu_3N crystallites, an electrical conduction path solely via Cu chains can be established through the sample, i.e. the percolation mechanism comes into play. The resistivity for the sample with 76.9% Cu is 4.04×10^{-5} Ωm, just with a little bit more excess Cu it drops to 1.0×10^{-6} Ωm, falling already into the range for good conductors. The sensitivity of the electrical resistivity to the content of Cu explains the large inconsistency in the reported values of electrical resistivity for copper nitride deposits.

The films obtained above at 150 W seriously deviate from the stoichiometry for Cu_3N. In order to obtain a nearly stoichiometric sample, the combination of power supply and the pressure of working gas (now we employ pure nitrogen gas) should be tested. Figure 8.5(a) and (b) display the XRD patterns for the samples deposited with various RF powers, and under two working pressures of 0.7 Pa and 0.9 Pa for comparison. It can be seen that the films

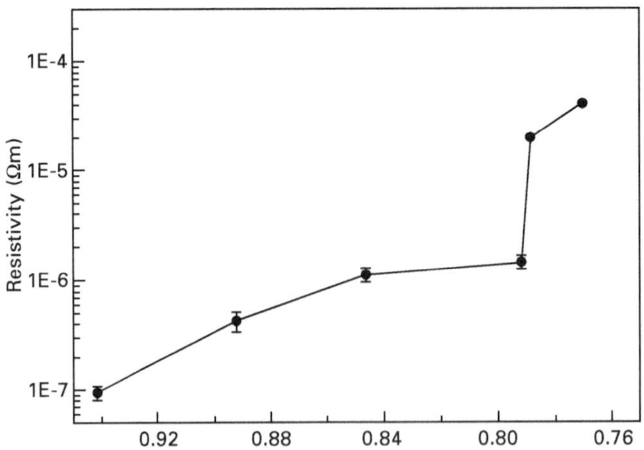

8.4 Electrical resistivity of the copper nitride deposits at room temperature as a function of Cu content.

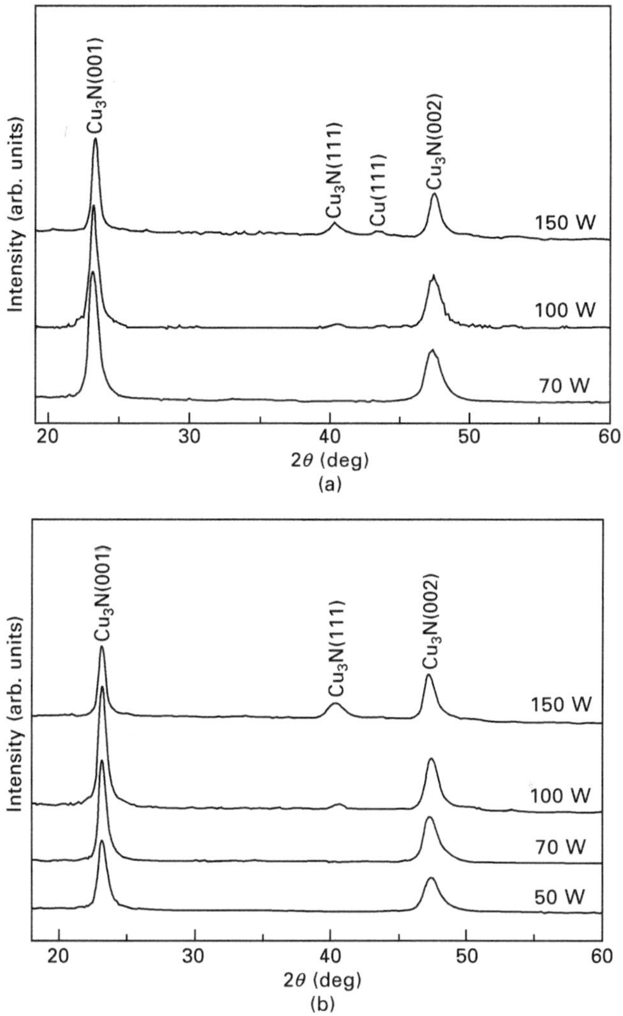

8.5 XRD patterns for the copper nitride deposits prepared at different RF powers under a working pressure of (a) 0.7 Pa and (b) 0.9 Pa, respectively.

always display the (001) and (002) reflections, and the (111) reflection appears when the supplied power is over 100 W. The presence of pure Cu crystallites is disclosed by the minor Cu-(111) reflection at $2\theta = 43.32°$. Clearly, the higher pressure and the lower RF power favor the formation of deposits of pure Cu_3N phase in a competitive way. At 0.7 Pa, the Cu-(111) reflection is visible in the deposit prepared with an RF power of 100 W; at 0.9 Pa, however, it remains absent when the RF power was raised to 150 W. This is quite easy to understand since both the higher pressure and lower RF power can reduce the energy of nitrogen ions bombarding the Cu target. By reactive sputtering of Cu target with nitrogen plasma, a very large projectile energy (large power and low pressure) is unfavorable for the formation of Cu–N bonds in the sputtered particles. Generally speaking, to obtain nearly stoichiometrical copper nitrides at a preset low substrate temperature, the gas pressure and the power should be chosen in such a way that the density of nitrogen ions is sufficiently high and the ion energy has to be low so as not to spoil the Cu–N bonded species emitted from the bombarded target. A strict specification of the processing parameters demands a thoughtful mass-resolved analysis of the bombardment products. Remarkably, in addition to the [001]-orientation, the larger RF power also provokes the occurrence of the [111]-orientation for the Cu_3N crystallites. From Fig. 8.5 one can say with confidence that with a power supply at 70 or 50 W, [001] oriented thin films of copper nitride without any trace of the Cu(111) reflection on the XRD pattern can be obtained.

Copper nitride even in the pure Cu_3N phase still reveals a minor deficiency of nitrogen. This is to say that in the stoichiometrical samples we presented here, the Cu content is still somewhat larger than 75.0%. In fact, from the energy-dispersive X-ray analysis and photoelectron spectroscopic data, the lowest Cu content is about 75.6%, to the accuracy of the methods. The nearly stoichiometrical copper nitride sample should be a typical deficit semiconductor, as seen from the temperature dependence of the electrical resistivity (see Fig. 8.6(a)). A slight increase in the content of Cu to ~76.0% can totally destroy this behavior for a deficit semiconductor, as the resistivity first drops irregularly with the decreasing temperature, and below 50 K it turns to increase steeply (not shown). In the Cu-rich sample with 78.8% of Cu, due to the presence of distinct Cu particles (Du *et al.* 2005), it shows a characteristic metallic behavior that the electrical resistivity increases linearly with the temperature rising from 5 K to 300 K (Fig. 8.6(b)). According to the equation $\rho(T) = \rho_0 [1 + \alpha(T - T_0)]$, the temperature coefficient α is determined to be 7.54×10^{-4}, much less than the value 1/232 for pure Cu. This can be explained by the fact that in the Cu-rich copper nitride, the conduction path is made of both Cu nanoparticles and the Cu_3N nanoparticles enclosed by the Cu-terminated {111}-planes. On the contrary, the stoichiometric Cu_3N sample manifests a typical semiconductor character, the temperature dependence of electrical resistivity (Fig. 8.6(b)) can be approximately fitted by the equation $\rho(T) = C \exp(-E_a/kT)$, with the activation energy E_a being approximately 8.59×10^{-2} eV. For comparison, E_a is 1.47×10^{-2} eV as reported by Wang *et al.* (1998).

The host material Cu_3N in the cubic anti-ReO_3 lattice is a typical semiconductor with an indirect band-gap of ~1.9 eV (Moreno-Armenta *et al.* 2004, Nosaka *et al.* 2001, Maruyama and Morishita 1996). Due to the weak Cu–N bonding that facilitates nitrogen re-emission from the compound,

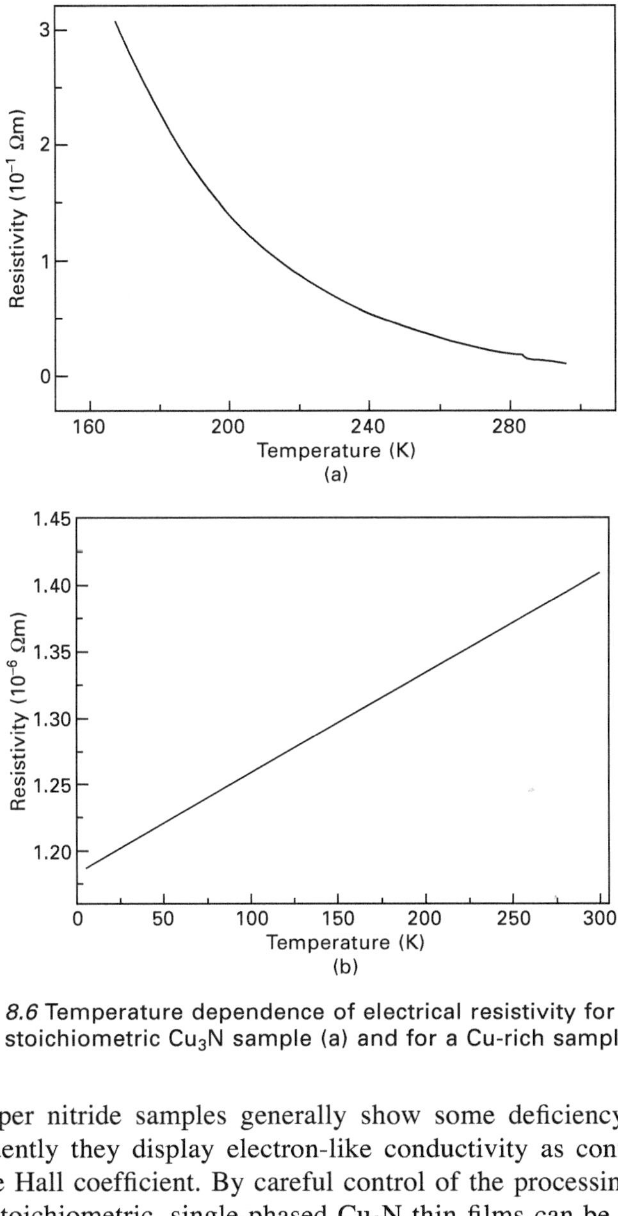

8.6 Temperature dependence of electrical resistivity for a stoichiometric Cu_3N sample (a) and for a Cu-rich sample (b).

the copper nitride samples generally show some deficiency of nitrogen; consequently they display electron-like conductivity as confirmed by the negative Hall coefficient. By careful control of the processing parameters, nearly stoichiometric, single-phased Cu_3N thin films can be obtained with a carrier density brought down to below $3.0 \times 10^{18}/cm^3$.

With the stoichiometrical Cu_3N sample at hand, a high-resolution TEM image can be obtained, which requires enormous expertise since the energetic electron beam in a TEM can readily destroy the Cu_3N lattice. Figure 8.7 displays the first TEM image for the stoichiometric Cu_3N, exhibiting well-aligned {001} fringes, which confirms the high crystallographic quality of the

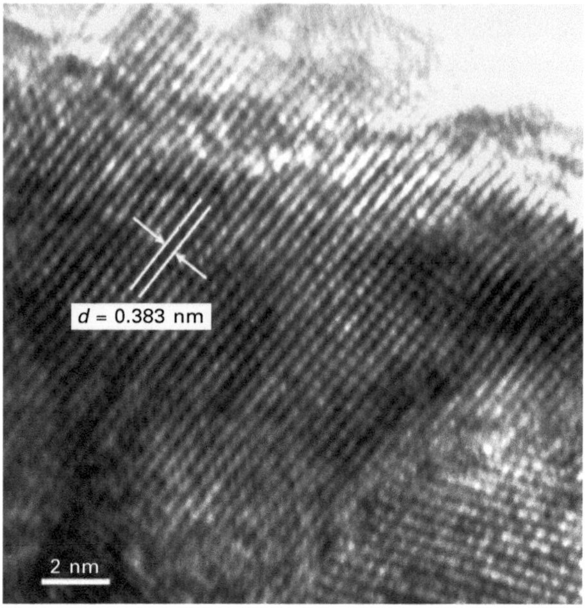

8.7 TEM image of the stoichiometric Cu_3N with a Cu content of 75.6%.

deposit. The planar distance for the {001}-planes, i.e., the lattice constant, measures 0.383 nm in the TEM image, which is in good agreement with the XRD measurements.

Due to the nitrogen deficiency, the deposits consist of distinct nanosized particles instead of being a single crystalline as claimed by some authors. This can be judged easily from the line width of the XRD patterns. From the SEM images (Fig. 8.8) we see that the sample with 76.9% Cu is sharply contrasted, where the size of the crystallites ranges from 40 to 60 nm. For the sample with 75.6% Cu, the crystallites are roughly 40 nm in size, and the image seems smeared. This is obvious since the nearly stoichiometric sample is an insulator – it causes a serious charging effect under the illumination of the electron beam in an SEM.

With the nearly stoichiometrical samples, many inconsistencies concerning the physical properties of Cu_3N can be resolved. First, the pure, nearly stoichiometric Cu_3N deposit is a typical wide-gap semiconductor, whose electrical resistivity at room temperature measures 2×10^{-2} Ωm. Smaller values for this quantity originate in the incorporation of Cu particles or Cu-rich {111} surfaces of the Cu_3N crystallites. It has a band-gap larger than 1.8 eV, as determined from the photoreflectance spectrum even for the slightly substoichiometric copper nitride thin film with 76.9% Cu (Du *et al.* 2005). Another important feature is the decomposition temperature. To study the

8.8 Surface morphology of copper nitride thin films with a Cu content of (a) 76.9% and (b) 75.6%.

thermal stability of the as-deposited films, the samples were annealed in the same chamber under the nitrogen environment at 0.9 Pa for 20 minutes. in a temperature range from 220°C to 420°C. The temperature at which Cu(111) reflection becomes noticeable on the XRD pattern of the post-annealed samples is taken as the decomposition temperature. Aannealing for 20 minutes. at 220°C and 300°C does not provoke any discernible changes to the diffraction pattern. Only at 350°C do two tiny peaks emerge at $2\theta \approx 43.32°$ and $2\theta \approx 50.55°$ (arrowed in Fig. 8.9) which can be assigned to the (111) and (002) reflections of the pure Cu phase (Fig. 8.9). The primarily absent Cu (111) and (002) reflections unambiguously signify the onset of thermal decomposition of Cu_3N. At 420°C, only strong Cu peaks were observable, the Cu_3N completely dissolved. Although from the current experimental data the precise value of the decomposition temperature cannot be fixed, we

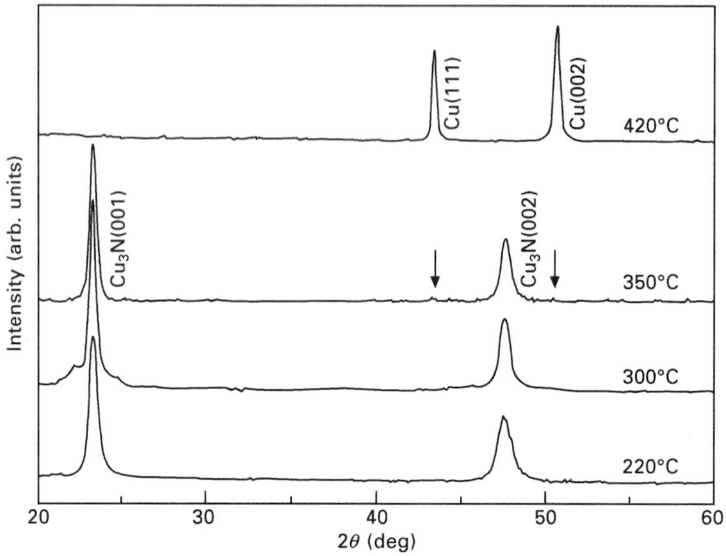

8.9 XRD patterns of the Cu_3N films after annealing at temperatures varying from 220°C to 420°C.

can say with confidence that the Cu phases emerge from the decomposed stoichiometrical Cu_3N, revealed by XRD, only at a temperature of ~350°C by annealing under the protection of nitrogen at a pressure of 0.9 Pa.

8.3 Nitrogen re-emission

The thermal instability of copper nitrides constitutes essential difficulties for their synthesis and structural characterization. By reactive magnetron sputtering deposition of copper nitride films, the effect of nitrogen re-emission on the film morphology may be directly observable, from which the effect on the growth process and on the microscopic structure of deposits can be inferred. The investigation of this topic is also helpful in understanding the failure mechanism of copper nitrides in usage, which may cause serious instability problems in practical applications.

To demonstrate the effect of nitrogen re-emission, copper nitride thin films were grown with mixed nitrogen (of 60% to 80% in flow rate) and argon as working gas, maintained at a pressure of 1.1–1.3 Pa. The power supply was set at 150 W. Under the given conditions, the deposits are substoichiometric with copper contents exclusively below 77.6%. Under SEM, the films grown for 15 minutes, which are roughly 0.3 μm thick, display a morphology with isolated round blisters (Fig. 8.10). These blisters are typically about 10 microns in dimension. The coverage by these protruding features is about 20%. In films deposited for 30 minutes, the protruding features turned into

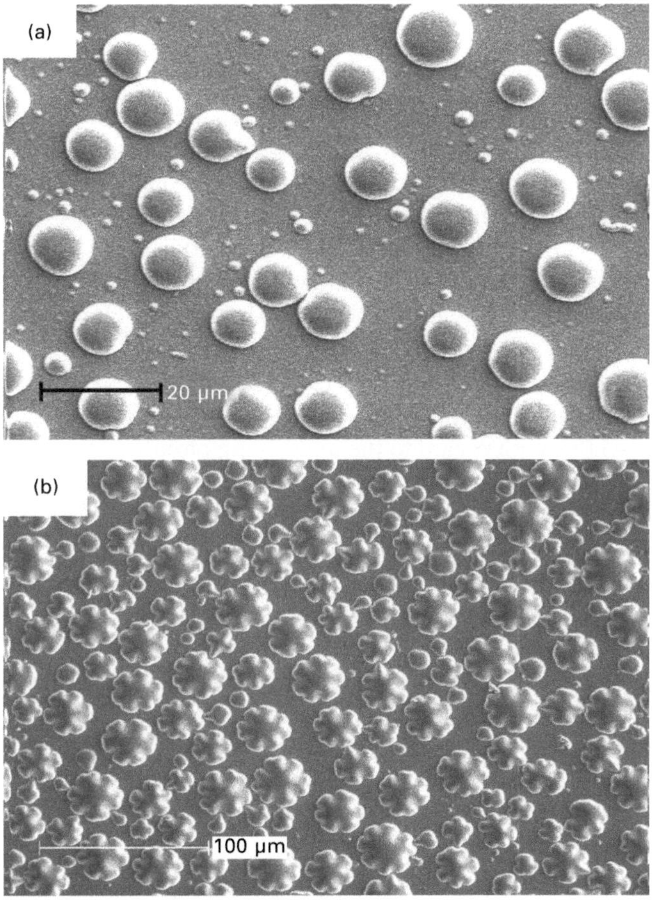

8.10 Effect of nitrogen re-emission on the morphology of copper nitride films. A 0.3 µm thick film exhibits distinct round blisters (a), while in a film grew to ~0.6 µm thick, rosettes with bifurcated rays were observed (b). The rosettes are generally quintilobed, often fivefold symmetry can be identified at the centers.

a ramified rosette-like structure. Such copper nitride rosettes are distributed homogeneously across the sample surface, and show a striking size and shape uniformity, with a lateral dimension around 23 microns (Fig. 8.10). By close inspection of the individual rosettes, it is observed that the rays bifurcate at an average angle of ~74°, slightly larger than the wedge angle of 71.53° for a tetrahedron. In some well-developed ones, fivefold symmetry can be confirmed at the center of the rosettes, though not as strict as in the geometrical sense.

At first glance, this morphology reminds us of island formation as frequently encountered in growing crystalline films, which can be generally modeled with an atomistic view. But we found that this is not the case. Both the smaller round caps and the larger rosettes are hollow, as confirmed by the presence of some cracked entities. When probing through the crack of a broken rosette, the energy-dispersive X-ray spectrometry detected a dominantly strong signal only for silicon and a weak signal for copper – the nitrogen is completely absent. This is to say that the protruding features are due to a complete local detachment of the film; and in that process some tiny amount of copper is left behind on the Si substrate (Ji *et al.* 2005, 2006a). Nevertheless, such protruding structures cannot have developed from a previously well-deposited coating due to the accumulated stress. This point can be clarified by briefly examining the necessary area expansion of the material for the formation of such hollow features. As a conservative estimation, consider a round cap 20 μm in lateral dimension and 1 μm high. This corresponds to an area expansion of ~5.0% with regard to the underlying circular base. Such a large area expansion coefficient is prohibitive for rigid materials such as the ionic copper nitride. Moreover, such a morphology is formed in the growth stage, post-growth degradation of an originally flat film due to nitrogen re-emission leaves behind a flat, but Cu-rich surface. The aging of films with a compact morphology in the ambient will not invoke any rosette structure.

The puzzle of the enormously expanded area of the rosettes with regard to the underlying base becomes immediately resolved with the aid of scanning electron micrographs at an enlarged magnification. We see that the deposits with rosette structures are composed of distinct crystallites, around 45 nm in dimension, just like a compact film, but the SEM images taken directly on a rosette reveals the peculiarity in the manner of crystallite stacking. The surface of the rosette structures displays ragged steps and terraces, and it is thinner than the flat portion of the deposits. The typical width of a terrace is about 100 nm (Fig. 8.11). Such a morphology has never been reported, to the best knowledge of the author. We are inspired to speculate that the crystallites in a rosette have experienced a rearrangement process. The area expansion in due course of relief formation is sustained through the gliding of nanocrystals, which is facilitated when the nanocrystals are clothed with the amorphized Cu-terminated {111}-planes; and the ongoing fast growth prevents the film from cracking.

In order to fully understand the formation mechanism for such protruding features, we take a close look at the morphological profile of the rosettes. They have the shape of a starfish with rays, but these rays generally do not meet at the same point in the central disk. Occasionally, a symmetrically developed rosette could be found, in which a perfect pentagram is discernible at the center, as shown in Fig. 8.12(a). The rays show further bifurcations, and the offshoots extend generally at an angle of about 74°, a little larger than 72° as required by fivefold rotational symmetry, which in turn is a little larger than the wedge angle of a tetrahedron ($\alpha = 70.53°$). This cannot be

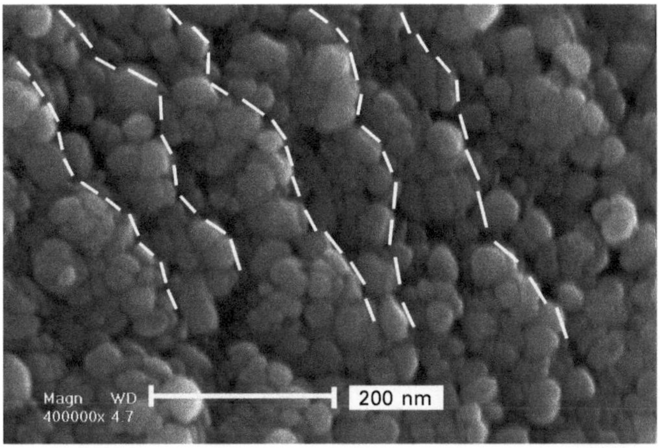

8.11 Scanning electron micrographs at a large magnification of the rosette. Typical crystallite size is 45 nm. The dashed lines are plotted to guide the eye.

a coincidence, bearing in mind that the tetrahedral crystallites do have the tendency to juxtapose at an angle of about 72° with a [110] orientation in order to fill the space, giving rise to a loosely fivefold symmetry. Fivefold symmetry evolving from this construction mechanism has been confirmed in nanostructures of cubic metals such as Au, Ag and Cu.

As we know, the {111}-planes of the cubic Cu_3N lattice comprise exclusively either Cu atoms or N atoms. When nitrogen re-emits from a Cu_3N crystallite, it leaves behind crystallites with a soft clothing of amorphous Cu layer in a habit, not necessarily tetrahedral, but with facets joining at the wedge angle ($\alpha \approx 70.53°$). The clothing of Cu_3N crystallites by Cu {111}-planes also builds a particular electrical conductance path, as verified from the percolation transition in the variation of its electrical resistivity versus the Cu content in the film. When driven by a sufficiently large stress, here initiated by the re-emitted nitrogen gas from beneath, the nanocrystals will undergo a gliding motion against each other along the {111}-planes, and conglomeration of a large quantity of such nanocrystals with facets joining at $\alpha \approx 70.35°$ results in fivefold symmetry at larger scales. Displayed in Fig. 8.12(b) is a piece of Cu_3N crystal in a specimen prepared with mechanical milling for transmission electron microscopic investigation – it is pentagonal with an edge length of 660 nm.

Now the picture becomes clear. An 'orogenic movement' model can be formulated to illustrate the formation of relief rosettes in a growing Cu_3N thin film (Fig. 8.5). Copper nitride keeps decomposing while film growth proceeds. The re-emitted nitrogen atoms recombine into molecules and together with entrapped nitrogen they agglomerate to form gas bubbles at the

8.12 Fivefoldness in the microstructure of a copper nitride deposit. (a) A rosette displaying perfect fivefold symmetry at the center; (b) a pentagonal assembly of the Cu_3N nanocrystals under transmission electron microscope.

film–substrate interface by virtue of a large lateral mobility, since solubility of nitrogen in both Si and Cu_3N is low. Adhesion to the local substrate is undermined where a gas bubble forms, and finally detachment occurs (Fig. 8.5(b)). By a delicate balance between expansion via nanocrystal glide and fast growth, the total area and the lateral dimension of the delaminated part grow in pace without cracking. As can be expected, at reduced growth rates a compact morphology of the film ensues. Gliding motion of the crystallites into ragged steps and terraces – similar to the movement of rocks in mountain building, i.e. the orogenic movement – provides the path for a complete strain relaxation; this explains the negligible residual stress in the sample despite the presence of the relief features, as concluded from the transmission electron microscopic and X-ray diffraction data about the lattice constant (Fig. 8.1). The possibility to glide along the Cu-terminated {111}-planes for the pyramidal Cu_3N crystallites is prerequisite for the formation of the relief morphology; this also explains its absence in the films of nearly stoichiometrical or heavily Cu-rich samples. As to the quintilobed structure of the rosettes that ramify at a branching-off angle of ~74°, they are the consequence of the preferred packing and rearrangement driven by an anisotropic stress of about two million {111}-faceted Cu_3N nanocrystals (Ji *et al.* 2007). Unlike in the nanorods of cubic metals where the core is fivefold twin crystals textured along the [110]-axis, the fivefold symmetry here is quite 'soft' in the sense that it arises rather as an integral effect at large scales.

8.4 Doping of Cu_3N by co-sputtering

The structure of Cu_3N is, in itself, rather interesting. It is an open structure that can, theoretically, accommodate one more metal ion in each unit cell. The incorporation of an excess metal atom can significantly alter the band structure of the material that continuous semiconducting-to-(semi)metallic transition can be expected, through which various interesting phenomena can be measured in the ternary compounds (Lovett 1977). Here, through the co-sputtering of Cu and Pd targets, we demonstrate the deposition of ternary Cu_3NPd_x thin films, where a vanishing temperature coefficient of resistivity (TCR) in the range from 240 K down to 5 K could be measured in $Cu_3NPd_{0.238}$ (Ji *et al.* 2006b). Such a phenomenon has not been found in any other single solids.

The inclusion of another metal atom in each unit cell of the Cu_3N lattice results in compounds like Cu_4N or the ternary Cu_3NPd (Hahn and Weber 1996). Remarkably, Cu_3NPd is a semimetal due to the intersection of the energy bands at the Fermi level. Consequently, a semiconducting-to-semimetallic transition is anticipated in the off-stoichiometric Cu_3NPd_x when 'x' increases continuously from zero to unity.

We prepared thin films of Cu_3N and Cu_3NPd_x ($0 < x < 0.350$) on the Si (001) wafers, a low substrate temperature (~100°C) in combination with a small power supply (<150 W) and a high working pressure (>0.7 Pa) is applied. The interposition of metal atoms at the cell centers of the cubic Cu_3N lattice severely deteriorates the covalent bonding in the resulting materials

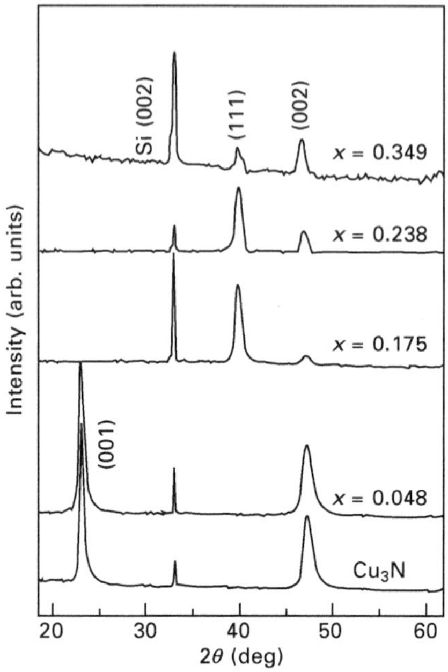

8.13 X-ray diffraction patterns of Cu_3NPd_x with varying '*x*' values.

such that the compounds Cu_4N and Cu_3NPd only exist in theory. In fact, the largest reported value of '*x*' never exceeds 0.989 in Cu_3NPd_x whereby the compounds exist in fine powders (Jacobs and Zachwieja 1991). We found that those Cu_3NPd_x thin film samples with $x > 0.5$ show a very poor crystallinity; therefore we restrict our discussion to the samples with $0.0 \leq x \leq 0.350$ which are well crystallized as confirmed by both X-ray diffraction pattern (Fig. 8.13) and transmission electron micrograph (Fig. 8.14). For samples containing only a tiny amount of Pd, e.g., $Cu_3NPd_{0.048}$, the X-ray diffraction pattern remains unchanged, like that of the host Cu_3N lattice showing only the (001) and (002) reflections. When the Pd concentration is large enough ($x \geq 0.175$), the (001) reflection was replaced by the strong (111) reflection which deteriorates both in intensity and in spectral profile. We can say that the Cu_3NPd_x thin films here concerned with $x < 0.35$ are crystalline. The addition of Pd atoms results in a slightly enlarged lattice constant which increases from $a = 0.383$ nm for the pure Cu_3N to $a = 0.385$ nm for $Cu_3NPd_{0.175}$ (Fig. 8.14). The Cu_3NPd_x samples obviously have a better electrical conductivity which in turn results in an improved image quality for the transmission electron micrograph (cf. Fig. 8.7). The fact that the Cu_3NPd_x is metallic at sufficiently large '*x*' values, while Cu_3N is semiconducting, may suggest that the interposition of Pd atoms alter the energy bands near the Fermi surface.

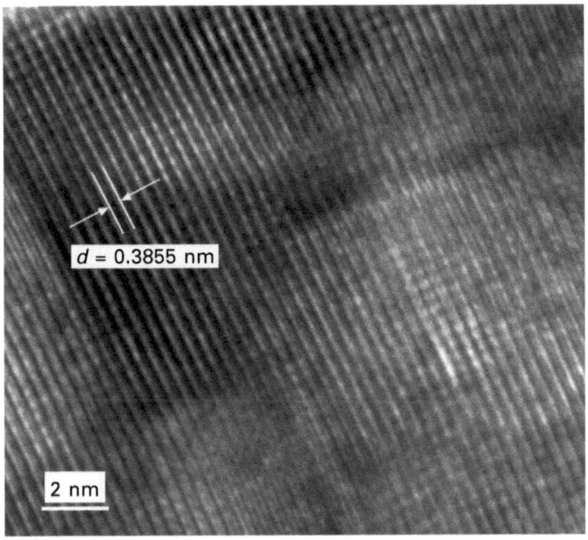

8.14 Transmission electron micrograph for the sample $Cu_3NPd_{0.175}$.

Figure 8.15 displays the temperature dependence of the electrical resisitivty for the nearly stoichiometric Cu_3N and for the different ternary Cu_3NPd_x compounds. Upon cooling from room temperature, the electrical resistivity in the intrinsic Cu_3N film rises rapidly (Fig. 8.15(a)). This observation is consistent with the wide band-gap nature of this semiconducting material. The insertion of Pd atoms confers Cu_3NPd_x an improved electrical conductivity owing to the narrowed band-gap. Even with a tiny amount of Pd insertion as in the samples with $x = 0.048$ and $x = 0.071$, the rapid onset of resistivity upon cooling is shifted to below 50 K, indicating a narrowed band-gap for the current samples (Fig. 8.15(b) and (c)). A further addition of Pd atoms initiates the semiconducting-to-semimetallic transition such that the TCR changes its sign. For $Cu_3NPd_{0.175}$, its electrical resistivity rises roughly linearly with decreasing temperature in a range as wide as 270 K, and the TCR is $-0.00039\ K^{-1}$ (Fig. 8.15(d)). Taking $T = 280$ K as the reference point, the resistance over such a broad temperature range changes only by ~5.0%. The resistivity for $Cu_3NPd_{0.349}$ also displays a good linearity from 50 K to 280 K, but now the TCR is $0.00117\ K^{-1}$ (Fig. 8.15(f)), indicating unambiguously that now the compound has changed to be metallic. In between these two cases of opposite tendencies for the variation of the electrical resistivity with temperature, we may expect a vanishing TCR over a considerable temperature range at some '*x*' value. This occurs in the sample $Cu_3NPd_{0.238}$. From 240 K down to 5 K, the TCR is nearly zero ($< 3.0 \times 10^{-6}$/K), and the relative variance of the electrical resistivity in this temperature range measures as small as 1.42×10^{-5} (Fig. 8.15(e)).

The temperature dependence of the electrical conductivity for solids is susceptible to many influencing factors; no hard-and-fast rules are available. For a semiconductor, the electrical conductivity can be formally written as $\sigma = e(n_e\mu_e + n_h\mu_h)$, where e and h denote electron and hole, while n and μ stand for the carrier density and carrier mobility, respectively. Both the carrier mobility and the carrier density, or the number of the carriers, are temperature dependent, yet the temperature dependence of the electrical conductivity is dominated by that of the latter, since the mobility changes only in a mild way with the lattice temperature while the number of carriers

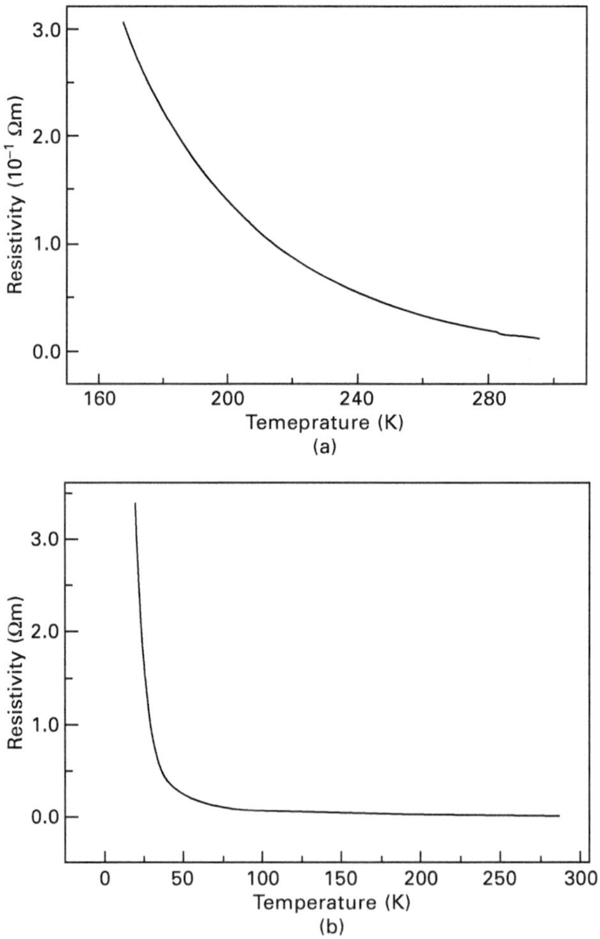

8.15 Temperature (in K) dependence of the electrical resistivity measured in crystalline Cu_3NPd_x thin films. (a) $x = 0.0$; (b) $x = 0.048$; (c) $x = 0.071$; (d) $x = 0.175$; (e) $x = 0.238$; and (f) $x = 0.349$. From (b) to (f), the electrical resistivity at room temperature is reduced by four orders of magnitude.

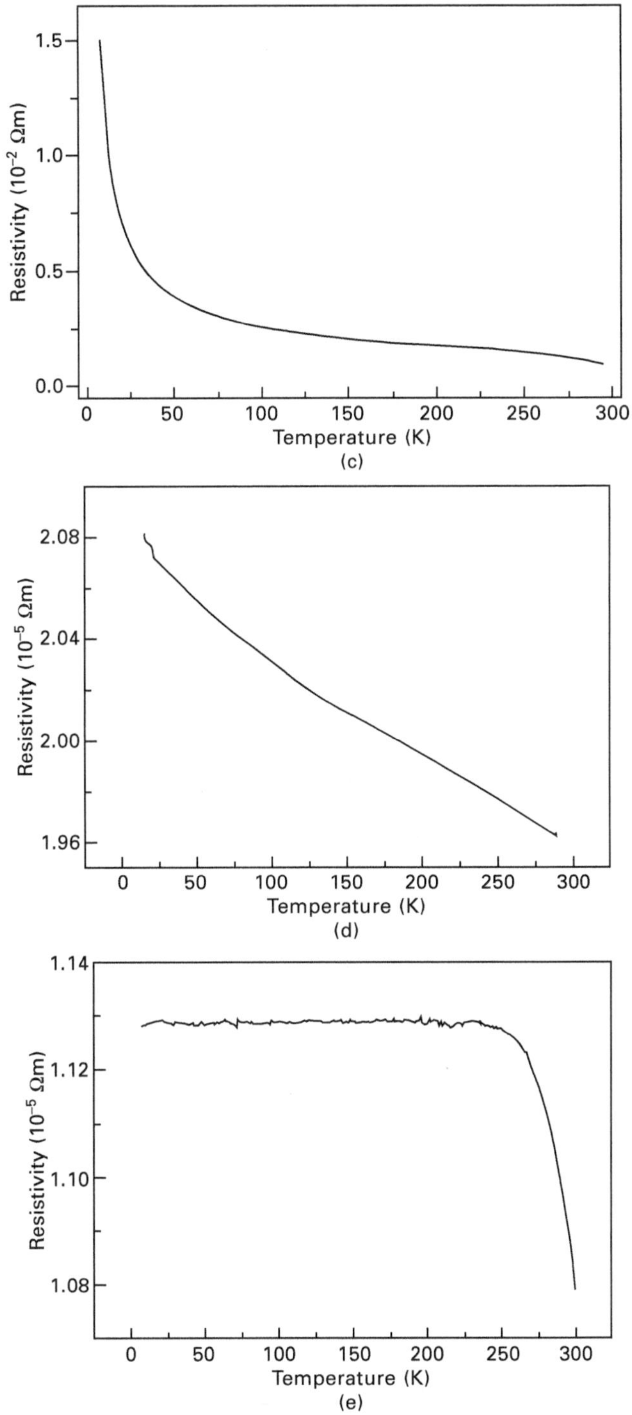

8.15 Continued.

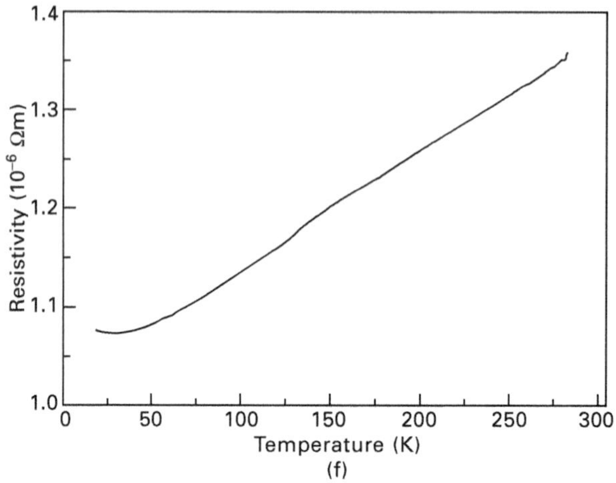

8.15 Continued.

can be an exponential-like function of $1/T$. With decreasing temperature, the number of carriers decreases while the mobility generally increases, but in different manners; consequently, the resulting electrical resistivity is exclusively temperature dependent. For narrow band-gap semiconductors and semimetals, the situation is more complicated than in the conventional semiconductors like Si or Ge. The carrier distribution function obeys the Fermi–Dirac statistics which in this case cannot be approximated by the Maxwell–Boltzmann formula. Moreover, the conduction band is Kane-type that the effective mass of electron, thus the mobility of electron, changes with the carrier density (Lovett 1977). The vanishing TCR occurs when this compound has narrowed its band-gap to be a semimetal for which the temperature dependence of the number of the carriers becomes rather mild-mannered in contrast to the situation in a semiconductor of a definite band-gap. It results from a delicate balance between the opposite changes of the number of carriers and of the carrier mobility with temperature, and reasonably it appears only at temperatures below 240 K – at higher temperatures the increasing number of carriers can easily compensate for the loss in carrier mobility to give rise to a rapidly decreasing electrical resistivity. The current discovery, though lacking a clear scenario of detailed microscopic processes due to the difficulty in calculating the band structure for the off-stoichiometric compounds such as $Cu_3NPd_{0.238}$ and in performing Hall measurement for some supplementary information, does demonstrate the possibility that a balanced change of carrier density and carrier mobility over a wide temperature range is in principle possible in a class of semimetallic materials. Considering the unlikelihood of a temperature-independent electrical resistivity for solids, the current result also bears some significance for solid state physics.

8.5 Conclusions

By carefully adjusting the working parameters including the applied power, gas pressure and the ratio of nitrogen in the working gas, and substrate temperature, thin films of noble-metal nitrides such as Cu_3N and Cu_3NPd_x can be successfully prepared by using the reactive magnetron sputtering method. For stoichiometric Cu_3N, the films are found to comprise distinct crystallites in a typical size of 40–60 nm. The peculiarity of growing noble-metal nitride films lies in the fact that the reaction of sputtered metal atoms or clusters with nitrogen occurs most probably only at the target surface, hence the matching of the density and energy of nitrogen ions with the sputtering yield for a chosen target material is the key factor to realizing stoichiometry in the resulting copper nitride films. A large ion energy is unfavorable since the higher sputtering yield of the metal atoms and the weakened combination with energetic nitrogen ions will give rise to a metal-rich deposit. A low substrate temperature is critical since at slightly elevated temperatures nitrogen re-emission occurs, which may even result in microsized protruding features. Semiconducting to (semi)metallic transition can be observed in doped Cu_3N structures wherein the excessive atoms are expected to sit at the cell centers of the Cu_3N lattice. A vanishing temperature coefficient of resistivity over a temperature of 200 K has been realized in $Cu_3NPd_{0.238}$. Many other surprising properties are anticipated in the noble-metal nitrides, which await discovery in the future.

8.6 References

Asano M, Umeda K, and Tasaki A (1990), Cu_3N thin film for a new light recording media, *Jpn. J. Appl. Phys.* 29, 1985–1986.

Baranov O, Romanov M, Wolter M, Kumar S, Zhong X X, and Ostrikov K (2010), Low-pressure planar magnetron discharge for surface deposition and nanofabrication, *Physics of Plasma* 17, 053509.

Berg S, and Nyberg T (2005), Fundamental understanding and modeling of reactive sputtering processes, *Thin Solid Films* 476, 215–230.

Borsa D M, Grachev S, Presura C, and Boerma D O (2002), Growth and properties of Cu_3N films and Cu_3N/γ-Fe_4N bilayers, *Appl. Phys. Lett.* 80, 1823–1825.

Cao Z X (2001), Electron cyclotron waveresonance plasma assisted deposition of cubic boron nitride thin films, *J. Vac. Sci. Technol. A* 19, 485–489.

Cao Z X (2002), Plasma enhanced deposition of silicon carbonitride films and property characterization, *Diamond & Relat. Mater.* 11, 16–21.

Cao Z X, and Oechsner H (2004), Effect of concurrent N_2^+-ion bombardment on the physical vapor deposition of nitrides thin films, *J. Vac. Sci. Technol. A* 22(2), 321–323.

Chambers S A (2010), Epitaxial growth and properties of doped transition metal and complex oxide films, *Adv. Mater.* 22(2), 219–248.

Cremer R, Witthaut M, Neuschutz D, Trappe C, Laurenzis M, Winkle O, and Kurz H (2000), Deposition and characterization of metastable Cu_3N layers for applications in optical data storage, *Mikro. Acta.* 133, 299–302.

Dorranian D, Dejam L, Sari A H, and Hojabri A (2010), Structural and optical properties of copper nitride thin films in a reactive Ar/N^2 magnetron sputtering system, *Euro. Phys. J.-Appl. Phys.* 50, 20503.

Du Y, Ji A L, Ma L B, Wang Y Q, and Cao Z X (2005), Electrical conductivity and photoreflectance of nanocrystalline copper nitride thin films deposited at low temperature, *J. Crystal Growth* 280, 490–494.

Hahn U, and Weber W (1996), Electronic structure and chemical-bonding mechanism of Cu_3N, Cui_3NPd, and related Cu(I) compounds, *Phys. Rev. B* 53, 12684–12693.

Hojabri A, Haghighian N, Yasserian K, and Ghoranneviss M (2010), The effect of nitrogen plasma on copper thin film deposited by DC magnetron sputtering, *IOP Conf. Ser.: Mater. Sci. Eng.* 12, 012004.

Jacobs H, and Zachwieja U (1991), Kupferpalladiumnitride, Cu_3Pd_xN mit $x = 0.020$ und 0.989, Perowskite mit 'bindender $3d^{10}$-$4d^{10}$-Wechselwirkung', *J. Less-common Met.* 170, 185–190.

Ji A L, Li C R, Du Y, Ma L B, Song R, and Cao Z X (2005), Formation of rosette pattern in copper nitride thin films via nanocrystals gliding, *Nanotech.* 16, 2092–2095.

Ji A L, Huang R, Du Y, Li C R, and Cao Z X (2006a), Growth of stoichiometric Cu_3N thin films by reactive magnetron sputtering, *J. Crystal. Growth* 295, 79–83.

Ji A L, Li C R, and Cao Z X (2006b), Ternary Cu_3NPd_x exhibiting invariant electrical resisitivity over 200 K, *Appl. Phys. Lett.* 89, 252120.

Ji A L, Du Y, Li C R, and Cao Z X (2007), Formation of symmetrical relief features in nanocrystalline copper nitride thin films, *J. Vac. Sci. Technol. B* 25 (1), 208–211.

Juza R, and Hahn H (1939), Copper nitride (in German), *Z. Anorg. Allg. Chem.* 241, 172–178.

Kim K J, Kim J H, and Kang J H (2001), Structural and optical characterization of Cu_3N films prepared by reactive RF magnetron sputtering, *J. Crystal Growth* 222, 767–772.

Lovett D R (1977), *Semimetals & Narrow-bandgap Semiconductors*, Pion Limited, London.

Ma G M, Alejandro M, and Noboru T (2004), *Ab initio* total energy calculations of copper nitride: the effect of lattice parameters and Cu content in the electronic properties, *Solid Stat. Sci.* 6, 9–14.

Maruyama T, and Morishita T (1996), Copper nitride and tin nitride thin films for write-once optical recording media, *Appl. Phys. Lett.* 69, 890–891.

Moreno-Armenta M G, Martínez-Ruiz A, and Takeuchi N (2004), *Ab initio* total energy calculations of copper nitride: the effect of lattice parameters and Cu content in the electronic properties, *Solid State Sciences* 6, 9–14.

Musil J, Baroch P, Vlcek J, Nam K H, and Han J G (2005), Reactive magnetron sputtering of thin films: present status and trends, *Thin Solid Films* 475, 208–218.

Norskov J K, and Hammer B (1995), Why gold is the noblest of all the metals, *Nature* 376, 238–240.

Nosaka T, Yoshitake M, Okamoto A, Ogawa S, and Nakayama Y (2001), Thermal decomposition of copper nitride thin films and dots formation by electron beam writing, *Appl. Surf. Sci.* 169–170, 358–361.

Reddy K V S, Reddy A S, Reddy P S, and Uthanna S (2007), Copper nitride films deposited by dc reactive magnetron sputtering, *J. Mater. Sci. – Materials in Electronics* 18, 1003–1008.

Terada S, Tanaka H, and Kubota K (1989), Heteroepitaxial growth of Cu_3N thin films, *J. Crystal Growth* 94, 567–568.

Wang D Y, Nakamine N, and Hayashi Y (1998), Properties of various sputter-deposited Cu–N thin films, *J. Vac. Sci. Technol. A* 16, 2084–2092.

9

Growth of graphene layers for thin films

B. H. HONG and H. R. JEON, Sungkyunkwan University, South Korea

Abstract: Graphene has received considerable attention due to its fascinating physical properties such as quantum electronic transport, a tunable band gap, extremely high mobility, high elasticity and electromechanical modulation. This chapter describes the direct synthesis of large-scale graphene films using chemical vapour deposition on thin nickel layers, and presents two different methods of patterning the films and transferring them to arbitrary substrates. Furthermore, we detail the roll-to-roll production and wet chemical doping of predominantly monolayer 30-inch graphene films grown by chemical vapour deposition onto flexible copper substrates.

Key words: graphene growth, chemical vapour deposition, half-integer quantum Hall effect, roll-to-roll production.

9.1 Introduction

In order to be used as nano-electronic devices, it is essential to grow graphene on a large scale. Among methods for graphene synthesis, graphene growth by chemical vapour deposition (CVD) on Ni and Cu has the advantage of growing graphene at large scale with uniform thickness. This chapter deals with graphene growth on Ni and Cu by CVD. First we describe large-scale graphene film synthesis using CVD on nickel layers, their pattern and their transfer to arbitrary substrates.[1] In next section we detail the roll-to-roll production and wet chemical doping of predominantly monolayer 30-inch graphene films grown by CVD onto flexible copper substrates.[2] Further, we detail how graphene electrodes were incorporated into a fully functional touchscreen panel device capable of withstanding high strain.

9.2 Large-scale pattern growth of graphene films for stretchable transparent electrodes

9.2.1 Direct synthesis of large-scale graphene films

Ni films were e-beam evaporated on SiO_2/Si substrates, and thermally annealed. After exposing the Ni surface to a CH_4, H_2 and Ar mixture in atmospheric conditions, the system was rapidly cooled to room temperature. This fast cooling rate is critical in suppressing formation of multiple layers and for separating graphene layers efficiently from the substrate in the later process.[3]

A scanning electron microscope (SEM) image of graphene films on a thin nickel substrate shows clear contrast between areas with different numbers of graphene layers (Fig. 9.1a). Transmission electron microscope (TEM) images (Fig. 9.1b) show that the film consists mostly of less than a few layers of graphene. After transfer of the film to a silicon substrate with a 300 nm thick SiO_2 layer, optical and confocal scanning Raman microscope images were made of the same area (Fig. 9.1c,d).[4] The brightest area in Fig. 9.2(d) corresponds to monolayers, and the darkest area is composed of more than ten layers of graphene. Bilayer structures appear to predominate in both TEM and Raman images for this particular sample, which was prepared from 7 minutes of growth on a 300 nm thick nickel layer. It is found that the average number of graphene layers, the domain size and the substrate coverage can be controlled by changing the nickel thickness and growth process, thus providing a way of controlling the growth of graphene for different applications.

Atomic force microscope (AFM) images often show the ripple structures caused by the difference between the thermal expansion coefficients of nickel and graphene (Fig. 9.1c).[5] It is believed that these ripples make the graphene films more stable against mechanical stretching,[6] making the films more expandable. Multilayer graphene films are preferable in terms of mechanical strength for supporting large-area film structures, whereas thinner graphene films have higher optical transparency. It is found that a ~300 nm thick nickel layer on a silicon wafer is the optimal substrate for the large-scale CVD growth that yields mechanically stable, transparent graphene films to be transferred and stretched after they are formed, and that thinner nickel layers with a shorter growth time yield predominantly mono- and bilayer graphene film for microelectronic device application.

9.2.2 Transfer processes for large-scale graphene films

The transfer of CVD-derived graphene films to non-specific substrates is enabled by $FeCl_3$ etching of the underlying Ni film. Use of buffered oxide etchant (BOE) or hydrogen fluoride solution removes silicon dioxide layers, so the patterned graphene and the nickel layer float together on the solution surface. After transfer to a substrate, further reaction with BOE or hydrogen fluoride solution completely removes the remaining nickel layers.

A dry-transfer process has been developed for the graphene film using a soft substrate such as polydimethylsiloxane (PDMS) stamp.[7] Here the PDMS stamp is first attached to the CVD-grown graphene film on the nickel substrate (Fig. 9.2d). The nickel substrate can be etched away using $FeCl_3$ as described above, leaving the adhered graphene film on the PDMS

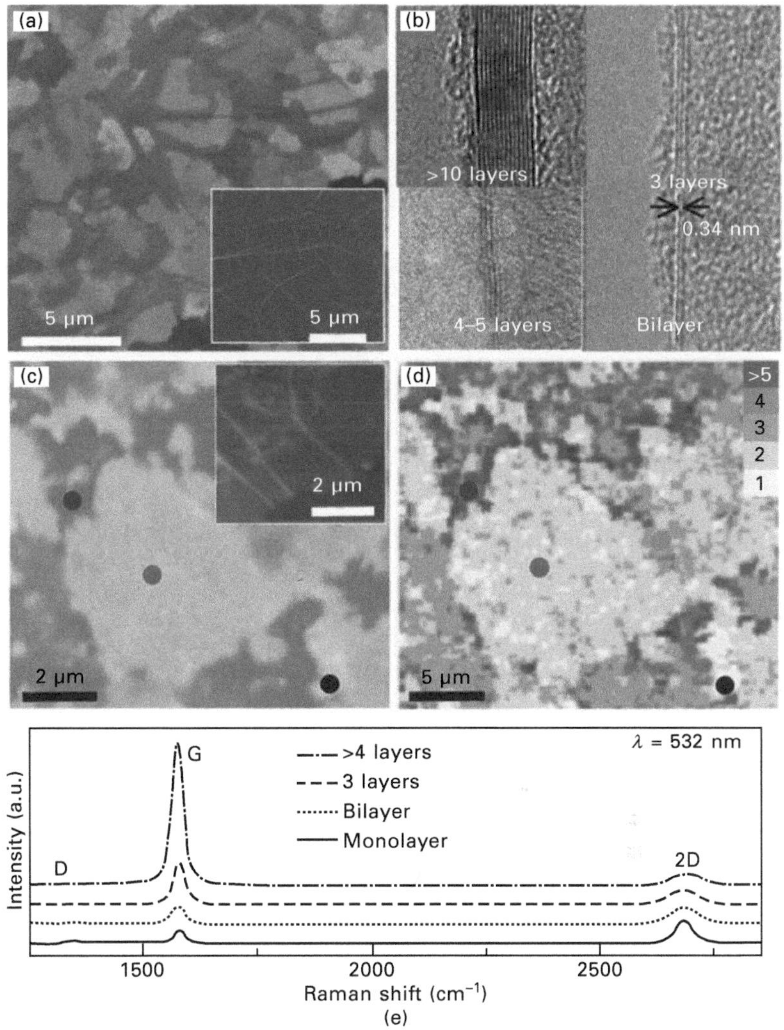

9.1 Various spectroscopic analyses of large-scale graphene films grown by CVD. (a) SEM images of as-grown graphene films on thin (300 nm) nickel layers and thick (1 mm) Ni foils (inset). (b) TEM images of graphene films of different thickness. (c) An optical microscope image of the graphene films transferred to a 300 nm thick silicon dioxide layer. The inset AFM image shows typical rippled structures. (d) A confocal scanning Raman image corresponding to (c). The number of layers is estimated from the intensities, shapes and positions of the G-band and 2D-band peaks. (e) Raman spectra (532 nm laser wavelength) obtained from the corresponding coloured spots in (c) and (d). a.u. = arbitrary units. (From Ref. [1].)

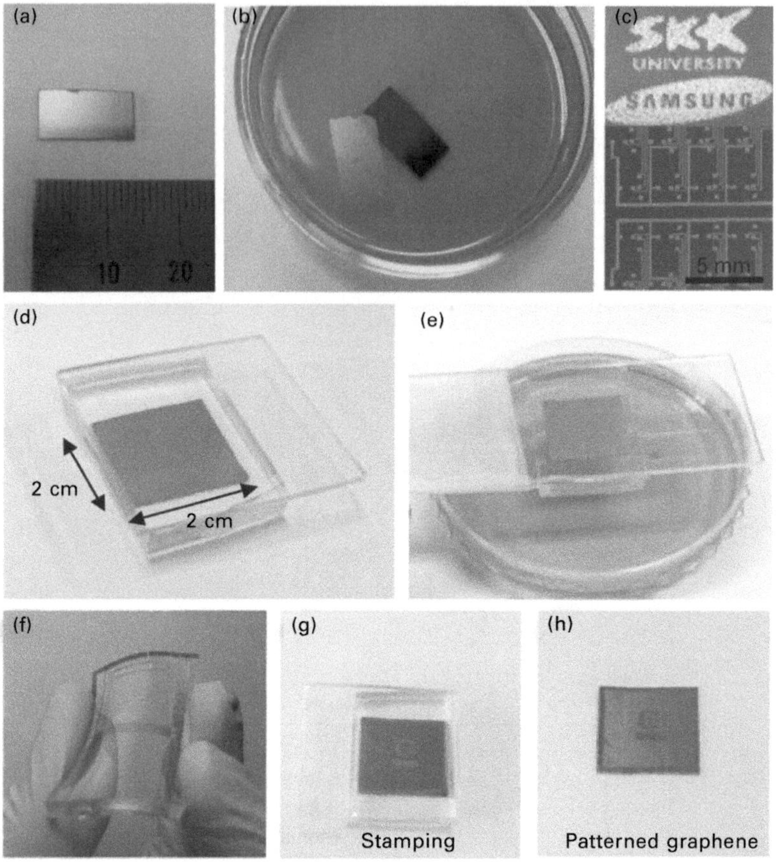

9.2 Transfer processes for large-scale graphene films. (a) A centimeter-scale graphene film grown on Ni(300 nm)/SiO_2(300 nm)/Si substrate. (b) A floating graphene film after etching the nickel layers in 1M $FeCl_3$ aqueous solution. After the removal of the nickel layers, the floating graphene film can be transferred by direct contact with substrates. (c) Various shapes of graphene films can be synthesized on top of patterned nickel layers. (d, e) The dry-transfer method based on a PDMS stamp is useful in transferring the patterned graphene films. After attaching the PDMS substrate to the graphene (d), the underlying nickel layer is etched and removed using $FeCl_3$ solution (e). (f) Graphene films on the PDMS substrates are transparent and flexible. (g, h) The PDMS stamp makes conformal contact with a silicon dioxide substrate. Peeling back the stamp (g) leaves the film on a SiO_2 substrate (h). (From Ref. [1].).

substrate (Fig. 9.2e). By using the pre-patterned nickel substrate (Fig. 9.2c), we can transfer various sizes and shapes of graphene film to an arbitrary substrate. This dry-transfer process turns out to be very useful in making large-scale graphene electrodes and devices without additional lithography processes (Fig. 9.2f–h). Microscopically, these graphene films often show linear crack patterns with an angle of 60°C or 120°C, indicating a particular crystallographic edge with large crystalline domains. In addition, the Raman spectra measured for graphene films on nickel substrates show a strongly suppressed defect-related D-band peak. This D peak grows only slightly after the transfer process (Fig. 9.1e), indicating overall good quality of the resulting graphene film. Further optimization of the transfer process with substrate control makes possible transfer yields approaching 99%.

9.2.3 Optical and electrical properties of the graphene films

For the macroscopic transport electrode application, the optical and electrical properties of 1×1 cm^2 graphene films were measured by ultraviolet-visible spectrometer and four-probe Van der Pauw methods, respectively (Fig. 9.3a,b). The transmittance is measured using an ultraviolet-visible spectrometer after transferring the floating graphene film to a quartz plate (Fig. 9.3a). In the visible range, the transmittance of the film grown on a 300 nm thick nickel layer for 7 minutes is ~80%, a value similar to those found for previously studied assembled films.[8,9] Because the transmittance of the graphene layer is ~2.3%,[10] this transmittance can be increased to ~93% by further reducing the growth time and nickel thickness, resulting in a thinner graphene film. Ultraviolet/ozone etching is also useful in controlling the transmittance under ambient conditions (Fig. 9.4a, upper inset). Indium electrodes were deposited on each corner of the square (Fig. 9.3a, lower inset) to minimize contact resistance. The minimum sheet resistance is ~280Ω per square, which is ~30 times smaller than the lowest sheet resistance measured on assembled films.[8,9] The values of sheet resistance increase with the ultraviolet/ozone treatment time, in accordance with the decreasing number of graphene layers (Fig. 9.3a)

For microelectronic application, the mobility of the graphene film is critical. To measure the intrinsic mobility of a single-domain graphene sample, the graphene samples are transferred from a PDMS stamp to a degenerate doped silicon wafer with a 300 nm deep thermally grown oxide layer. Monolayer graphene samples were readily located on the substrate from optical contrast[10] and their identification was subsequently confirmed by Raman spectroscopy.[4] Electron beam lithography was used to make multi-terminal devices (Fig. 9.3b, lower inset). Notably, the multi-terminal electrical measurements showed that the electron mobility is ~3750 cm^2 V^{-1} s^{-1} at a carrier density of $\sim 5 \times 10^{12}$ cm^{-2} (Fig. 9.3b). For a high magnetic field of 8.8*T*, the half-integer quantum Hall effect (Fig. 9.3b) is observed, indicating that the quality of CVD-grown graphene is comparable to that of mechanically cleaved graphene.

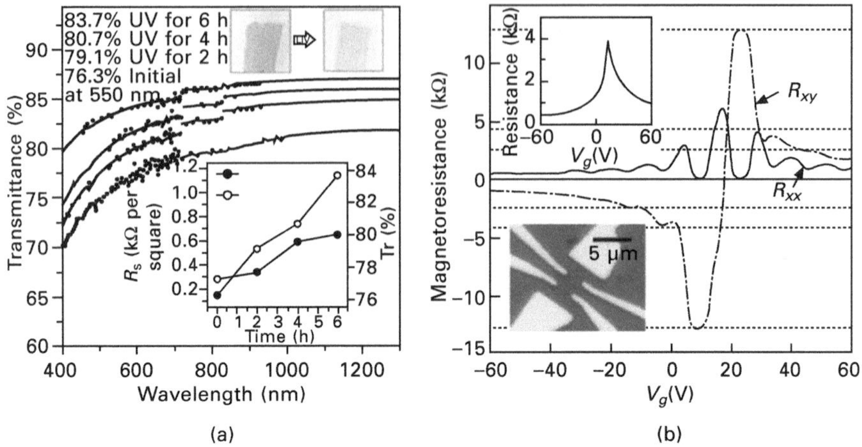

9.3 Optical and electrical properties of graphene films. (a) Transmittance of the graphene films on a quartz plate. The discontinuities in the absorption curves arise from the different sensitivities of the switching detectors. The upper inset shows the ultraviolet (UV)-induced thinning and the consequent enhancement of transparency. The lower inset shows the changes in transmittance, T_r, and sheet resistance, R_s, as functions of ultraviolet illumination time. (b) Electrical properties of monolayer graphene devices showing half-integer quantum Hall effect and high electron mobility. The upper inset shows a four-probe electrical resistance measurement on a monolayer graphene Hall bar device (lower inset) at 1.6 K. A gate voltage, V_g, is applied to the silicon substrate to control the charge density in the graphene sample. The main panel shows longitudinal (R_{xx}) and transverse (R_{xy}) magnetoresistances measured in this device for a magnetic field $B = 8.8T$. The monolayer graphene quantum Hall effect is clearly observed, showing the plateaux with filling factor $\upsilon = 2$ at $R_{xy} = (2e^2/h)^{-1}$ and zeros in R_{xx}. (Here e is the elementary charge and h is Plank's constant.) Quantum Hall plateaux (horizontal dashed lines) are developing for higher filling factors. (c) Variation in resistance of a graphene film transferred to a ~0.3 mm thick PDMS/PET substrate for different distances between holding stages (that is, for different radii). The left inset shows the anisotropy in four-probe resistance, measured as the ratio R_y/R_x, of the resistances parallel and perpendicular to the binding direction, y. The right inset shows the bending process. (d) Resistance of a graphene film transferred to a PDMS substrate isotropically stretched by ~12%. The left inset shows the case in which the graphene film is transferred to an unstretched PDMS substrate. The right inset shows the movement of holding stages and the consequent change in shape of the graphene film. (From Ref. [1].)

In addition to the good optical and electrical properties, the graphene film has excellent mechanical properties when used to make flexible and stretchable electrodes (Fig. 9.3c,d).[7] The foldability of the graphene films, transferred

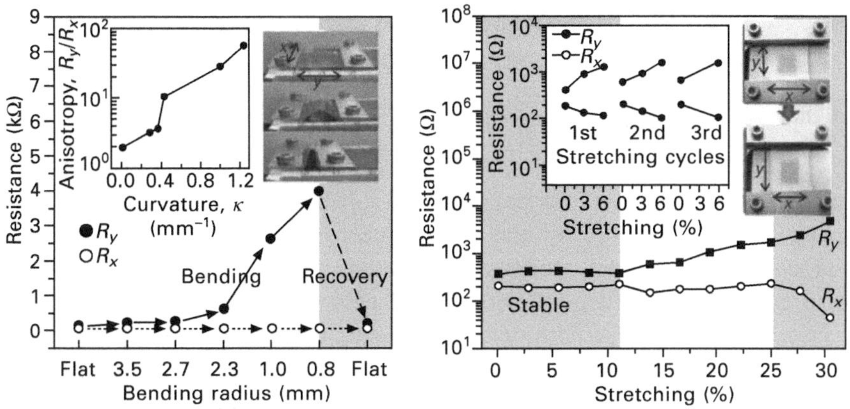

(c) (d)

9.3 Continued

to a polyethylene terephthalate (PET) substrate (thickness, ~100 μm) coated with a thin PDMS layer (thickness, ~200 μm; Fig. 9.3c), is evaluated by measuring resistances with respect to bending radii. The resistances show little variation up to the bending radius 2.3 mm (approximate tensile strain of 6.5%), and are perfectly recovered after unbending. Notably, the original resistance can be restored even for the bending radius of 0.8 mm (approximate tensile strain of 18.7%), exhibiting extreme mechanical stability in comparison with conventional materials used in flexible electronics.[11]

The resistance of graphene films transferred to pre-strained and unstrained PDMS substrates was measured with respect to uniaxial tensile strain ranging from 0 to 30% (Fig. 9.3d). Similar to the results in the folding experiment, the transferred film on the unstrained substrate recovers its original resistance after stretching by ~6% (Fig. 9.3d, left inset). However, further stretching often results in mechanical failure. Thus, it was tried transferring the film to pre-strained substrates[12] to enhance the electromechanical stabilities by creating ripples similar to those observed in the growth process (Fig 9.1c, inset). The graphene transferred to a longitudinally strained PDMS substrate does not show much enhancement, owing to the transverse strain induced by Poisson's effect.[13] To prevent this problem, the PDMS substrate was isotopically stretched by ~12% before transferring the film to it (Fig. 9.3d). Surprisingly, both longitudinal and transverse resistance (R_y and R_x) appear stable up to ~11% stretching and show only one order of magnitude change at ~25% stretching. It is supposed that further uniaxial stretching can change the electronic band structure of graphene, leading to the modulation of sheet resistance. These electrochemical properties thus show graphene films to be not only the strongest[14] but also the most flexible and stretchable conducting transparent materials so far measured.[10]

9.3 Roll-to-roll production of 30-inch graphene films for transparent electrodes

9.3.1 Synthesis of 30-inch graphene films and roll-to-roll transfer of graphene films

Figures 9.4a–c present photographs of the roll-based synthesis and transfer process. An 8-inch-wide tubular quartz reactor (Fig. 9.4a) is used in the CVD system, allowing a monolayer graphene film to be synthesized on a roll of copper foil with dimensions as large as 30 inches in the diagonal direction (Fig. 9.4c). A temperature gradient usually exists that depends on the radial position inside the tubular reactor. In our preliminary work, this sometimes resulted in inhomogeneous growth of the graphene on the copper foils. To solve this problem, a ~7.5-inch quartz tube wrapped with a copper foil was inserted and suspended inside the 8-inch quartz tube. In this way, the radial inhomogeneity in the reaction temperature could be minimized. In the first step of synthesis, the roll of copper foil is inserted into a tubular quartz tube and then heated to 1000°C with flowing 8 standard cubic centimetres per minute) (sccm) H_2 at 90 mtorr. After reaching 1000°C, the sample is annealed for 30 minutes without changing the flow rate or pressure. The copper foils are heat-treated to increase the grain size from a few micrometers to ~100 μm, as we have found that the copper foils with larger grain size yield higher-quality graphene films, as suggested by Li and colleagues.[15] The gas mixture of CH_4 and H_2 is then flowed at 460 mtorr at rates of 24 and 8 sccm for 30 minutes, respectively. Finally, the sample is rapidly cooled to room temperature (~10°C s^{-1}) with flowing H_2 under a pressure of 90 mtorr.

After growth, the graphene film grown on copper foil is attached to a thermal release tape by applying soft pressure (~0.2 MPa) between two rollers. After etching the copper foil in a plastic bath filled with copper etchant, the transferred graphene film on the tape is rinsed with deionized water to remove residual etchant, and is ready to be transferred to any kind of flat or curved surface on demand. The graphene film on the thermal release tape is inserted between the rollers together with a target substrate (Fig. 9.4b). By repeating these steps on the same substrate, multilayered graphene films can be prepared that exhibit enhanced electrical and optical properties, as demonstrated by Li and colleagues using wet-transfer methods at the centimeter scale.[16] Figure 9.4(c) shows the 30-inch multilayer graphene film transferred to a roll of 188 μm thick polyethylene terephthalate (PET) substrate. Figure 9.4(d) shows a screen-printing process used to fabricate four-wire touchscreen panels[17] based on graphene/PET transparent conducting films. After printing electrodes and dot spacers, the upper and lower panels are carefully assembled and connected to a controller installed in a laptop computer (Figs 9.4e and f),

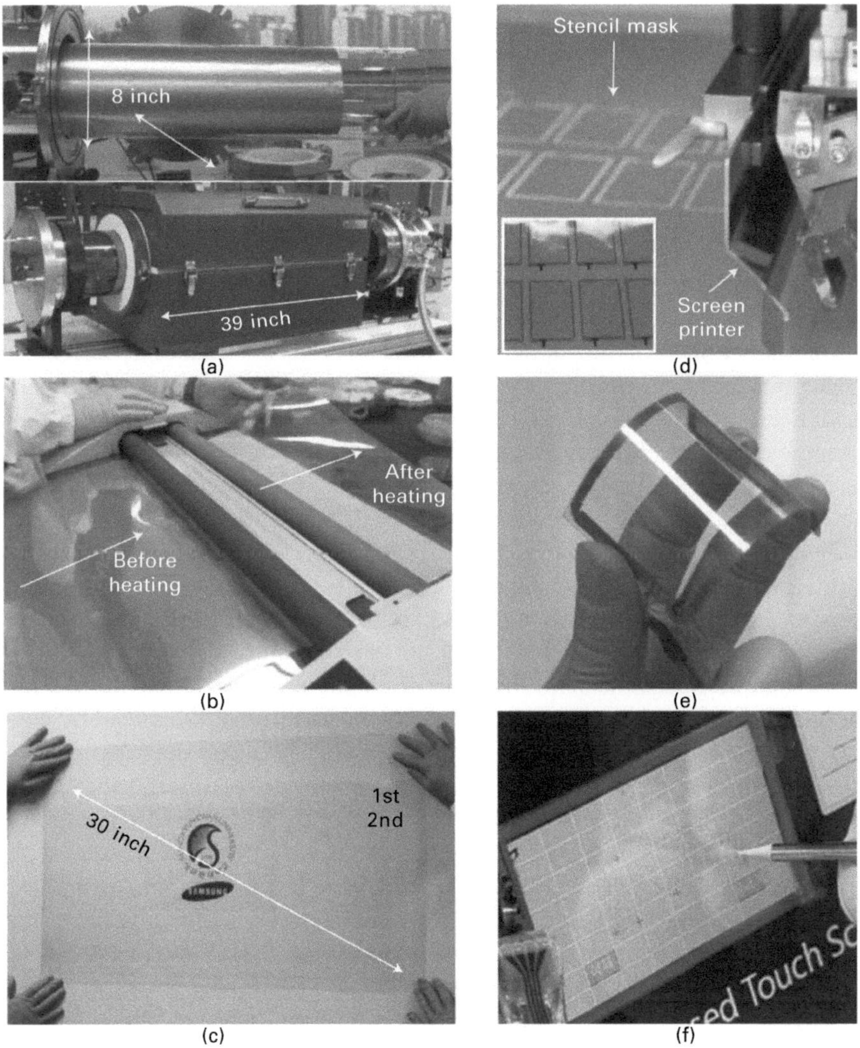

9.4 Photographs of the roll-based production of graphene films. (a) Copper foil wrapping around 7.5 inch quartz tube to be inserted into an 8 inch quartz reactor. The lower image shows the stage in which the copper foil reacts with CH_4 and H_2 gases at high temperatures. (b) Roll-to-roll transfer of graphene films from a thermal release tape to a PET film at 120°C. (c) A transparent ultra-large-area graphene film transferred on a 35 inch PET sheet. (d) Screen printing process of silver paste electrodes on graphene/PET film. The inset shows 3.1 inch graphene/PET panels patterned with silver electrodes before assembly. (e) An assembled graphene/PET touch panel showing outstanding flexibility. (f) A graphene-based touchscreen panel connected to a computer with control software. (From Ref. [2].)

which shows extraordinary flexibility. The scalability and processability of CVD graphene and the roll-to-roll methods presented here are expected to enable the continuous production of graphene-based electronic devices at large scales.

The graphene films seem to be predominantly composed of monolayers when analysed using Raman spectra (Fig. 9.5a). However, atomic force microscope (AFM) and transmission electron microscope (TEM) images often show bilayer and multilayer islands. As the graphene layers are transferred one after another,[16] the intensities of the G- and 2D-band peaks increase together, but their ratios do not change significantly. This is because the hexagonal lattices of the upper and lower layers are randomly oriented, unlike the graphite, so the original properties of each monolayer remain unchanged, even after stacking into multilayers;[18,19] this is clearly different from the case of multilayer graphene exfoliated from graphite crystals.[4] The randomly stacked layers behave independently without significant change in the electronic band structures, and the overall conductivity of the graphene films appears to be proportional to the number of stacked layers.[16] The optical transmittance is usually reduced by ~2.2–2.3% for an additional transfer, implying that the average thickness is approximately a monolayer (Fig. 9.5b).[10]

The unique electronic band structure of graphene allows modulation of the charge carrier concentrations, depending on an electric field induced by gate bias[20] or chemical doping,[21] resulting in enhancement of sheet resistance. We tried various types of chemical doping methods, and found that nitric acid is very effective for *p*-doping of graphene films. Figure 9.5(c) shows Raman spectra of the graphene films before and after doping with 63 wt% HNO_3 for 5 minutes. The large peak shift ($\Delta u = 18\ cm^{-1}$) indicates that the graphene film is strongly *p*-doped. The shifted G peak is often split near the randomly stacked bilayer islands, as shown in Fig. 9.5(c). We hypothesize that the lower graphene layer, which is screened by top layers, experiences a reduced doping effect, leading to G-band splitting. In X-ray photoelectron spectra (XPS), the C 1*s* peaks corresponding to sp^2 and sp^3 hybridized states are shifted to lower energy, similar to the case for *p*-doped carbon nanotubes.[21] However, multilayer stacking results in blueshifted C1*s* peaks. We suppose that weak chemical bonding such as π-π stacking interaction causes descreening of nucleus charge, leading to an overall increase in core electron binding energies. We also find that the work functions of graphene films as estimated by UV photoelectron spectroscopy (UPS) are blueshifted by ~130 meV with increasing doping time (Fig. 9.5d, inset). The multiple stacking also changes the work functions (Fig. 9.5d, inset) which could be very important in controlling the efficiency of photovoltaic[22] or light-emitting devices based on graphene transparent electrodes.[23]

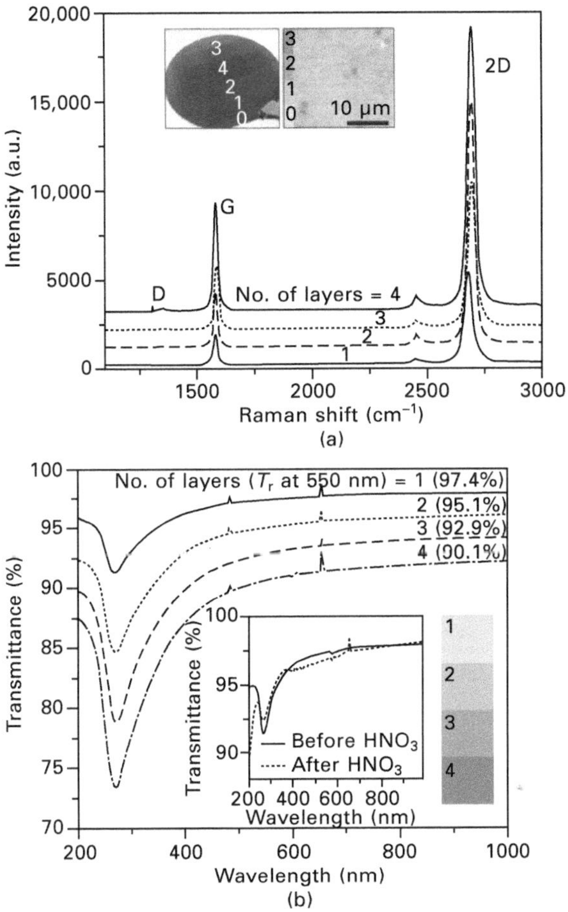

9.5 Optical characterizations of the graphene films prepared using layer-by-layer transfer on SiO_2/silicon and PET substrates. (a) Raman spectra of graphene films with different numbers of stacked layers. The left inset shows a photograph of transferred graphene layers on a 4 inch SiO_2 (300 nm)/silicon wafer. The right inset is a typical optical microscope image of the monolayer graphene, showing >95% monolayer coverage. A PMMA-assisted transfer method is used for this sample. (b) UV-vis spectra of roll-to-roll layer-by-layer transferred graphene films on quartz substrates. The inset shows the UV spectra of graphene films with and without HNO_3 doping. The right inset shows optical images for the corresponding number of transferred layers (1×1 cm^2). The contrast is enhanced for clarity. (c) Raman spectra of HNO_3-doped graphene films, showing ~18 cm^{-1} blueshift both for G and 2D peaks. D-band peaks are not observed before or after doping, indicating that HNO_3 treatment is not destructive to the chemical bonds of graphene. (d) XPS peaks of monolayer graphene films transferred on SiO_2/Si substrates, showing typical redshift and broadening of carbon 1*s* peaks (C1*s*) caused by *p*-doping. The inset shows work function changes ($\Delta\Phi$) with respect to doping time (lower *x*-axis), measured by UPS. (From Ref. [2].)

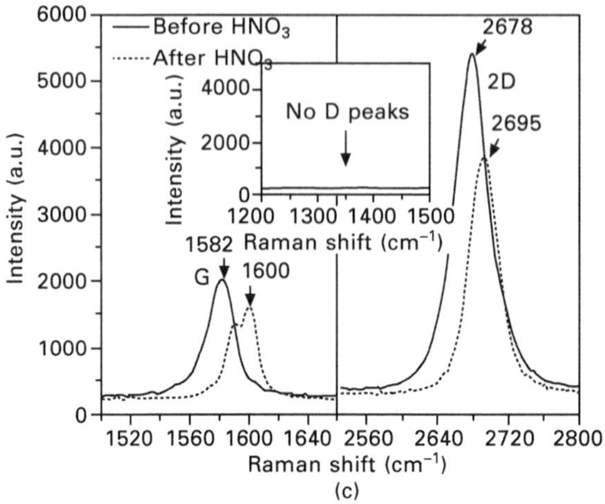

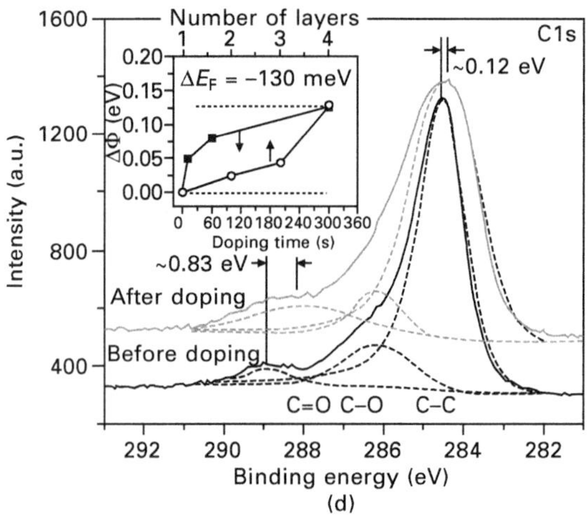

9.5 Continued

9.3.2 Optical and electrical properties of graphene films

The electrical properties of graphene films formed using layer-by-layer staking methods were also investigated. Usually, the sheet resistance of graphene film with 97.4% transmittance is as low as ~125 Ω/sq (Fig. 9.6a) when it is transferred by a soluble polymer support such as polymethyl methacrylate (PMMA).[16,24,25] The transferrable size achievable using a wet transfer method is limited to less than a few inches of wafer because of the weak mechanical strength of spin-coated PMMA layers. However, the scale of roll-to-roll dry transfer assisted by a thermal release tape is in principle unlimited. In the process of roll-to-roll dry transfer, the first layer sometimes shows approximately two to three times larger sheet resistance than that of the PMMA-assisted wet transfer method. As the number of layers increases, the resistance drops faster compared to the wet transfer method (Fig. 9.6a). We postulate that the adhesion of the first layer with the substrate is not strong enough for complete separation of the graphene films from the thermal release tape. As a result, there can be mechanical damage on the graphene films, leading to an increase in the overall sheet resistance. Because additional layers are not directly affected by the adhesion with the substrate surface, the sheet resistance of multilayers prepared by the roll-to-roll method does not differ much from that for the wet transfer case. The *p*-doping with HNO_3 clearly enhances the electrical properties of graphene

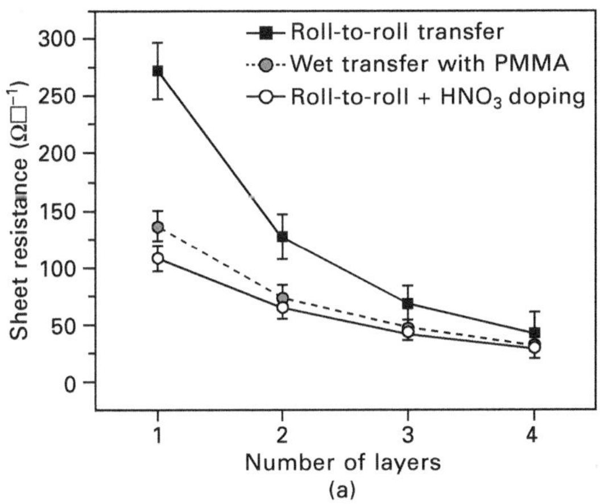

9.6 Electrical characterization of layer-by-layer transferred and HNO_3-doped graphene films. (a) Sheet resistances of transferred graphene films using roll-to-roll (R2R) dry transfer method combined with thermal release tapes and a PMMA-assisted wet transfer method. (b) Comparison of sheet resistance from this research and transmittance plots taken from other references. The dashed arrows indicate the expected sheet resistances at lower transmittance. (c) Electrical properties of a monolayer graphene Hall bar device in vacuum. Four-probe resistivity (left inset) is measured as a function of gate voltage in the monolayer graphene Hall bar shown in the right inset at room temperature (solid curve) and $T = 6$ K (dashed curve). The QHE effect at $T = 6$ K and $B = 9$ *T* is measured in the same device. The longitudinal resistivity ρ_{xx} and Hall conductivity σ_{xy} are plotted as a function of gate voltage. The sequence of the first three half-integer plateaux corresponding to $\upsilon = 2$, 6 and 10, typical for single-layer graphene, are clearly seen. The Hall effect mobility of this device is $\mu_{Hall} = 7350$ cm^{-2} V^{-1} s^{-1} at 6 K (~5100 cm^{-2} V^{-1} s^{-1} at 295 K). Scale bar (inset), 3 μm. (d) Electromechanical properties of graphene-based touchscreen devices compared with ITO/PET electrodes under tensile strain. The inset shows the resistance changes with compressive and tensile strain applied to the upper and lower graphene/PET panels, respectively. (From Ref. [2].)

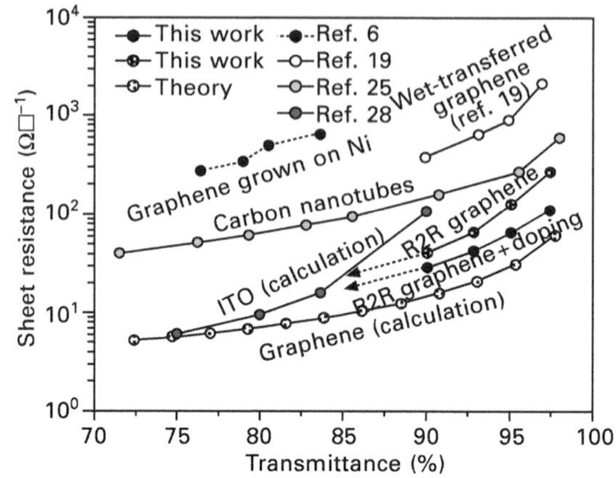

(b)

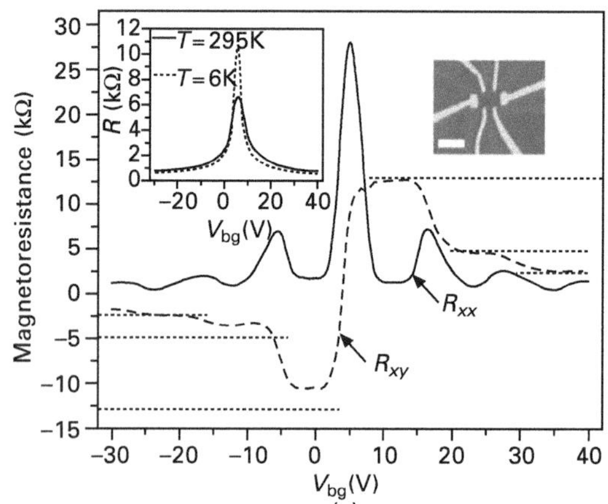

(c)

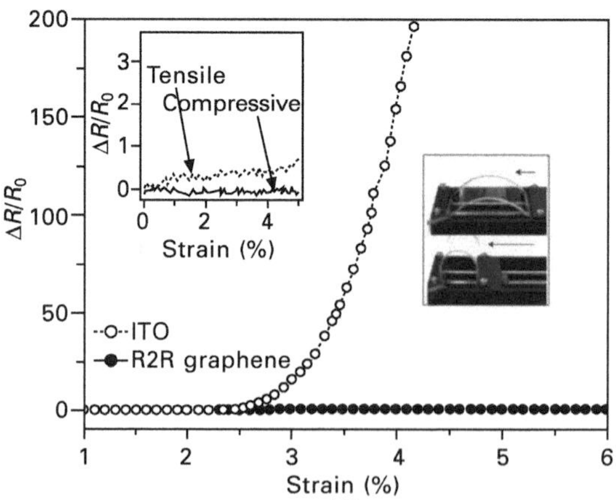

(d)

9.6 Continued

films, and is more effective in roll-to-roll processes. The sheet resistance of the *p*-doped four-layer graphene film with ~90% optical transmittance is as low as ~30 Ω/sq, which is superior to common transparent electrodes such as indium tin oxide (ITO) and carbon-nanotube films (Fig. 9.6b).[26]

Standard electron-beam lithography has been used to fabricate graphene Hall bars on conventional 300 nm SiO_2/Si substrates (Fig. 9.6c). The left inset of Fig. 9.6c shows the four-terminal resistance of such samples as a function of backgate voltage (V_{bg}) at both room temperature (solid curve) and at a low temperature ($T = 6K$) and zero magnetic field. We observe the graphene specific gate bias dependence of the resistance with a sharp Dirac peak and an effective Hall mobility of 7350 $cm^2\ V^{-1}\ s^{-1}$ at low temperatures. This allows the observation of the quantum Hall effect (QHE)[27] at 6K and a magnetic field of $B = 9T$ (Fig. 9.6c, right). The fingerprint of single-layer graphene, the half-integer quantum Hall effect, is observed with plateaux at filling factors of $u = 2,6$ and 10 at $R_{xy} = 1/2$, 1/6 and 1/10 (h^2/e^2), respectively. Although the sequence of the plateaux remains for both the electron side and the hole side, there is a slight deviation from the fully quantized values on the hole side.

Finally, the electromechanical properties of graphene/PET touchscreen panels were tested (Fig. 9.6d). Unlike an ITO-based touch panel, which easily breaks under ~2–3% strain, the graphene-based panel resists up to 6% strain; this is limited not by the graphene itself, but by the printed silver electrodes (Fig. 9.6d).[28]

9.4 Conclusions

In conclusion, we showed a method for graphene growth on nickel and copper layers using chemical vapour deposition and further described a method to transfer graphene films to stretchable substrates. This will lead to numerous applications including use in large-scale flexible, stretchable, foldable transparent electronics. In addition, given the scalability and processability of roll-to-roll and CVD methods and the flexibility and conductivity of graphene films, we anticipate that the commercial production of large-scale transparent electrodes, replacing ITO, will be realized in the near future.

9.5 References

1. Hong, B. H. *et al.* Large-scale pattern growth of graphene films for stretchable transparent electrodes. *Nature* **457**, 706–710 (2009).
2. Hong, B. H. *et al.* Roll-to-roll production of 30-inch graphene films for transparent electrodes. *Nature Nanotech.* **5**, 574–578 (2010).
3. Yu, Q. *et al.* Graphene segregated on Ni surfaces and transferred to insulators. *Appl. Phys. Lett.* **93**, 113103 (2008).
4. Ferrari, A. C. *et al.* Raman spectrum of graphene and graphene layers. *Phys. Rev. Lett.* 187401 (2006).

5. Obraztsov, A. N., Obraztsova, E. A., Tyurnina, A. V. and Zolotukhin, A. A. Chemical vapor deposition of thin graphite films of nanometer thickness. *Carbon* **45**, 2017–2021 (2007).
6. Khang, D.-Y. *et al.* Individual aligned single-wall carbon nanotubes on elastomeric substrates. *Nano Lett.* **8**, 124–130 (2008).
7. Yang, P. *et al.* Mirrorless lasing from mesostructured waveguides patterned by soft lithography. *Science* **287**, 465–467 (2000).
8. Li, X. *et al.* Highly conducting graphene sheets and Langmuir–Blodgett films. *Nature Nanotechnol.* **3**, 538–542 (2008).
9. Eda, G., Fanchini, G. and Chhowalla, M. Large-area ultrathin films of reduced graphene oxide as a transparent and flexible electronic material. *Nature Nanotechnol.* **3**, 270–274 (2008).
10. Nair, R. R. *et al.* Fine structure constant defines visual transparency of graphene. *Science* **320**, 1308 (2008).
11. Lewis, J. Material challenge for flexible organic devices. *Mater. Today* **9**, 38–45 (2006).
12. Sun, Y., Choi, W. M., Jiang, H., Huang, Y. Y. and Rogers, J. A. Controlled buckling of semiconductor nanoribbons for strechable electronics. *Nature Nanotechnol.* **1**, 201–207 (2006).
13. Khang, D.-Y., Jiang, H., Huang, Y. and Rogers, J. A. A stretchable form of single crystal silicon for high-performance electronics on rubber substrates. *Science* **311**, 208–212 (2006).
14 Lee, C., Wei, X., Kysar, J. W. and Hone, J. Measurement of the elastic properties and intrinsic strength of monolayer graphene. *Science* **321**, 385–388 (2008).
15. Li, X. *et al.* Large-area synthesis of high-quality and uniform graphene films on copper foils. *Science* **324**, 1312–1314 (2009).
16. Li, X. *et al.* Transfer of large-area graphene films for high-performance transparent conductive electrodes. *Nano Lett.* **9**, 4359–4363 (2009).
17. Hecht, D. S. *et al.* Carbon nanotube film on plastic as transparent electrode for resistive touch screens. *J. Soc. Inf. Display* **17**, 941–946 (2009).
18. Hass, J. *et al.* Why multilayer graphene on 4H-SiC(000-1) behaves like a single sheet of graphene. *Phys. Rev. Lett.* **100**, 125504 (2008).
19. Sprinkle, M. *et al.* First direct observation of a nearly ideal graphene band structure. *Phys. Rev. Lett.* **103**, 226803 (2009).
20. Das, A. *et al.* Monitoring dopants by Raman scattering in an electrochemically top-gated graphene transistor. *Nature Nanotech.* **3**, 210–215 (2008).
21. Geng, H.-Z. *et al.* Effect of acid treatment on carbon nanotube-based flexible transparent conducting films. *J. Am. Chem. Soc.* **129**, 7758–7759 (2007).
22. Schrivera, M., Reganb, W., Losterb, M. and Zettl, A. Carbon nanostructure-aSi:H photovoltaic cells with high open-circuit voltage fabricated without dopants. *Solid State Commun.* **150**, 561–563 (2010).
23. Wu, J. *et al.* Organic light-emitting diodes on solution-processed metal nanowire mesh transparent electrodes. *ACS Nano* **4**, 43–48 (2010).
24. Reina, A. *et al.* Large area, few-layer graphene films on arbitrary substrates by chemical vapor deposition. *Nano Lett.* **9**, 30–35 (2009).
25. Cai, W. W. *et al.* Large area few-layer graphene/graphite films as transparent thin conducting electrodes. *Appl. Phys. Lett.* **95**, 123115 (2009).
26. Lee, J.-Y., Connor, S. T., Cui, Y. and Peumans, P. Solution-processed metal nanowire mesh transparent electrodes. *Nano Lett.* **8**, 689–692 (2008).

27. Cao, H. L. *et al.* Electronic transport in chemical vapor deposited graphene synthesized on Cu: Quantum Hall effect and weak localization. *Appl. Phys. Lett.* **96**, 122106 (2010).
28. Cairns, D. R. *et al.* Strain-dependent electrical resistance of tin-doped indium oxide on polymer substrates. *Appl. Phys. Lett.* **76**, 1425–1427 (2000).

10
Epitaxial growth of graphene thin films on single crystal metal surfaces

J. CORAUX, Institut Néel, France and A. T. N'DIAYE, C. BUSSE and T. MICHELY, Universität zu Köln, Germany

Abstract: Epitaxial growth of graphene on metals has been known about for five decades and has encountered renewed interest since 2006–2007 when it was realized that graphene mass production is a major hurdle in the development of graphene-based applications. With the help of model systems, i.e. prepared in clean conditions (ultra-high vacuum) and at clean surfaces (single-crystalline metallic ones), much progress has been made towards the understanding of graphene growth on metals, from the elementary processes governing growth towards the tailoring of the morphology of the graphene samples.

Key words: graphene, metals, chemical vapour deposition, defects, structure, edges, multilayer, growth processes.

10.1 Introduction

Graphene is the name given to an atomically thin layer of sp^2-hybridized carbon (Fig. 10.1). Its unconventional properties were explored by researchers from 2004, after it was isolated by A. Geim and colleagues. in Manchester by mechanical exfoliation of graphite (Novoselov, 2004). Such samples have fuelled exceptionally sustained research owing to their high structural quality, and they are set to keep revealing a wealth of remarkable properties resulting from the high surface to volume ratio, mechanical properties, chemical inertness, electronic band structure, etc., of graphene (Geim, 2009). A consensus has emerged as to the necessity for alternative preparation methods, because of large deviations in the quality of samples obtained by mechanical exfoliation, and because this technique is very fastidious, together precluding the reliable and efficient production of graphene. Accordingly alternative preparation methods have been explored, among which are epitaxial growth on silicon carbide and on metals. In contrast to silicon carbide, single-crystalline metal substrates can yield high-quality graphene with one layer uniformly covering a surface (some applications then require the transfer of graphene onto an adequate support). Another direction in graphene research focuses towards understanding properties, possibly new

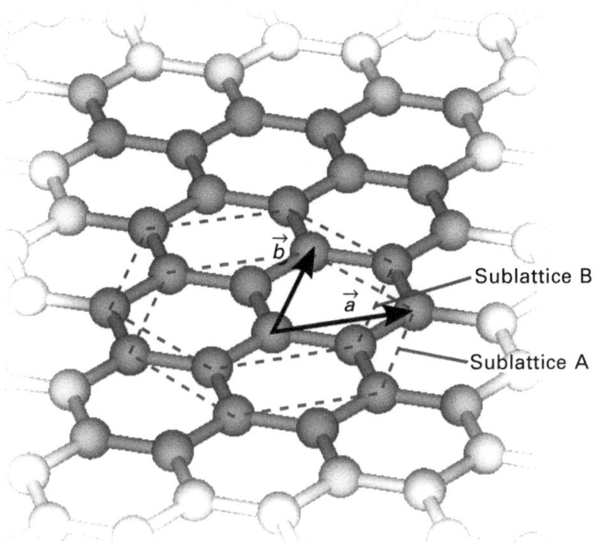

10.1 Schematics of graphene's honeycomb structure showing the two carbon hexagonal sublattices in different shades. The unit cell is defined by the two lattice vectors represented with black arrows.

ones, resulting from its interaction with an environment, in particular with a metal, which has obvious relevance when considering the graphene/electrode contact. Graphene prepared directly on metal surfaces, which has been known for decades (Banerjee, 1961; Karu, 1966; Irving, 1967; Presland, 1969), provides a system ideally suited to address such issues.

Growth processes are often monitored by fine inspection of a system's structure. Therefore we first describe the structure of graphene on metals including the defects that are commonly encountered. We then focus on the growth of graphene. Throughout the discussion, we lay emphasis on the strong influence of the metal–graphene interaction upon growth processes and consequently graphene's structure. Based on our studies of graphene growth on iridium and other contributions in the literature, we review conditions for obtaining high-quality graphene.

10.2 Structure of graphene on metals

The C–C bond is one of the strongest in nature (3.61 eV for a single bond), stronger than the bond between carbon and a noble metal. This explains why *sp*2 carbon (more stable than *sp*3) and the noble metal coexist as two separate phases, at least at room temperature and ambient (or below) pressure. As we shall discuss in Section 10.3, graphene or multilayer graphene can be obtained. The epitaxial relationship is more or less well defined: while the texture orientation is unique ([0001] direction of graphene or graphite parallel to the dense-packed metal surface normal), the azimuthal (in-plane) orientation can be multiple, either because different variants can be formed, or because the azimuthal orientation scatters around an energetically favourable situation. Graphene is the stiffest material known to date (Lee, 2008), which has important consequences for its structure: biaxial strains in graphene resulting from its epitaxial growth are restricted to a few tenths of a percent, which renders graphene's lattice incommensurate with that of the substrate in the general case. Due to the flexibility of graphene, delaminations (wrinkles) are possible in response to thermal mismatch between the substrate and graphene. Graphene can also bend up or down at substrate edges.

10.2.1 Commensurate or not

Among transition metals, Co and Ni dense-packed surfaces exhibit only small lattice mismatches with graphene, of 1.8 and 1.2% respectively (for Fe, this is only 0.9%, but presumably due to the large solubility of C in Fe at ambient conditions, the control of the number of graphene layers on Fe is almost impossible). Considering graphene's high Young's modulus (Lee, 2008), stretching the graphene to match the metal lattice parameter would, however, imply that the metal can sustain a considerable amount of stress, actually more than its breaking strength. This is in apparent contradiction with experimental observations pointing to a commensurate graphene layer on the metal. This contradiction may be explained if one considers a possible change of graphene's mechanical properties upon its bonding on the metal (which is strong on Co or Ni). Electron diffraction revealed a (1×1) (commensurate) superstructure for graphene/Ni(111). In this superstructure half the carbon atoms sit on top of the Ni atoms in the topmost substrate layer, and the remaining carbon atoms sit on top of the Ni atoms in the third (relative to the topmost Ni layer) substrate layer. The former carbon atoms are slightly higher than the latter (Gamo, 1997) (Fig. 10.2a).

The situation is more complex in the case of a larger lattice mismatch between graphene and the substrate. This is the case for a number of transition metals, like Cu(111) (4% lattice mismatch with graphene), Ir(111), Pt(111), or Ru(0001) (~10%). With such substrates the coincidence of the graphene and metal surface lattices can only be local and partial: if at a given location the centres of carbon rings approximately prolong the fcc(111) or hcp(0001) atomic arrangement of the metal atoms below, a few nearest neighbour distances away, this coincidence is lost. This gives rise to a periodic lattice with the symmetry of the less symmetric surface in the {graphene,metal} system. In the case of graphene on a metallic surface with hexagonal symmetry, the superlattice exhibits hexagonal symmetry too. Such lattices

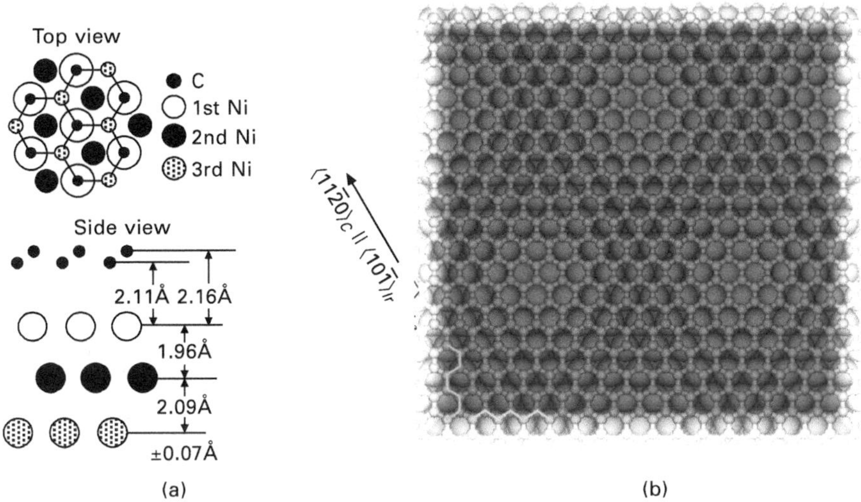

10.2 (a) Structural model for graphene on Ni(111) derived from a fit of I-V LEED spectra (from Gamo, 1997, © Elsevier); (b) ball model for a moiré between graphene and a Ir(111) plane. Section of zigzag and armchair carbon rows are highlighted in the bottom left corner.

are encountered in various systems, such as metal/metal epitaxial systems or oxide layers on metals (see, for instance Wiederholt (1995) and Ritter (1998), respectively). Coincidence site lattices are often described following a moiré pattern (Fig. 10.2b), by analogy with the optical beating fringes appearing when looking through two thin, transparent, superimposed tissue veils (historically composed of fibres from angora goats, a type of textile whose name was adapted from Arabic to French as 'moiré' and to English as 'mohair') with similar fibre structure. For convenience and consistency with the literature, we hereafter use the term 'moiré'.

Interesting geometric properties are associated with moirés (Amidror, 2007). It can be shown that in the case when the metal surface dense-packed directions align with <11–20> directions in graphene (zigzag directions), the lattice parameter of the moiré ($a_{\text{moiré}}$) can be written in a simple way as a function of the lattice parameter of the metal (a_m) and of graphene (a_C): $1/a_{\text{moiré}} = 1/a_C - 1/a_m$. This implies that $a_{\text{moiré}}$ is inversely proportional to the lattice mismatch between graphene and the metal. For ~ 10% lattice mismatches between graphene and the metal, $a_{\text{moiré}}$ is in the range of 2–3 nm, for graphene on Cu(111), $a_{\text{moiré}}$ was recently found to be 6.6 nm (Gao, 2010; Zhao, 2011). Note that the geometrical parameters of the moiré (orientation, lattice parameter) are very sensitive to faint variations of those of the metal and of the graphene, which allows a fine description of the graphene or metal structure (Coraux, 2008; N'Diaye, 2008b, 2009a).

Standard techniques that have been used for identifying graphene's structure are surface science ones, most prominently low-energy electron diffraction (LEED), also scanning tunnelling microscopy (STM), and to a much lesser extent reflection high-energy electron diffraction (RHEED) or surface X-ray diffraction (SXRD). As an illustration, Fig. 10.3 shows characteristic signatures for graphene on lattice mismatched transition metals using LEED, STM and RHEED. The LEED pattern displays a sixfold symmetry, appearing as six groups of spots. The finer structure of these groups reveals two prominent spots (the outer one for the metal, the inner one for graphene) and satellites spots that arise from the moiré (N'Diaye, 2008b). STM captures the atomic structure of graphene (dark spots are centres of carbon rings) modulated by a larger scale superstructure with a periodicity in the range of 2–3 nm, which is the moiré pattern (Land, 1992). The RHEED diagrams show sets of crystal truncation rods whose streaky character point to the flatness (except for the moiré corrugation) of the surface. Such patterns are observed periodically every 60°, in agreement with the sixfold symmetry of the surface. Besides the zero-order central rod, strong first-order rods are related to the graphene (inner) and metal (outer) lattices (see solid arrows in Fig. 10.3c). These main rods are surrounded by satellite ones, which are related, similar to the satellite spots in LEED, to the moiré (dotted arrows in the figure).

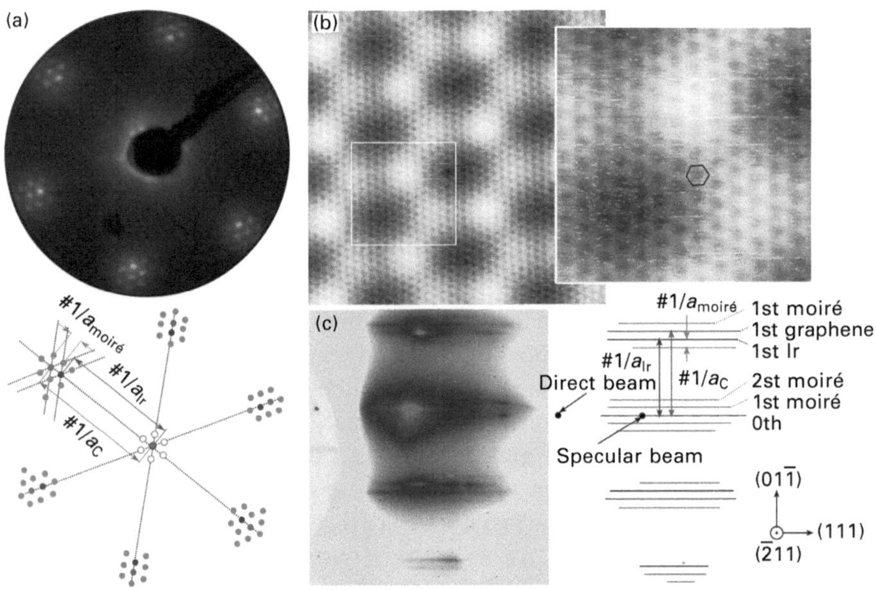

10.3 (a) LEED pattern measured with 80 eV electrons, (b) STM topographs (left: 8.8 × 8.8 nm^2, right: 2.5 × 2.5 nm^2), and (c) RHEED pattern recorded with 10 keV electrons, along the [$\bar{2}$11] azimuth, for graphene on Ir(111).

Finally, note that the above discussion applied to graphene on metal surfaces with a threefold symmetry. Much less literature addresses graphene prepared on, for example, (100) surfaces of fcc metals (Hamilton, 1980; Zhao, 2011) or (110) surfaces of body-centred cubic metals. On these metals graphene/metal moirés are also found, but their symmetry is decreased compared to that of moirés with graphene on threefold symmetric surfaces.

10.2.2 Height of the graphene sheet

The distance between the metallic surface and the graphene surface is characteristic of the metal/graphene interaction. Extreme cases for this interaction are van der Waals binding, such as in between the graphite plane, and strong hybridization of the metal d bands with the π bands of graphene. While the first case mostly preserves the conical character of graphene π bands, the second deeply modifies them, causing band gap opening at the K point in the Brillouin zone and bending of the bands. In both cases, charge transfer between graphene and the metal is *a priori* expected.

In the case of dominating van der Waals interactions, as in graphite, the graphene–metal distance is expected to be close to the graphite interplane distance, i.e. around 0.345 nm. In the opposite situation, the carbon and metal atoms form bonds with a covalent character, therefore the distance is expected to be much shorter (e.g. 0.21 nm for graphene/Ni(111); Gamo, 1997).

The graphene–metal distance remains poorly characterized at experimental level. This is mainly ascribed to the limitations of the techniques that are commonly employed: STM has so far been unable to disentangle topographic and electronic contributions to the apparent height measurements in the graphene–metal system (Marchini, 2007; Vazquez de Parga, 2008); atomic force microscopy (AFM) was only used in air for graphene on metals, so that the sensitivity of the technique does not allow for a sufficiently accurate determination of the height. The only techniques which proved relevant up to now are LEED I-V measurements and SXRD which were applied to graphene on Ni(111), Ru(0001), and Pt(111) (Gamo, 1997a; Martoccia, 2008; Sutter, 2009a; Moritz, 2010). The analysis of the SXRD data relies on the choice of structural models *a priori*, which introduces an (unknown) uncertainty in the value of the graphene–metal height. The LEED I-V analysis is based on the simulation of electron reflectivity using a dynamical diffraction framework, which allows a partial agreement between the simulations and the experiment, thus imposing careful interpretation of the simulations.

Concerning theory studies, a number of reports provide estimates of the graphene–metal distance (Bertoni, 2004; Nemec, 2006, 2008; Giovannetti, 2008; Wang, 2009; Khomyakov, 2009; Ran, 2009). Yet, noticeable deviations are found, for example for graphene/Pd(111) (Nemec, 2006; Giovannetti,

2008). It can be argued that the geometry used for the calculations is frequently over-simplified. The approximations employed to perform the *ab initio* calculations yield different results: the local density spin approximation is known to result in overbinding, while the generalized gradient approximation is expected to result in large graphene–metal distances as suggested by the over-estimation of the graphene–graphene distance in graphite. Most important, for a long time van der Waals interactions were not included in the calculations and are only now starting to be considered (Vanin, 2009; Lazic, 2010; Busse, 2011).

The difficulty in assessing the graphene–metal height also concerns the modulation of this height in the case of graphene–metal systems with a moiré. So far, with the exception of one experimental report (Martoccia, 2008), only demanding first principle calculations were employed in this respect (N'Diaye, 2006, 2008a, Feibelman, 2008; Wang, 2008; Brugger, 2009). From these works it seems that graphene height is modulated between 0.30 and 0.38 nm on Ir(111), a system for which graphene–metal hybridization is *a priori* weak (Pletikosic, 2009). On Ru(0001), the height was claimed to vary between 0.22 and 0.37 nm (Wang, 2008): for the regions where the height is lower, there is a strong covalent-like graphene–Ru interaction.

10.2.3 Orientation variants, small-angle twins and dislocations

The epitaxial relationship between graphene and the metal is well defined along the direction perpendicular to the graphene and metal surface planes: [0001] for graphite parallel to [111] (fcc metal) or [0001] (hcp metal) direction. This is not always the case in the plane of the graphene and metal surfaces, in other words, there exist several preferred azimuthal orientations for graphene on metal (orientation variants) and the orientation of graphene may fluctuate around these preferred situations. It was recently argued that the weaker the graphene–metal interaction, the less well defined should the graphene azimuthal orientation be (Sutter, 2009a).

The formation of several orientation variants was highlighted for graphene/Ir(111) for specific preparation conditions (Loginova, 2009a, 2009b), for which four orientation variants were identified (Fig. 10.4a). The azimuthal twin between each variant is large, for several to several tens of degrees. Obviously, each variant represents a local minimum for the total energy of the graphene–metal system. It seems, however, that the configuration corresponding to the dense-packed metal rows aligned to the carbon zigzag edges is more energetically favourable as it always forms first and can be exclusively obtained for graphene/Ir(111) in some growth conditions. Also consistent with this larger stability is the fact that this rotation variant is more inert against oxidation (van Gastel, 2009). Rotational variants were

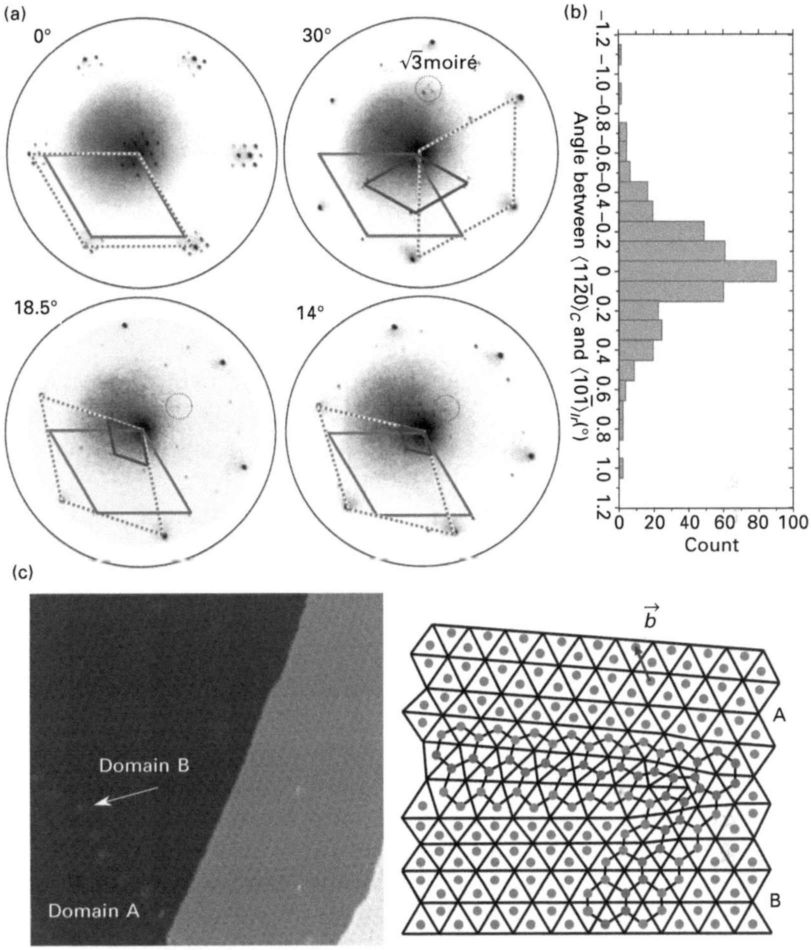

10.4 (a) Micro-LEED pattern for each of the four rotational variants for graphene on Ir(111), corresponding to a rotation of 0°, 30°, 18.5° and 14° of the carbon zigzag rows with respect to the dense-packed rows of Ir(111). The dark grey, dotted, and light grey rhombus highlight the Ir(111), graphene, and moiré unit cells, respectively (from Loginova, 2009a, © The American Physical Society, http://prb.aps.org/abstract/PRB/v80/i8/e085430). (b) Distribution of angles between the graphene zigzag rows (<11-20>) and the dense-packed rows (<10-1>) of an Ir(111) surface for the energetically preferred 0° rotational variant. (c) (left) STM topograph (108 nm × 108 nm^2) showing three domains with a different orientation of graphene on Ir(111) and (right) the atomic structure of defects, such as the one indicated with an arrow on the left panel, which the grain boundaries consist of (adapted from Coraux, 2008, © The American Chemical Society). Note that the observed superstructure is the graphene/Ir(111) moiré and that it amplifies small-angle twins. A and B mark two of the domains; the arrow highlights one of the typical defects forming the grain boundary.

also found for graphene/Pt(111) (Sasaki, 2000; Sutter, 2009a), but it has not been possible so far to achieve a single orientation, suggesting that all variants have comparable stability.

Depending on the growth conditions (see Section 10.3), the azimuthal orientation of graphene might be more or less well-defined around each orientation variant (N'Diaye, 2008b). The scatter in this orientations is characterized by a few tenths of a degree only for graphene/Ir(111) if graphene is prepared above 1000°C (Fig. 10.4b), and several tens of degrees if graphene is prepared at 600°C (Coraux, 2008). The smallest scatters can be efficiently characterized by analysing the moiré scatters, which largely amplify the carbon lattice ones (Coraux, 2008; N'Diaye, 2008a).

At the boundary between graphene domains that are twinned (scattered twins or orientation variants), the carbon lattice tends to minimize the number of dangling bonds through the formation of edge dislocations, that is pairs of a heptagon and a pentagon (Coraux, 2008) as shown in Fig. 10.4(c). Such defects are known to accommodate twin boundaries (Bollmann, 1964).

10.2.4 Other defects

Deviations from the pure two-dimensional honeycomb structure of graphene might be considered as defects. A detailed review of defects would include a discussion of their origin, stability and electronic properties, and is beyond the scope of this chapter. Defects can be intrinsic to the graphene itself, which is the case for vacancies, substitutional atoms, or graphene edges (Fig. 10.5a). They can also be extrinsic, such as being imposed by the graphene environment, which is the case for graphene undulations induced by those of the substrate (e.g., step edges) or thermally induced delaminations (Fig. 10.5b). Intrinsic defects for epitaxial graphene are so far poorly characterized. Extrinsic ones are better explored. Graphene wrinkles (linear delaminations) present after cooling samples down to room temperature following growth are due to the mismatch of thermal expansion coefficients. The wrinkle formation actually sets in at a well-defined temperature (whose value depends on the growth temperature) when cooling samples down (N'Diaye, 2009a), corresponding to a situation when strain accumulation becomes unfavourable as compared to a loss of graphene/substrate binding energy and accompanying bending energy associated with graphene delamination. Such defects were first investigated for epitaxial graphene on SiC (Cambaz, 2008; Biedermann, 2009; Sun, 2009), and further for graphene on metals (Kim, 2009; Obraztsov, 2007; Chae, 2009; Loginova, 2009a; N'Diaye, 2009a; Sutter, 2009a). Their nucleation occurs abruptly within a fraction of a second, presumably at the location of defects in graphene or around substrate step edges where the graphene lattice is already bent (Chae, 2009). Their propagation is to the contrary progressive (Sutter, 2009a). The formation of wrinkles (when

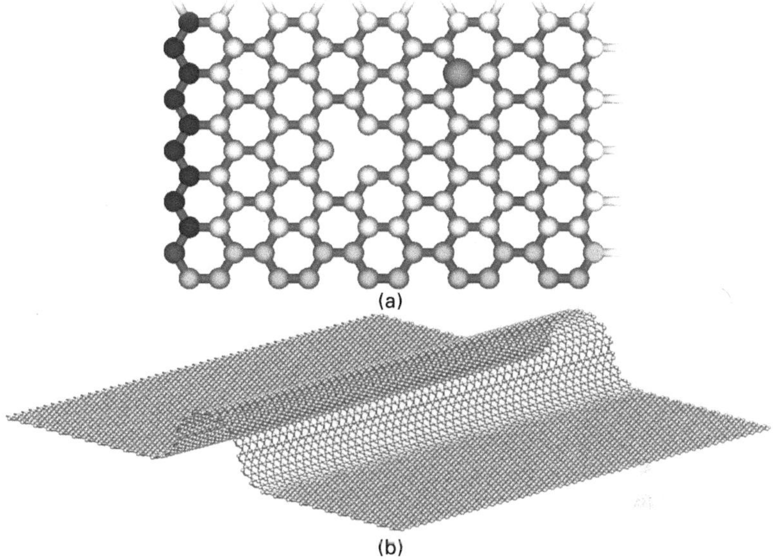

10.5 (a) Vacancies (missing atom in the carbon lattice), substitutional atoms (replacing a carbon one, shown with a larger radius), and graphene edges (black: zigzag, grey: armchair); (b) wrinkle formed in graphene following delamination from the substrate (not shown).

cooling down the sample from growth temperature) and their suppression (upon heating up the sample again) were shown to take place at distinct temperatures, pointing to a hysteresis.

The bending of the graphene lattice by the substrate step edges (the graphene lattice is not interrupted by the step edges) was mainly addressed for graphene in epitaxy on metals, for both Ru(0001) (Pan, 2007) and Ir(111) (Coraux, 2008) substrates. The graphene lattice was shown to be continuous across the substrate step edges. On Ir(111) the radius of curvature could even be evaluated by taking benefit of the graphene/Ir(111) moiré and its high sensitivity to sub-Ångström displacements. It was found to be close to that of thinnest single wall carbon nanotubes, i.e. approximately 0.27 nm.

10.3 Growth of graphene on a metal

In discussing graphene growth on metals, one may consider several situations, first depending on the nature of the carbon source and the way it is provided, then depending on the affinity of carbon with the metal substrate. Both aspects can be linked. For instance, carbon solubility in Ni is of the order of 1 at% at 1000°C, a temperature which is often used for graphene growth on Ni, and drastically drops at lower temperature, so that graphene growth can proceed from segregation of carbon dissolved in bulk Ni, which may be eased by slow cooling of Ni from ~ 1000°C down to room temperature (after Ni has been enriched with carbon at high temperatures if required) (Kim, 2009). On the contrary, Ir can store much less carbon in its bulk (Arnoult, 1972), which implies that graphene growth proceeds only by impinging carbon adspecies supplied to the surface.

A variety of carbon sources can be employed for the growth, and it is probable that others will be used in the future. The most convenient source is a gas of carbon-containing molecules (Fig. 10.6a). Such molecules are usually hydrocarbons (Oshima, 1997). Until 2009, the gases were injected at low-pressure in ultra-high vacuum systems (UHV). In 2009, gas mixtures (e.g. $CH_4/Ar/H_2$) were also employed at ambient pressure (Kim; 2009; Reina, 2009a), opening valuable perspectives for graphene production. In both cases (UHV and ambient pressure) the carbon-containing molecules are cracked at the metal surface; for this the metal surface is made catalytically active by heating it. This process can be considered an heterogeneous chemical reaction between a gas phase and a solid one, which is why the technique is often referred to as chemical vapour deposition (CVD). Another option for graphene growth is the use of an atomic carbon source (Loginova, 2008). Finally, in the case of metals having a non-negligible carbon solubility, carbon segregation from the bulk towards the surface might be an alternative route (Sutter, 2008).

In the following we focus on the growth processes disregarding the way the carbon ad-species have been formed at the metal surface (CVD, atomic

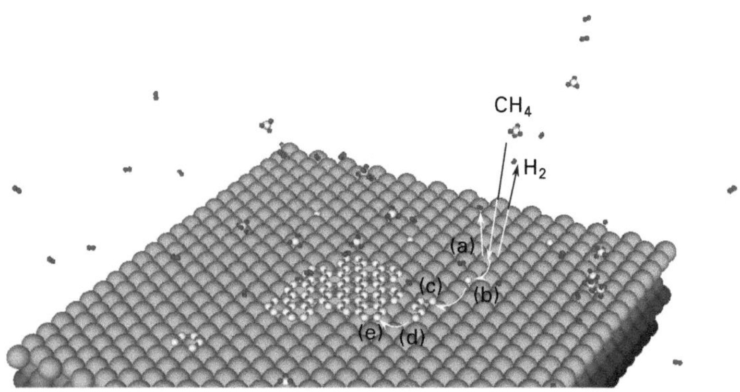

10.6 Schematics for the different elementary processed during CVD growth of graphene on a metal, with a methane precursor as an illustration. (a) A methane molecule is adsorbed at the surface, cracked into a carbon adatom, leaving two hydrogen molecules released in vacuum. (b) The carbon adatom diffuses at the surface (or sub-surface). (c) The carbon adatom incorporates in a carbon chain, forming a pentamer (on metal surfaces like Ir(111) or Ru(0001), Loginova, 2008). (d) The carbon chain diffuses at the surfaces and (e) is incorporated in the graphene flake.

carbon source, segregation). This is legitimate if the nature of the carbon source only affects graphene growth through the rate of adatom formation. This is at least the case in prototypical systems that are suited to address elementary growth processes, for instance growth by CVD with ethene or by an atomic carbon source at single-crystalline metal surfaces (Loginova, 2009b).

10.3.1 Growing graphene nanoflakes

A derivative of the CVD growth for graphene consists in (i) adsorbing carbon-containing molecules at the metallic surface, at such a temperature that the molecules do not desorb to vacuum, and (ii) increasing the sample temperature for promoting hydrocarbon decomposition and eventually graphene growth (Coraux, 2009).

The surface is first saturated with molecules, and the resulting adlayer prevents the strong bonding of further molecules at the surface. In practice, ethene molecules are transformed into ethyledine ones at Pt(111) and Ir(111) (Nieuwenhuys, 1976) surfaces at room temperature. The amount of carbon in the resulting adlayer is less than that in a full graphene layer, due to steric effects.

In the next step the substrate is heated up. In the course of increasing the sample temperature, dehydrogenation of the molecules proceeds, leaving pure-carbon adspecies being mobile at the surface. Nanometer-scale carbidic islands develop from this carbon sea. The electronic interaction between the carbidic islands and the metal surface is strong, all the more as the islands are small, as has been shown for graphene on Ir(111). This is because in small islands the proportion of edge atoms is large: such low coordinated atoms have a tendency to form bonds with the metal atoms underneath. This strong bonding was detected through energy shifts in the carbon and metal atoms core levels. As a result of the enhanced bonding of graphene edges to the substrate, the small graphene islands were shown to be dome-shaped and characterized by a small distance to the metal surface, noticeably below that for a plain graphene sheet on Ir(111), for island radii smaller than 1–2 nm, and all the more as the radii is small (Fig. 10.7) (Lacovig, 2009). Such small islands, as imaged by STM (Fig. 10.8a), are found at the terraces of the metallic surface, and occasionally bond to substrate step edges. They are characterized by a poorly defined height and rounded edges. Further increase of the temperature modifies their morphology (Fig. 10.8b–e). Their height becomes well defined as seen by STM for islands radii larger than a few nanometres.

Such islands with well-defined height are bound by straight edges that are precisely oriented along the <11-20> directions of graphene, consistent with the prominent proportion of zigzag portions composing these edges. At

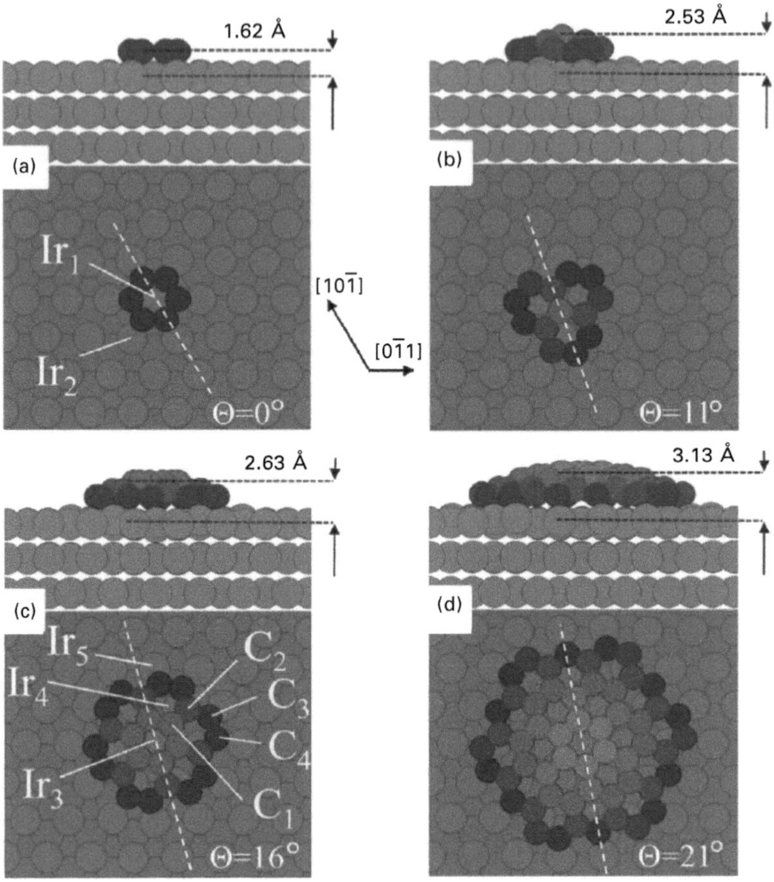

10.7 Calculated structural models for graphene nanoislands on Ir(111) as a function of their (small) size, on a side view and on a top view: for (a) one, (b) three, (c) seven, and (d) 19 carbon rings. Note that the average height of the islands increases with their size (from Lacovig, 2009, © The American Physical Society, http://prl.aps.org/abstract/PRL/v103/i16/e166101).

sufficiently high growth temperature, it was shown for graphene on Ir(111) that the island edges align to the <1-10> direction of Ir(111) (N'Diaye, 2008b). Graphene islands were also reported on Pt(111) (Land, 1992), Ir(111) (N'Diaye, 2006; Coraux, 2009), and Co(0001) (Eom, 2009). The occurrence of straight edges for graphene is the evidence that carbon mobility is active at edges. In the explored growth temperature range (700–1300°C), these effects are rather slow as the islands are most often not in their equilibrium, hexagonal shape. Carbon diffusion at edges could proceed through that of carbon pentagons continuously formed at graphene edges during growth (Frenklach, 2004; Whitesides, 2010). Pentagon collisions result in hexagons,

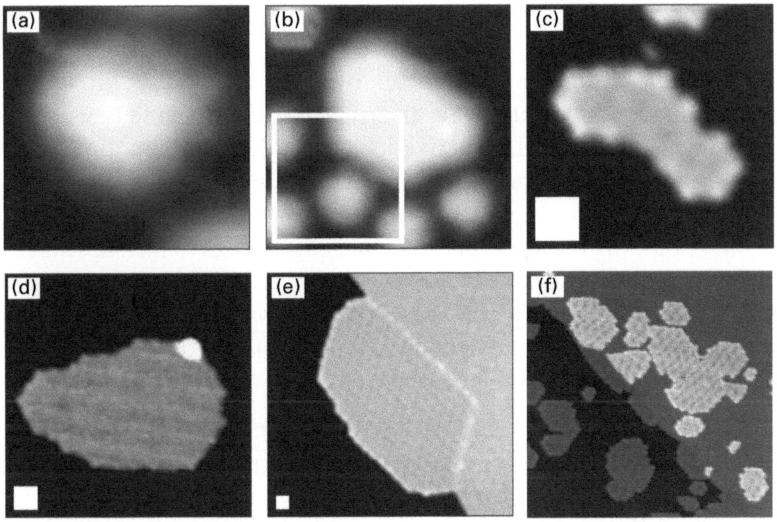

10.8 (a–e) STM topographs of typical carbidic (below 600°C) and graphene (higher temperatures) islands on Ir(111) as a function of the growth temperature: (a) 600°C (3.2 × 3.2 nm^2), (b) 700°C (6.2 × 6.2 nm^2), (c) 850°C (18 × 18 nm^2), (d) 1050°C (32 × 32 nm^2), and (e) 1200°C (64 × 64 nm^2). The size of the white box in (b–e) corresponds to the size of (a) (from Coraux, 2009, © Institute of Physics). (f) STM topograph (73 × 73 nm^2) of a graphene island grown on Ir(111) and presumably formed upon the coalescence of smaller islands.

i.e. graphene building blocks (Whitesides, 2007). On the contrary, edge smoothing through a 2D carbon gas surrounding the edges, that would be formed via carbon atom detachment at edges (more readily at regions where the 2D pressure is higher, i.e. at convex edges), is not a prominent process at the temperature of interest.

The temperature to which the substrate is heated up determines the size of the graphene islands: their average diameter can be adjusted between a few and several tens of nanometres for 600°C and 1200°C respectively on Ir(111). Through this temperature range, the island density drops by almost three orders of magnitude, consistent with the fact that the total amount of carbon at the surface is not modified. As the growth temperature increases, the proportion of graphene islands bound to substrate step edges increases. The density and size of the graphene islands vary because of a thermally induced ripening process. Considering the strength of the C–C bond in graphene, Ostwald ripening, i.e. the preferential growth of large islands at the expense of the dissolution of smaller ones through their higher pressure of carbon adatoms, cannot account for the observed efficient ripening below 1300°C. The presence of large islands with irregularly shaped (though zigzag) edges, exhibiting large vacancies, for growth temperature of 850°C during a few seconds (Fig. 10.8f), suggests that large islands are formed upon the coalescence of smaller ones that are mobile at the surface (Coraux, 2009). This process is known as Smoluchowski ripening. It is probably effective due to the incommensuracy of the graphene and metal lattices. For a strictly incommensurate island the activation energy for lateral motion would vanish, as for every carbon moving away from its optimum binding site, another one finds it. The existence of a migration barrier is probably linked to the stronger binding of the graphene island edge atoms to the substrate. As larger graphene islands display sizes consisting of an integer number of moiré units it is likely that the graphene edges bind to specific energetically preferred substrate sites. Prolonged heating results in islands becoming more compact, presumably due to efficient carbon mobility at graphene edges.

10.3.2 Growing plain graphene sheets

We now consider the situation where carbon atoms, whatever their source, are provided at the hot metallic surface.

From graphene islands to plain sheets

Graphene islands preferentially bind to the substrate step edges. The size (density) of the islands is the larger (lower) the higher the growth temperature. This is evidence that graphene nucleation along the steps is homogeneous and not at a specific defect site. The nucleation at step edges was observed by STM (Fig. 10.9a) and low-energy electron microscopy (LEEM) on Ir(111) (Coraux, 2009), Ru(0001) (Sutter, 2008), and Pt(111) (Sutter, 2009a), in a wide range of growth temperatures, and for different types of carbon sources. The graphene islands nucleate preferentially at the ascending side of the step edge. As discussed in the next subsection, this is presumably promoted by the fact that carbon atoms attach via σ bonds to the ascending substrate step.

Providing more carbon adatoms at the surface results in the growth of the islands. This growth is always anisotropic, faster at the lower terrace than at the upper one. For graphene on Ru(0001), there is almost no growth at the upper terrace (Sutter, 2008). On Ir(111) and Pt(111) (Coraux, 2009; Loginova, 2009b; Sutter, 2009a) the onset of growth is delayed (Fig. 10.9b). Probably it is delayed by an energetically costly transient state, for which the carbon–metal bond is broken before a carbon–carbon bond is formed across the substrate step edge. The growth of graphene over steps does not interrupt the graphene lattice. The carbon rows are continuous and bent vertically to adapt to the the terrace flow (Pan, 2007; Coraux, 2008); in other words, graphene rests like a blanket on the substrate.

Growing islands will eventually meet. STM showed that the moiré displays continuous rows through the location where the islands coalesced,

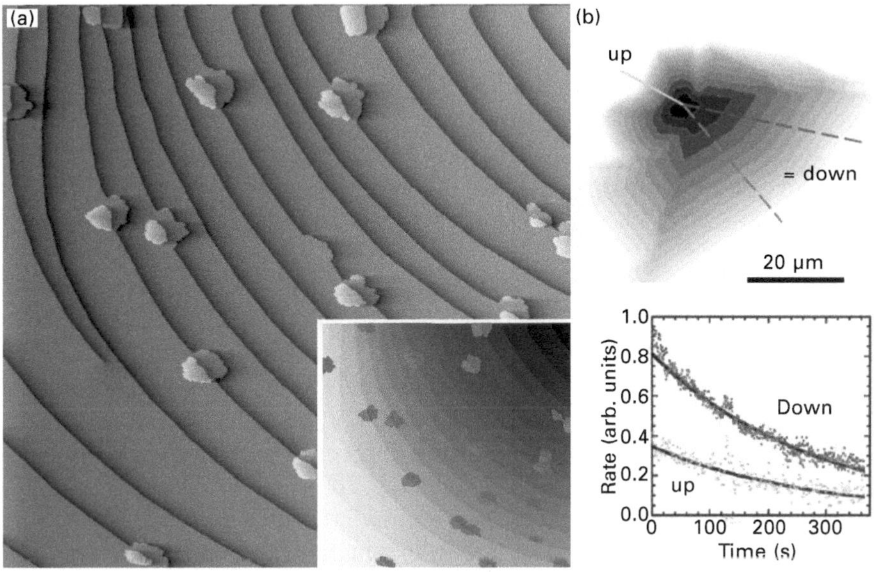

10.9 (a) Differentiated STM topograph (1 × 1 µm²) of graphene islands nucleated at Ir(111) atomic step edges (the islands are highlighted in dark and light gray if nucleated at the upper or lower terrace respectively) (from Coraux, 2009, © Institute of Physics). (b) Contours of a growing graphene island on Pt(111) at the vicinity of an atomic substrate step edge, as a function of the growth time, as images with LEEM (top panel) and growth rate derived from such a sequence of images at the upper and lower terrace around the step edge (from Sutter, 2009a, © The American Physical Society, http://prb.aps.org/abstract/PRB/v80/i24/e245411).

which is evidence that the carbon rows are also continuous (Coraux, 2009). This observation is puzzling at first glance, as there is no reason why the graphene lattices from the two lattices would be in registry. Rather we would expect a shift by a fraction of a lattice constant. Lattice distortions, which could be adapted either locally or at a larger scale, are therefore expected. Dislocations (heptagon-pentagon pairs) are also found where the islands met to accommodate for small tilts between the islands.

Prolonged growth will result in a lateral expansion of the graphene flakes which will meet and eventually cover the whole metallic surface. This is discussed in more detail on page 245.

Graphene interaction with the substrate step edges

The nucleation of graphene at substrate step edges and the partial or total hindering of uphill growth over step edges are evidence for the interaction of graphene edges with substrate steps. This interaction is probably lowering the edge energies of both graphene and the metal by increasing the low coordination of edge atoms.

Graphene growth was also observed to induce reshaping of the substrate step edges. The magnitude of this reshaping varies. On Ir(111) or Ru(0001) (Coraux, 2009; Loginova, 2010), it is in the form of a reorientation of the metal step edge that is in contact with one of the graphene island edges, so that the metal edge transforms into segments aligning the zigzag direction of graphene (Fig. 10.10a). On Pt(111) and Pd(111), the effects are much stronger: a whole graphene island can penetrate the metal terraces so that all graphene edges are bound to metal edges, with the latter aligning graphene's zigzag edges (Nakagoe, 2002; Fujita, 2005; Land, 1992; Kwon, 2009) (Fig. 10.10b). Such processes can only be effective if flow of metal atoms develops at the step edges. The preferential orientation of the reshaped metal edges is probably driven by maximization of the number of carbon–metal σ-bonds.

So far only in the cases of Ni(111) (Helveg, 2004) and Ru(0001) (Loginova, 2010) supports has it been shown that a growth front of graphene is possible also at the location where it is attached to a metallic step edge. Accordingly graphene grows at the expense of a retraction of the substrate step edge (Fig. 10.11). It was argued (Loginova, 2010) that the metal atoms are then displaced away from the step edge and can incorporate in the topmost metal layer, causing the formation of a dislocation network pattern in this layer. Metal step edge retraction may also result from the equilibrium between the step edge and the 2D metal gas surrounding it at the growth temperature, allowing for an efficient exchange between step atoms and adatoms.

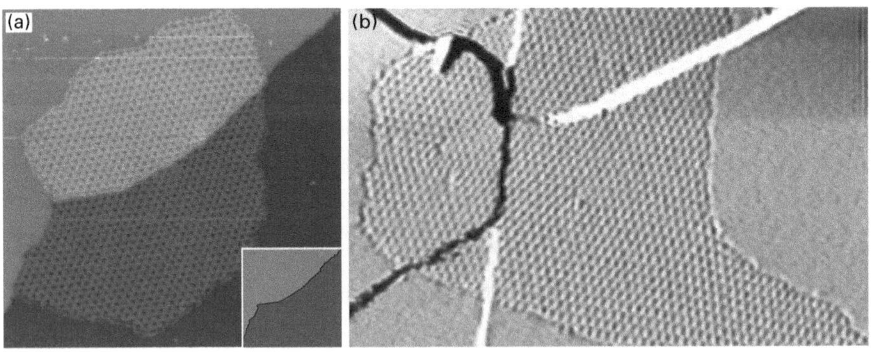

10.10 STM topographs evidencing the interaction of metal step edges upon graphene growth, on Ir(111) (a) and on Pd(111) (b) (from Kwon, 2009, © The American Chemical Society). The superstructure is the moiré of graphene/Ir(111) and graphene/Pd(111). The inset in (a) highlights the structure of the substrate step edge reshaped by the graphene island. The interaction of graphene with Pd(111) steps is such that the graphene island is incorporated into a Pd terrace. Image sizes are 125 × 125 nm^2 for (a) and 380 × 250 nm^2 for (b).

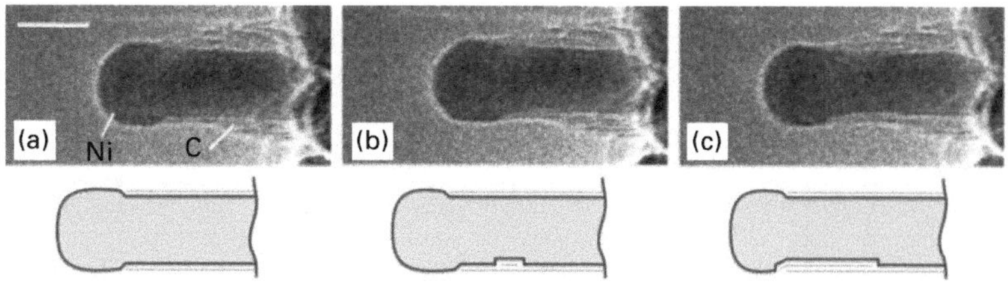

10.11 *In situ* observation of carbon nanotube growth at the surface of a catalytic (Ni) particle with transmission electron microscopy, where methane is provided with increasing doses from (a) to (c). Scale bar is 5 nm. The schematics highlight the graphene growth at the expense of the retraction of the substrate step edges (from Helveg, 2004, © Nature Publishing Group).

Elementary processes during the growth

As we already discussed, graphene growth starts at the step edges. It was shown that the amount of graphene does not depend on the size of the terraces surrounding the step edges (Coraux, 2009). In addition, at steps graphene islands nucleate with random separations giving rise to earlier coalescence. These observations indicate the absence of any significant gradients induced in the concentration of the adspecies from which the islands grow. Graphene island growth is not diffusion limited.

On the contrary, it was proposed that the growth is interface limited because carbon adatoms experience a large energy barrier for attaching to graphene step edges (Loginova, 2008). A large carbon supersaturation is needed at the surface so that several atom carbon clusters, being mobile at the surface, can be formed. Such clusters have been calculated to possess a much lower energy barrier for attachment at graphene edges, and are thus considered as building blocks for graphene growth. They are as large as heptamers on Ru(0001) (Loginova, 2008) and Ir(111) (Loginova, 2009b). These results are qualitatively very similar for two different carbon sources, CVD with ethylene or atomic carbon flux.

The analysis of the surface coverage with graphene as a function of the carbon dose provides information about the stability of carbon adspecies at the bare metal surfaces and regions covered with graphene. For growth on Ir(111), it was shown that the experimental data are well reproduced by a mere exponential law (Fig. 10.12), with a rapid increase at initial stages followed by a slowing down of the growth rate towards zero as full coverage is approached (Coraux, 2009). Quantitative analysis showed that the exponential behaviour is well reproduced by assuming (i) the absence of desorption of carbon adspecies from the Ir surface areas, (ii) the absence of C incorporation to the bulk and (iii) no sticking of molecules or C adspecies to the graphene areas. Therefore, under these conditions graphene growth is self-limiting

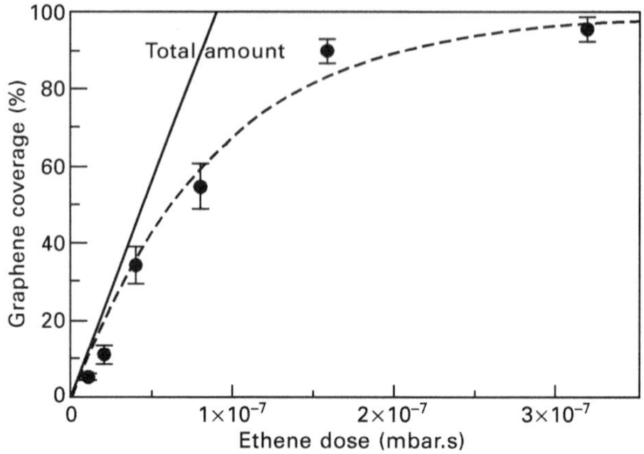

10.12 Graphene coverage as a function of the ethene dose during CVD growth of graphene on Ir(111), as estimated from STM topographs, and exponential law fit (adapted from Coraux, 2009, © Institute of Physics).

and terminates upon monolayer completion. In the opposite situation where the bulk of the support has the capability to store carbon and supply mobile species, like it is the case from 1500°C on Ir(111) or 800–900°C on metals like Ni or Ru, the graphene growth is not self-limiting. A second and even more layers may form underneath the monolayer graphene.

On the formation and stability of rotational variants

Graphene rotational variants are found at Ir(111) and Pt(111) surfaces (see Section 10.2.3), and will presumably also be found at the surface of other metals weakly interacting with graphene, typically those for which the average graphene–metal distance is close to that between planes in graphite. On Ir(111) the variant with graphene zigzag edges aligned to Ir dense-packed rows grows is energetically preferred and dominates the growth; at high growth temperatures and fluxes, other variants may nucleate at its edges (Loginova, 2009b). For graphene on Pt(111) the different variants nucleate independently and at the same time (Sutter, 2009a). On Ir(111), it was suggested that the nucleation of the other variants is heterogeneous, taking place at defects at the first variant edges. Once the other variants are nucleated, they grow faster than the first variant. It was argued that this is due to different carbon attachment kinetics at graphene edges for the different variants, as carbon attachment is the limiting step in graphene growth (Lovinova, 2009b).

The growth of other variants can be suppressed if the development of graphene edges not aligned to the dense-packed rows of the substrate can be avoided, or at least largely postponed. For this, a fraction of the surface might first be covered by graphene islands having well-defined zigzag edges following the recipe described in Section 10.3.2 (pre-adsorption of ethene at room temperature followed by high temperature annealing). Accordingly a high density of islands is achieved, and the growth can be continued by CVD at high temperature, yielding exclusively the more stable and energetically preferred variant (van Gastel, 2009). In practice, the first step may be performed at 1200°C and the second at 800°C, which results in a single crystallographic orientation of graphene across the whole sample surface. The as-grown graphene has a low density of wrinkles (see Section 10.2.4) due to the low growth temperature of 800°C during graphene layer completion. The lower growth temperature induces less thermal lattice mismatch during cooling and consequently less strain relieving defects.

The different variants have distinct stabilities, as suggested by the preferential high temperature oxygen etching of the graphene variants of which the zigzag edges do not align to dense-packed rows in the metal (van Gastel, 2009; Starodub, 2010). This difference in reactivity was employed as an alternative route to achieve macroscopic graphene samples with a single orientation on Ir(111), via cycles of CVD growth with ethylene and oxygen etching of the undesired variants (van Gastel, 2009) (Fig. 10.13).

10.3.3 Graphene multilayers on metals

Graphene multilayers are commonly obtained on metals which can store a non-negligible amount of carbon in their bulk. This is the case for instance for Ru, Ni or Co. Relatively slow cooling rates (typically 10°C/s, but this figure depends on the amount of C stored in the metal) are usually employed to promote the diffusion of carbon towards the bulk of the metal, leaving only a limited amount of carbon close to the metal surface, which favours the growth of few-layer and even single layer graphene (Yu, 2008; Reina, 2009b). Too slow cooling rates leave too much time for carbon to diffuse towards the metal bulk, resulting in negligible amounts of carbon near the surface and accordingly no graphene growth by segregation. On the contrary, fast cooling rates inhibit carbon diffusion towards the bulk, so that large amounts of carbon are available close to the surface of the metal: this favours the growth of multilayer and defective graphene. LEEM showed that an additional graphene layer starts to grow after each layer is completed (Sutter, 2009b, 2009c). So far the question of how the second layer of graphene grows after the first is completed remains open: on Pt(111) it was shown that the second layer does not grow between the first layer and the topmost Pt layer (Sutter, 2009a), pushing the first layer upwards, which suggests that carbon atoms are escaping the bulk towards the graphene surface via defects (e.g. holes) present in the first graphene layer. LEEM studies showed that the graphene growth by segregation is mostly a bulk-diffusion limited process (McCarty,

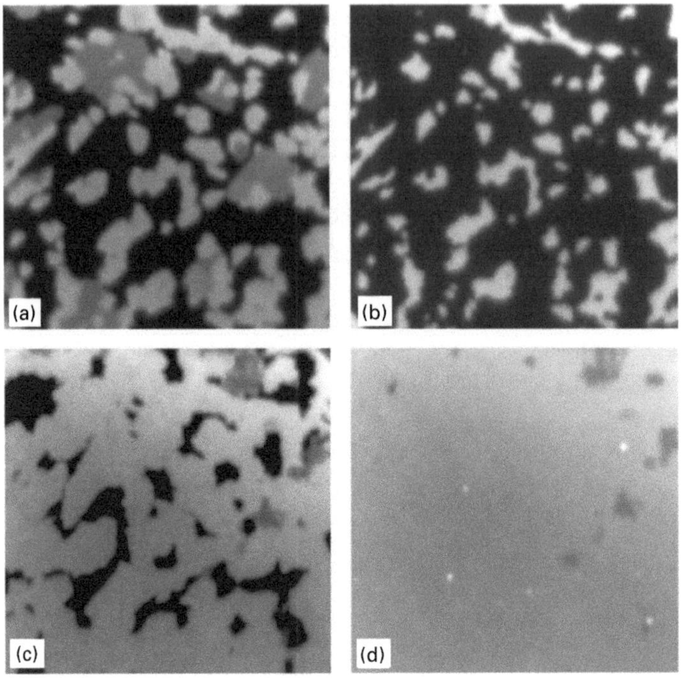

10.13 Repeated sequences (a–d) of CVD growth of graphene on Ir(111) and selective high temperature (856°C) oxygen etching (at a partial pressure of 5×10^{-8} mbar), as observed *in situ* by photoemission electron microscopy (field of view, 102 μm). The rotational variant with carbon zigzag rows parallel to dense-packed rows of Ir(111) appear brighter, while the other variant appears in grey. Ir(111), due to its high work function, appears in black. The procedure leads to the selective etching of the second variant and to the prominent formation of the first one with a 99% yield (from van Gastel, 2009, © American Institute of Physics).

2009). Multilayers and the number of layers they contain are readily identified by electron reflectivity as was done for multilayer graphene on SiC (Ohta, 2008). The in-plane morphology of multilayer graphene is much less studied. The stacking of the multilayers and azimuthal orientation between layers are, for instance, not reported or discussed. Diffraction experiments, such as (micro) low-energy electron diffraction (Sutter, 2008), should provide valuable hints in this respect. STM was used to show that the second layer in a bilayer might exhibit nanometre scale ripples at the surface or be flat depending on the location on the sample (Sutter, 2008, 2009c).

At the surface of metals with low carbon solubility such as Ir or Pt, the formation of multilayer graphene is generally absent, since once the surface is covered with graphene, the graphene growth reaction stops. At high temperatures also in these metals carbon can be stored in the bulk, yet in much lower proportions than in Ni or Ru. Cooling down leads to the segregation of this carbon exclusively at the location of defects in the first graphene layers (e.g. holes, substrate step bunches), promoting the local growth of multilayer graphene (Sutter, 2009a). It seems that only small multilayer domains (with extensions of a few microns), rather than large surfaces, can be obtained that way (Sutter, 2009a; Starodub, 2010; Meyer zu Heringdorf, private communication).

10.4 Future trends

An important direction for research is the growth of graphene on thin metallic films. This is the first step towards cheap graphene production on a large scale (Bae, 2009) which may be followed by the transfer of the graphene onto suitable supports following the etching of the metallic film (Kim, 2009; Reina, 2009a). So far polycrystalline thin films were employed in this context, yet the electronic properties of the samples are remarkably high. Controlling the number of graphene layers, ultimately down to a single one, and the crystallinity of the lattice to further increase the electronic performance will require the optimization of the growth conditions (Li, 2010) and the use of metallic thin films with much better structural quality. Educated choices for the nature of the metallic films should also allow better control of the number of layers (Li, 2009) and reducing the growth temperatures which are prohibitively high at present (~1000°C). In this latter respect the plasma-enhanced growth, which provides carbon species in an already active form at the surface, is a promising approach, which is not only applied on metal surfaces (Ismach, 2010; Rümmeli, 2010). Graphene growth on metals from metallic carbon sources is also an interesting alternative route (Hofrichter, 2009; Juang, 2009; Xi, 2011).

The experimental investigation of the interaction between graphene and the metallic surface has made significant progress in the last few years, noticeably concerning the charge transfer between the two materials and the carbon–metal hybridization, thanks to transport measurements (Huard, 2008; Blake, 2009), photoemission spectroscopy (Oshima, 1997; Grüneis, 2008; Varykhalov, 2008; Pletikosic, 2009; Lacovig, 2009; Sutter, 2009b), and *ab initio* calculations (Nemec, 2006, 2008; Giovannetti, 2008; Wang, 2009; Khomyakov, 2009; Ran, 2009). Yet many questions remain open. The effort to include van der Waals interactions in *ab initio* simulations has just been initiated (Vanin, 2009) and is mandatory for quantitative understanding of graphene–metal interaction in the case of weakly bound systems. The identification of general trends explaining the different interactions, from chemisorption to physisorption is in progress at the moment (Khomyakov, 2009). The connection between the strength of the graphene–metal/carbon–metal interaction and the growth and structure of graphene is not clearly understood so far, and should become more obvious once growth and structure of graphene have been studied at elementary scales on a wider variety of substrates.

The moiré lattices between graphene and transition metal surfaces were shown to be quite efficient patterns for the self-organization of a variety of nanometer size clusters. It was initially proven on Ir(111) substrates (N'Diaye, 2006, 2009b) and later on Ru(0001) (Donner, 2009; Pan, 2009; Zhang, 2009). Novel magnetic, catalytic, or optical effects are being considered in these systems.

10.5 Sources of further information and advice

Few reviews address graphene growth on metals (Wintterlin, 2009; Oshima, 1997), on SiC (de Heer, 2007), and by chemical ways (Park, 2009). The unconventional electronic properties of graphene were reviewed in a comprehensive article (Castro Neto, 2009). More accessible reviews on this aspect also exist (Geim, 2007; Katsnelson, 2007). A point of view on the prospects of graphene research and applications was proposed by Geim (2009).

Worth mentioning in the context of this chapter, which is devoted to graphene on metals, are spintronics effects in this system. Efficient spin-polarization and filtering were predicted (Karpan, 2007, 2008; Yazyev, 2009) and have started to be explored at the experimental level (Dedkov, 2008a). Rashba effects developing at the graphene–metal interface were recently the matter of noticeable interest (Dedkov, 2008b; Varykhalov, 2008; Rader, 2009; Rashba, 2009; Kuemmeth, 2009).

10.6 Acknowledgements

Johann Coraux acknowledges the Alexander von Humboldt Foundation for a research grant.

10.7 References

Amidror I. (2007), *The Theory of the Moiré Phenomenon*, Dordrecht, Springer.

Arnoult W. J., McLellan R. B. (1972), The solubility of carbon in rhodium, ruthenium, iridium and rhenium, *Scr. Metall.* 6, 1013.

Bae S., *et al.* (2009), 30-inch roll-based production of high quality graphene films for flexible transparent electrodes, *Sci. Technol.* 5, 1.

Banerjee B. C., Hirt T. J., Walker Jr. P. L. (1961), Pyrolitic carbon formation from carbon suboxide, *Nature* 192, 450.

Bertoni G., Calmels L., Altibelli A., Serin V. (2004), First-principles calculation of the electronic structure and EELS spectra of the graphene/Ni(111) interface, *Phys. Rev.* B 71, 075402.

Biedermann L. B., Bolen M. L., Capano M. A., Zemlyanov D., Reifenberger R. G. (2009), Insights into few-layer epitaxial graphene growth on 4H-SiC(0001) substrates from STM studies, *Phys. Rev. B* 79, 125411.
Blake P., *et al.* (2009), Influence of metal contacts and charge inhomogeneity on transport properties of graphene near neutrality point, *Solid State Comm.* 149, 1068.
Bollmann W. (1964), Some basic problems regarding subgrain boundaries, *Discuss. Faraday Soc.* 38, 26.
Brugger T., *et al.* (2009), Comparison of electronic structure and template function of single-layer graphene and a hexagonal boron nitride nanomesh on Ru(0001), *Phys. Rev. B* 79, 045407.
Busse C., *et al.* (2011), Graphene on Ir(111): a weakly bonded system, Submitted.
Cambaz Z. G., Yushin G., Osswald S., Mochalin V., Gogotsi Y. (2008), Non catalytic synthesis of carbon nanotubes, graphene and graphite on SiC, *Carbon* 46, 841.
Castro Neto A. H., Guinea F., Peres N. M. R., Novoselov K. S., Geim A. K. (2009), The electronic properties of graphene, *Rev. Mod. Phys.* 81, 109.
Chae S. J., *et al.* (2009), Synthesis of large-area graphene layers on poly-nickel substrates by chemical vapor deposition: wrinkle formation, *Adv. Mater.* 21, 2328.
Coraux J., N'Diaye A. T., Busse C., Michely T. (2008), Structural coherency of graphene on Ir(111), *Nano Lett.* 8, 565.
Coraux J., *et al.* (2009), Growth of graphene on Ir(111), *New J. Phys.* 11, 023006.
Dedkov Y. S., Fonin M., Laubschat C. (2008a), A possible source of spin-polarized electrons: the inert graphene/Ni(111) system, *Appl. Phys. Lett.* 92, 052506.
Dedkov Y. S., Fonin M., Rüdiger U., Laubschat C. (2008b), Rashba effect in the graphene/Ni(111) system, *Phys. Rev. Lett.* 100, 107602.
de Heer W. A., *et al.* (2007), Epitaxial graphene, *Solid State Comm.* 143, 92.
Donner K., Jakob P. (2009), Structural properties and site specific interactions of Pt with graphene/Ru(0001) moiré overlayer, *J. Chem. Phys.* 131, 164701.
Eom D., *et al.* (2009), Structure and electronic properties of graphene nanoislands on Co(0001), *Nano Lett.* 9, 2844.
Feibelman P. J. (2008), Pinning of graphene to Ir(111) by flat Ir dots, *Phys. Rev. B* 77, 165419.
Frenklach M., Ping J. (2004), On the role of surface migration in the growth and structure of graphene layers, *Carbon* 42, 1209.
Fujita T., Kobayashi W., Oshima C. (2005), Novel structures of carbon layers on a Pt(111) surface, *Surface and Interface Analysis* 37, 120.
Gao L., Guest J. R., Guisinger N. P. (2010), Epitaxial graphene on Cu(111), *Nano Lett.* 10, 3512.
Gamo Y., *et al.* (1997), Atomic structure of monolayer graphite formed on Ni(111), *Surf. Sci.* 374, 61.
Geim A. K. (2009), Graphene: status and prospects, *Science* 324, 1530.
Geim A. K., Novoselov K. S. (2007), The rise of graphene, *Nature Mater.* 6, 183.
Giovannetti G., *et al.* (2008), Doping graphene with metal contacts, *Phys. Rev. Lett.* 101, 026803.
Grüneis A., Vyalikh D. V. (2008), Tunable hybridization between electronic states of graphene and a metal surface, *Phys. Rev. B* 77, 193401.
Hamilton J. C., Blakely J. M. (1980), Carbon segregation to single-crystal surfaces of Pt, Pd and Co, *Surf. Sci.* 91, 199.
Helveg S., *et al.* (2004), Atomic-scale imaging of carbon nanofibre growth, *Nature* 427, 426.

Hofrichter J., *et al.* (2009), Synthesis of graphene on silicon dioxide by a solid carbon source, *Nano Lett.* 10, 36.

Huard B., Stander N., Sulpizio J. A., Goldhaber-Gordon D. (2008), Evidence of the role of contacts on the observed electron-hole asymmetry in graphene, *Phys. Rev. B* 78, 121402(R).

Irving S. M., Walker Jr. P. L. (1967), Interaction of evaporated carbon with heated metal substrates, *Carbon* 5, 399.

Ismach A., *et al.* (2010), Direct chemical vapor deposition of graphene on dielectric surfaces, *Nano Lett.* 10, 1542.

Juang Z.-Y., *et al.* (2009), Synthesis of graphene on silicon carbide substrates at low temperature, *Carbon* 47, 2026.

Karpan V. M., *et al.* (2007), Graphite and graphene as perfect spin filters, *Phys. Rev. Lett.* 99, 176602.

Karpan V. M., *et al.* (2008), Theoretical prediction of perfect spin filtering at interfaces between close-packed surfaces of Ni or Co and graphite or graphene, *Phys. Rev. B* 78, 195419.

Karu A. E., Beer M. (1966), Pyrolytic formation of highly crystalline graphite films, *J. Appl. Phys.* 37, 2179.

Katsnelson M. I. (2007), Graphene: carbon in two dimensions, *Materials Today* 10, 20.

Khomyakov P. A., Giovannetti G., Rusu P. C., Brocks G., van den Brink J., Kelly P. J. (2009), First-principle study of the interaction and charge transfer between graphene and metals, *Phys. Rev. B* 79, 195425.

Kim K. S., *et al.* (2009), Large-scale pattern growth of graphene films for stretchable transparent electrodes, *Nature* 457, 706.

Kuemmeth F., Rashba E. I. (2009), Spin-polarized photoemission from graphene: joint effect of sublattice interference and spin-orbit coupling to substrate, *Phys. Rev. B.* 80, 241409(R).

Kwon S. Y., *et al.* (2009), Growth of semiconducting graphene on palladium, *Nano Lett.* 9, 3985.

Lacovig P., *et al.* (2009), Growth of dome-shaped carbon nanoislands on Ir(111): the intermediate between carbidic clusters and quasi free-standing graphene. *Phys. Rev. Lett.* 103, 166101.

Land T. A., Michely T., Behm R. J., Hemminger J. C., Comsa G. (1992), STM investigation of single layer graphite structures produced on Pt(111) by hydrocarbon decomposition, *Surf. Sci.* 264, 261.

Lazic P., Atodiresei N., Alai M., Caciuc V., Blügel S., Brako R. (2010), JuNoLo – Jülich nonlocal code for parallel post-processing evaluation of vdW-DF correlation energy, *Comput. Phys. Commun.* 181, 371.

Lee C., Wei X., Kysar J. W., Hone J. (2008), Measurement of the elastic properties and intrinsic strength of monolayer graphene, *Science* 321, 385.

Li X., *et al.* (2009), Large-area synthesis of high quality and uniform graphene films on copper foils, *Science* 324, 1312.

Li X., *et al.* (2010), Graphene films with large domain size by a two-step chemical vapor deposition process, *Nano Lett.* 10, 4328.

Loginova E., Bartelt N. C., Feibelman P. J., McCarty K. F. (2008), Evidence for graphene growth by C cluster attachment, *New J. Phys.* 10, 093026.

Loginova E, Nie S, Thürmer K., Bartelt N. C., McCarty K. F. (2009a), Defects of graphene on Ir(111): rotational domains and ridges, *Phys. Rev. B* 80, 085430.

Loginova E., Bartelt N. C., Feibelman P. J., McCarty K. F. (2009b), Factors influencing the growth of graphene on metals, *New J. Phys.* 11, 063046.
Loginova E., *et al.* (2010), Graphene growth by metal etching on Ru(0001), *Phys. Rev. B* 85, 235422.
Marchini S., Günther S., Wintterlin J. (2007), Scanning tunneling microscopy of graphene on Ru(0001), *Phys. Rev. B* 76, 75429.
Martoccia D., *et al.* (2008), Graphene on Ru(0001): a 25 × 25 supercell, *Phys. Rev. Lett.* 101, 126102.
McCarty K., Feibelman P. J., Loginova E., Bartett N. C. (2009), Kinetics and thermodynamics of carbon segregation and graphene growth on Ru(0001), *Carbon* 47, 1806.
Moritz W., *et al.* (2010), Structure determination of the coincidence phase of graphene on Ru(0001), *Phys. Rev. Lett.* 104, 136102.
Nakagoe O., Tgaki N., Matsumoto Y. (2002), Thermal decomposition of acetylene on Pt(111) studied by scanning tunneling microscopy, *Surf. Sci.* 514, 414
N'Diaye A. T., Bleikamp S., Feibelman, P. J., Michely T. (2006), Two-dimensional Ir cluster lattice on a graphene moiré on Ir(111), *Phys. Rev. Lett.* 97, 215501.
N'Diaye A. T., Bleikamp S., Feibelman, P. J., Michely T. (2008a), Erratum: two-dimensional Ir cluster lattice on a graphene moiré on Ir(111), *Phys. Rev. Lett.* 101, 219904.
N'Diaye A. T., Coraux J., Plasa T. N., Busse C., Michely T. (2008b), Structure of epitaxial graphene on Ir(111), *New J. Phys.* 10, 043033.
N'Diaye A. T., *et al.* (2009a), *In situ* observation of stress relaxation in epitaxial graphene, *New J. Phys.* 11, 113056.
N'Diaye A. T., Gerber T., Busse C., Myslivecek J., Coraux J., Michely T. (2009b), A versatile fabrication method for cluster superlattices, *New J. Phys.* 11, 103045.
Nemec N., Tomanek D., Cuniberti G. (2006), Contact dependence of carrier injection in carbon nanotubes: an *ab initio* study, *Phys. Rev. Lett.* 96, 076802.
Nemec N., Tomanek D., Cuniberti G. (2008), Modeling extended contacts for nanotube and graphene devices, *Phys. Rev. B* 77, 125420.
Nieuwenhuys B. E., Hagen D. I., Rovida G., Somorjai G. A. (1976), LEED, AES and thermal desorption studies of chemisorbed hydrogen and hydrocarbons (C_2H_2, C_2H_4, C_6H_6, C_6H_{12}) on the (111) and stepped [6(111) × (100)] iridium crystal surfaces; comparison with platinum, *Surf. Sci.* 59, 155.
Novoselov K. S., *et al.* (2004), Electric field effect in atomically thin carbon films, *Science* 306, 666.
Obraztsov A. N., Obraztsov E. A., Tyurnina A. V., Zolotukhin A. A. (2007), Chemical vapor deposition of thin graphite films of nanometer thickness, *Carbon* 45, 2017.
Ohta T., *et al.* (2008), Morphology of graphene thin film growth on SiC(0001), *New J. Phys.* 10, 023034.
Oshima C., Nagashima A. (1997), Ultra-thin epitaxial films of graphite and hexagonal boron nitride on solid surfaces, *J. Phys.: Condens. Matter* 9, 1.
Pan Y., Shi D.-X., Gao H.-J. (2007), Formation of graphene on Ru(0001) surface, *Chinese Phys.* 16, 3151.
Pan Y., Gao M., Huang L., Liu F., Gao H.-J. (2009), Directed self-assembly of monodispersed platinum clusters on graphene moiré template, *Appl. Phys. Lett.* 95, 093106.
Park S., Ruoff R. S. (2009), Chemical methods for the production of graphene, *Nature Nanotech.* 4, 217.
Pletikosic I., *et al.* (2009), Dirac cones and minigaps for graphene on Ir(111), *Phys. Rev. Lett.* 102, 056808.
Presland A. E. B., Walker Jr. P. L. (1969), Growth of single-crystal graphite by pyrolysis of acetylene over metals, *Carbon* 7, 1.

Rader O., Varykhalov A., Sanchez-Barriga J., Marchenko D., Rybkin A., Shikin A. M. (2009), Is there a Rashba effect in graphene on 3d ferromagnets?, *Phys. Rev. Lett.* 102, 057602.

Ran Q., Gao M., Guan X., Wang Y., Yu Z. (2009), First-principle investigation on bonding formation and electronic structure of metal-graphene contacts, *Appl. Phys. Lett.* 94, 103511.

Rashba E. I. (2009), Graphene with structure-induced spin-orbit coupling: spin-polarized states, spin zero modes, and quantum Hall effect, *Phys. Rev. B* 79, 161409(R).

Reina A., *et al.* (2009a), Large area, few-layer graphene films on arbitrary substrates by chemical vapor deposition, *Nano Lett.* 9, 30.

Reina A., *et al.* (2009b), Growth of large-area single- and bi-layer graphene by controlled carbon precipitation on polycrystalline Ni surfaces, *Nano Research.* 2, 509.

Ritter M., Ranke W., Weiss W. (1998), Growth and structure of ultrathin FeO films on Pt(111) studied by STM and LEED, *Phys. Rev. B* 57, 7240.

Rümmeli M. H., *et al.* (2010), Direct low-temperature nanographene CVD synthesis over a dielectric insulator, *ACS Nano* 4, 4206.

Sasaki M., Yamada Y., Ogiwara Y., Yagyu S. I., Yamamoto S. (2000), Moiré contrast in the local tunneling barrier height images of monolayer graphite on Pt(111), *Phys. Rev. B* 61, 15653.

Starodub E., Bartelt N. C., McCarty K. F. (2010), Oxidation of graphene on metals, *J. Phys. Chem. C* 114, 539.

Sun G. F., Jia J. F., Xue Q. K., Li L. (2009), Atomic-scale imaging and manipulation of ridges on epitaxial graphene on 6H-SiC(0001), *Nanotechnology* 20, 355701.

Sutter P. W., Flege J.-I., Sutter E. A. (2008), Epitaxial graphene on Ru(0001), *Nature Mater.* 7, 406.

Sutter P., Sadowsky J. T., Sutter E. (2009a), Graphene on Pt(111): Growth and substrate interaction, *Phys. Rev. B* 80, 245411.

Sutter P., Hybersten M. S., Sadowski J. T., Sutter E. (2009b), Electronic structure of few-layer epitaxial graphene on Ru(0001), *Nano Lett.* 9, 2654.

Sutter E., Acharya D. P., Sadowsky J. T., Sutter P. (2009c), Scanning tunneling microscopy of epitaxial bilayer graphene on ruthenium (0001), *Appl. Phys. Lett.* 94, 133101.

van Gastel R., *et al.* (2009), Selecting a single orientation for millimeter sized graphene sheets, *Appl. Phys. Lett.* 95, 121901.

Vanin M., *et al.* (2009), Graphene on metals: a van der Waals density functional study, *Phys. Rev. B* 81, 081408.

Varykhalov A., *et al.* (2008), Electronic and magnetic properties of quasifreestanding graphene on Ni, *Phys. Rev. Lett.* 101, 157601.

Vazquez de Parga A. L., *et al.* (2008), Periodically rippled graphene: growth and spatially resolved electronic structure, *Phys. Rev. Lett.* 100, 56807.

Wang B., Bocquet M.-L., Marchini S., Günther S., Wintterlin J. (2008), Chemical origin of a graphene moiré overlayer on Ru(0001), *Phys. Chem. Chem. Phys.* 10, 3530.

Wang Q. J., Che J. G. (2009), Origins of distinctly different behaviors of Pd and Pt contacts on graphene, *Phys. Rev. Lett.* 103, 066802.

Whitesides R., Kollias A. C., Domin D., Lester Jr. W. A., Frenklach M. (2007), Graphene layer growth: collision of migrating five-member rings. *Proc. Combust. Inst.* 31, 539.

Whitesides R., Frenklach M. (2010), Detailled kinetic Monte Carlo simulations of graphene edge growth, *J. Phys. Chem. A* 114, 689.

Wiederholt T., *et al.* (1995), Formation of two-dimensional sulfide phases on Al(111): an STM study, *Surf. Sci.* 324, 91.

Wintterlin J., Bocquet M.-L. (2009), Graphene on metal surfaces, *Surf. Sci.* 603, 1841.
Xi M., *et al.* (2011), Single-layer graphene nearly 100% covering an entire substrate, arXiv:1006.5085.
Yazyev O. V., Pasquarello A. (2009), Magnetoresistive junctions based on epitaxial graphene and hexagonal boron nitride, *Phys. Rev. B* 80, 035408.
Yu Q., Lian J., Siriponglert S., Li H., Chen Y. P., Pei S.-S. (2008), Graphene segregated on Ni surfaces and transfered to insulators, *Appl. Phys. Lett.* 93, 113103.
Zhang H., Fu Q., Cui Y., Tan D., Bao X. (2009), Fabrication of metal nanoclusters on graphene grown on Ru(0001), *Chinese Sci. Bull.* 54, 2446.
Zhao L., *et al.* (2011), Influence of copper crystal surface on the CVD growth of large area monolayer graphene, *Solid State Comm* 151, 509.

11
Electronic properties and adsorption behaviour of thin films with polar character

N. NILIUS, Fritz-Haber-Institut der MPG, Germany

Abstract: Thin films of polar nature are distinctively different from their non-polar counterparts. They feature characteristic properties in their surface morphology, electronic structure and adsorption behaviour, which originate from the effort of such systems to lower the polarity-induced electrostatic energy. This chapter provides insight into the general concepts of polarity and typical polarity healing mechanisms. A main focus is the polarity of thin oxide films and the chapter deals with the various observable consequences of uncompensated surface dipoles in such systems. Furthermore, the distinct adsorption characteristic of polar films is discussed, putting special emphasis on the peculiar growth modes, self-assembly phenomena and charging effects that accompany residual polarity. Finally, brief reference is made to potential applications of polar materials.

Key words: polar thin films, adsorption behaviour, charge transfer, self-assembly effects, scanning tunnelling microscopy, density functional theory.

11.1 Introduction to oxide polarity

Thin films with polar character form a particularly interesting class of materials, as they possess a number of unusual structural and electronic properties and a unique adsorption and reaction behaviour with respect to non-polar systems (Noguera, 1996; Goniakowski *et al.*, 2008). The decisive property of a polar system is a macroscopic surface dipole that orients perpendicular to the surface and originates from a distinct crystallographic structure of the material. Surface polarity usually arises from the presence of elementary cells in the atomic lattice that carry an uncompensated dipole moment. As electrostatic dipoles are related to the occurrence of charged species in the lattice, surface polarity is restricted to highly ionic materials and does not occur in covalently bound or metallic systems. Prototypical polar materials are therefore sulfides, oxides and halides that are characterized by a high level of ionicity. In this chapter, we will concentrate exclusively on polar oxide films, whereas information on halides and sulfides can be found in the literature.

P. W. Tasker introduced a simple three-level scheme to classify polar systems (Tasker, 1979). Type I materials consist neither of charged atomic layers parallel to the surface nor of unit cells with a dipole moment and are consequently non-polar (Fig. 11.1a). Type II systems are built of atomic planes that carry a net charge. However those charges add up to a vanishing dipole moment within each unit cell, rendering the ideal system non-polar again. One recognizes immediately, however, that polarity can be introduced to such systems if the top-most surface plane does not coincide with the boundaries of the unit cell and a layer of uncompensated charges resides at the surface (Fig. 11.1b, dashed line). The polarity of type II systems is therefore not intrinsic to a certain crystallographic orientation, but depends on the surface termination. A typical example is the $Al_2O_3(0001)$ surface, which is non-polar for the Al-termination as a charge-compensated unit cell that includes the surface can be constructed, but polar for the O-termination. The same consideration holds for the O- (non-polar) versus the Ti-terminated (polar) surface of $TiO_2(110)$. Type III systems are intrinsically polar according to Tasker's scheme, because they comprise stacks of oppositely-charged atomic planes as well as elementary cells that have a non-vanishing dipole moment along the surface normal (Fig. 11.1c). Prominent examples are the rocksalt crystals with (111) orientation, e.g. MgO(111), NiO(111) and FeO(111), and the (0001) planes of wurzite, for instance ZnO(0001).

The specific properties of polar systems become evident in thicker films. Each unit perpendicular to the surface carries a dipole moment μ of:

$$\mu = \sigma \cdot d \qquad [11.1]$$

with σ being the charge density in the layer and d the layer distance. For polar films consisting of thousands of charge-separated planes, those dipole moments add up to a macroscopic quantity that finally diverges in the bulk material (Fig. 11.2). The electrostatic energy V_{total} to stabilize the dipole

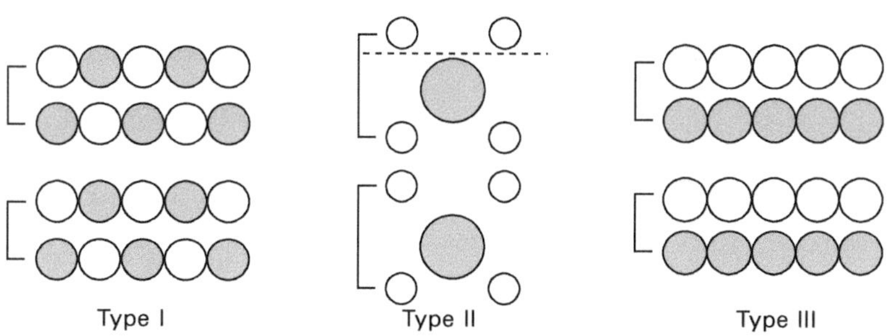

11.1 Classification of ionic systems according to the Tasker scheme. (a) and (c) are intrinsically non-polar and polar surfaces, respectively. The polarity of (b) depends on the termination. A non-polar surface is obtained when cutting the crystal between two dipole-compensated unit cells, whereas polarity arises for a cut through the unit cell (dashed line)

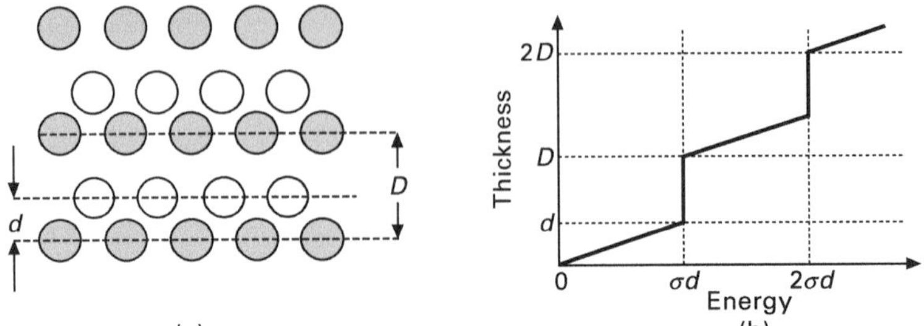

11.2 (a) Vertical cut through a polar system. The layer distance *d* and the size of the unit cell *D* are indicated. (b) Dependence of the electrostatic energy on the thickness of the polar slab.

moment in a slab of *N* unit cells can be approximated with the help of a plate capacitor model:

$$V_{\text{total}} = \frac{1}{\varepsilon_0 \cdot \varepsilon_r} N \cdot \sigma \cdot d \qquad [11.2]$$

whereby ε_0 and ε_r are the vacuum permittivity and the dielectric constant of the material, respectively (Kittel, 1996). To get an idea of the energies involved, the electrostatic potential is calculated for a free-standing MgO(111) bilayer ($\varepsilon_r \sim 10$). The system consists of a positively-charged Mg^{2+} and a negative-charged O^{2-} plane ($\sigma = 2|e|/7.8$ Å^2) separated by the bulk lattice parameter of 1.8 Å and already has an electrostatic energy of 8.3 eV. Apparently, the values for thicker slabs easily reach the lattice energy in ionic crystals of 20–30 eV, rendering polar materials energetically unstable. Bulk-like structures with uncompensated polarity should therefore not exist in nature.

The latter statement seems to be in conflict with the experimental observation of systems with polar character. Moreover, polar terminations are often characterized by a lower surface free energy than their non-polar counterparts. For example, MgO cubes terminated by non-polar (100) surfaces were found to develop polar (111) facets in a humid environment, although this should be accompanied by a dramatic increase of the electrostatic energy (Hacquart and Jupille, 2007). Similarly, a (111)-type NiO surface has been identified as the one with the lowest free energy even with respect to the non-polar (100) and (110) planes (Barbier *et al.*, 2000). This discrepancy can be solved by considering the various polarity-healing mechanisms that are able to remove the electrostatic dipole and therewith the polarity of the respective surface (Goniakowski *et al.*, 2008). From a purely mathematical viewpoint, polarity cancellation is achieved by adjusting a distinct charge density at the surface that compensates the bulk dipole. For simple systems, the required surface charge density is given by:

$$\sigma_{\text{surf}} = \sigma \frac{d}{D} \qquad [11.3]$$

with D being the height of the unit cell and d the interlayer distance in the crystal. As an example, layers with 25% and 50% of the bulk charge density need to be formed on the wurzite (0001) (d/D = 0.25) and the rocksalt (111) surface (d/D = 0.5) in order to remove the polarity in the whole system. In simple terms, the polarity annihilation might be understood in these cases as the result of an additional dipole that is created between the upper and lower surface of the slab and has exactly the same size but an opposite sign as the bulk moment.

Several mechanisms have been identified that enable an adjustment of the required charge density at the surface (Noguera, 1996; Goniakowski *et al.*, 2008). A first possibility is the restructuring of the top-most surface layers, creating planes with fractional atom filling and hence modified charge distribution (Fig. 11.3a). The best studied example is the octopolar reconstruction of the rocksalt (111) surface, which was predicted by Wolf in 1992 and later verified experimentally through X-ray diffraction measurements on NiO(111) and MgO(111) (Barbier *et al.*, 2000; Finocchi *et al.*, 2004). The octopolar reconstruction comprises two surface layers with 75% and 25% atom filling with respect to the ideal plane, which adds up to the required 50% charge density at the surface according to Eq. 11.3. The dipole compensation in this case is so efficient that the free energy of the reconstructed surface drops below the value for comparable non-polar oxide planes (Barbier *et al.*, 2000). Alternative reconstructions have been identified for other lattice geometries and different chemical environments. For instance, a spinel-like termination was found to develop on NiO(111) at low O_2 partial pressures (Barbier and Renaud, 1997; Barbier *et al.*, 1999), while several non-stoichiometric surface compositions were revealed for $SrTiO_3$(111) in a Sr- or O-rich environment (Bottin *et al.*, 2003). For the Zn-terminated ZnO(0001) surface, polarity healing has been investigated at the local scale with the help of scanning tunnelling microscopy (STM) (Parker *et al.*, 1998; Dulub *et al.*, 2003; Ostendorf *et al.*, 2008). The studies revealed triangular-shaped islands and pits of 1 ML height that cover the

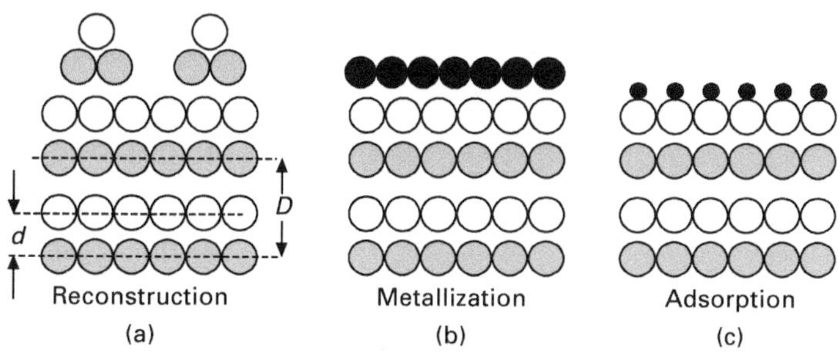

11.3 Sketch of the three main polarity-healing mechanisms in ionic systems.

complete oxide surface (Fig. 11.4a). According to density functional theory (DFT) calculations, the edges of those triangular structures are saturated with O atoms, rendering the surface oxygen-rich with respect to the bulk (Kresse *et al.*, 2003). As discussed above, the polarity of wurzite(0001) is cancelled by creating a surface layer with 25% extra charges, which corresponds to a local Zn:O stoichiometry of 3:4 with respect to the bulk value of 1:1. Exactly this ratio is adjusted by introducing the triangular islands and pits into the ZnO surface. Each equilateral triangle comprises $\frac{1}{2} \cdot n \cdot (n + 1)$ anions and $\frac{1}{2} \cdot n \cdot (n - 1)$ cations (n being the side length) and has an excess of n oxygen ions. The desired 3:4 surface stoichiometry is now realized by forming islands with seven edge atoms ($n = 7$), a structure that is indeed frequently found on the surface (Fig. 11.4b). For larger islands, additional atoms have to be removed from the layer underneath to reach the ideal Zn:O surface ratio (Fig. 11.4c,d). The patched Zn-terminated ZnO(0001) surface is therefore an ideal example for polarity healing via surface reconstruction (Dulub *et al.*, 2003).

11.4 (a) Empty-states STM image of Zn-terminated ZnO(0001) (50 × 50 nm^2). The terraces are covered with triangular islands and pits that are involved in healing the surface polarity. (b–d) Structure models of three triangular islands of different size. The island borders are saturated with oxygen. The resulting non-stoichiometry produces a charged surface layer that compensates the bulk dipole moment. Reprinted with permission from Dulub *et al.*, 2003, Copyright (2010) by the American Physical Society.

Alternatively, dipole compensation might be achieved via purely electronic effects, leaving the atomic structure of the surface unchanged (Fig. 11.3b). One proposed mechanism is the creation of surface states, the electron filling of which is adjusted to reach the required charge density for polarity healing (Goniakowski *et al.*, 2008). The formation of a partly-filled electronic state is always connected with the metallization of the oxide surface. This particular mechanism has been predicted for bulk Al_2O_3(0001) (Wang *et al.*, 2000) and thin, unreconstructed MgO(111) films (Goniakowski *et al.*, 2007). The formation of dipole-compensating surface states becomes particularly easy when the polar oxide is capped by a metal film, as demonstrated for Pd and Cu over-layers on MgO(111) and ZnO(0001), respectively (Goniakowski and Noguera, 2002; Meyer and Marx, 2004).

A third way to compensate the polarity of oxide materials is the binding of ad-species that become charged upon adsorption (Fig. 11.3c). The prototype adsorbate to heal surface polarity is hydrogen, which forms hydroxyl groups consisting of a surface oxygen ion and a positively charged H^+ ion. The hydroxylation of oxide surfaces is often triggered by the heterolytic splitting of water, which renders this compensation mechanism especially efficient in an ambient environment. Hydroxylation was predicted to occur spontaneously on most rocksalt (111) surfaces (Pojani *et al.*, 1997), on ZnO(0001) and on Al_2O_3(0001) (Wang *et al.*, 2000). It has been revealed experimentally for instance on MgO(111) (Poon *et al.*, 2006; Hacquart and Jupille, 2007), NiO(111) (Rohr *et al.*, 1994; Kitakatsu *et al.*, 1998) and ZnO(0001) (Wang, 2008) by detecting the O–H vibrational bands. Also combined mechanisms are reported, where molecular adsorption induces the formation of a partly-filled surface state at the Fermi level, which in turn removes the surface polarity (Wang *et al.*, 2005). Adsorbate-mediated polarity healing, in general, is responsible for the unique binding properties of polar systems and their enhanced chemical reactivity with respect to non-polar materials (Sun *et al.*, 2009).

11.2 Polar oxide films

Whereas for bulk materials the polarity needs to be healed in order to avoid a divergence in the electrostatic energy, thin films grown on metal and semiconductor supports can be stabilized even in a polar state. This difference to bulk materials relies on two effects. First, the electrostatic energy might be kept below the lattice energy of the film, as the number of polar units is small (see Eq. 11.2). As a consequence, reconstruction of the surface can be avoided and the film keeps its polar nature. Second, the substrate contributes to a reduction of the film dipole, especially when using a polarizable metal support. In this case, the required charge density that heals the polarity according to Eq. 11.3 is provided by the substrate and localized at the metal–oxide interface (Goniakowski and Noguera, 2002, 2009). The interplay between a polar film and a metal support shall be demonstrated in the following for a simple bilayer structure, consisting of a cationic and an anionic plane (Fig. 11.5).

The preferred adsorption geometry of the film is given by the electron affinity of the substrate. Whereas the interface is formed by positively charged ions on electronegative metals (e.g. Au and Pt), a negatively charged oxygen layer sits on top of electropositive materials (e.g. Mg and Al) (Goniakowski and Noguera, 2009). In the first case, electrons accumulate in the topmost substrate plane, leading to the following sequence of charged layers: metal surface(–)/cations(+) / anions(–). In the second scenario, the charge distribution changes to metal(+)/anions(–)/cations(+), as the electron density in the metal surface is depleted by the polar film. In both cases, the tri-layer structure has a reduced vertical dipole moment and hence low polarity. This substrate-mediated effect on the polarity might be enhanced by a real charge transfer, whereby electrons flow out of the polar film on electronegative metals but into the ad-layer on electropositive supports. The resulting charge-driven dipole always aligns in opposite direction to the film moment and further quenches the polarity of the metal–oxide system. The various electrostatic interactions that occur between a metal surface and a polar film, as well as the dipole moments involved are summarized in Fig. 11.5.

Thanks to efficient polarity stabilization schemes, thin oxide films with residual dipole moment can be prepared on metal surfaces, even if the respective bulk oxides are thermodynamically unstable. Well-studied examples for polar oxide films are MgO(111) on Ag(111) and Au(111) (Kiguchi *et al.*, 2003; Arita *et al.*, 2004; Mantilla *et al.*, 2008; Myrach *et al.*, 2011), ZnO(0001) on Ag(111) (Tusche *et al.*, 2007), CoO(111) on Ir(100) (Giovanardi *et al.*, 2006) as well as FeO(111) on Pt(111) (Vurens *et al.*, 1988; Ritter *et al.*, 1998; Rienks *et al.*, 2005). STM topographic images of the respective films are displayed in Fig. 11.6. The residual polarity of the various systems depends

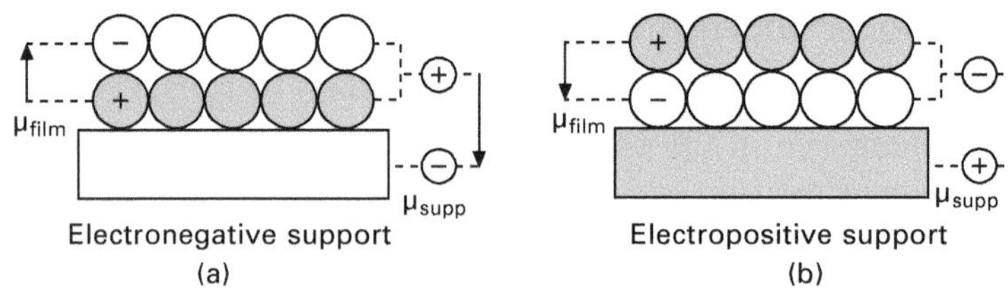

11.5 Sequence of oxide layers at the interface to an electronegative (a) and an electropositive (b) support. In both cases, the intrinsic dipole of the film is oriented in opposite direction to the interface dipole that results from a polarization/charge transfer between oxide and metal support. This compensation effect partly quenches the polarity of the combined system.

11.6 STM topographic images of polar oxide films prepared on metal supports. (a) ZnO/Ag(111) (9×9 nm^2, 1.2 V). Reprinted with permission from Tusche *et al.* (2007), Copyright (2010) by the American Physical Society. (b) FeO/Pt(111) (9×9 nm^2, 0.65 V). (c) CoO/Ir(100) (4.5×4.5 nm^2, 0.15 V). Reprinted with permission from Giovanardi *et al.* (2006), Copyright (2010) by the American Physical Society. (d) VO/Rh(111) (6×6 nm^2, 2.0 V). Reprinted with permission from Parteder *et al.* (2008), Copyright (2010) by Elsevier.

largely on the film structure and the nature of the metal–oxide interactions and is rather different in all cases. Rocksalt MgO(111) is only polar in the limit of ultrathin films, but becomes non-polar with increasing thickness due to the formation of 3D oxide islands (Myrach *et al.*, 2011). In ZnO(0001)/Ag(111), on the other hand, the polarity is suppressed already in the first atomic planes, which adopt a hexagonal boron nitride structure with Zn and O ions lying in the same layer (Tusche *et al.*, 2007). The wurzite structure of bulk ZnO is only restored in thicker films, which are, however, subject to considerable surface roughening that quenches the reappearing surface dipole. In rocksalt CoO(111) and spinel Co_3O_4(111), polarity healing is achieved by a substantial decrease in the interlayer distance between the top-most O and Co planes and a concomitant reduction of the ionicity of the surface species (Giovanardi *et al.*, 2006). A similar means to reduce the surface polarity has been identified for FeO(111) on Pt(111). Using X-ray diffraction techniques, the Fe-O layer separation in the film was determined as 0.68 Å, which is 50% smaller than the bulk value and substantially decreases the surface dipole (Kim *et al.*, 1997). As a result of this vertical contraction, the FeO layer is expanded within the surface plane and the Fe nearest-neighbour distance increases from its bulk value of 3.0 Å to 3.1 Å in the film. A similar tetragonal distortion has been revealed for other polar films, e.g. for VO/Rh(111) (Schoiswohl *et al.*, 2005; Parteder *et al.*, 2008) and TiO_x/Pt(111) (Sedona *et al.*, 2005), indicating the universal nature of this polarity healing mechanism.

In all these experiments, changes in the oxide stoichiometry and parasitic adsorption processes have been excluded as means to heal the oxide polarity. Thin oxide films that can be prepared on various metal supports are therefore well suited to study the adsorption behaviour of polar materials. However, before this issue is addressed in Section 11.4, experimental techniques to quantify the amount of oxide polarity will be introduced in the next Section.

11.3 Measuring polarity of thin oxide films

In most experiments, the polar nature of thin films is concluded from indirect evidence, for instance from a reconstruction of the surface (Dulub *et al.*, 2003; Ostendorf *et al.*, 2008), unusual electronic properties (Kiguchi *et al.*, 2003; Arita *et al.*, 2004) or an adsorption behaviour that strongly deviates from the non-polar case (Rohr *et al.*, 1994; Wang *et al.*, 2005; Poon *et al.*, 2006). Such indirect indications give no information on the nature of the actual polarity healing mechanism and on the size of the remaining surface dipole. However, this kind of data can be obtained with STM, a technique that is usually employed to probe the surface topography. In the following, two spectroscopic applications of STM are discussed that enable detection of the local surface potential as a measure for the polarity. The techniques are based on the evaluation of (i) energy positions of field emission resonances (FER) and (ii) effective barrier heights for electron-tunnelling into the polar film. They are demonstrated using the example of the FeO(111) film mentioned above (Rienks *et al.*, 2005), but can be applied to any other polar system with sufficient conductivity to perform STM.

The FeO film has a bilayer structure, consisting of a hexagonal O layer at the surface and a hexagonal Fe plane at the interface to the Pt(111) (Fig. 11.7). It has a polar character due to the ionic nature of the $Fe^{\delta+}$ and $O^{\delta-}$ species and belongs to type III materials in the Tasker scheme (Tasker, 1979). The FeO film has an in-plane lattice parameter of 3.1 Å, being 11% larger than that of Pt(111) (2.76 Å). This mismatch leads to the formation of a coincidence lattice with 25 Å edge length and a crystallographic relation of $(\sqrt{91} \times \sqrt{91})R \pm 5.2°$ with respect to the Pt support (Ritter *et al.*, 1998). Within the coincidence cell, three stacking domains are distinguishable that

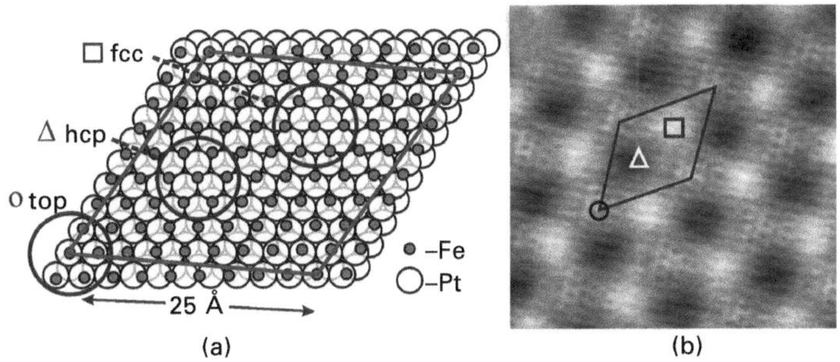

11.7 (a) Structure model of the coincidence cell formed between FeO and Pt(111). The O-top layer is omitted for the sake of clarity. (b) STM topographic image taken at 65 mV (7 × 7 nm^2). The different stacking domains are assigned in accordance to the DFT calculations presented in Giordano *et al.* (2007b).

differ in their Fe binding geometry on the Pt(111) surface (Fig. 11.7a). In the top domain, the interfacial Fe atoms bind on top of Pt, while the O atoms in the layer above occupy fcc hollow sites. In the fcc (hcp) domains, Fe binds to fcc (hcp) hollow sites while O sits in hcp (top) positions of the Pt(111) lattice. The different binding configurations give rise to a distinct contrast of the FeO film in low-bias STM images (Fig. 11.7b). The discernible regions have been assigned to the underlying stacking domains with the help of model calculations based on DFT (Giordano *et al.*, 2007b) and a quantum-chemical scattering approach (Galloway *et al.*, 1996).

The binding of Fe and O atoms either to Pt hollow or top sites modulates the interlayer distance and hence the surface dipole in the three oxide domains. The resulting change in polarity is an observable quantity and produces a strong bias-dependent contrast of the FeO film in STM images taken at high sample bias (Fig. 11.8b). At 4.25 V, one region of the coincidence cell (marked with a circle) is imaged with a particularly bright contrast. The contrast is not related to a geometric effect, as the low-bias corrugation of the film that reflects its true morphology is very small (0.3 Å). It is, in fact, of electronic origin and caused by a strong increase in the conductance through the STM junction at this particular bias. To verify this statement, the differential conductance (dI/dV) through the FeO/Pt(111) system has been measured directly as a function of bias voltage, as shown in Fig. 11.8a. In correspondence to the topographic image, high dI/dV intensity at 4.25 V is observed in the domain marked by the circle. However, with increasing bias the dI/dV maximum moves to an adjacent domain, being flagged by the square, and finally appears in the triangle region at 4.9 V sample bias.

These bias-dependent conductance changes are caused by electron transport through confined states that develop in the classical part of an STM junction and become accessible for electrons at elevated bias (Binnig *et al.*, 1982;

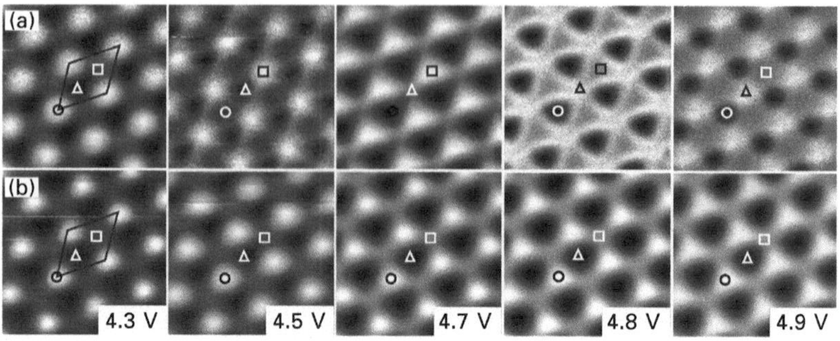

11.8 (a) Conductance and (b) topographic images of FeO/Pt(111) taken as a function of the bias voltage (9 × 9 nm^2). The contrast change reflects the varying contributions of field emission resonances to the electron transport through the different stacking domains. Reprinted with permission from Rienks *et al.* (2005), Copyright (2010) by the American Physical Society.

Becker *et al.*, 1985). Those states may be considered as eigenstates in a triangular potential, being confined by the sample surface on one side and the down-sloping vacuum barrier on the other (Fig. 11.9b). Electrons penetrating this classical region are able to form standing waves, if multiples of half their electron wavelength match the distance between the two boundaries. In this case, propagating and reflected waves interfere constructively and quasi-bound electronic states with a high transmission probability develop in the STM junction. These states are termed field emission resonances (FER) and dominate the STM image contrast at elevated bias. Interestingly, they also contain information on the local surface potential Φ and might therefore be used to probe oxide polarity (Fig. 11.9b). The interplay between FER and Φ can be explained with a simple picture. The electrostatic energy due to uncompensated polarity (Eq. 11.2) produces an offset on the surface potential that shifts the FER to higher energy and modulates their availability for electron transport through the oxide film. Up-shifted FER are therefore indicative for polar regions, while down-shift states occur in areas with compensated polarity.

Already this crude model is sufficient to connect the contrast changes observed in the dI/dV maps of FeO/Pt(111) to spatial modulations of the surface polarity (Rienks *et al.*, 2005). The domain that turns bright at 4.25 V is the one with lowest surface potential, as the first FER becomes available at relatively low bias (Fig. 11.8a). Based on a simple hard-sphere model of the film, this region is assigned to the FeO top domain, in which Fe and O atoms occupy top and hollow sites of the Pt surface, respectively, and the Fe-O interlayer distance and hence the surface dipole are small. At 4.5 V, the FER become available in the square regions, which have a slightly higher Φ as indicated by the up-shifted resonances. The region marked by the

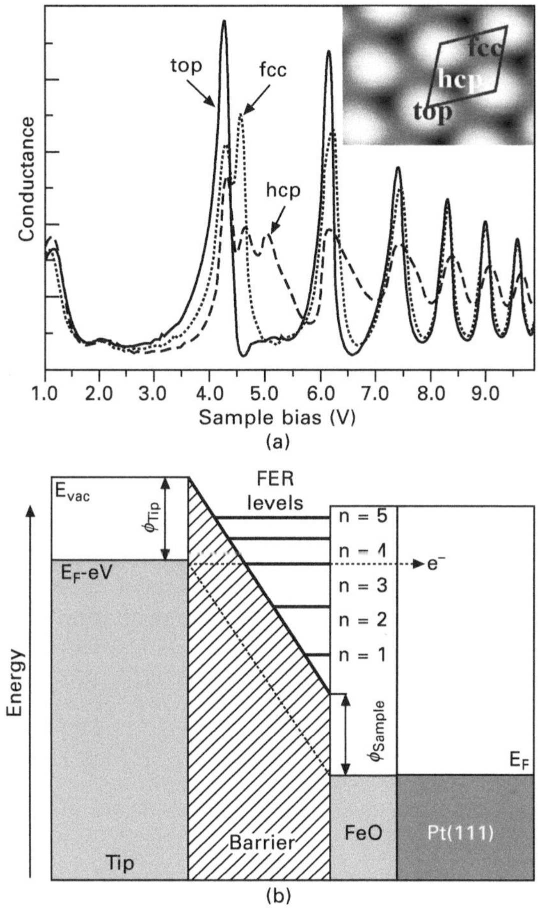

11.9 (a) Conductance spectra obtained with closed feedback loop on the three stacking domains of the FeO film, as indicated in the STM image shown in the inset. Each maximum marks the position of a field emission resonance (FER) in the tip-sample junction. (b) Potential diagram visualizing the formation of FER at high bias voltage and their dependence on the local work function. Reprinted with permission from Rienks *et al.* (2005), Copyright (2010) by the American Physical Society.

triangle turns bright only at 4.9 V and consequently has the highest Φ and hence the highest degree of polarity. The hard-sphere model connects this region to the hcp domains, as the vertical distance between the hollow-bound Fe atoms and the top-bound O atoms is expected to be largest there. The sequential availability of FER in the different domains is not so evident in the topographic images, which contain information on the integral conductance probed over a large bias window (Fig. 11.8b).

The changes in the electrostatic potential due to the modulations in the polarity can even be quantified for the three stacking domains of FeO/Pt(111) (Rienks *et al.*, 2005). For this purpose, the position of the first six FER was detected with dI/dV spectroscopy performed in a large bias window (1–10 V) (Fig. 11.9a). In correspondence to the dI/dV maps, the series of FER starts in the top domain at 4.25 V (see circle in Fig. 11.8a), whereas the fcc (square) and hcp (triangle) domains follow at 4.7 and 5.1 V, respectively. It should be noted that due to the limited spatial resolution of dI/dV spectroscopy at high bias, low-lying FER of neighbouring domains show up in the spectrum of a selected region as well. Consequently, the dI/dV peaks of fcc and hcp domains are split into two and three maxima with the lower ones arising from conductance contributions of the adjacent top and fcc domains, respectively.

The local surface potential Φ is determined from fitting the experimental FER positions to a quantum-mechanical model that describes the resonances as eigenstates in a triangular potential. The well is confined by the sample surface and the vacuum barrier that slopes down with the tip-induced electric field F (Fig. 11.9b) (Kolesnychenko *et al.*, 2000). Here, F is assumed to be constant during spectroscopy due to the enabled feedback loop of the STM. The position of the nth FER calculates to:

$$eV_n = \Phi + \left(\frac{3\pi\hbar e}{2\sqrt{2m}}\right)^{2/3} F^{2/3} n^{2/3}$$

with m being the free electron mass (Kolesnychenko *et al.*, 2000). A fit of all experimental FER except the first one, whose energy position is altered by other effects, yields the surface potential. The derived Φ values increase from 3.5 eV in the top region to 3.65 and 3.85 eV in the fcc and hcp domains, in good agreement with the results of the dI/dV maps. Conclusively, the hcp and top domains are the most and least polar ones, corroborating the predictions from the hard sphere model.

Complementary information on the surface polarity is obtained by measuring the effective barrier height Φ_{eff} that is experienced by electrons tunnelling into the FeO film (second approach). The barrier height is measured via dI/dz spectroscopy, detecting the current response I to a change in the tip-sample distance z (Chen, 1993). The relationship between those quantities can be derived from a one-dimensional model for electron tunnelling through a square-shaped barrier:

$$\Phi_{eff} = \frac{\hbar^2}{8m}\left(\frac{d \ln I}{dz}\right)^2$$

Although the barrier height is not identical to the real surface potential, it is a monotonous function of Φ (Olesen *et al.*, 1996). It will thus be largest on the most polar oxide region, as the electrons have to overcome a substantial surface dipole. Figure 11.10 shows two approaches to probe the effective barrier height of the FeO thin film. In Fig. 11.10(a), the logarithmic current response to a linear distance ramp is recorded and Φ_{eff} is extracted from the slope of the ln(I)-Δz curves. In Fig. 11.10(b), the $d(\ln I)/dz$ signal is mapped directly with lock-in technique by adding a small z-modulation to the STM feedback loop. In both cases, spatial variations in the effective barrier height are detected on the FeO surface. Here, the top domains exhibit the smallest slope in ln(I)-Δz curves and the lowest intensity in $d(\ln I)/dz$ maps, indicating a low barrier height. The hcp domains, on the other hand, are characterized by a large Φ_{eff} value, in agreement with their more polar nature. The two techniques discussed above, namely evaluating the FER energies and probing the effective barrier height, therefore reveal similar modulations in the local surface potential and provide unmatched insight into the polarity of the three FeO stacking domains.

The experimental results obtained on the polar FeO/Pt(111) film have been corroborated by DFT calculations (Giordano *et al.*, 2007b). Due to the large size of the real FeO-Pt coincidence cell, they were performed with a smaller computational cell that still contains Fe atoms in characteristic top, fcc and hcp binding configurations (Fig. 11.11a). While the hcp and fcc regions display similar properties in the simulations, the top domain sticks out in various aspects. It is characterized by the largest interfacial separation from

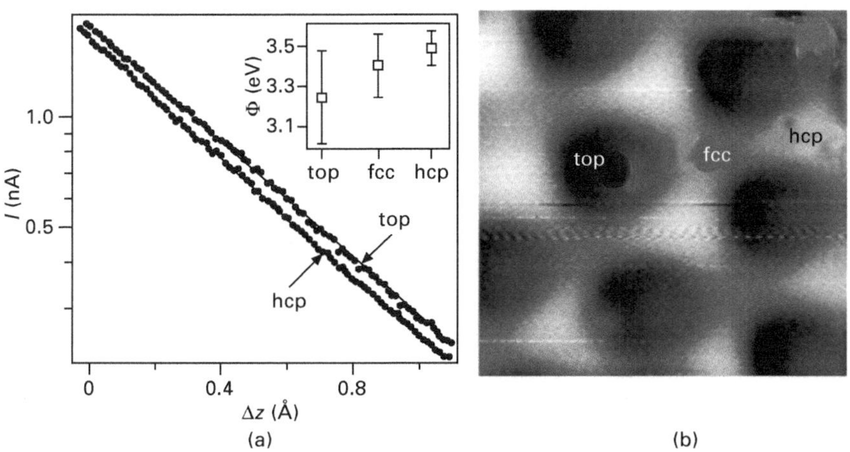

11.10 (a) Tunnel current versus tip-sample distance recorded for the top and hcp domain of the FeO coincidence cell. From the current slope, the effective barrier height is determined and displayed in the inset. (b) $d(\ln I)/dz$ image of FeO/Pt(111) taken with lock-in technique and closed feedback loop. (Image size 5.7 × 5.7 nm^2, V_s = 4.5 V, Δz_{rms} = 1 Å.) Reprinted with permission from Rienks *et al.* (2005), Copyright (2010) by the American Physical Society.

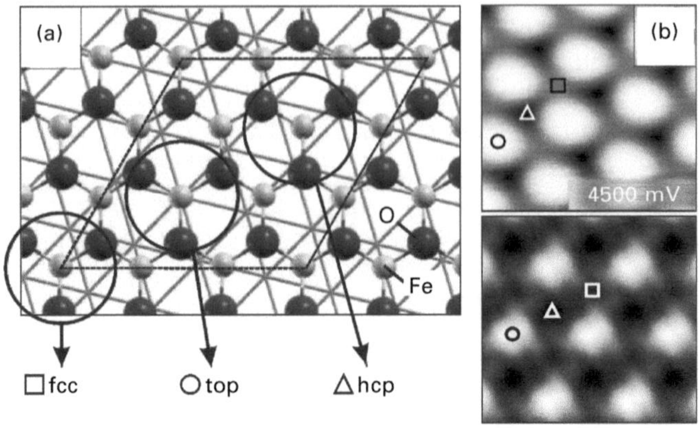

11.11 (a) Top view of a model structure for the FeO/Pt(111) system that contains all relevant Fe binding configurations but is much smaller than the real coincidence cell. (b) STM image of the polar FeO film taken at +4.5 V sample bias (top, 7 × 7 nm^2) and corresponding surface potential map calculated with the model shown in (a) (bottom). The map depicts the -Φ signal, and regions with low surface potential. Reprinted with permission from Giordano *et al.* (2007b), Copyright (2010) by the American Physical Society.

the Pt(111) surface and therefore mimics the properties of a free-standing FeO layer. Additionally, the vertical separation between Fe and O planes (0.5 Å) is 15–20% smaller in the top than in adjacent hcp and fcc domains (0.8 Å). The top region therefore has the smallest $Fe^{\delta+}$-$O^{\delta-}$ surface dipole and hence the lowest surface potential. Its Φ value is 0.29 and 0.23 eV lower than those of the hcp and fcc regions, respectively, in good agreement with the measured potential modulations in the FeO coincidence cell of 0.35 eV.

Based on the computed surface potentials, STM images have been simulated for bias voltages in the field emission mode (Giordano *et al.*, 2007b). As discussed above, the contrast in this regime is governed by the availability of FER, which in turn depends on the local surface potential. A small Φ gives rise to a bright contrast, as low-lying FER with high electron transmissibility promote the electron transport, while regions with large Φ appear dark. Calculated Φ maps, depicted with inverted contrast, indeed reproduce the high-bias STM images of the FeO/Pt system (Fig. 11.11b). In agreement with our earlier interpretation, the brightest region corresponds to the one with the lowest surface potential and is assigned to the top domain that has the lowest polarity of all FeO stacking domains.

Measurements of the surface potential with an STM have proven to be an adequate means to probe the polarity of FeO thin films (Rienks *et al.*, 2007). Similar results were obtained for other polar systems, e.g. MgO(111) films grown on Au(111) (Myrach *et al.*, 2011). Also here, a considerable surface dipole develops from the stacking of a surface O^{2-} layer on top of an interfacial Mg^{2+} layer. This layer sequence is compatible with the electronegative character of gold, which promotes a charge transfer out of the cationic plane to reduce the vertical dipole moment (Fig. 11.5a). Similar to the FeO/Pt(111) system, the surface polarity is not homogeneous on the MgO(111) surface, but for a different reason. The oxide polarity quickly vanishes with film thickness, as three-dimensional MgO islands with zero dipole develop on the surface. The initial building blocks that are responsible for polarity healing are nano-pyramids of a few layers height (Wolf, 1992), which are readily observed in STM images of the oxide film (Fig. 11.12a). Their effect on the oxide polarity has been deduced from the barrier height images again (Fig. 11.12b). As expected, the Φ_{eff} value turns minimal above the surface protrusions, demonstrating the effective quenching of the surface dipole due to the island growth. It should be noted that this morphology-driven removal of the polarity bears similarities to the octopolar reconstruction that is the dominant healing mechanism on bulk MgO(111) (Barbier and Renaud, 1997; Barbier *et al.*, 2000).

In general, the STM is able to provide spatially resolved information on the distinct properties of polar surfaces. The technique is not only suited to characterize the various polarity healing mechanisms from a topographic point of view, but enables a local determination of the surface potential and its correlation to the oxide polarity. Although STM-based approaches do not reach the same quantitative accuracy in determining surface potentials as,

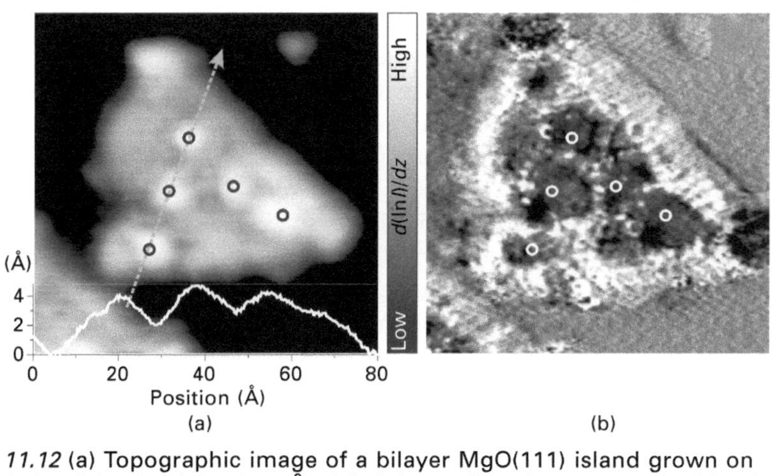

11.12 (a) Topographic image of a bilayer MgO(111) island grown on Au(111) (–0.25 V, 9 × 9 nm^2). A height profile taken across the island is shown in the inset. (b) Effective barrier height ($d(\ln I)/dz$) image of the same surface region. Particularly low values of the barrier height are revealed on the ad-structures that cover the oxide island, indicating their importance in quenching the surface dipole.

for instance, Kelvin–Probe spectroscopy, they feature an unmatched spatial resolution with respect to alternative techniques.

11.4 Adsorption properties of polar films

Residual polarity gives rise to unusual adsorption and chemical properties of thin oxide films (Goniakowski *et al.*, 2008; Sun *et al.*, 2009). The difference to non-polar systems lies in the electrostatic contribution to the surface free energy that originates from the uncompensated surface dipole (see discussion in Section 11.1). The binding behaviour of adsorbates is therefore not only governed by the usual physisorption and chemisorption effects, but includes changes in the electrostatic energy of the system as adsorbates might reduce its polarity. Given the magnitude of the energies involved, the binding potential of polar surfaces can be substantially higher compared to non-polar ones. A direct manifestation of this effect is the wetting growth of metals on polar surfaces, whereas mainly three-dimensional deposits form on non-polar oxide materials (Goniakowski and Noguera, 2002; Meyer and Marx, 2004).

Two mechanisms have to be considered in conjunction with polarity healing via adsorbates. In a first scenario, the ad-species become charged upon adsorption and alter the electron density on the surface. As discussed in Section 11.1, depolarization of the system takes place when the surface charge density equals the bulk density times the ratio between interlayer distance d and unit cell height D:

$$\sigma_{\mathrm{surf}} = \sigma \frac{d}{D} \qquad [11.4]$$

This condition can now be fulfilled by adsorbing the required number of charged species to the surface. The most prominent example of this mechanism is the attachment of protons (H^+) that often originate from the heterolytic splitting of water to polar surfaces (hydroxylation). For rocksalt (111), every surface site needs to be occupied by a hydroxyl group in order to quench the polarity, although $d/D = 0.5$. The reason is that each H^+ carries only half the charge of an oxide ion (Mg^{2+}, O^{2-}). In the case of wurzite (0001), on the other hand, 50% surface coverage would be sufficient. The hydroxylation of polar oxide surfaces has been intensively studied with infrared reflection absorption and high resolution electron energy loss spectroscopy (HREELS), as discussed for instance in Rohr *et al.* (1994), Poon *et al.*, (2006) and Wang (2008). At the local scale, proton attachment was investigated with the STM on Cr_2O_3/Cr(110) films (Maurice *et al.*, 2001) and more recently on FeO/Pt(111) (Merte *et al.*, 2009; Knudsen *et al.*, 2010). Although dipole removal via hydroxylation is observed most frequently, any other adsorbate that is easily ionized or polarized can be used instead. Examples for the adsorption of charged ad-species on polar oxide films are given in Section 11.4.1.

A second mechanism for dipole quenching is the metallization of polar surfaces. In this case, the adsorbates create electron bands that cross the Fermi energy of the system and can be filled with the required number of electrons to satisfy Eq. 11.4. Those electronic states might be intrinsic to the ad-species or induced into the polar material upon adsorption. The first case is typically realized with metallic adsorbates that possess states around the Fermi level. For instance, a single Pd layer attached to a Mg-terminated MgO(111) slab was found to charge up negatively by filling a Pd-like interface state (Goniakowski and Noguera, 2002). The computed charge density in the Pd layer hereby adopts 50% of the MgO bulk value in accordance with Eq. 11.4. Further thickening of the metal film does not change this situation, indicating that a single layer is sufficient to cancel the oxide polarity. Experimentally, the wetting growth of metals on polar oxide materials has been studied for Cu on Zn(0001) (Koplitz *et al.*, 2003; Dulub *et al.*, 2005; Kroll and Köhler, 2007). Especially, on the O-terminated surface, a layer-by-layer growth of Cu is observed that reflects, however, not only the impact of oxide polarity but also the generally higher adhesion for this surface termination. A similar result was obtained for the polar FeO/Pt(111) film, where extended 2D islands form upon Pd deposition (Fig. 11.13a). It should be noted that metals usually develop 3D particles when deposited

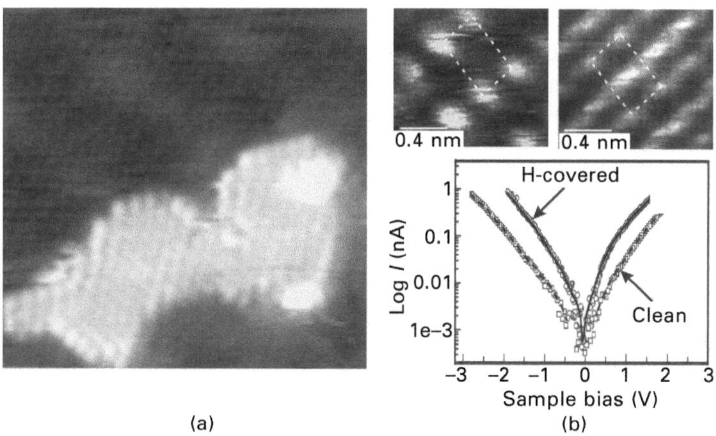

11.13 (a) STM image of a planar Pd island grown on a FeO/Pt(111) film (9.0×9.0 nm^2). The two-dimensional growth mode is triggered by the polarity of the system. (b) STM images of a clean (left) and a hydrogen covered ZnO(10-10) surface (right). The dashed rectangle marks the (1×1) surface unit cell. H adsorption leads to a metallization of the ZnO(10-10) surface, as shown by the vanishing band gap in I-V curves taken on the H-covered surface. Reprinted with permission from Wang *et al.* (2005), Copyright (2010) by the American Physical Society.

on non-polar materials in order to minimize the contact area with the inert oxide surface (Bäumer and Freund, 1999).

Induction of a metallic surface state in the substrate itself has been demonstrated with STM conductance spectroscopy and HREELS for the ZnO(10-10) surface (Wang *et al.*, 2005; Yin *et al.*, 2006). In those experiments, the oxide band gap was found to disappear after adsorbing one H atom per unit cell of the ZnO surface (Fig. 11.13b). This result was explained with the transfer of the hydrogen electron into the initially empty *4s*-band of zinc that subsequently becomes half-filled. Increasing the dosage to two H atoms per unit cell renders the surface non-metallic again, as a second electron is added to the surface band that consequently shifts below the Fermi level. A similar metallization effect has so far not been revealed for the polar ZnO(0001) surface.

The detailed adsorption properties of polar systems are discussed for the FeO/Pt(111) film in the following section. This particular system is selected again because it has already been introduced in the previous sections. Furthermore, the FeO film exhibits spatial modulations of the surface polarity, which induces a template effect in the arrangement of the adsorbates. It is therefore an example where uncompensated polarity induces self-assembly phenomena on an oxide surface.

11.4.1 Adsorption of metal atoms on polar FeO films

As discussed in Section 11.3, the polarity of the FeO film arises from the interplay between an $Fe^{\delta+}$ interface and an $O^{\delta-}$ surface plane (Vurens *et al.*, 1988; Kim *et al.*, 1997). The Fe-O layer separation is not constant, but varies between 0.52 and 0.78 Å due to different Fe-O stacking configurations on the Pt(111) surface. Large interlayer distances occur for the fcc and hcp domains, while small separations are found in the top region of the FeO-Pt coincidence cell (see Fig. 11.7a for a structure model). As discussed above, such modulations in the interlayer distance affect the vertical dipole strength and therefore the polarity of the film, whereby the hcp and top regions turn out to be the most and least polar ones, respectively (Rienks *et al.*, 2005). The same modulations should govern the spatial distribution of adsorbates on the FeO surface as well.

A first demonstration of this effect was provided in a low-temperature STM experiment, where single Au atoms were deposited onto the film at 10 K sample temperature (Nilius *et al.*, 2005). The incoming atoms perform a transient diffusion on the FeO surface due to their initial thermal energy, and are able to reach their preferred adsorption sites. After thermalization, they can be imaged in the STM as 1.0 Å high, circular protrusions. The adatoms are not randomly distributed on the surface, but exhibit a large tendency to attach to the hcp domains of the coincidence cell. From the five adatoms shown in Fig. 11.14(a), only the uppermost one binds to a top region while the four lower ones sit in hcp domains. At a nominal Au coverage of 0.01 ML, which is close to one atom per coincidence cell, more than 70% of the hcp domains are occupied with Au atoms, whereas adjacent fcc and top regions remain nearly adsorbate-free (occupancy below 10%) (Fig. 11.14c). The site-specific adsorption behaviour of Au therefore leads to the formation of a hexagonal adatom array, the lattice parameter of which matches the size of the FeO/Pt(111) coincidence cell. The long-range order in the adatom arrangement becomes evident in a 2D pair-distribution function $n(x, y)$, calculated from more than 700 atom positions (Fig. 11.15). The highest probability to find a neighbouring Au atom is in the hcp domain of the next FeO/Pt coincidence cell; however, also second and third neighbouring hcp domains show substantial occupancy due to the perfect ordering of the adsorbates. A distinct atom distribution is discernible even within each hcp region, as marked by seven, hexagonally-arranged spots with 3.1 Å mutual distance. This fine structure reflects the few atomic binding sites in the centre of each hcp domain that are actually populated by Au. Based on the pair distribution function, the modulation of Au binding energies within the coincidence cell ΔE can be estimated by assuming a Boltzmann distribution for the occupation probability: $n(x, y) \propto \exp(^{-\Delta E}/_{kT})$ (Silly *et al.*, 2004; Kulawik

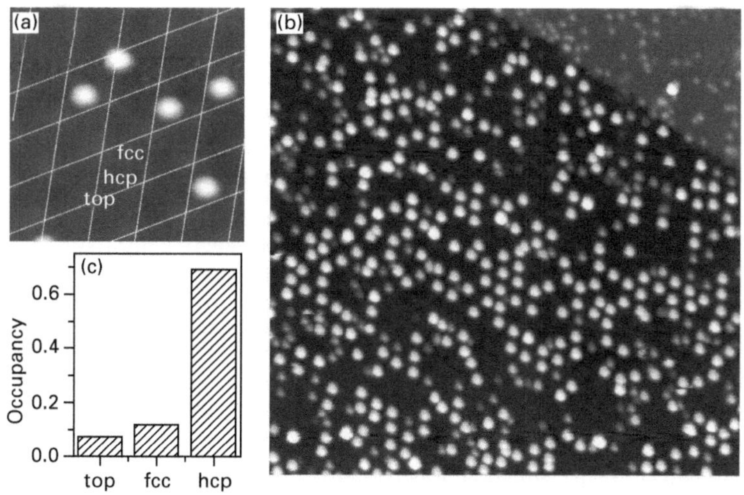

11.14 (a) STM topographic image of Au atoms on FeO/Pt(111) (0.5 V, 13 × 13 nm^2). Four out of five adatoms occupy hcp domains in the FeO coincidence cell. (b) Overview STM image demonstrating the ordering of Au atoms into a hexagonal lattice (0.5 V, 60 × 60 nm^2). (c) Histogram for the probability of finding an Au adatom on the different FeO stacking domains. The plot is based on the evaluation of ~700 atom positions. Reprinted with permission from Nilius *et al.* (2005), Copyright (2010) by the American Physical Society.

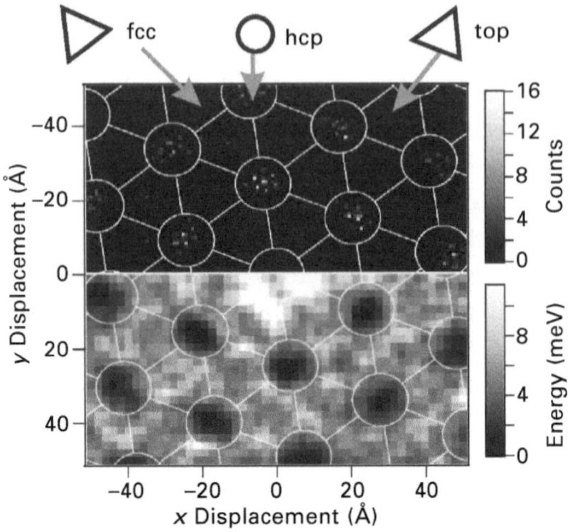

11.15 Two-dimensional pair-distribution function of Au atoms on the FeO film (top). Most events occur in the hcp domains; the sharp maxima within those regions reflect the atomic binding sites of the adsorbates. The bottom part depicts the potential landscape for Au adsorption, being calculated by assuming a Boltzmann distribution according to the Au binding energy. Reprinted with permission from Nilius *et al.* (2005), Copyright (2010) by the American Physical Society.

et al., 2005). As the self-assembly is observed for temperatures as high as 50 K, this temperature is used to calculate a lower bound for ΔE. The analysis yields an energetic preference of (10 ± 2) meV for Au binding to the hcp domains with respect to adjacent fcc and top regions. It is interesting to note that Au atoms remain essentially monomeric upon adsorption onto the oxide film even at relatively high coverage. No aggregation sets in before the majority of hcp domains is filled, which suggests a repulsive, most likely Coulomb-type interaction between the adatoms that inhibits cluster formation (Nilius *et al.*, 2005).

The self-assembly of Au atoms on the FeO/Pt(111) surface can be traced back to the polar nature of the film. The adatoms preferentially attach to the hcp domains having the largest surface dipole (Giordano *et al.*, 2007b). The uncompensated polarity in this region enables an efficient polarization of the Au, which in turn leads to an electrostatic coupling between the adatoms and the dipole field of the polar film. This binding contribution will be smaller on the adjacent fcc and top domains, where the dipole strength is reduced.

A more detailed picture of the Au ordering effect on the FeO is obtained from DFT calculations (Giordano *et al.*, 2008). In this study, the Au interaction with O-top, Fe-top, bridge and hollow sites has been analysed for all three stacking domains of the coincidence cell. The largest adsorption energy of 0.6 eV was computed for the O-top sites in hcp domains, where the binding is 30% stronger than on Fe sites in the hcp and O sites in the fcc and top domains. This binding preference explains the observed self-assembly of Au adatoms on the FeO film. It should be noted that theory finds an even more favourable adsorption configuration, in which an interfacial Fe atom flips above the O-plane (Fig. 11.16). The hence under-coordinated Fe atom is able to bind the surface Au species with ~1.5 eV, much stronger than in the regular geometry. Due to a concomitant negative charging of the Au adatom, also the film polarity is locally quenched in this configuration, as the resulting $O^{\delta-}$-$Fe^{\delta+}$-$Au^{\delta-}$ stacking carries no dipole moment (Fig. 11.16b). A similar binding configuration has been predicted for other electronegative metals, such as Pt (Goniakowski *et al.*, 2009). However, the inverted adsorption geometry could not be confirmed experimentally, as neither the height nor the electronic properties of a potential Au-Fe ad-species was compatible with the STM data. Apparently, the energy barrier to break a Pt-Fe bond at the interface and swap the local Fe-O stacking sequence is not overcome at the low temperature of the experiment (Nilius *et al.*, 2005).

The experimentally confirmed Au binding to O atoms in the hcp domains leads to a 20% increase of the Fe-O layer distance around the adsorption site, which enhances the surface dipole and therewith the electrostatic Au–FeO interaction (Giordano *et al.*, 2008). The local lattice distortion is initiated by a charge transfer from the Au atom into the Pt crystal. The Coulomb interaction of the Au^+ lifts the $O^{\delta-}$ ion underneath above the surface plane, but repels the three adjacent $Fe^{\delta+}$ ions, thereby inducing a polaronic distortion of the oxide lattice. The formation of charged adsorbates combined with a polaronic distortion of the surrounding lattice is a common binding scenario on ionic oxide films (Pacchioni *et al.*, 2005; Giordano *et al.*, 2007a). The cationic nature of Au atoms on the FeO film is further supported by state density calculations (Fig. 11.17b). The Au *6s* orbital of hcp-bound adatoms is located at +0.3 eV above the Fermi level and thus empty, in contrast to its half-filled nature in gas-phase gold. This finding indicates that the *6s* electron has been transferred into the support, rendering the Au positively

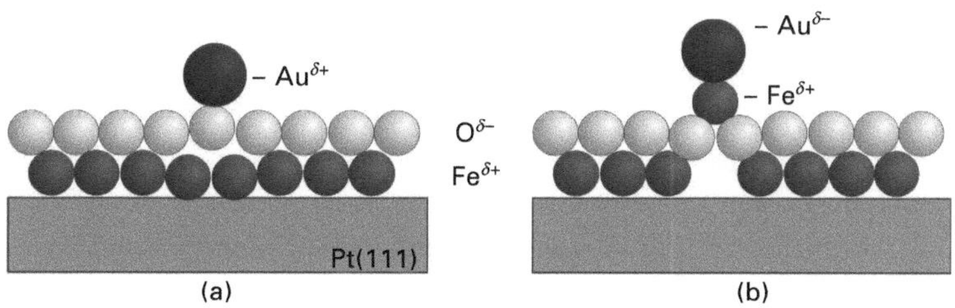

11.16 (a) Experimentally observed and (b) energetically favoured binding configuration of Au on FeO/Pt(111). In the optimum geometry, a Fe atom flips above the oxide plane in order to increase the interaction with the Au. The energy barrier for this restructuring is not known.

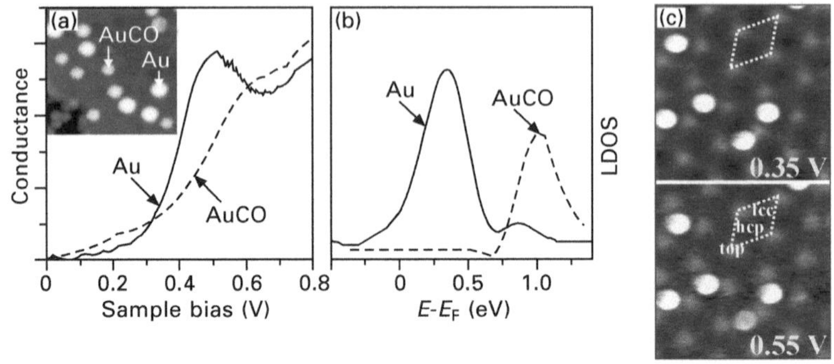

11.17 (a) Conductance spectra and (b) calculated state-density of Au and AuCO species adsorbed on hcp-domains in the FeO/Pt(111) film. An STM image of the probed species is shown in the inset. The Au-related peak at 0.5 V (exp.) and 0.3 eV (theory) corresponds to the Au *6s* orbital, while the AuCO state at 1.0 eV is of CO $2\Pi^*$ character and not detected in STM. (c) Conductance maps of differently bound Au atoms (13×13 nm^2). Whereas the upper atoms are attached to hcp domains and appear bright in the conductance map taken at the position of the Au *6s* orbital (0.55 V), the lower species sits in an adjacent domain and shows no *6s*-related dI/dV intensity (lower panel). Below the resonance position, all adatoms are imaged with the same dI/dV contrast (upper panel). Reprinted with permission from Giordano *et al.* (2008), Copyright (2010) by the American Physical Society.

charged. Again, the formation of cationic gold is closely related to the polar nature of the FeO film. Only attachment of a positively charged species is able to heal the oxide polarity, as this neutralizes locally the $Fe^{\delta+}$–$O^{\delta-}$ surface dipole (Fig. 11.16a). Adsorption of Au^+ is therefore an efficient means to lower the surface free energy of the polar system and fully comparable with the common polarity healing mechanism via hydroxylation (H^+ attachment). Energetically, the formation of cationic gold is supported by the unusually high work function of the FeO/Pt(111) system that impedes charge transfer out of the support (Giordano *et al.*, 2007b, 2008). It should be noted at this point that Au^+ is an unusual charge state for the strongly electronegative gold. In fact, Au becomes anionic on most non-polar oxide films due to a charge transfer into the adsorbate (Pacchioni *et al.*, 2005; Nilius, 2009). The positive charging in the present case can only be rationalized if the polar character of the FeO film is taken into account.

The charge state of Au atoms on the FeO/Pt(111) system has been verified experimentally by probing the position of the Au *6s* orbital with respect to the Fermi level with STM conductance spectroscopy. The hcp-bound atoms exhibit a pronounced dI/dV peak at +0.5 V that is readily assigned to the unfilled Au *6s* orbital with help of the DFT results (Fig. 11.17a) (Nilius *et al.*, 2005; Giordano *et al.*, 2008). Further evidence for the *6s* character of this state comes from CO adsorption experiments. After CO exposure onto the Au–FeO film, the +0.5 V conductance peak vanishes, because the Au *6s* level shifts out of the accessible spectral range upon formation of the Au–CO bond (Fig. 11.17a). The origin of this up-shift is the Pauli repulsion that is exerted by the CO orbitals onto the Au states (Persson, 2005). However, even bare Au atoms do not display the +0.5 V dI/dV peak if they are not bound to an hcp domain. As concluded from conductance spectra and images, Au atoms in fcc and top regions of the coincidence cell do not possess an unoccupied *6s* orbital (Fig. 11.17c). Apparently, those adatoms have a deviating electronic structure and are not charged, which explains their lower binding energy with respect to the hcp-bound species. The self-assembly of Au therefore results from an interplay between the spatially modulated oxide polarity and the concomitant charging response of the gold, which leads to a binding enhancement only for Au atoms on the FeO–hcp domains.

As the self-assembly of gold on the FeO/Pt(111) film is partially induced by a charge-mediated binding mechanism, it is not surprising that atoms with a deviating electronic structure do not show the same ordering effect. To a certain extent, palladium can be considered as the counterpart to gold, because the Pd affinity levels are far away from the Fermi energy and hardly involved in any charge transfer processes. More precisely, the Pd *5s* orbital is well above E_{Fermi} and empty, while the *4d* states are below and completely filled. The inhibited electron transfer is expected to hamper the self-ordering of Pd atoms on the FeO surface. Indeed, Pd deposition at 10 K results in a random distribution of adatoms on the oxide film, as evident in the STM image shown in Fig. 11.18. Furthermore, the adsorbates tend to aggregate

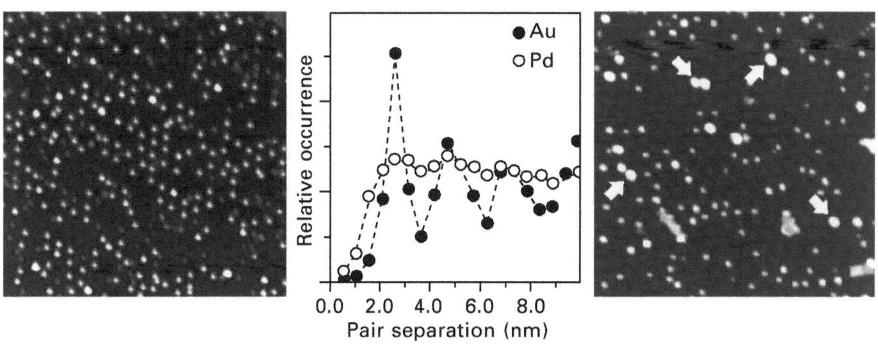

(a) (b) (c)

11.18 STM images of (a) Au and (c) Pd atoms on FeO/Pt(111) (0.25 V, 50 × 50 nm²). Arrows in (c) mark small Pd clusters. (b) Radial pair-distribution function for the Au and Pd adatoms on the FeO surface, calculated from hundreds of atom positions in different STM images. In contrast to Au, no ordering effect is observed for Pd adatoms. Reprinted with permission from Giordano *et al.* (2008), Copyright (2010) by the American Physical Society.

even at small Pd exposure and low deposition temperature (see arrows in Fig. 11.18c). The uncorrelated spatial arrangement becomes manifest in the radial pair-distribution function calculated for Pd, which does not display any maxima, in sharp contrast to the results obtained for Au (Fig. 11.18b).

The absence of self-ordering effects for Pd on the polar FeO film has several reasons, being elucidated in a DFT study (Giordano *et al.*, 2008). In general, Pd atoms experience a much higher binding energy than Au atoms (1.4 eV versus 0.6 eV). The largest binding contribution comes from the formation of covalent bonds with either a Fe or an O ion in the surface. Polarization interactions in the spatially modulated FeO dipole field, on the other hand, are comparatively small. As a consequence, the adsorption landscape of Pd adatoms is rather smooth on the FeO surface and local variations in binding strength are only of the order of 5% (compared to 30% for the Au). In particular, the different polarity of the top, fcc and hcp region does not give rise to prominent changes in the adsorption strength. In addition, the amplification of polar oxide properties due to charge transfer processes out of the adatoms, as observed for Au, does not take place for Pd that remains neutral on all binding sites of the FeO coincidence cell. It is the sum of all these effects that prevents the ordered arrangement of Pd atoms on the FeO film and even facilitates their aggregation into small aggregates on the surface (Giordano *et al.*, 2008).

11.4.2 Adsorption of molecules on polar FeO films

Charging is, however, not essential to induce the self-assembly of adsorbates on polar oxide films. Especially in the absence of strong chemical forces, the modulated surface-dipole within the FeO coincidence cell might sufficiently perturb the adsorption landscape to trigger ordering effects. Self-assembly phenomena that are exclusively driven by the spatially varying oxide polarity are therefore expected for closed-shell, but highly polarizable electronic systems that are unable to form strong covalent bonds with the support. Ideal candidates in this respect are molecules with a conjugated π-electronic system that couple to the inert oxide surface mainly via polarization interactions (Witte and Wöll, 2004). Therefore, magnesium phthalocyanine (MgPc) molecules have been chosen to study the adsorption behaviour of a polarizable, but chemically inactive species on the FeO/Pt(111) film (Lin and Nilius, 2008).

Figure 11.19(a) shows an STM topographic image taken after MgPc deposition onto the surface at 300 K. Close inspection of such images reveals that also the MgPc preferentially adsorbs on the hcp domains of the coincidence cell, which show up with bright contrast under the selected experimental conditions. Particularly attractive binding sites for the molecules are those hcp domains that are situated directly next to a rising step edge. A

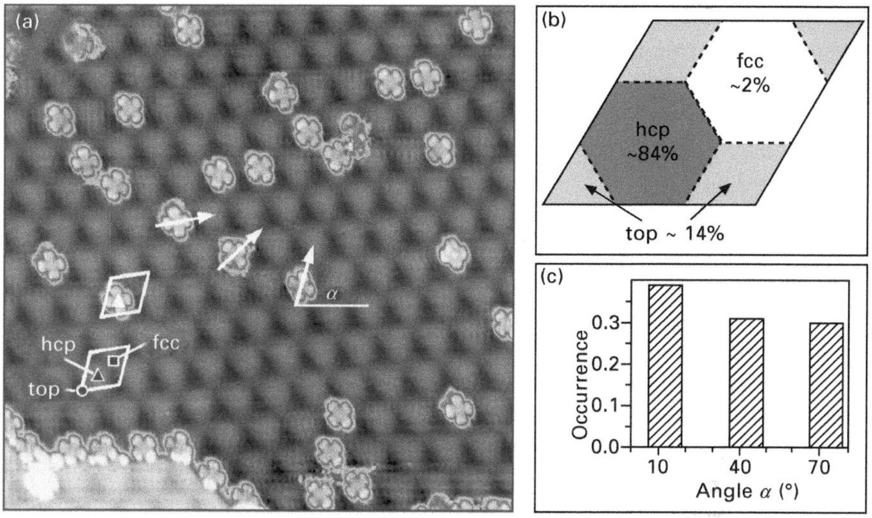

11.19 (a) Topographic image of MgPc molecules on FeO/Pt(111) adsorbed at 300 K (35 × 35 nm^2). The coincidence cell is indicated by a rhomboid, the arrows mark the three rotational orientations of the molecules. (b) Relative occurrence of MgPc in different domains of the FeO coincidence cell. (c) Angular orientation of the MgPc symmetry axes with respect to a horizontal line. More than 100 molecules have been included in the statistics shown in (b) and (c). Reprinted with permission from Lin and Nilius (2008), Copyright (2010) by the American Chemical Society.

statistical evaluation of dozens of STM images revealed that roughly 84% of the MgPc bind to hcp domains, while only 14% and 2% attach to top and fcc regions, respectively (Fig. 11.19b) (Lin and Nilius, 2008). Assuming a Boltzmann distribution according to the MgPc binding energy, adsorption to the FeO hcp domain is preferred with 90 meV over the fcc and with 46 meV over the top region. This result demonstrates the crucial importance of polarization interactions between MgPc molecules and the dipole field of the polar oxide. It should be noted that an electrostatic coupling to the vertical FeO dipole requires the induction of a dipole moment in the initially planar MgPc molecule, for instance via a small vertical displacement of the central $Mg^{\delta+}$ ion against the Pc-cage (Ruan *et al.*, 1999; Janczak and Kubiak, 2001).

The binding preference to the FeO hcp domain is almost one order of magnitude larger for MgPc molecules than for Au adatoms (90 meV versus 10 meV) (Lin and Nilius, 2008; Nilius *et al.*, 2005). This finding reflects the superior polarizability of the highly flexible, organic molecule with respect to a single atom. Furthermore, while polarization in the spatially modulated dipole field is the decisive interaction mechanism for MgPc, this contribution competes with true chemical bonding in the case of Au. Because the latter effect is less site-specific, the spatial modulations in the Au binding energy are weaker, and well-defined Au super-lattices only develop at cryogenic temperature. A self-assembly of the MgPc, on the other hand, is observed even at 300 K.

The ordering of MgPc on the FeO film affects not only the spatial distribution, but also the rotational alignment of the organic molecule (Lin and Nilius, 2008). The molecules always bind with one of their two equivalent axes oriented parallel to a close-packed FeO direction. Reflecting the hexagonal symmetry of the oxide support, three molecular orientations are distinguishable on the surface, separated by angles of 30° and 60° (Fig. 11.19c). A rotation by 90°, on the other hand, reproduces the initial binding situation of the fourfold symmetric adsorbate. The self assembly of MgPc on the FeO/Pt(111) system therefore has two aspects: while the spatial distribution is governed by the modulated surface potential within the coincidence cell, the angular alignment is fixed by the interaction of the molecule with the atomic FeO lattice. In general, the adsorption of MgPc to the polar FeO film is a rare example, where self-assembly is not induced by inter-molecular coupling (see, for instance, Dmitriev *et al.*, 2003; Barth *et al.*, 2005), but by a template effect of the oxide support.

11.5 Conclusion and future trends

The adsorption properties of polar oxide surfaces have been discussed for atomic and molecular adsorbates on the FeO/Pt(111) film. In both examples, the uncompensated surface dipole of the oxide film influences considerably the binding behaviour, predominantly by inducing polarization and electrostatic interactions between the ad-species and the oxide support. The binding occurs preferentially in regions with large dipole strength, triggering the formation of long-range ordered adsorbate structures on the oxide surface. Similar behaviour has been reported for comparable systems, e.g. for Cr and Fe aggregates on Fe_3O_4/Pt(111) and RhO_2/Rh(111) films (Berdunov *et al.*, 2006; Gustafson *et al.*, 2004). For many other polar oxides, elucidation of their adsorption behaviour including potential self-assembly effects is still lacking.

In conclusion, polar materials have a number of interesting morphological, electronic and chemical properties that arise from the electrostatic energy stored in those systems. As demonstrated in this chapter, the adsorption behaviour of polar oxides strongly deviates from that of non-polar systems, as the ad-species are partly involved in quenching the energetically unfavourable surface dipole. The resulting adsorption characteristics can be exploited in various technologically relevant applications. The strong adhesion between polar supports and metal adsorbates might be used, for example, to produce ultrathin, homogeneous metal coatings for optical and electronic devices. The observed self-assembly effects on polar surfaces are an interesting starting point to produce ordered arrangements of single atoms, molecules and metal particles. The fabrication of regular arrays of defined entities is highly desirable for applications in molecular electronics, heterogeneous catalysis and chemical sensing. However, the role of the oxide polarity in triggering such ordering phenomena needs to be explored in more detail. Another potential interest in polar systems arises from their unusual chemical properties, which originate from the high surface free energy of such structures. This energy contribution might be temporally released by restructuring the surface during a chemical reaction and can be transferred to the reactants. In a recent study, the polar FeO film was found to be 50 times more reactive than the widely-used Pt catalyst in CO oxidation reactions (Sun *et al.*, 2009). The understanding of the interplay between polarity and chemical properties is, however, just at the beginning and more work is required in the future. However, thin films with polar character will certainly remain in the focus of research in the next few years.

11.6 Sources of further information and advice

For more detailed information on the topic, the reader is referred to the books and review articles listed below:

Noguera C (1996), *Physics and Chemistry at Oxide Surfaces*, Cambridge University Press.

Noguera C (2000), 'Polar oxide surfaces', *Journal of Physics – Condensed Matter* **31**, R367.

Goniakowski J, Finocchi F and Noguera C (2008), 'Polarity of oxide surfaces and nanostructures', *Reports on Progress in Physics* **71**, 016501.

Dillmann B, Rohr F, Seiferth O, Klivenyi G, Bender M, Homann K, Yakovkin IN, Ehrlich D, Bäumer M, Kuhlenbeck H and Freund H-J (1996), 'Adsorption on a polar oxide surface: Cr_2O_3(0001)/Cr(110)', *Faraday Discussion* **105**, 295.

11.7 Acknowledgements

My special thanks go to Hans-Joachim Freund for his constant support and to the members of my group that have contributed to this work: Emile Rienks, Xiao Lin and Philipp Myrach. I am grateful to Gianfranco Pacchioni, Livia Giordano and Jacek Goniakowski for their theoretical insights into polar materials.

11.8 References

Arita R, Tanida Y, Entani S, Kiguchi M, Saiki K and Aoki H (2004), 'Polar surface engineering in ultrathin MgO(111)/Ag(111): possibility of a metal-insulator transition and magnetism', *Phys Rev B*, **69**, 235423.

Barbier A and Renaud G (1997), 'Structural investigation of the NiO(111) single crystal surface', *Surf Sci*, **392**, L15–L20.

Barbier A, Renaud G, Mocuta C and Stierle A (1999), 'Structural investigation of the dynamics of the NiO(111) surface by GIXS', *Surf Sci* **433**, 761.

Barbier A, Mocuta C, Kuhlenbeck H, Peters KF, Richter B and Renaud G (2000), 'Atomic structure of the polar NiO(111)-p(2 × 2) surface', *Phys Rev Lett*, **84**, 2897–2900.

Barth JV, Costantini G and Kern K (2005), 'Engineering atomic and molecular nanostructures at surfaces', *Nature*, **437**, 671–679.

Bäumer M and Freund HJ (1999), 'Metal deposits on well-ordered oxide films', *Prog. Surf Sci* **61**, 127–198.

Becker RS, Golovchenko JA and Swartzentruber BS (1985), 'Electron interferometry at crystal-surfaces', *Phys Rev Lett*, **55**, 987–990.

Berdunov N, Mariotto G, Balakrishnan K, Murphy S and Shvets IV (2006), 'Oxide templates for self-assembling arrays of metal nanoclusters', *Surf Sci*, **600**, L287–L290.

Binnig G, Rohrer H, Gerber C and Weibel E (1982), 'Tunneling through a controllable vacuum gap', *Appl Phys Lett*, **40**, 178–180.

Bottin F, Finocchi F and Noguera C (2003), 'Stability and electronic structure of the (1 ×1) $SrTiO_3$(110) polar surfaces by first-principles calculations', *Phys Rev B*, **68**, 035418.

Chen JC (1993), *Introduction Into Scanning Tunneling Microscopy*, New York, Oxford University Press.

Dmitriev A, Spillmann H, Lin N, Barth JV and Kern K (2003), 'Modular assembly of two-dimensional metal-organic coordination networks at a metal surface', *Angew Chem Int Ed*, **42**, 2670–2673.

Dulub O, Diebold U and Kresse G (2003), 'Novel stabilization mechanism on polar surfaces: ZnO(0001)-Zn', *Phys Rev Lett*, **90**, 016102.

Dulub O, Batzill M and Diebold U (2005), 'Growth of copper on single crystalline ZnO: Surface study of a model catalyst', *Top Catal*, **36**, 65–76.

Finocchi F, Barbier A, Jupille J and Noguera C (2004), 'Stability of rocksalt (111) polar surfaces: beyond the octopole', *Phys Rev Lett*, **92**, 136101.

Galloway HC, Sautet P and Salmeron M (1996), 'Structure and contrast in scanning tunneling microscopy of oxides: FeO monolayer on Pt(111)', *Phys Rev B*, **54**, 11145–11148.

Giordano L, Martinez U, Sicolo S and Pacchioni G (2007a), 'Observable consequences of formation of Au anions from deposition of Au atoms on ultrathin oxide films', *J Chem Phys*, **127**, 144713.

Giordano L, Pacchioni G, Goniakowski J, Nilius N, Rienks EDL and Freund HJ (2007b), 'Interplay between structural, magnetic, and electronic properties in a FeO/Pt(111) ultrathin film', *Phys Rev B*, **76**, 075416.

Giordano L, Pacchioni G, Goniakowski J, Nilius N, Rienks EDL and Freund HJ (2008), 'Charging of metal adatoms on ultrathin oxide films: Au and Pd on FeO/Pt(111)', *Phys Rev Lett*, **101**, 026102.

Giovanardi C, Hammer L and Heinz K (2006), 'Ultrathin cobalt oxide films on Ir(100)-(1 × 1)', *Phys Rev B*, **74**, 125429.

Goniakowski J and Noguera C (2002), 'Microscopic mechanisms of stabilization of polar oxide surfaces: transition metals on the MgO(111) surface', *Phys Rev B*, **66**, 085417.

Goniakowski J and Noguera C (2009), 'Polarization and rumpling in oxide monolayers deposited on metallic substrates', *Phys Rev B*, **79**, 155433.

Goniakowski J, Noguera C and Giordano L (2007), 'Prediction of uncompensated polarity in ultrathin films', *Phys Rev Lett*, **98**, 205701.

Goniakowski J, Finocchi F and Noguera C (2008), 'Polarity of oxide surfaces and nanostructures', *Rep Progr Phys*, **71**, 016501.

Goniakowski J, Noguera C, Giordano L and Pacchioni G (2009), 'Adsorption of metal adatoms on FeO(111) and MgO(111) monolayers: effects of charge state of adsorbate on rumpling of supported oxide film', *Phys Rev B*, **80**, 125403.

Gustafson J, Mikkelsen A, Borg M, Lundgren E, Kohler L, Kresse G, Schmid M, Varga P, Yuhara J, Torrelles X, Quiros C and Andersen JN (2004), 'Self-limited growth of a thin oxide layer on Rh(111)', *Phys Rev Lett*, **92**, 126102.

Hacquart R and Jupille J (2007), 'Hydrated MgO smoke crystals from cubes to octahedra', *Chem Phys Lett*, **439**, 91–94.

Janczak J and Kubiak R (2001), 'X-ray single crystal investigations of magnesium phthalocyanine. The 4+1 coordination of the Mg ion and its consequence', *Polyhedron*, **20**, 2901–2909.

Kiguchi M, Entani S, Saiki K, Goto T and Koma A (2003), 'Atomic and electronic structure of an unreconstructed polar MgO(111) thin film on Ag(111)', *Phys Rev B*, **68**, 115402.

Kim YJ, Westphal C, Ynzunza RX, Galloway HC, Salmeron M, van Hove MA and Fadley CS (1997), 'Interlayer interactions in epitaxial oxide growth: FeO on Pt(111)', *Phys Rev B*, **55**, 13448–13451.

Kitakatsu N, Maurice V and Marcus P (1998), 'Local decomposition of NiO ultra-thin films formed on Ni(111)', *Surf Sci*, **411**, 215–230.

Kittel C (1996), *Introduction to Solid State Physics*, New York, Wiley.

Knudsen J, Merte LR, Grabow LC, Eichhorn FM, Porsgaard S, Zeuthen H, Vang RT, Lægsgaard E, Mavrikakis M and Besenbacher F (2010), 'Reduction of FeO/Pt(111) thin films by exposure to atomic hydrogen', *Surf Sci* **604**, 11–20.

Kolesnychenko OY, Kolesnichenko YA, Shklyarevskii O and van Kempen H (2000), 'Field-emission resonance measurements with mechanically controlled break junctions', *Physica B*, **291**, 246–255.

Koplitz LV, Dulub O and Diebold U (2003), 'STM study of copper growth on ZnO(0001)-Zn and ZnO(0001)-O surfaces', *J Phys Chem B*, **107**, 10583–10590.

Kresse G, Dulub O and Diebold U (2003), 'Competing stabilization mechanism for the polar ZnO(0001)-Zn surface', *Phys Rev B*, **68**, 245409.

Kroll M and Köhler U (2007), 'Small Cu-clusters on ZnO(0001)-Zn: nucleation and annealing behavior', *Surf Sci* **601**, 2182.

Kulawik M, Rust HP, Heyde M, Nilius N, Mantooth BA, Weiss PS and Freund HJ (2005), 'Interaction of CO molecules with surface state electrons on Ag(111)', *Surf Sci*, **590**, L253–L258.

Lin X and Nilius N (2008), 'Self-assembly of MgPc molecules on polar FeO thin films', *J Phys Chem C*, **112**, 15325–15328.

Mantilla M, Jedrecy N, Lazzari R and Jupille J (2008), 'Oxidation of Mg/Ag(111) investigated using scanning tunneling microscopy: towards atomically smooth MgO nanostructures', *Surf Sci*, **602**, 3089–3094.

Maurice V, Cadot S and Marcus P (2001), 'Hydroxylation of ultra-thin films of alpha-Cr_2O_3(0001) formed on Cr(110)', *Surf Sci*, **471**, 43–58.

Merte LR, Knudsen J, Grabow LC, Vang RT, Laegsgaard E, Mavrikakis M and Besenbacher F (2009), 'Correlating STM contrast and atomic-scale structure by chemical modification: vacancy dislocation loops on FeO/Pt(111)', *Surf Sci*, **603**, L15–L18.

Meyer B and Marx D (2004), 'Density-functional study of Cu atoms, monolayers, films, and coadsorbates on polar ZnO surfaces', *Phys Rev B*, **69**, 235420.

Myrach P, Benedetti S, Valeri S, Nilius N and Freund HJ (2011), 'Investigating polarity healing mechanisms at the atomic scale: MgO on Au(111)', forthcoming.

Nilius N (2009), 'Properties of oxide thin films and their adsorption behavior studied by scanning tunneling microscopy and conductance spectroscopy', *Surf Sci Rep*, **64**, 595–659.

Nilius N, Rienks EDL, Rust HP and Freund HJ (2005), 'Self-organization of gold atoms on a polar FeO(111) surface', *Phys Rev Lett*, **95**, 066101.

Noguera C (1996), *Physics and Chemistry at Oxide Surfaces*, Cambridge, Cambridge University Press.

Olesen L, Brandbyge M, Sorensen MR, Jacobsen KW, Laegsgaard E, Stensgaard I and Besenbacher F (1996), 'Apparent barrier height in scanning tunneling microscopy revisited', *Phys Rev Lett*, **76**, 1485–1488.

Ostendorf F, Torbrugge S and Reichling M (2008), 'Atomic scale evidence for faceting stabilization of a polar oxide surface', *Phys Rev B*, **77**, R041405.

Pacchioni G, Giordano L and Baistrocchi M (2005), 'Charging of metal atoms on ultrathin MgO/Mo(100) films', *Phys Rev Lett*, **94**, 226104.

Parker TM, Condon NG, Lindsay R, Leibsle FM and Thornton G (1998), 'Imaging the polar (0001) and non-polar (1010) surfaces of ZnO with STM', *Surf Sci*, **415**, L1046–L1050.

Parteder G, Allegretti F, Surnev S and Netzer FP (2008), 'Growth of cobalt on a VO(111) surface: template, surfactant or encapsulant role of the oxide nanolayer?', *Surf Sci*, **602**, 2666–2674.

Persson M (2005), 'Adsorption-induced constraint on delocalization of electron states in an Au chain on NiAl(110)', *Phys Rev B*, **72**, R081404.

Pojani A, Finocchi F, Goniakowski J and Noguera C (1997), 'A theoretical study of the stability and electronic structure of the polar {111} face of MgO', *Surf Sci*, **387**, 354–370.

Poon HC, Hu XF, Chamberlin SE, Saldin DK and Hirschmugl CJ (2006), 'Structure of the hydrogen stabilized MgO(111)-(1 × 1) surface from low energy electron diffraction (LEED)', *Surf Sci*, **600**, 2505–2509.

Rienks EDL, Nilius N, Rust HP and Freund HJ (2005), 'Surface potential of a polar oxide film: FeO on Pt(111)', *Phys Rev B*, **71**, 241404.

Rienks EDL, Nilius N, Giordano L, Goniakowski J, Pacchioni G, Felicissimo MP, Risse T, Rust HP and Freund HJ (2007), 'Local zero-bias anomaly in tunneling spectra of a transition-metal oxide thin film', *Phys Rev B*, **75**, 205443.

Ritter M, Ranke W and Weiss W (1998), 'Growth and structure of ultrathin FeO films on Pt(111) studied by STM and LEED', *Phys Rev B*, **57**, 7240–7251.

Rohr F, Wirth K, Libuda J, Cappus D, Baumer M and Freund HJ (1994), 'Hydroxyl driven reconstruction of the polar NiO(111) surface', *Surf Sci*, **315**, L977–L982.

Ruan CY, Mastryukov V and Fink M (1999), 'Electron diffraction studies of metal phthalocyanines, MPc, where M=Sn, Mg, and Zn (reinvestigation)', *J Chem Phys*, **111**, 3035–3041.

Schoiswohl J, Surnev S, Sock M, Eck S, Ramsey MG, Netzer FP and Kresse G (2005), 'Reduction of vanadium-oxide monolayer structures', *Phys Rev B*, **71**, 165437.

Sedona F, Rizzi GA, Agnoli S, Xamena F, Papageorgiou A, Ostermann D, Sambi M, Finetti P, Schierbaum K and Granozzi G (2005), 'Ultrathin TiOx films on Pt(111): a LEED, XPS, and STM investigation', *J Phys Chem B*, **109**, 24411–24426.

Silly F, Pivetta M, Ternes M, Patthey F, Pelz JP and Schneider WD (2004), 'Creation of an atomic superlattice by immersing metallic adatoms in a two-dimensional electron sea', *Phys Rev Lett*, **92**, 016101.

Sun YN, Qin ZH, Lewandowski M, Carrasco E, Sterrer M, Shaikhutdinov S and Freund HJ (2009), 'Monolayer iron oxide film on platinum promotes low temperature CO oxidation', *J Catal*, **266**, 359–368.

Tasker PW (1979), 'Stability of ionic-crystal surfaces', *J Phys C*, **12**, 4977–4984.

Tusche C, Meyerheim HL and Kirschner J (2007), 'Observation of depolarized ZnO(0001) monolayers: formation of unreconstructed planar sheets', *Phys Rev Lett*, **99**, 026102.

Vurens GH, Salmeron M and Somorjai GA (1988), 'Structure, composition and chemisorption studies of thin ordered iron-oxide films on platinum(111)', *Surf Sci*, **201**, 129–144.

Wang XG, Chaka A and Scheffler M (2000), 'Effect of the environment on alpha-Al_2O_3 (0001) surface structures', *Phys Rev Lett*, **84**, 3650–3653.

Wang Y (2008), 'High resolution electron energy loss spectroscopy on perfect and defective oxide surfaces', *Z Phys Chem*, **222**, 927–964.

Wang Y, Meyer B, Yin X, Kunat M, Langenberg D, Träger F, Birkner A and Wöll C (2005), 'Hydrogen induced metallicity on the ZnO(1010) surface', *Phys Rev Lett*, **95**, 266104.

Witte G and Wöll C (2004), 'Growth of aromatic molecules on solid substrates for applications in organic electronics', *J Mater Res*, **19**, 1889–1916.

Wolf D (1992), 'Reconstruction of NaCl surfaces from a dipolar solution to the madelung problem', *Phys Rev Lett*, **68**, 3315–3318.

Yin XL, Birkner A, Hanel K, Lober T, Kohler U and Wöll C (2006), 'Adsorption of atomic hydrogen on ZnO(1010): STM study', *Phys Chem Chem Phys*, **8**, 1477–1481.

12
Polarity controlled epitaxy of III-nitrides and ZnO by molecular beam epitaxy

X. Q. WANG, Peking University, P. R. China and
A. YOSHIKAWA, Chiba University, Japan

Abstract: This chapter discusses the polarity controlled epitaxy of III-nitrides and ZnO by molecular beam epitaxy. The chapter first explains what the lattice polarity is and reviews the available detection methods for the polarity. After a brief introduction of the polarity behavior at heteroepitaxy and homoepitaxy, the chapter discusses the polarity controlled epitaxy of GaN, AlN, InN and ZnO, paying special attention to InN.

Key words: polarity control, epitaxy, III-nitrides, InN, ZnO.

12.1 Introduction

III-nitrides and ZnO are very promising materials which have many applications in for example, solid state lighting, full color display, laser printers, and high density information storage. These materials all have a wurtzite lattice structure, with lattice polarity along the *c*-direction, the most usual growth direction in their epitaxy. Polarity plays an important role in the epitaxy since the growth mode/behaviors, surface morphology, crystalline quality, doping ability, chemical and thermal stability are all influenced by the polarity. In this chapter, we introduce the polarity controlled epitaxy of III-nitrides and ZnO grown by using molecular beam epitaxy, paying attention to how to control the polarity on non-polar sapphire substrate and the effect of the polarity on sample properties. Section 12.2 briefly introduces what the lattice polarity is and how to detect it. Section 12.3 illustrates the polarity control in heteroepitaxy and homoepitaxy, in particular the former. Section 12.4 briefly demonstrates the polarity controlled epitaxy of GaN and AlN. This section is not discussed in detail since GaN and AlN have been studied extensively. We discuss in detail polarity controlled epitaxy of InN in Section 12.5 including its p-type doping because research into this has recently become very topical and InN is the most mysterious material of the III-nitrides. Section 12.6 shows the epitaxy of ZnO, where the polarity control issue is more complicated than for III-nitrides. Finally, we present our conclusions in Section 12.7.

12.2 Lattice polarity and detection methods

Crystals with wurtzite lattice structure such as III-nitrides and ZnO are polar crystal along the *c*-direction, and lacking inversion symmetry along this direction. These crystals or films are either +*c* polarity or –*c* polarity. Figure 12.1 shows a lattice structure of a hexagonal crystal, for example, GaN. It is shown that the lattice polarity can be easily clarified, that there is only one dangling bond along the +*c* polarity for a Ga atom, while three dangling bonds are available toward the –*c* direction. Lattice polarity is an important issue in the wurtzite semiconductor epitaxy since the growth behavior and properties of these semiconductors (thermal and chemical stability, optical properties, surface properties, etc.) are greatly affected by the polarity. For example, III-nitrides are usually grown on +*c*-polarity where the growth is under step-flow mode and atomically flat surfaces are obtained. On the other hand, the –*c*-polarity usually leads to rough surfaces. Thus, it is important to perform epitaxy under controlled polarity.

Several methods have been established to detect the lattice polarity. They include chemical etching, convergent beam electron diffraction (CBED), coaxial impact collision ion scattering spectroscopy (CAICISS) (Wang and Yoshikawa, 2004), surface reconstruction revealed from reflection high-energy electron diffraction (RHEED), and circular photogalvanic effect (CPGE). Chemical etching is a simple method to detect the polarity, but destroys the sample. Due to the chemical nature of different polarities, the chemical stability varies. For example, +*c*-polarity III-nitrides are more chemically stable and more difficult to be chemically etched than the –*c*-polarity ones. And the step-flow-like morphology usually keeps well for +*c*-polar samples except that etching pits appear. On the other hand, hexagonal islands are

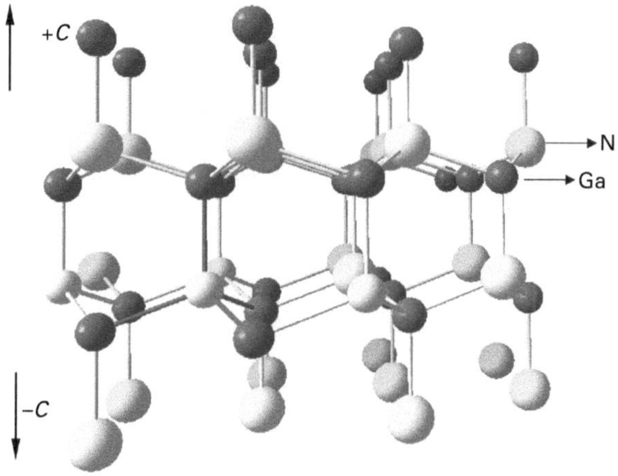

12.1 Schematic illustration of wurtzite GaN in different polarities.

usually observed in the –*c*-polarity case, which is accompanied by roughening surface. This different behavior after chemical etching is shown in Fig. 12.2 for InN, where the +*c* and –*c*-polarity are In- and N-polarity, respectively.

CBED is a powerful method to detect polarity in sample, which works in the transmission electron microscope (TEM) (Ponce *et al.*, 1996). In the measurement, the CBED pattern observed from the sample is compared with the simulated CBED pattern by considering the lattice structure and sample thickness as well. The CBED measurement is a unique method to detect polarity at the atomic level, and demonstrates significant advantages in analyzing domain structures at the micro- to nano-scale. The disadvantage is that the measurement is complicated and the sample preparation is also rather difficult. The surface reconstruction revealed from RHEED has been widely used to determine the polarity of III-nitrides and ZnO (Smith *et al.*, 1998). This measurement can be taken *in-situ* in molecular beam epitaxy (MBE). For example, the N-polar GaN surface shows the reconstruction of 1×1, 3×3, 6×6, and 6×12 while that of Ga-polar shows the reconstruction of 2×2, 5×5, 6×4 and pseudo-1×1. As for the AlN, it has been suggested that the reconstructions of 1×1 and 2×6 showed Al-polarity while that of 1×1, 1×3, 3×3, and 6×6 indicated N-polarity (Lebedev *et al.*, 1999). One problem concerning this method is that it is difficult to find surface reconstruction for semiconductor layers since it requires rather flat and metal-rich surfaces.

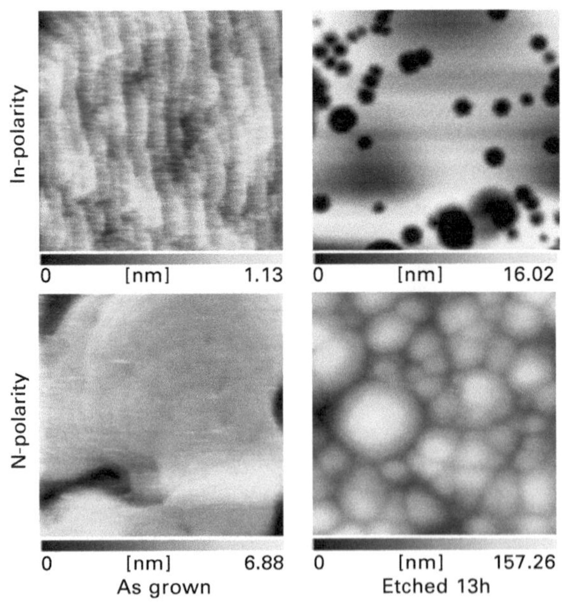

12.2 Atomic force microscopy images of as-grown and 13 h-etched InN layers with different polarities. The scanned area for each image is 1 × 1 μm.

CAICISS is a derivative of the generic technique of ion scattering spectroscopy, in particular of low energy ion scattering spectroscopy. In CAICISS measurement, a simple 180° backscattering geometry enables us to determine the atomic positions at both an outermost surface and slightly deeper layers. CAICISS spectra, including information on the surface structure and lattice polarity, is obtained by analyzing the intensity of scattered signals as a function of incident angle. The polarity is determined by comparing the experimental CAICISS spectra to the simulated ones shown in Fig. 12.3, for example in the case of InN. As shown in the figure, three peaks at 23°, 47° and 72° are observed in simulated CAICISS spectra for +*c*-polarity while six peaks at 15°, 23°, 32°, 51°, 67° and 74° are seen for –*c*-polarity.

CPGE is a newly reported method to detect the lattice polarity (Zhang *et al.*, 2009). In CPGE, a net current was generated without adding any external bias under the irradiation of a circularly polarized light. The origin of the CPGE can be attributed to the lack of spatial inversion symmetry. The former degeneration band splits into two bands with opposite spin indices as symmetry reduces. When the semiconductor layer is radiated by circularly polarized light, an asymmetrical distribution of excited electrons in momentum space is generated due to the optical selection rules, which leads to a net current in the sample.

As a wurtzite semiconductor, III-nitrides and ZnO are lacking in inversion symmetry. Along the *c*-direction, the lattice of +*c*-polarity can be regarded as a complete inversion of that of –*c*-polarity. Thus the spin-orbit interaction in the samples with contrary polarities should be opposite, resulting in an opposite spin splitting. Therefore, the sign of photocurrent due to CPGE should be opposite for samples with different polarities because of the opposite

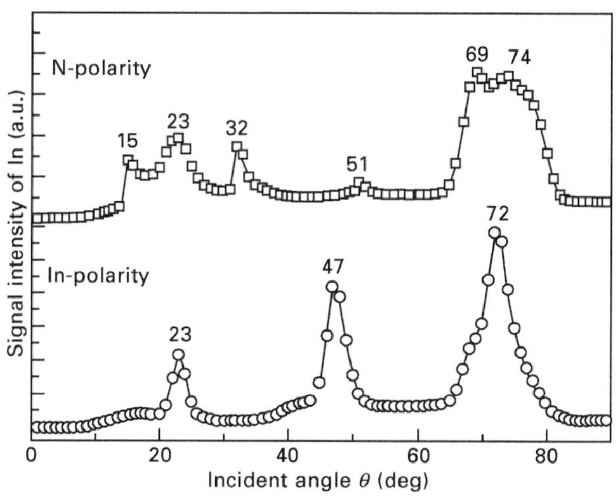

12.3 Simulated CAICISS spectra of InN.

spin splitting. It has been confirmed by experimental observation that both CPGE photocurrents and their dependences on the incident phase angle are opposite for InN layers with different polarities, as shown in Fig. 12.4. It is found that the above phenomenon is independent of the conduction type, i.e. both n- and p-type InN shows the same sign of CPGE current and the same dependence on the incident phase angle as well. Therefore, polarity can be easily determined by investigating CPGE photocurrents in the measured samples. It is worth noting that the CPGE measurement system is very easy to construct at low cost and the measurement itself is also very simple and non-destructive to samples. The disadvantage is that measurement is at the macro-scale and it is very difficult to detect samples at the micro- or nano-scale.

12.3 Polarity issues at heteroepitaxy and homoepitaxy

Although free-standing GaN, AlN and ZnO substrates are commercially available now, they are seldom used in epitaxy yet because they are several tens to hundreds times more expensive than sapphire, which is the most popular substrate up to now. It means that most epitaxy is heteroepitaxy. Sapphire itself is non-polar along the *c*-direction. Its atomic structure consists of an Al atom bilayer and an O single layer. Thus, it is very important to control the surface of sapphire just in front of epitaxy. Next, the effect of surface treatment on rotation domains/polarity is shown, using as an example ZnO epitaxy on sapphire by MBE.

Table 12.1 shows the surface treatment methods for sapphire substrate before epitaxy of ZnO buffer layer, where etching, TC, H*, O*, Ga and N* means chemical etching by a mixture of H_2SO_4:H_3PO_4 = 3:1 at 110°C, thermal cleaning (TC) under ultra-high vacuum (UHV), atomic hydrogen treatment, oxygen radical pre-treatment, Ga atomic layer pre-deposition and nitridation by using nitrogen radical, respectively (Yoshikawa *et al.*, 2004). Figure 12.5 shows the XRD ϕ-scans for (102) plane of ZnO epilayers grown with the different sapphire surface pre-treatments listed in Table 12.1. The XRD ϕ-scan for (113) Al_2O_3 is also shown for comparison. The dominant domain in sample A is 30°-rotated compared to the 'main' domains observed in the other samples B–I whose peaks are located at the same φ angles as those of (113) plane of Al_2O_3. These two domains have the epitaxy relationship $[11\bar{2}0]$ZnO//$[11\bar{2}0]Al_2O_3$ and $[10\bar{1}0]$ZnO//$[11\bar{2}0]Al_2O_3$, respectively. The latter is often called the main domain and the former is called the 30°-rotated domain. There are three kinds of possible surface structures on clean sapphire: terminated by O layer, single Al layer, and Al bi-layer. The oxygen layer is sixfold symmetric while the Al layer is threefold symmetric. On the other hand, both Zn and O are sixfold symmetric in ZnO (the same is also true for

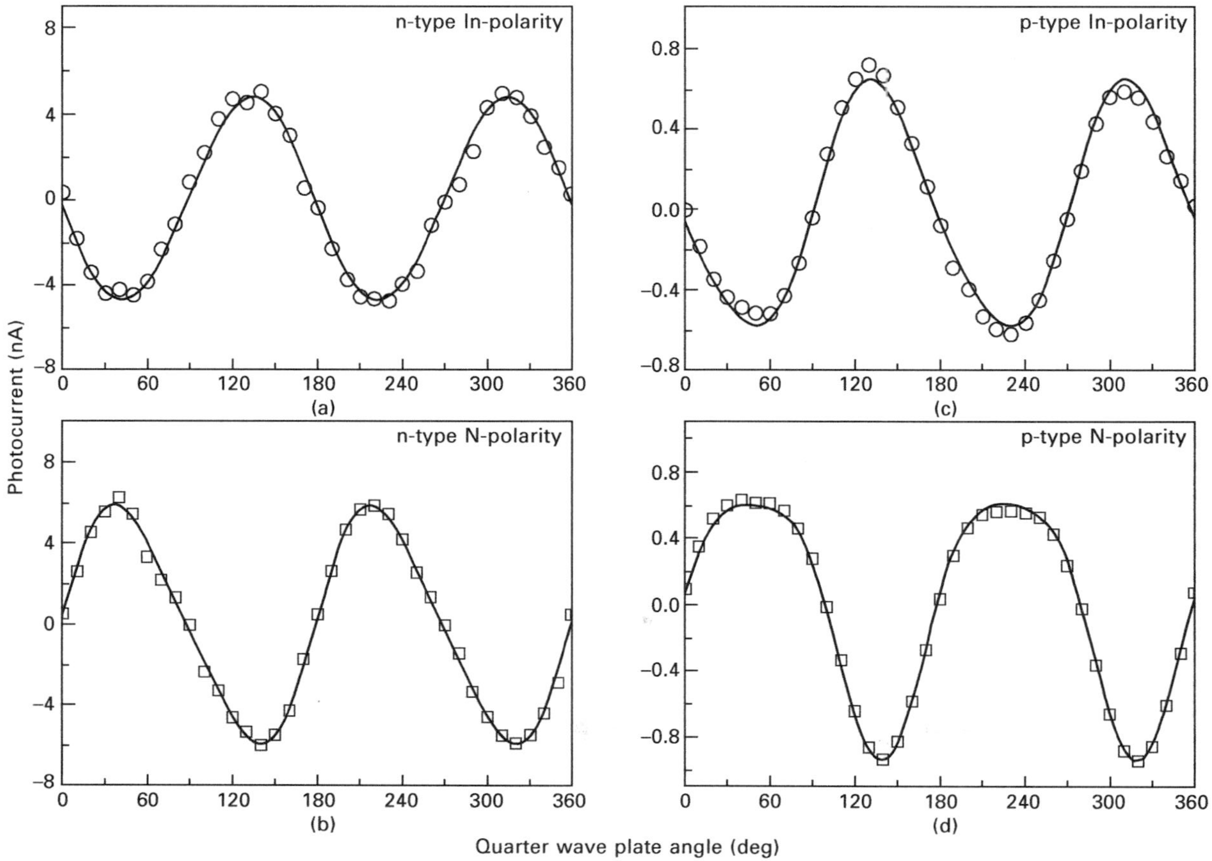

12.4 Photocurrents observed in four kinds of InN layers with different polarities as a function of quarter-wave plate angle φ. Samples in (a)–(d) are n-type In-polarity, n-type N-polarity, p-type In-polarity and p-type N-polarity, respectively. Incidence angles are +25° in all measurements.

Table 12.1 The different pre-treatment methods of sapphire substrate before the growth of ZnO buffer layer

Samples	Etching time (mins)	TC time (mins)	H* time (mins)	O* time (mins)	Ga thickness (ML)	N* time (mins)
A	30					
B	30	30				
C	30	30	10			
D	30	30	30			
E	30	30	30	30		
F	30	30			2	
G	30	30	10		2	
H	30	30	30	30	2	
I	30	30	10			60

ML = monolayer.

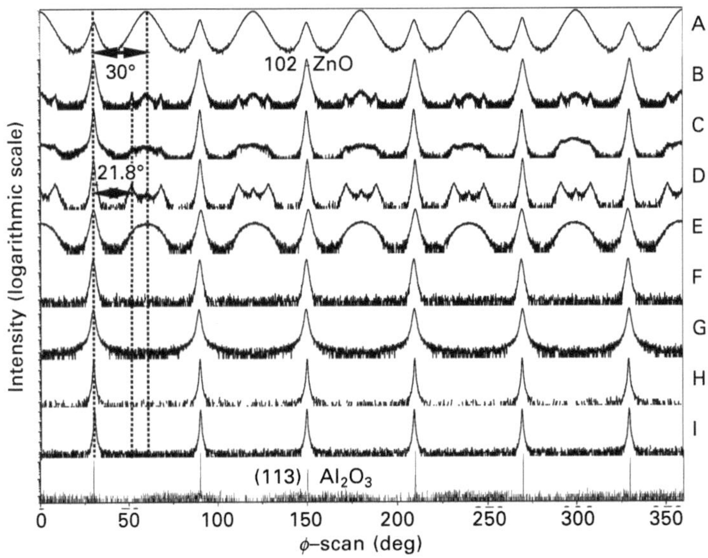

12.5 XRD ϕ-scans for (102) ZnO epilayers grown with several different procedures of sapphire surface treatments. Three kinds of rotation domains were observed after several surface treatments, but single domain was obtained after Ga pre-exposure and sapphire nitridation. XRD ϕ-scan for (113) Al_2O_3 is also shown here for comparison. The surface pre-treatment conditions are shown in Table 12.1.

III-nitrides). The main domain is usually formed when the Zn atom forms bonds with O atoms on the sapphire surface with epitaxial relationship of $[10\bar{1}0]$ZnO//$[11\bar{2}0]Al_2O_3$ and lattice mismatch of about 18.2%, as shown in Fig. 12.6. On the other hand, the 30° domain is formed due to the bonding of O atoms in ZnO with Al atoms resulting in an epitaxial relationship of $[11\bar{2}0]$ZnO//$[11\bar{2}0]Al_2O_3$ with a mismatch of about 31.8%. It is clear that

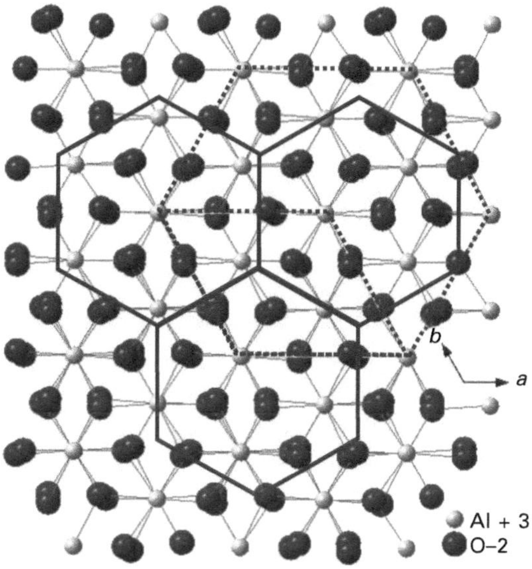

12.6 Atomic structure of sapphire (perpendicular to *c*-axis) and schematic illustration of the epitaxial relationship between ZnO and sapphire. The dotted and solid lines show the bonding configurations of two rotation domains, respectively.

the main domain is dominant in samples B–I in Fig. 12.5, though the 30°-rotated domain is dominant in sample A. The third rotation domain, which is 21.8°-rotated compared to the main domain, is observed in samples B, C, and D, where it is stronger in longer H*-treated sample D. But if the surface treatment for sample D is followed by O* pre-treatment, those peaks become broad and finally the 21.8° domain cannot be identified though the integrated intensity around both 21.8°- and 30° -rotated domains becomes stronger. The 21.8°-rotated domain is most likely due to the Al-rich surface, which has been confirmed by CAICISS and scanning force microscopy (SFM) measurements (Barth and Reichling, 2001). The sapphire surface is Al-rich after TC and it becomes Al-richer after H* pre-treatment due to the loss of O atoms. Figure 12.5 shows that the intensity of the diffraction peak of the 21.8° domain becomes stronger with increasing exposure period of H*. It shows evidence that this rotation domain is ascribed to the Al layer of sapphire. When ZnO buffer layer growth begins, O atoms form bonds with Al atoms first due to the Al-rich surface and relatively stronger bonds of Al–O in comparison to Zn–O. Hence, the rotation domains of Al layer influence the following ZnO layer and results in the 21.8° domain. On the other hand, the surface is terminated at the O-plane after O* pre-treatment and the Al metallic plane cannot be formed on the surface, which leads to the elimination of the 21.8° domain as shown in Fig. 12.5.

Ga pre-exposure on the O*-treated surface affects the ZnO epitaxy dramatically as shown in Fig. 12.5, and the rotation domains were completely suppressed resulting in a single domain, which is independent of treatment condition, i.e., TC, TC+H* and TC+H*+O*. This is because Ga acts as the template following ZnO layer deposition since it can migrate more easily than the Al onto the surface and can uniformly cover it. When the growth of the ZnO buffer layer begins, O atoms will bond with Ga atoms, resulting in a single rotation domain. Furthermore, sapphire nitridation is also very effective in eliminating the rotation domains as shown in Fig. 12.5. Following sapphire nitridation, a very thin N-polarity AlN layer is formed and this layer acts as the template following ZnO layer growth. The AlN layer is a single domain with hexagonal structure, which leads to the single domain ZnO. In addition, the ZnO shows O-polairty on the Ga-treated surface while Zn-polarity is formed on the nitrided sapphire surface. This will be discussed in detail in Section 12.6.

The important point to be emphasized here is that the samples with multiple domains usually show mixed polarity and the treatment of surfaces has great influence on the epitaxy of ZnO and nitrides. Thus, it is important to control the surface to eliminate the multiple domains and thus to be able to control the polarity in heteroepitaxy.

The polarity control epitaxy on polar substrate is much easier than on non-polar substrate. Due to the stronger bonds between metal atoms and N or O atoms, the polarity usually follows that of the substrate. For example, +*c*-polar nitride is grown on +*c*-polarity SiC while –*c*-polarity nitride is obtained on –*c*-polaritySiC. In the case of homoepitaxy of III-nitrides and ZnO, it is quite simple that the polarity usually remains the same during epitaxy. The only exception happens when we do doping, for example, p-type doping of III-nitrides by using Mg as a dopant. Figure 12.7 shows the TEM image and the polarity of InN layers. This multiple-InN layer structure sample consists of four 390 nm thick InN layers grown at different [Mg] levels, three 110 nm thick undoped spacer layers, a non-doped InN layer and cap layer, respectively. [Mg]s in four Mg:InN layers from bottom to top are 1.0×10^{18} (1st layer), 5.6×10^{18} (2nd), 2.9×10^{19} (3rd) and 1.8×10^{20} cm^{-3} (4th). It is shown that the polarity was inverted from In- to N-polarity above the V-shaped domains, which happens at [Mg] ~ 2.9×10^{19} cm^{-3}. Further detailed investigation shows that the polarity inversion happens at [Mg] ~1 $\times 10^{19}$ cm^{-3} (Wang *et al.*, 2007b). The observed N-polarity at position E in Fig. 12.7 indicates that polarity inversion does not happen with further Mg doping in the N-polarity case. Investigation of the N-polarity sample does not show any polarity inversion and thus this kind of polarity inversion only happens in the In-polarity case. Very similar phenomena have also been observed in GaN (Green *et al.*, 2003).

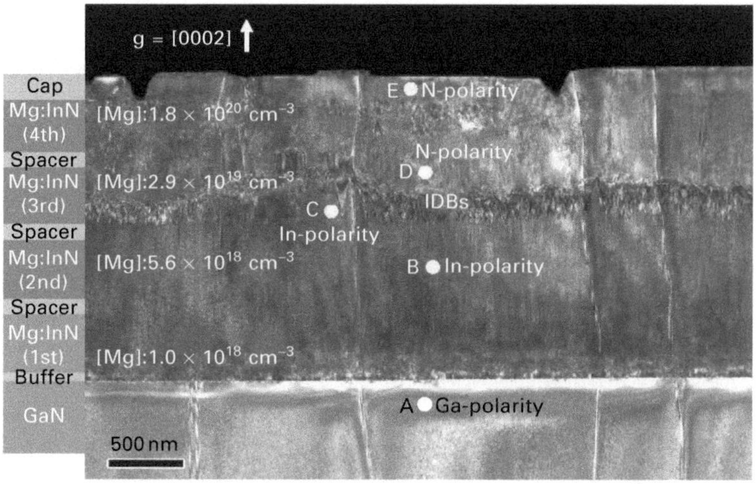

12.7 Cross-sectional dark field TEM images of a multiple-layer-structure InN film recorded with g = [0002]. The sample structure is shown on the left.

12.4 Polarity controlled epitaxy of GaN and AlN

The most popular epitaxy methods for GaN and AlN are metal–organic vapor phase epitaxy (MOVPE/also called MOCVD) and MBE. The polarity of GaN and AlN can be controlled in both methods. In MOCVD growth, due to the near quasi-equilibrium, the growth window to get $+c$-polar nitrides is rather broad, while it is difficult to get $-c$-polar samples, in particular to get a flat surface for $-c$-polarity, since the growth window for $-c$-polarity is rather narrow and the sample often includes $+c$-polarity domains, which lead to rough surface with hexagonal grains with size up to hundreds of micrometers. In MBE growth, it is relatively easy to control the polarity by controlling the buffer layer growth.

If GaN is grown directly on sapphire substrate, it is mainly Ga-polarity, which is due to the Al-rich surface of sapphire substrate after thermal cleaning under UHV (Xu *et al.*, 2002). However, sapphire nitridation is commonly used in order to improve the crystal quality of GaN and/or AlN. During the nitridation, N atoms replace O atoms and a thin AlN or AlNO layer is formed with an epitaxial relationship of $[10\bar{1}0]$AlN//$[11\bar{2}0]Al_2O_3$ (Namkoong *et al.*, 2002). The nitridation temperature has a large effect on the properties of the formed AlN thin layer and influences the properties of the grown GaN. Low temperature nitridation (200–400°C) is often used since the AlN layer formed is much purer and includes less NO. But this process usually takes a much longer time than high temperature nitridation though the AlN layer formed by the lattice process often includes NO in a higher atom ratio. As far as sample quality is concerned, the AlN layer formed at low temperature is homogeneous and smooth, which acts as a high quality template and results in a less defective and smooth buffer GaN layer, leading to high quality GaN film. On the other hand, the AlN layer formed at high temperature is rough and inhomogeneous, which leads to a defective GaN buffer layer including inversion domains, resulting in low quality GaN film.

It should be noted that the AlN formed by nitridation is mainly N-polarity. On this N-polar AlN ultra-thin layer, GaN buffer layer grown at 600°C has been proved to be mainly N-polarity (90%) by *in-situ* CAICISS measurement independent of the III/V ratios. This means that with Ga-rich growth or even Ga bilayer, it is difficult to invert the polarity. However, the behavior of the AlN layer is different from that of GaN. It has been found that the polarity of AlN layer grown at low temperature depends critically on both the surface stoichiometry in the initial growth stage and during epitaxy. The N-polar AlN buffer or epitaxial layer has been grown under N-rich conditions while the Al-polar buffer has been obtained under Al-rich growth conditions. The N-polar AlN layer even changes to Al-polar after the growth conditions change from N-rich to Al-rich as shown in Fig. 12.8(a). Even at the high temperature (820°C) growth of AlN, the polarity of AlN is inverted from N-polarity to Al-polarity by changing the growth conditions from N-rich to Al-rich. This polarity inversion is due to the formation of Al bilayer in the Al-rich growth condition as shown in Fig. 12.8(b). In the case of Ga-rich growth for GaN epitaxy, Ga bilayer is also easily formed. However, it is also easily broken by the strike of N atoms. The Al bilayer is different. Once the Al bilayer is formed, it is more difficult to break than the Ga bilayer and results in Al-polarity. This is also confirmed by the polarity inversion of GaN through an insertion of Al bilayer. Figure 12.9 shows the polarity dependence of GaN and AlN on growth conditions. The original polarity of GaN is not influenced by Ga/N ratios while that of AlN is maintained in N-rich growth conditions. Al-polar AlN is usually obtained in Al-rich growth conditions. So in the case of GaN epitaxy, N-polarity is easily obtained by using low temperature GaN buffer layer while Ga-polarity is achieved by using AlN buffer layer grown under Al-rich conditions.

The growth behavior and properties of GaN in different polarities differ largely. Growth is usually performed under Ga-rich conditions to improve sample quality, which is independent of polarity. However, the Ga-bilayer always remains at the growth front for Ga-polarity while a single Ga atom layer exists on the top surface during epitaxy of N-polarity. The Ga-polar GaN usually shows better crystal quality and smoother surface than that of N-polar GaN. The surface of Ga-polar GaN is essentially featureless between steps while the N-polar surface exhibits an island structure. The Ga-polar GaN surface is more stable than that of the N-polar one in vacuum. The growth rate of Ga-polar GaN is sometime higher than that of the N-polar one,

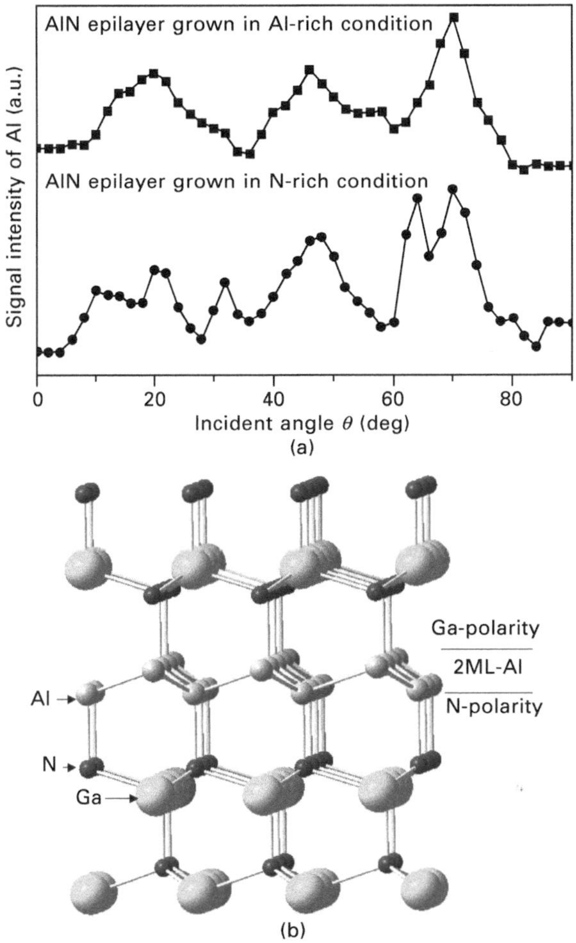

12.8 (a) CAICISS spectrum of N-polar AlN epilayer grown in N-rich conditions, and the spectrum of AlN epilayer after the growth condition is switched to Al-rich. The thickness of an AlN layer grown in Al-rich conditions is about 30 nm. (b) The schematic atomic structure of polarity inversion by an Al bilayer.

which is one of the reasons for the formation of pits with inversion domains. The impurity incorporation and doping efficiency are also influenced by the polarity. P-type doping using Mg is much easier on the Ga-polar GaN than on that of the N-polar one. Li *et al.* (2000) reported that the p-type conductivity with a free-hole concentration up to 5×10^{17} cm^{-3} was realized on Ga-polar GaN while N-polar GaN was highly resistive or semi-insulating. In the case of n-type doping, Ng and Cho (2002) found that there was no significant difference for the incorporation of Si into GaN of either polarity. However, the background impurities such as C and O were found to be more

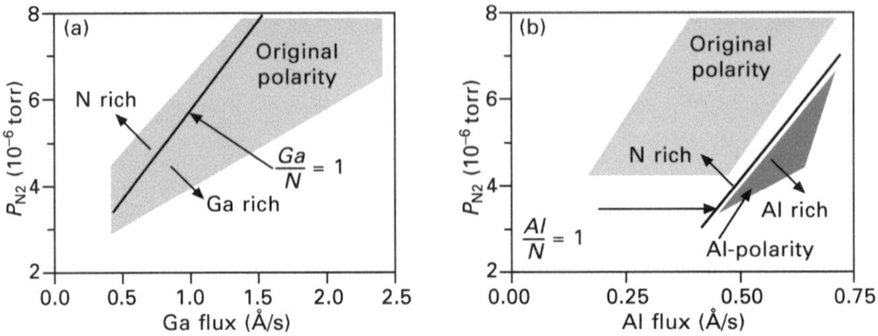

12.9 Schematic diagrams showing effect of surface stoichiometry on polarity control processes of (a) GaN and (b) AlN in MBE growth.

easily incorporated into N-polar GaN. Sumiya *et al.* (2000) also reported that the impurities related to C, O and Al were more readily incorporated into N-polar GaN film. Oxygen was incorporated at a rate ten times faster into the N-polar GaN than into the Ga-polar GaN, as reported by Park *et al.* (2001). The effect of polarity on the AlN layer is very similar to that on the GaN.

12.5 Polarity controlled epitaxy of InN

InN epitaxy is the most difficult among the III-nitrides, and the effect of polarity on it is more serious than on GaN and AlN. This is mainly because the dissociation temperature of InN is very low, much lower than the re-evaporation temperature of In metal. One of the most obvious effects of polarity on InN epitaxy is the maximum epitaxy temperature (Xu and Yoshikawa, 2003). As shown in Fig. 12.10, the maximum epitaxy temperature is about 600°C for N-polarity, which is about 100°C higher than for In-polarity. Above these temperatures, InN epilayers start to decompose and In droplets appear on the surface. Once those In droplets appear on the surface, they do not re-evaporate and the only way to remove them is nitrogen radical irradiation. It means that the growth process is limited by the dissociation of InN itself but not by the re-evaporation of excess In on the surface. This is different from the other III-nitrides such as GaN. The growth regimes are also different between InN and GaN. As shown in Fig. 12.11, there are two regimes for InN, i.e. In-rich and N-rich regimes. Under In-rich regimes, there is no way to avoid the formation of In droplets. Under N-rich regimes, the surface is rough and the quality is not good. The two growth regimes of InN epitaxy are different from those of GaN, where three growth regimes exist, i.e. Ga-rich, intermediate (Ga-rich but free of Ga droplets) and N-rich.

The growth mode and surface morphology are also greatly affected by polarity. Step-flow growth mode is obtained in the In-polarity growth regime

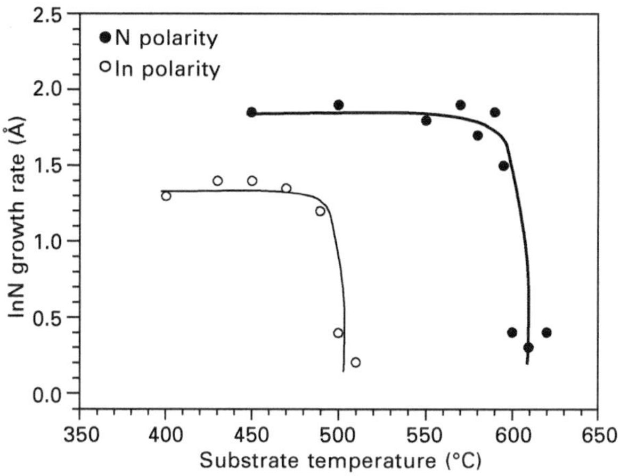

12.10 Growth rate of InN films with different polarities as a function of growth temperature.

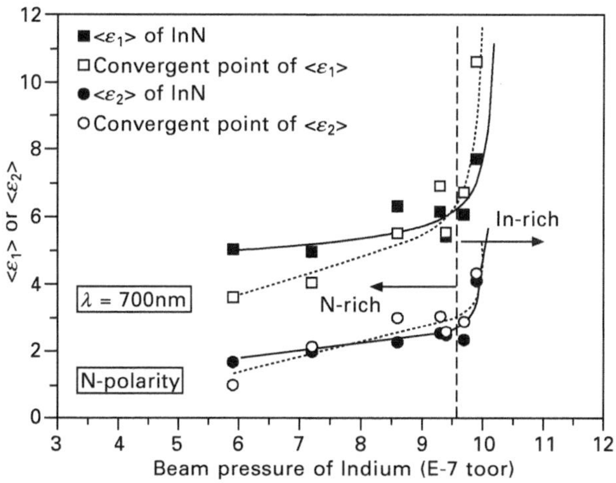

12.11 Growth regime diagram illustrated by $<\varepsilon_1>$ and $<\varepsilon_2>$ at InN thickness of 100 nm and the convergent points. Solid and dotted lines show guidelines for values of $<\varepsilon_1>$ and $<\varepsilon_2>$ at InN thickness of 100 nm and at convergent point respectively. As shown in the figure, there are two growth regimes, In-rich and N-rich.

but not in the N-polarity regime (Wang *et al.*, 2006). Figure 12.12 shows the typical surface morphology of InN with In- and N-polarities, respectively. An atomically flat surface with a step and terrace feature formed in step-flow growth mode can be observed for In-polar InN. The step height is 0.28–0.29 nm, which coincides with 1/2 of the *c* lattice constant of InN, showing that the step height is a single monolayer. The surface roughness is 0.9 nm in

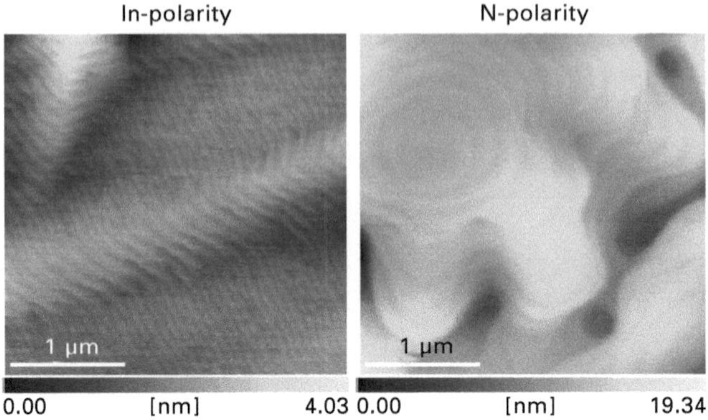

12.12 AFM images of surface morphology of InN at In- and N-polarities, respectively.

a 10 μm × 10 μm area. This surface morphology is very similar to that of homoepitaxial GaN grown by MBE on MOVPE-grown GaN. On the other hand, grain-like morphology with step-flow features within each grain is observed in N-polar samples. The step height is usually 2 ML or multiple MLs.

Thus, it is better to grow InN epilayers under In-polarity in order to achieve an atomically flat surface. In addition, it was observed that the following factors are necessary: (1) the growth temperature should be as high as possible; (2) growth should be kept under slightly In-rich conditions; and (3) keeping the screw-component threading dislocation as low as possible. It should be noted here that it is impossible to avoid the formation of In droplets to get a flat surface at such low temperatures as we mentioned above. However, a regular growth interruption under N beam irradiation can be used to eliminate the In droplets, though one should carefully control the irradiation time since longer irradiation is found to result in a rough surface.

The control of polarity for InN is basically achieved by using GaN buffer layers with controlled polarity. The reason is that the lattice matched substrate for InN is not available and sapphire is the most popular substrate in InN epitaxy. Since the lattice mismatch between InN and sapphire is very big (~25%), GaN is usually used as a buffer layer or template for InN epitaxy, which shows better crystalline quality than the InN layer grown on sapphire by only using low temperature InN buffer layer (Chen *et al.*, 2006). Then, the In- and N-polarity InN layers are grown on Ga- and N-polarity GaN buffer layer/template. For the Ga-polar layer/template, either MOVPE-grown GaN epilayer or MBE-grown GaN/AlN double buffer layers can work well (Heying *et al.*, 1999; Ide *et al.*, 2003). As for the N-polar GaN layer, it is grown on GaN buffer layer using nitridation process by MBE as we discussed in the last section.

A summary of MBE-grown high quality and high purity InN epilayers under different polarities is shown in Table 12.2. The In-polar samples are grown on 3.7 μm thick MOVPE-grown GaN templates and the N-polar samples are grown by the whole MBE process on the 500 nm thick N-polar GaN epilayers. The N-polar samples show better crystalline structural quality, which is due to the higher maximum epitaxial temperature. The N-polarity growth is preferable from the viewpoint of better matching with epitaxy temperatures among other III-nitrides. From the viewpoints of surface flatness and sharp interface, however, In-polarity is definitely a better choice. As shown in Table 12.2, the lowest n_e is at 10^{17} cm^{-3} levels for both polarities, where the In-polar samples show a little lower residual electron density and higher mobility. Although the residual donor levels are still high, the data imply that the extrinsic donor impurities in the InN layers are pretty low considering the contribution from high density edge-type threading dislocations.

It is important to investigate the origin of fairly high residual electron concentration in InN. Since the Fermi stabilization energy is located in the conduction band for InN, unintentionally doped impurities and native defects tend to be 'shallow' donors. It is believed that the nitrogen vacancy would be a major native defect due to the lowest formation energy among native defects. In addition, oxygen and hydrogen are also expected to be a significant source of free electrons in InN. Thus, the growth should be performed under growth chamber conditions as pure as possible and under controlled surface stoichiometry as well. Furthermore, the edge-type threading dislocations (ETDs) should be reduced to as low as possible. It has been found that this type of threading dislocation is the dominant one in the InN layer which basically originates from large lattice misfit with the substrate. Its density is about 1–2 orders higher than that of the screw-type one. It is known that the edge-type threading dislocations in GaN introduce deep levels in the forbidden band. However, it was theoretically confirmed that they generate donor levels above the conduction band bottom in InN,

Table 12.2 Comparison of MBE grown In-polarity and N-polarity InN epilayers

	In-polarity InN	N-polarity InN
Growth temperature	Less than ~500°C	Less than ~600°C
Surface flatness/ morphology	Atomically flat, monolayer steps	Columnar/grain-like rough, bilayer steps
XRD-FWHM (002)	200–540 arcsec	200–250 arcsec
XRD-FWHM (102)	800–1100 arcsec	1000–650 arcsec
Residual electron density at RT	2–30 × 10^{17} cm^{-3}	4–50 × 10^{17} cm^{-3}
Electron mobility at RT	1500–2200 cm^2/Vs	1500–2150 cm^2/Vs

though their exact levels are slightly different depending on their detailed core structures (Takei and Nakayama, 2009). Then it is considered that the edge-type threading dislocation is one of residual donor species in InN when the residual impurities and point defects are reduced below 10^{17} cm^{-3} levels. This is also confirmed by experimental observation as shown in Fig. 12.13, where the electron concentrations (n_e) and mobilities (μ_e) of dozens of In-polar InN films with thicknesses of 0.5–5 μm are shown as a function of full width at half maximum (FWHM) measurements of (102) InN ω scans. These n_e's are net for the bulk, i.e, the surface/interface charges of $3–5 \times 10^{13}$ cm^{-2} has been removed.

As shown in Fig. 12.13, n_e increases from 4.0×10^{17} to 2.4×10^{18} cm^{-3} with increasing $FWHM_{(102)}$ from 950 to 2200 arcsec, indicating the increase of n_e with increasing ETD density since the $FWHM_{(102)}$ is correlated with the ETDs. The dashed line indicates the contribution of dangling bonds to the n_e in the InN epilayers, assuming that each dangling bond at the ETDs acts as a singly ionizable donor, i.e., two electrons are supplied every monolayer along each ETD. It is shown that the dependence of the dashed line agrees well with that of the n_e on $FWHM_{(102)}$ and even the values are almost exactly the same as the experimental data too. This confirms the theoretical prediction that ETDs act as donors and are probably a dominant source of n_e in high purity InN epilayers. The discrepancy between the measured n_e and the density of dangling bonds is probably caused by the contributions from impurity- and point-defects-generated carrier concentrations not being taken into account. In addition, μ_e decreases from 2150 to 1000 cm^2/Vs with increasing $FWHM_{(102)}$ from 950 to 2200 arcsec, showing that the μ_e is also influenced by the ETDs.

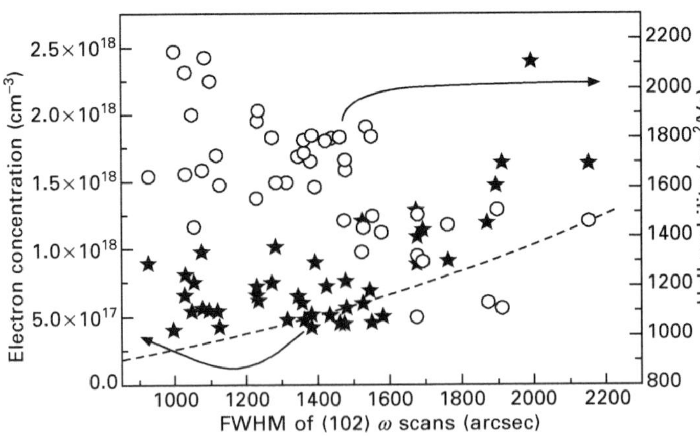

12.13 Electron concentrations and Hall mobilities of many InN films as a function of FWHMs of (102) ω scans. The estimated density of dangling bonds originating from edge-type threading dislocations is shown by the dashed line.

P-type doping is one of the most important topics. Although the residual donor levels are still fairly high, it is considered that extrinsic donor impurity levels are almost low enough to try p-type doping. Mg is a popular and effective dopant in p-type doping of GaN and AlN, and it is also the most studied one in InN. It was found that the effects of Mg doping on properties of InN with different polarities are very similar (Wang *et al.*, 2007a). The sticking coefficient of Mg on In-polarity InN is almost unity, which is probably attributed to the growth temperature being low and thus Mg is effectively adsorbed and incorporated into InN. The Mg concentrations for the N-polarity InN at the same Mg beam flux are almost the same as those for the In-polarity case, indicating that the Mg sticking coefficient in InN is independent of polarity.

An important factor in evaluating the doping difficulty is the activation energy of Mg acceptors. To evaluate it, photoluminescence (PL) properties of Mg-doped InN are investigated. Figure 12.14 shows PL spectra and corresponding PL intensity measured at 16 K as a function of [Mg]. It is shown that PL intensity changes in more than five orders of magnitude. At [Mg] ~ 10^{18} cm^{-3}, the intensity is already lower by about four orders in magnitude than that of the undoped one, and no PL emission is detected at [Mg] ~5.6–29 × 10^{18} cm^{-3}, where it is proven to be p-type region as shown

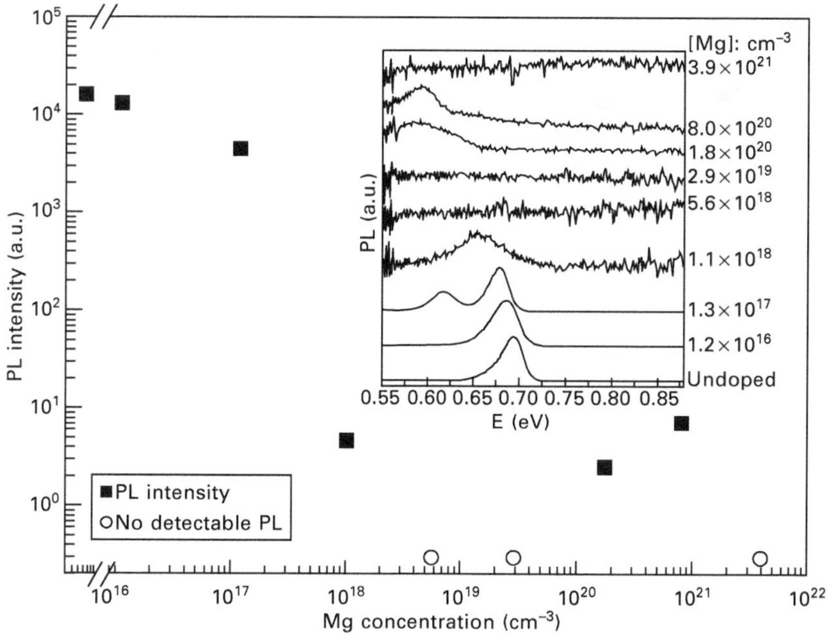

12.14 PL intensity and PL spectra (inset) of InN layers at 16 K as a function of Mg concentrations.

later. Weak PL emission is observed again with increasing [Mg]. Further, no PL emission can be detected at [Mg] ~3.9×10^{21} cm^{-3} again, which is probably because of unknown luminescence-killer processes in these poor quality samples. The exact reason for weak emission in InN:Mg at [Mg] ~$5.6–29 \times 10^{18}$ cm^{-3} as well as p-type InN reported later are not currently clear, but one possible reason is the long diffusion length of minority carriers (electrons) in p-InN resulting in their easy annihilation at killer defects and/or the introduction of high-density complex defect levels by Mg-overdoping resulting in about five orders of magnitude higher carrier relaxation paths. In the InN:Mg sample at [Mg] ~1.3×10^{17} cm^{-3}, two emission peaks are clearly observed at around 0.67 (labelled as I_{bb} in Fig. 12.15) and 0.61 (I_{fa}) eV, respectively. Excitation power dependent PL study of this sample is shown in Fig. 12.15. It is clear that intensity of I_{bb} is linearly increased with increasing excitation power while that of I_{fa} tends to be saturated. This indicates that I_{bb} originates from the band-to-band transitions while I_{fa} comes from the free-to-acceptor transitions, where the activation energy for Mg acceptors is estimated to be about 61 meV.

The observation of such shallow Mg acceptor levels implies that the

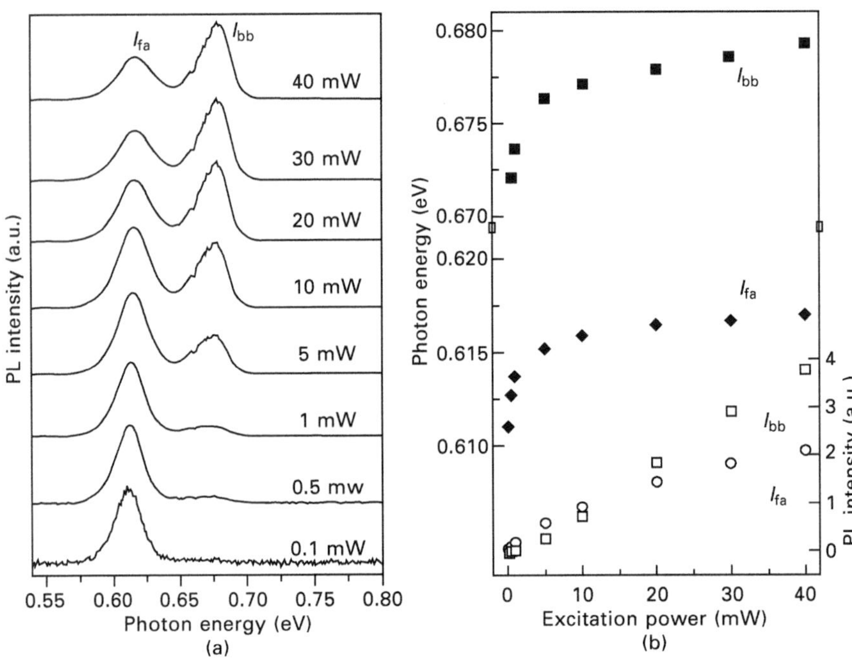

12.15 (a) 16 K PL spectra of partly compensated InN:Mg under different excitation powers. (b) Intensity and energy of I_{fa} (free to acceptor recombination) and I_{bb} (band to band) as a function of excitation power.

p-type doping may not be as difficult as we imagined previously. It seems that the problem of how to detect the p-type conduction is more serious at this moment. The p-type conduction has not been directly confirmed by conventional Hall effect measurements. This is due to the existence of surface electrons as well as the interface accumulation layer with an electron density of $3–5 \times 10^{13}$ cm^{-2}, which is independent of bulk layer conductivity (Mahboob *et al.*, 2004). Therefore, the most commonly used method to detect p-type conduction is currently electrolyte-based capacitance voltage (ECV) and thermal power (Jones *et al.*, 2006; Yim *et al.*, 2007; Wang *et al.*, 2007a).

ECV measurements are usually performed by using an electrolyte, for example KOH or NaOH solutions, to form a Helmholtz-layer contact on InN. This Helmholtz layer behaves as a very thin insulator and the total structure is like a 'metal-insulator-semiconductor (MIS) structure' in principle. To successfully modulate the surface Fermi level, the electric field intensity in the Helmholtz layer must be strong enough to deplete the surface electrons. The principle of how to detect the buried p-type region under high density surface electron is shown in Fig. 12.16. As shown in Fig. 12.16(a), the surface Fermi level in InN is located inside the conduction band due to high-density surface electrons. Since the InN is highly conductive in this case, the situation for

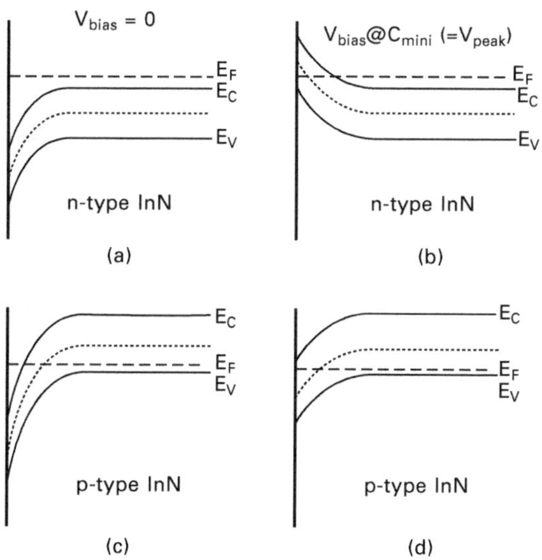

12.16 Energy band diagram for ECV characterization on detecting the conduction type of bulk InN region located below high-density surface electrons. The vertical thick lines indicate the Helmholtz layer in the electrolyte. (a) and (c) are for the zero bias voltage for n-type and p-type samples, respectively. Correspondingly, (b) and (d) are for those under voltage biased for detecting the minimum capacitance.

measuring the capacitance value is almost the same as that for the electrical double layer capacitor. The measured capacitance value corresponds only to the Helmholtz layer itself and is the largest. Figure 12.16(b) shows the situation where the surface Fermi level is effectively modified into the inside forbidden gap by applying bias voltage. The measured capacitance value in this case is the synthetic capacitance value for serial connection of the Helmholtz layer and the depletion layer in InN. By further increasing the applied bias voltage (V_{bias}), the measured capacitance value finally approaches that for the much smaller side capacitance, i.e. the depletion layer in InN. The net donor (or acceptor in the p-type sample) concentration can be estimated by the minimum capacitance value or the slope for the relationship between the capacitance and applied voltage. Figures 12.16(c) and (d) show the similar surface Fermi level modulation for p-type InN samples where the surface is also covered by high-density electrons, i.e., Figs 12.16(c) and (d) are for under zero bias voltage and for the surface Fermi level modified into the forbidden gap, respectively. As stated above, the minimum capacitance value reflects the net donor/acceptor concentration inside the InN bulk layer.

The important point for detecting the conduction type of the semiconductor sample by ECV is as follows. The applied bias voltage necessary for modulating the surface Fermi level to detect the minimum capacitance is different between n- and p-type samples as shown in Figs 12.16(a) and (c). The reason is that the Fermi level position inside the layer is different for n- and p-type samples. The Mott–Schottky plot is usually prepared by $1/C^2$ vs V_{bias} and the bulk doping level and type can be estimated from the peak height of $1/C^2$ and the position of V_{bias}. During the ECV measurements, it was found that interface states existed on the surface of InN and contributed to the measured C_{min} and thus the directly estimated N_{ae} or N_{de} are overestimated. The effect of the interface states on ECV results is estimated by the difference between theoretically predicted and measured capacitance, where the average ΔC of 634 and 1320 nF/cm^2 were estimated for In- and N-polarity, respectively. The ΔC can be regarded as a contribution from the interface states, and their densities corresponding to ΔC are roughly estimated to be 4.0×10^{12} and 8.4×10^{12} cm^{-2} for In- and N-polarity. The difference between the densities is probably due to different surface properties of InN with opposite polarities.

Figure 12.17 summarizes the p-type doping regions of Mg-doped InN by ECV measurements. In the In-polarity case, N_{ae} increases from 7.0×10^{17} to 1.3×10^{19} cm^{-3} with increasing [Mg] from 1.2×10^{18} to 2.9×10^{19} cm^{-3}. For the over-doped InN at [Mg] $\geq 1.8 \times 10^{20}$ cm^{-3}, they change to n-type and N_{de} increases from 2.2×10^{18} to 2.4×10^{20} cm^{-3} with increasing [Mg] from 1.8×10^{20} to 3.9×10^{21} cm^{-3}. For InN:Mg at [Mg] $< 1.2 \times 10^{18}$ cm^{-3}, compensation of residual donors has been observed and N_{de} decreases with increasing [Mg]. Similar tendency of N_{ae}'s and N_{de}'s as a function of [Mg]

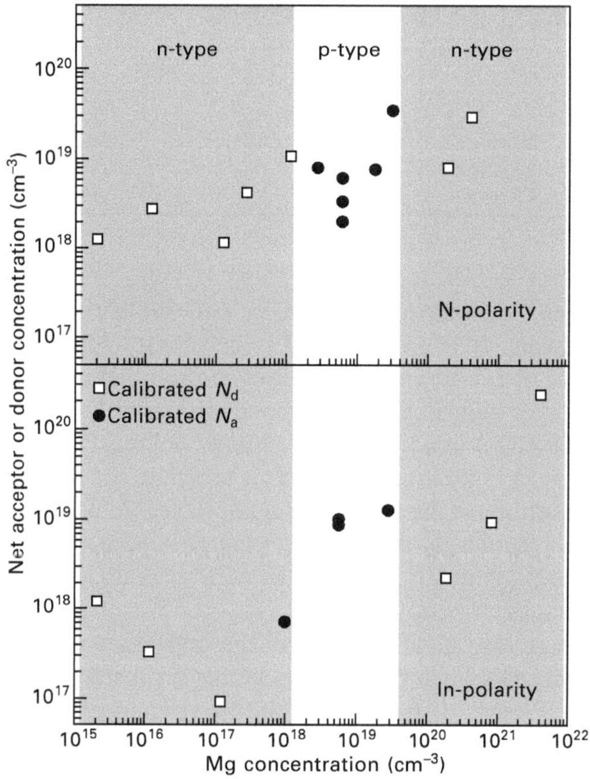

12.17 Conduction regions and calibrated net acceptor/donor concentrations of InN:Mg layers for In- and N-polarity as a function of Mg concentration.

has also been observed in the N-polarity case. Here, it should be noticed that there is a relatively large error for the estimated N_{ae}'s and N_{de}'s due to the complexity of the ECV measurement. Anyway, we can conclude that the p-type conduction is achieved at [Mg] ~ 10^{18}–3×10^{19} cm^{-3} while the InN:Mg at other regions are n-type conduction. In addition, the p-type conduction is also confirmed by thermal electric measurement for In-polar samples grown at [Mg] ~6×10^{18} cm^{-3}.

12.6 Polarity controlled epitaxy of ZnO

As has already been discussed in Section 12.2, surface treatment of sapphire substrate is important for polarity controlled epitaxy of high quality ZnO epilayers. During the ZnO epitaxy, the initial stage growth of the ZnO buffer layer determines the polarity of the ZnO film while the growth condition of the ZnO epilayer has no influence. Table 12.3 shows the polarities of single

Table 12.3 Growth conditions and polarities of single-domain ZnO epilayers

Samples	TC	H*	O*	Ga	N*	Polarity
F	30 mins			2 ML		O-polarity
G	30 mins	10 mins		2 ML		O-polarity
H	30 mins	30 mins	30 mins	2 ML		O-polarity
I	30 mins	10 mins			60 mins	Zn-polarity

ML = monolayer.

domain ZnO epilayers. As shown in the table, ZnO samples using Ga pre-exposure are all O-polar. In this case, the surface is terminated at Ga atoms no matter what kind of pre-treatment methods are used. When the growth of the ZnO buffer layer begins, the Ga–O bond is formed first followed by the formation of the Zn–O bond. This Ga–O bond belongs to the structure of Ga_2O_3 and results in O-polarity, acting as a template for the subsequent ZnO growth and resulting in the O-polar ZnO epilayer. In addition, the Ga pre-exposure greatly enhances the sample quality, where the FWHM of ω-rocking curve (002) symmetric plane is as narrow as 67 arcsec in X-ray diffraction (XRD) measurement.

Sapphire nitridation also greatly improves the crystalline quality of ZnO since the formed thin AlN layer reduces the lattice mismatch between ZnO and sapphire. The FWHMs of (002) and (102) ω-scans decrease from 912 arcsec to 95 arcsec and 2870 arcsec to 445 arcsec, respectively. In the case of the polarity, the ultra-thin AlN layer formed after nitridation is N-polar. However, the following ZnO buffer layer is found in Zn polarity, as shown in Table 12.3. It means that the polarity is inverted from N-polarity to Zn-polarity at the initial stage of ZnO buffer layer growth. At the initial growth stage, it seems that the Zn atom should bond with the N atom first because the cation–anion bond would be easily formed. In this case, if the Zn–N bond has a hexagonal structure, the polarity should not be inverted. However, if the Zn–N bond has the same structure as that of Zn_3N_2, it is possible to invert the polarity because the Zn_3N_2 may play the same role as the Mg_3N_2 which has been found to invert the polarity during GaN epitaxy. Unfortunately, there is no experimental evidence of the formation of Zn_3N_2. In our *in-situ* RHEED monitoring of the ZnO buffer layer growth, a halo pattern was observed shortly before the appearance of a ZnO RHEED pattern. This indicates that an amorphous/chaotic layer including Al, N, O and Zn is formed at the interface. This is reasonable because the O atom may replace the N atom partly at the interface due to the stronger bond of Al–O in comparison to that of Al–N. On this layer, the Zn-polar ZnO is more thermodynamically stable due to the low growth temperature, resulting in Zn-polar film (Ohkubo *et al.*, 1999). Here, the formation of the amorphous/chaotic layer is due to the AlN layer under it being terminated by the N atom plane. Therefore, if we deposit only one ML Ga, the top layer is not N but Ga. In this case, the defective layer should not be formed because the Ga–O bond will form at the interface. During the growth, the halo pattern does not appear from the RHEED *in-situ* investigation. Hence, the polarity follows the N-polarity of ultra-thin AlN and O-polar ZnO layer is grown. These polarities are confirmed by both CAICISS and chemical etching measurements.

Zn-polar ZnO layers usually show a higher growth rate, about 1.4 times to that of O-polar layers. The reason for the different growth rates is the different dangling bond configurations of the growing surfaces. Since each O atom on an O-polar ZnO surface has only one dangling bond along the *c*-axis while each O atom has three dangling bonds on the Zn-polar one, the Zn sticking coefficient on the O atom plane of the Zn-polar ZnO is higher than that of the O-polar one. Therefore, the growth rate of the Zn-polar ZnO is higher than that of the O-polar one. The Zn-polar samples also show a slightly better crystalline quality. The FWHM values of (002) and (102) ω scans for Zn-polar ZnO were 119 arcsec and 486 arcsec while those for O-polar one were 73 arcsec and 673 arcsec, respectively. Of course, the quality is also related to the buffer layer or interface modification.

The crystalline quality and surface morphology can be further improved by using a thin GaN interlayer. A quite thin (3 nm) GaN layer grown under heavily N-rich conditions has been used to greatly improve the surface flatness and also the crystalline quality. Figure 12.18 describes the detailed growth processes of the ZnO epilayers grown on nitridated sapphires with and without the 3 nm thick GaN interlayer based on *in-situ* RHEED investigation. It is clear that the RHEED pattern of $[10\bar{1}0]Al_2O_3$ changed to $[11\bar{2}0]$AlN at the same e-beam azimuth after sapphire nitridation. Then, during the growth of the GaN layer, the streaky RHEED pattern of AlN changed into a spotty pattern, indicating the three-dimensional growth of GaN. This spotty pattern was maintained even after high-temperature thermal annealing as shown in Fig. 12.18(c). During the subsequent growth of the ZnO buffer layer, the RHEED spots became gradually elongated at first and remained so until the end of buffer layer growth as shown in Fig. 12.18(d). This RHEED pattern changed to a streaky pattern when the temperature was increased to 600°C and became sharper as shown in Fig. 12.18(e) after the temperature was increased to 680°C, i.e., the epilayer growth temperature. Thus, higher temperature annealing of the ZnO buffer layer is not necessary in this case. This streaky pattern became more intense after several minutes of epilayer growth and was maintained until the end of growth as shown in Fig. 12.18(f), indicating a flat surface. In the case of the sample grown without the LT–GaN layer, however, a spotty RHEED pattern was observed at the end of the growth of the ZnO buffer layer as shown in Fig. 12.18(g). From Fig. 12.18(h), we found that these RHEED spotty patterns did not change into streaky patterns when the temperature

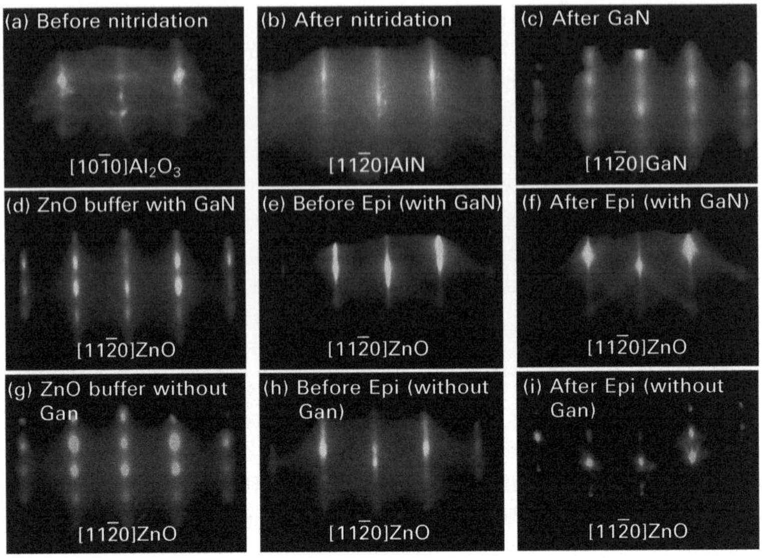

12.18 RHEED patterns along [10$\bar{1}$0]Al_2O_3 e-beam azimuth recorded during growth. (a) the (0001) Al_2O_3 surface before nitridation; (b) the surface after nitridation, it is clear that a smooth AlN layer was formed with an epitaxial relationship of [11$\bar{2}$0]AlN//[10$\bar{1}$0]Al_2O_3; (c)–(f), growth with the LT-GaN layer; (g)–(i), growth without the LT–GaN layer. (c) the thin GaN layer after thermal annealing; (d) the finished ZnO buffer layer (at 400°C); (e) the ZnO buffer layer before epitaxy (at 680°C); (f) the finished ZnO epilayer grown with the GaN layer; (g) the finished ZnO buffer layer grown without the GaN layer (at 400°C); (h) the ZnO buffer layer before epitaxy (at 680°C); (i) the finished ZnO epilayer grown without the GaN layer.

was increased to 680°C. They became more spotty after several minutes growth and were maintained until the end of growth, indicating a rough surface. The surface morphology is shown in Fig. 12.19, where several large triangular terraces can be clearly seen. The step height is confirmed to be 0.26 nm by the line scan of the surface. This value corresponds to the distance between the neighboring Zn layers (single layer or half a unit cell) in the ZnO crystal structure. These step edges are along the (10$\bar{1}$0), (01$\bar{1}$0) and ($\bar{1}$100) planes and the steps in the following layer are rotated by 60° as shown in the figure. This type of triangular terrace with single atomic steps has also been observed on the surface of Zn-polar ZnO bulk by scanning tunneling microscopy (STM) and was proved to be favorable to the surface stability. Small hexagonal pits can also be found in the surface as denoted in the AFM image. The surface is very flat with an rms roughness of 0.13 nm in a scanned area of 3×3 μm. This surface morphology was different from that of the O-polar ZnO epilayer, where a hexagonal morphology with double-atomic-layer step (0.52 nm) is usually observed.

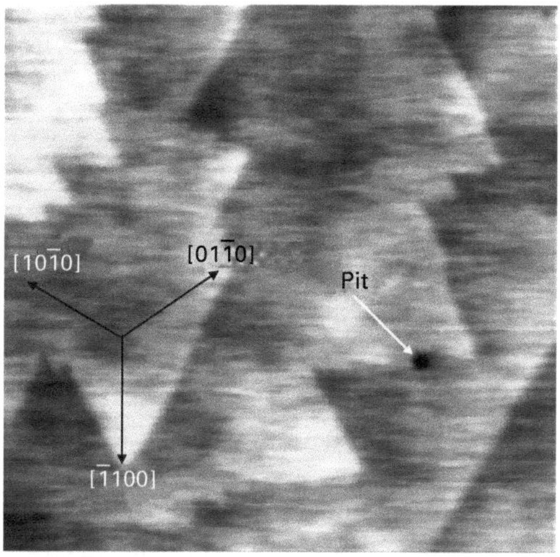

12.19 AFM surface image of the ZnO epilayer grown with the LT–GaN layer in a scanned area of 3 × 3 μm. Triangular terraces can be observed.

The crystalline quality was also improved in comparison with that of the ZnO epilayer grown without the LT–GaN layer. The FWHM values of (002) and (102) symmetric and asymmetric ω-scans were 41 and 378 arcsec, respectively. These narrow FWHM values indicate that the densities of both the edge-type and screw-type threading dislocations are low, which was also reflected by the high mobility (about 148 $cm^2 V^{-1} s^{-1}$ at room temperature).

12.7 Conclusions

Polarity does have great influence on the epitaxy and properties of III-nitrides and ZnO. Several methods such as CBED, chemical etching, CAICISS, and CPGE can effectively detect the polarity, with each having their own particular advantages in different specific areas. C-plane sapphire is the most commonly used substrate. On this non-polar substrate, the skillful modification of growing surface is very important to improve quality and control the polarity. Sapphire nitridation is a very popular and successful method for growing high quality, single polarity III-nitrides layers by MBE. The ultra-thin AlN layer with the thickness of several monolayers is formed by nitridation and it exhibits N-polarity. On this N-polar AlN ultra-thin layer, high temperature-grown AlN layers lead to the Al-polar layers and thus result in the Ga-polar GaN if the AlN layer is used as the buffer layer for the GaN. On the other hand, GaN epilayer with a low temperature GaN buffer layer on nitrided sapphire leads to a N-polar GaN layer. Basically, Ga-polar samples show better morphology and most nitride-based devices are under Ga-polarity. However, N-polarity shows its advantage on electronic devices such as HEMTs due to the different polarizations from the Ga-polar one.

InN is still the most mysterious among the III-nitrides. The growth of InN is also the most difficult. The polarity of InN is controlled by the downside GaN or AlN buffer layer. The effect of polarity on InN growth is more serious, in particular, the maximum growth temperature of the N-polarity sample is 100°C higher than that of In-polarity. This leads to a better match for epitaxy of alloys and quantum structures with GaN. On the other hand, In-polar samples show better surface morphology and relatively lower residual electron concentration. P-type doping has been performed on InN layers, where the doping efficiency is almost independent of polarity due to the low growth temperature. The activation energy of Mg in InN is about 61 meV. The p-type conduction has been confirmed by ECV and thermal power measurement. Further investigation on demonstration of InN pn junction should be carried on.

On non-polar *c*-plane sapphire substrate, polarity controlled epitaxy of ZnO is performed. Ga pretreatment and sapphire nitridation are very effective in improving sample quality. On nitrided sapphire, Zn-polar ZnO is obtained because Zn-polarity is more thermal-dynamically stable at relatively low temperatures. On the other hand, 1 ML Ga pre-exposure is enough to prevent the formation of defective layers and results in the O-polar ZnO. A 3 nm thick chaotic GaN interlayer is found to greatly improve the quality and surface flatness of ZnO.

12.8 References

Barth C and Reichling M (2001), 'Imaging the atomic arrangements on the high-temperature reconstructed α-Al_2O_3(0001) surface', *Nature* 414, 54.

Chen T C P, Thomidis C, Abell J, Li W and Moustakas T D (2006), 'Growth of InN films by RF plasma-assisted MBE and cluster beam epitaxy' *J. Crystal Growth* 288, 254.

Green D S, Haus E, Wu F, Chen L, Mishra U K and Speck J S (2003), 'Polarity control during molecular beam epitaxy growth of Mg-doped GaN', *J. Vac. Sci. Technol. B*, 21, 1804–1811.

Heying B, Tarsa E J, Elsass C R, Fini P, DenBaars S P and Speck J S (1999), 'Dislocation mediated surface morphology of GaN', *J. Appl. Phys.*, 85, 6470.

Ide T, Shimizu M, Kuo J, Jeganathan K, Shen X Q and Okumura H (2003), 'Surface morphology of GaN layer grown by plasma-assisted molecular beam epitaxy on MOCVD grown GaN template', *Phys. Status Solidi. C*, 0, 2549.

Jones R E, Yu K M, Li S X, Walukiewicz W, Ager J W, Haller E E, Lu H and Schaff W J (2006), 'Evidence for p-type doping of InN', *Phys. Rev. Lett.* 96, 125505.

Lebedev V, Schroter B, Kipshidze G and Richter W (1999), 'The polarity of AlN films grown on Si(111)', *J. Crystal Growth* 207, 266.

Li L K, Jurkovic M J, Wang W I, Van Hove J M and Chow P P (2000), 'Surface polarity dependence of Mg doping in GaN grown by molecular beam epitaxy', *Appl. Phys. Lett.* 76, 1740.

Mahboob I, Veal T D, McConville C F, Lu H and Schaff W J (2004), 'Intrinsic electron accumulation at clean InN surfaces', *Phys. Rev. Lett.* 92, 036804.

Namkoong G, Doolittle W A, Brown A S, Losurdo M, Capezzuto P and Bruno G (2002), 'Role of sapphire nitridation temperature on GaN growth by plasma assisted molecular beam epitaxy: Part I. Impact of the nitridation chemistry on material characteristics', *J. Appl. Phys.* 91, 2499.

Ng H M and Cho A Y (2002), 'Investigation of Si doping and impurity incorporation dependence on the polarity of GaN by molecular beam epitaxy', *J. Vac. Sci. Technol. B*, 20, 1217.

Ohkubo I, Ohtomo A, Ohnishi T, Mastumoto Y, Koinuma H and Kawasaki M (1999), 'In-plane and polar orientations of ZnO thin films grown on atomically flat sapphire', *Surf. Sci.* **443**, L1043.

Park A J, Holbert L J, Ting L, Swartz C H, Moldovan M, Giles N C, Myers T H, Van Lierde P, Tian C, Hockett R A, Mitha S, Wickenden A E, Koleske D D and Henry R L (2001), 'Controlled oxygen doping of GaN using plasma assisted molecular beam epitaxy', *Appl. Phys. Lett.* 79, 2740.

Ponce F A, Bour D P, Young W T, Saunders M and Steeds J W (1996), 'Determination of lattice polarity for growth of GaN bulk single crystals and epitaxial layers', *Appl. Phys. Lett.* 69, 337.

Smith A R, Feenstra R M, Greve D W, Shin M S, Skowronski M, Neugebauer J and Northrup J E (1998), 'Determination of wurtzite GaN lattice polarity based on surface reconstruction' , *Appl. Phys. Lett.* 72, 2114.

Sumiya M, Yoshimura K, Ohtsuka K and Fuke S (2000), 'Dependence of impurity incorporation on the polar direction of GaN film growth', *Appl. Phys. Lett.* 76, 2098.

Takei Y and Nakayama T (2009), 'Electron carrier generation by edge dislocations in InN films: first-principles study', *J. Crystal Growth* 311, 2767.

Wang X and Yoshikawa A (2004), 'Molecular beam epitaxy growth of GaN, AlN and InN', *Prog. Cryst. Growth Charact. Mater.,* 48/49, 42.

Wang X, Che S B, Ishitani Y and Yoshikawa A (2006), 'Step-flow growth of In-polar InN by molecular beam epitaxy', *Jpn. J. Appl. Phys. Part 2, Letters*, 45, L730.

Wang X, Che S B, Ishitani Y and Yoshikawa A (2007a), 'Systematic study of p-type doping control of InN with different Mg concentrations with both In and N polarities', *Appl. Phys. Lett.* 91, 242111.

Wang X, Che S B, Ishitani Y, Yoshikawa A, Sasaki H, Shinagawa T and Yoshida S (2007b), 'Polarity inversion in high Mg-doped In-polar InN epitaxial layers', *Appl. Phys. Lett.* 91, 81912.

Xu K and Yoshikawa A (2003), 'Effect of film polarities on InN grown by molecular beam epitaxy', *Appl. Phys. Lett.* 83, 251–253.

Xu K, Yano N, Jia A W, Yoshikawa A and Takahashi K (2002), 'Polarity control of GaN grown on sapphire substrate by RF-MBE', *J. Crystal Growth,* 237, 1003.

Yim J W L, Jones R E, Yu K M, Ager J W, Walukiezicz W, Schaff W J and Wu J (2007), 'Effects of surface states on electrical characteristics of InN and InGaN', *Phys. Rev. B.* 76, 041303(R).

Yoshikawa A, Wang X Q, Tomita Y, Roh O, Iwaki H and Ishitani Y (2004), 'Rotation-domains suppression and polarity control of ZnO epilayers grown on skillfully treated *C*-Al_2O_3 surface', *Phys. Stat. Solidi B* 241, 620.

Zhang Q, Wang X Q, He X W, Yin C M, Xu F J, Shen B, Chen Y H, Wang Z G, Ishitani Y and Yoshikawa A (2009), 'Lattice polarity detection of InN by circular photogalvanic effect', *Appl. Phys. Lett.* 95, 031902.

13
Understanding substrate plasticity and buckling of thin films

F. FOUCHER, Centre de Biophysique Moléculaire – CNRS, France and C. COUPEAU, J. COLIN, A. CIMETIÈRE, and J. GRILHÉ, Université de Poitiers, France

Abstract: The aim of this chapter is to study the effect of crystalline substrate plasticity on thin film buckling. A general description of buckling is given first within the framework of the Föppl–von Kármán theory of thin plates. The model is then used to describe the buckling induced by plasticity of the crystalline substrate. The localization of the buckling structures on the areas of dislocation emergence is explained. Finally, the model is generalized in order to obtain all the equilibrium states of a deformed thin film on a plastically strained crystal.

Key words: thin film buckling, crystalline plasticity, atomic force microscopy, Föppl–von Kármán theory of thin plates.

13.1 Introduction

Thin films are used in a wide range of domains in order to confer new properties to materials. To be efficient, the film must be fully adhesive to the substrate surface. The thin films may, however, peel off from their substrate after deposition, forming buckling structures. This phenomenon is due to the relaxation of the internal stresses in the film that can reach several GPa in compression. Several studies have been carried out in order to understand the underlying mechanisms of buckling since this phenomenon is associated with a loss of properties initially conferred to the system.

The first observations of buckling structures in thin films were carried out at the beginning of the 1980s by Matuda *et al.* (1981). The authors realized that the mechanical properties of thin films, such as Young's modulus, internal stress or adhesion energy, i.e. properties that are relatively difficult to determine experimentally, could be estimated through the phenomenon of buckling. In 1992, Hutchinson and Suo described the mechanical processes involved during buckling and delamination. Their model was based on the theory of thin plates developed at the beginning of the twentieth century by Föppl (1907) and von Kármán (1910).

In most of the present applications, the film thicknesses rarely exceed one micrometer, leading to the formation of micrometric buckles. Their observation is thus enhanced by the use of high magnification devices such as scanning electron microscopy (SEM) and atomic force microscopy (AFM). Many types of morphologies are observed: circular, straight-sided (linear structure with a sinusoidal profile) or telephone cords (streamers forming zigzags). Whatever the shape, the deflection is one order of magnitude higher than the width, giving the buckles a flat aspect. It is worth noting that AFM images are generally strongly dilated in the vertical direction in such a way that this flat aspect is not shown. A couple of examples of buckling structures are displayed in Fig. 13.1.

The formation of buckling structures results from two interacting phenomena: delamination and buckling. Evidence for this interaction comes from buckles propagating in a film with a thickness gradient. Since the energy for buckling is directly proportional to the film thickness, the buckles stop when their elastic energy becomes lower than the adhesion energy, as shown in Fig. 13.2.

Most of the common structures can be modelled based on the Föppl–von Kármán (FvK) theory of thin plates (Föppl, 1907; von Kármán, 1910). Analytical solutions for circular and straight-sided buckles have thus been obtained (Hutchinson and Suo, 1992). However, the experiments show that the most widely observed structures are the telephone cord patterns (Figs 13.1(b) and 13.2). Parry *et al.* (2006) demonstrated that the predominance of these structures is explained by the isotropy of the internal stresses generally observed in thin films. This kind of structure is relatively complex and can only be modelled using numerical methods (Crosby and Bradley, 1999; Audoly, 2003; Parry *et al.*, 2004, 2006; Jagla, 2006, 2007). Experimentally, the formation of straight-sided buckles was observed during uniaxial compression of coated substrates (Coupeau *et al.*, 1999a) as shown in Fig. 13.3.

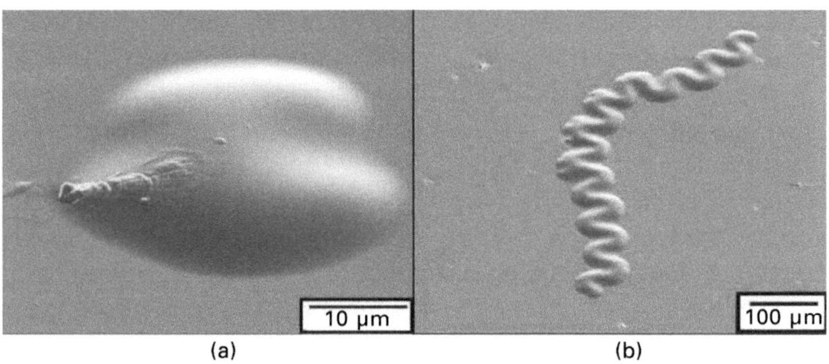

13.1 Diamond-like thin carbon film deposited on a glass substrate. SEM images of buckling structures with (a) a circular shape and (b) a telephone cord shape (after Moon *et al.*, 2002).

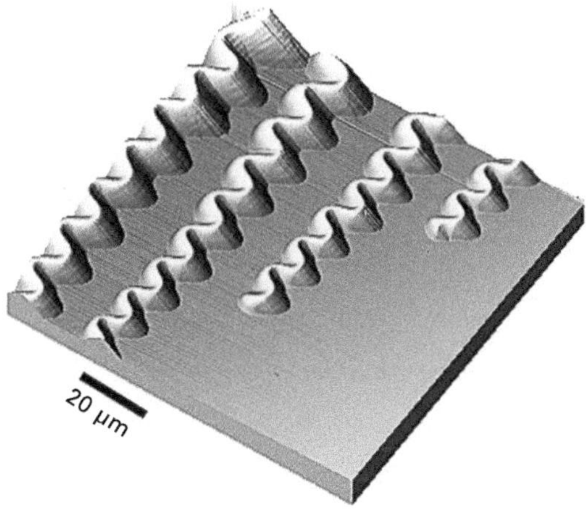

13.2 AFM image of telephone cord buckling patterns observed on a Y_2O_3 thin film of varying thickness (30–50 nm) deposited on a GaAs substrate. The structures terminate in the bottom of the image when the film thickness is too small (after Paumier *et al.*, 2003).

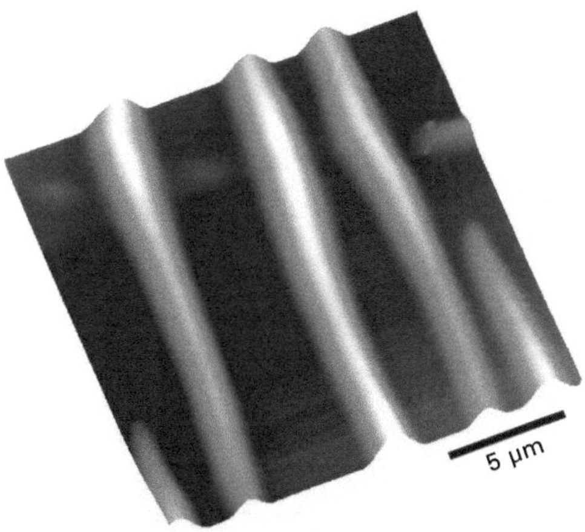

13.3 Straight-sided buckles observed by AFM on a 100 nm thick film of nickel deposited on a uni-axially strained polycarbonate substrate (after Coupeau *et al.*, 1999a). The structures are oriented perpendicular to the applied stress.

This chapter will focus on the effect of crystalline substrate plasticity on the generation of buckling structures. In the first part, general experimental observations of the phenomenon are described and the mechanisms involved during the film buckling and the crystal plastic deformation are explained. The second part deals with the modelling of the phenomenon in the framework of the FvK theory of thin plates to obtain the equilibrium shapes of the structures and of the critical stress for buckling. The localization of the buckling structures on the area of crystal strain is discussed and the asymmetry of the buckles observed on crystalline substrate is explained. The model is generalized for situations where several slip systems are activated under the same buckle and the equilibrium shape of the film is determined whatever the initial internal stress and the substrate deformation. Finally, it is shown in the last part of the chapter how buckling structures can be used to characterize the mechanical properties of thin films.

13.2 Experimental observations

13.2.1 Buckling structures

The first observations of spontaneously buckling structures in thin films were made on carbon thin films deposited on glass substrates (Matuda *et al.*, 1981). It was shown that spontaneous buckling was driven by the relaxation of the internal stresses in the film. Internal stresses are due to the method of deposition itself (intrinsic stresses) but also to the thermal expansion mismatch between the film and the substrate (thermal stresses), as given by Hutchinson *et al.* (2000):

$$\sigma_{\text{ther}} = \frac{E_{\text{f}}}{1-\nu_{\text{f}}}\Delta\alpha_{\text{t}}\Delta T \qquad [13.1]$$

where E_{f} is the Young's modulus of the film, ν_{f} its Poisson's ratio, $\Delta\alpha_{\text{t}}$ the difference in coefficients of linear thermal expansion (α film – α substrate) and ΔT the drop in temperature from the deposit temperature to room temperature. These stresses are mostly compressive and isotropic and favour the formation of more or less complex structures, especially telephone cord buckles (c.f. Figs. 13.1b and 13.2).

During uniaxial compressive tests of metallic thin films deposited on polymeric substrate, the formation of straight-sided structures perpendicular to the compression axis was observed (Coupeau *et al.*, 1999a) (cf Fig. 13.3). During these mechanical tests, the elastic deformation of the substrate is totally transmitted to the film. This deformation creates high stress anisotropy in the film, given by Colin *et al.* (2000):

$$\Delta\sigma^0_{\text{xx}} = \frac{E_{\text{f}}}{E_{\text{s}}}\frac{\nu_{\text{f}} - \nu_{\text{s}}}{1 - \nu_{\text{f}}^2}\sigma^{\text{exp}}_{\text{yy}} \qquad [13.2]$$

$$\Delta\sigma^0_{\text{yy}} = \frac{E_{\text{f}}}{E_{\text{s}}}\frac{1 - \nu_{\text{f}}\nu_{\text{s}}}{1 - \nu_{\text{f}}^2}\sigma^{\text{exp}}_{\text{yy}} \qquad [13.3]$$

The stress $\sigma^{\text{exp}}_{\text{yy}}$ corresponds to the stress applied to the sample along the (Oy) axis, E_{s} and ν_{s} are the Young's modulus and the Poisson's ratio of the substrate, respectively. In the film, the compressive stress along the (Oy) axis is strongly increased while, for most metallic thin films deposited on polymeric substrate, the stress is relaxed along the (Ox) axis since $\nu_f < \nu_s$. This anisotropy easily explains the formation of straight-sided buckles perpendicular to the applied stress as shown in Fig. 13.3, instead of telephone cords (Parry *et al.*, 2006).

13.2.2 Buckling induced by substrate plasticity

Similar compressive tests were carried out by replacing the polymeric substrate by a crystalline substrate (Foucher *et al.*, 2006). In this case the substrate is plastically strained during compression. Crystalline plasticity is driven by the activation and gliding of dislocations along particular crystal planes, which lead to the shearing of the sample. A slip system is described by a Burgers vector and a slip plane, the Burgers vector representing the magnitude and the direction of the lattice distortion induced by the dislocation in the crystal lattice. The primary slip systems (with the higher Schmidt's factor) will be activated as a function of the compression axis, crystal orientation and crystalline structure. For instance, assuming a parallelepiped sample with a cubic crystalline structure (lattice parameter *a*), oriented along the planes {100} and compressed along the [010] direction, the activated slip systems will be given by {110}<110>, corresponding to dislocations with a Burgers vector *a*/2<110> gliding in the planes {110}. On the observed (001) surface, two of the slip systems will create steps as shown in Fig. 13.4.

During plastic strain, the dislocations emerge in groups and form sub-micrometric steps at the crystal surface. The number and height of these steps increase with the increasing strain, as shown by Coupeau and Grilhé (2001). The total height of a surface step is given by $H = N \cdot h_e$ with h_e the height of an elementary step, formed by the emergence of one single dislocation,

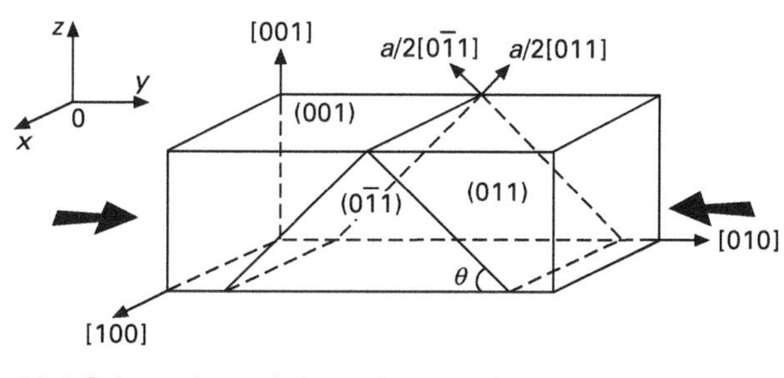

13.4 Orientation of the primary slip systems creating step structures on the surface (001) of a cubic lattice crystal during uni-axial strain. The stress is applied along the (Oy) axis corresponding to the [010] crystal plane (large arrows). The slip planes are symmetric and form an angle θ with the surface (001).

and N the number of emerged dislocations. AFM is particularly suited for characterizing the mechanisms taking place within the crystal through the study of these nanostructures (Coupeau *et al.*, 1999b; 2004; Coupeau and Grilhé, 2001).

In the case of coated crystals, the dislocations are locked at the interface between the film and the substrate (Coupeau *et al.*, 2000; Foucher and Coupeau, 2007; Foucher *et al.*, 2006). The formation of steps is observed on the film surface corresponding to the pile-up of several dislocations at the interface. Note that the film is elastically strained and not sheared by the dislocations. With increasing strain of the sample, the step height increases and the formation of straight-sided buckles following the surface steps is finally observed, as shown in Fig. 13.5.

Figure 13.5 shows that the size of the buckles is directly related to the film thickness, the thicker the film the larger the buckles. The buckles are only observed above the steps and are characterized by a vertical shift between their edges corresponding to the height of the underlying step, as clearly shown by the profile in Fig. 13.6.

13.3 Modelling

13.3.1 The Föppl–von Kármán theory of buckling

The modelling of the buckling phenomenon is based on the FvK theory of thin plates (Hutchinson and Suo, 1992). A buckled part of film is considered initially separated from the substrate. For a straight-sided buckle, the film is considered to be initially separated over an infinite length along the (Ox) axis, and over a width $2b$, between $y = -b$ and $y = b$, along the (Oy) axis. The film strip is then modelled as a thin plate clamped at its edges whose displacements are imposed by the substrate. In the case of spontaneous

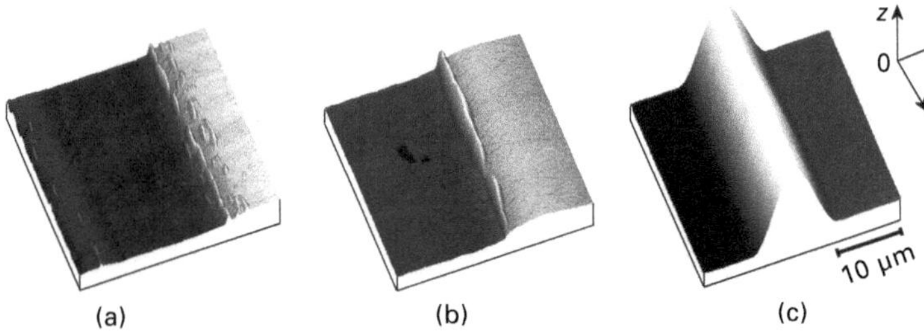

13.5 Nickel thin films deposited on LiF single crystals plastically strained at −2% (film thickness (a) 20 nm, (b) 75 nm and (c) 150 nm). AFM images show the localization of the buckles on the steps formed during the emergence of dislocations from the substrate. The size of the buckles is proportional to the film thickness.

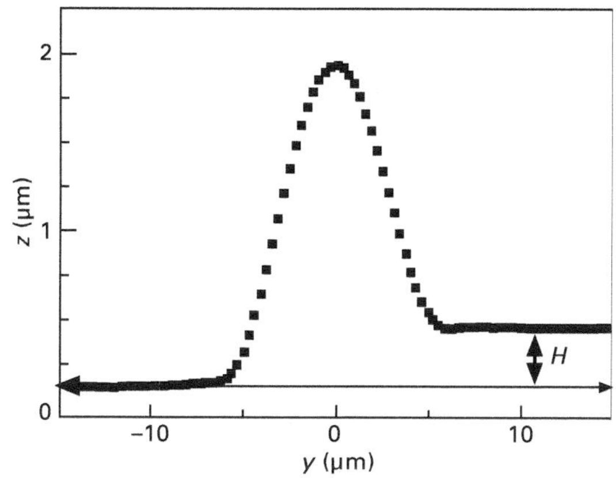

13.6 Nickel thin film 150 nm thick deposited on LiF single crystal strained at –2%. AFM profile of a straight-sided buckle formed above a step structure of height *H*.

buckling, these displacements are related to the internal stresses after deposition. These initial stresses are supposed to be isotropic, they are labelled $\sigma_{yy}^0 = \sigma_{xx}^0 = -\sigma_0$ in the middle plane of the film (using this notation, σ_0 is positive in compression). The change in the stresses after buckling is given by $\Delta\sigma_{ii} = \sigma_{ii} + \sigma_0$.

Föppl (1907) and von Kármán (1910) developed the theory of thin plate buckling in the general framework of linear elasticity. Since buckling is associated with high out-plane displacement and very small in-plane displacements (within the plate thickness), the second-order terms for the out of plane displacement in the Hooke's equations are considered. In this case, the FvK equilibrium equation that must be verified by the out of plane displacement w, is given by:

$$\frac{D}{h}\nabla^2\nabla^2 w = \sigma_{xx}\frac{\partial^2 w}{\partial x^2} + \sigma_{yy}\frac{\partial^2 w}{\partial y^2} - 2\sigma_{xy}\frac{\partial^2 w}{\partial x \partial y} \qquad [13.4]$$

with ∇^2 the Laplacian, h the film thickness and D the bending stiffness given by Hutchinson and Suo (1992):

$$D = \frac{h}{12}\frac{E_f}{1 - \nu_f^2} \qquad [13.5]$$

For a straight-sided buckle, the shape is invariant along the (Ox) axis, i.e. w does not depend on x, and the FvK equation can be written as:

$$\frac{d^4 w}{dy^4} + \alpha^2\frac{d^2 w}{dy^2} = 0 \qquad [13.6]$$

with

$$\alpha^2 = h\frac{(\sigma_0 - \Delta\sigma_{yy})}{D} \qquad [13.7]$$

In the framework of elasticity, it is shown that the transverse displacement in the buckle v must satisfy the following equation:

$$\frac{1-\nu_f^2}{E_f}(\sigma_{yy} + \sigma_0) = \frac{dv}{dy} + \frac{1}{2}\left(\frac{dw}{dy}\right)^2 \qquad [13.8]$$

The profile of the buckle is given by the parametric equation $(y + v(y), w(y))$. Solving Eqs 13.6 and 13.8 leads to two solutions. The first one is the planar fundamental solution corresponding to $w = 0$ and the second one is the buckled solution given by:

$$v_b(y) = \frac{\pi^2}{12}\frac{\delta^2}{b}\sin\left(\frac{2\pi}{b}y\right) \qquad [13.9]$$

$$w_b(y) = \frac{\delta}{2}\left[1 + \cos\left(\frac{\pi}{b}y\right)\right] \qquad [13.10]$$

with δ the maximal deflection of the buckle:

$$\delta = h\sqrt{\frac{4}{3}\left(\frac{\sigma_0}{\sigma_c} - 1\right)} \qquad [13.11]$$

The fundamental solution is valid whatever the value of the internal stress σ_0, but the buckled solution is only valid when σ_0 is higher than a critical value σ_c (also taken positive in compression) given by:

$$\sigma_c = \frac{\pi^2}{12}\frac{E_f}{1-\nu_f^2}\left(\frac{h}{b}\right)^2 \qquad [13.12]$$

The profile of the buckle $(y + v_b(y), w_b(y))$ is plotted in Fig. 13.7. It corresponds to a sinusoid slightly deformed by the transversal displacement v_b which leads to a widening of the upper part and a shrinking of the lower part of the structure.

In the case of buckling on polymeric substrates induced during compressive tests, the deformation of the substrate imposes displacements on the edges of the considered film strip. These displacements can be seen as an increase of the internal stress intensity now given by:

$$\sigma_{yy}^0 = -\sigma_0 + \Delta\sigma_{yy}^0 \qquad [13.13]$$

with $\Delta\sigma_{yy}^0$ given by Eq. 13.3.

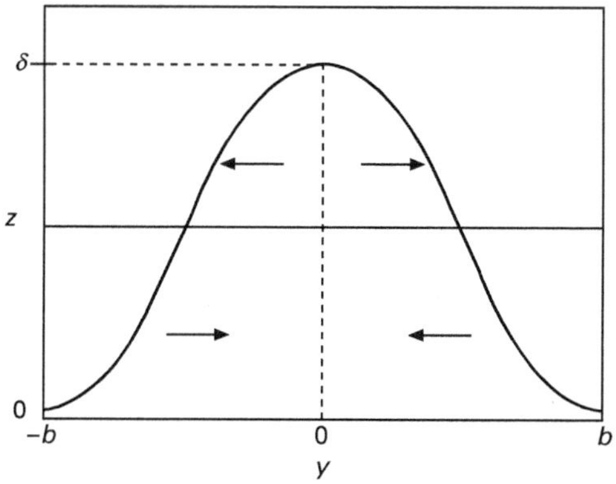

13.7 Theoretical profile of a straight-sided buckle given by $(y + v_b(y), w_b(y))$, of width $2b$ and height δ. The arrows show the orientation of the transversal displacement v_b in the upper and lower parts of the buckle.

In this case, the critical stress refers to the stress $-\sigma_{yy}^0$ $(-\sigma_{yy}^0 > \sigma_c)$ and the stress σ_0 needs to be replaced by $-\sigma_{yy}^0$ in Eqs 13.7–13.11. The uniaxial compressive tests induce an increase of the transversal stress and thus favour the formation of straight-sided buckles.

13.3.2 Buckling on crystalline substrates

Dislocations emerge on strained crystalline substrates at the film/substrate interface leading to the formation of steps above which straight-sided buckles are localized (cf Fig. 13.4). The presence of the step beneath the buckle modifies the boundary conditions on the edges of the thin film strip. For a thin film strip of initial width $2b$, the formation of dislocations induces an out of plane displacement $w(b) = H = N \cdot h_e$ and an in-plane displacement $v(b) = -H/\tan\theta$, where θ is the angle between the slip plane and the crystal surface $(0° < \theta \leq 90°)$ as shown in Fig. 13.8. Note that the imposed displacements are relatively small with respect to the strip width $(H << b)$.

The imposed displacements lead to the following compatibility equation derived from linear elasticity:

$$\frac{1}{2}\int_{-b}^{+b}\left(\frac{dw}{dy}\right)^2 dy = 2b\frac{1-\nu_f^2}{E}\Delta\sigma_{yy} + \frac{H}{\tan\theta} \qquad [13.14]$$

The transversal displacement is now given by:

$$v(y) = -\frac{1}{2}\int\left(\frac{dw}{dy}\right)^2 dy + \frac{1-\nu_f^2}{E}\Delta\sigma_{yy} \times y \qquad [13.15]$$

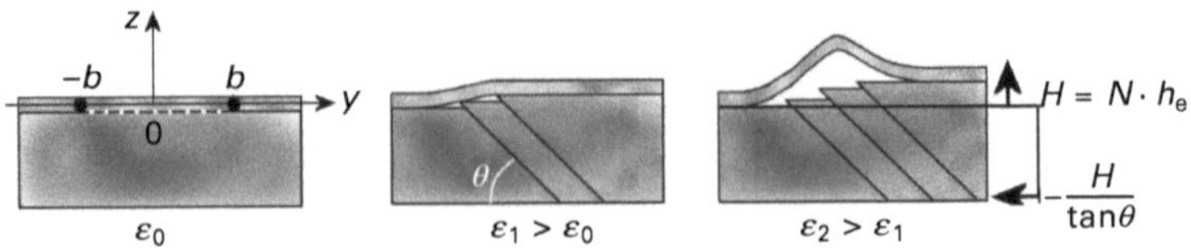

13.8 Evolution of a delaminated strip of the film of initial width 2*b* above a step structure formed during plastic strain of a crystalline substrate. The underlying step induces an out of plane displacement *H* and an in-plane displacement $-H/\tan\theta$ on the right side of the thin film strip. Above a critical strain (or stress), the formation of a straight-sided buckle is observed above the step.

Solving the FvK equation 13.2 leads to two solutions (Foucher *et al.*, 2006). The first one corresponds to an unbuckled fundamental solution (similar to the planar solution in the classical case) and the second one to a buckled solution. The unbuckled solution is given by:

$$w_1(y) = \frac{H}{2}\left(\frac{\sin\alpha y - \alpha y\cos\alpha b}{\sin\alpha b - \alpha b\cos\alpha b} + 1\right) \quad [13.16]$$

$$v_1(y) = \rho(b-y) + \alpha\frac{H^2}{32}\left(\frac{8\cos\alpha b.\sin\alpha y - \sin 2\alpha y - 3\sin 2\alpha b}{(\sin\alpha b - \alpha b\cos\alpha b)^2}\right) - \frac{H}{\tan\theta} \quad [13.17]$$

with

$$\rho = -\frac{1-v_f^2}{E_f}\sigma_0 + \alpha^2\frac{h^2}{12} + \alpha^2\frac{H^2}{16}\left(\frac{2+\cos 2\alpha b}{(\sin\alpha b - \alpha b\cos\alpha b)^2}\right) \quad [13.18]$$

α^2 is given by Eq. 13.7 and can be obtained from the compatibility equation (Eq. 13.14). The profile of the film is finally given by the parametric curve $(y + v_1(y), w_1(y))$. This profile is plotted in Fig. 13.9 for various step heights H. It is shown that H must be smaller than a critical value H_c otherwise the out of plane displacement would be associated with negative values (H_2 in Fig. 13.9). This critical value is associated with $\alpha = \pi/b$ and is given by:

$$H_c = \frac{4b}{3\tan\theta} - \frac{2}{3}\sqrt{\frac{12b^2(1-v_f^2)}{E_f}\sigma_0 - h^2\pi^2 + \frac{4b^2}{\tan^2\theta}} \quad [13.19]$$

When $H > H_c$, the fundamental solution w_1 is no longer acceptable due to the presence of the substrate.

The second solution, labelled buckled solution, for $\alpha = \pi/b$ is valid for

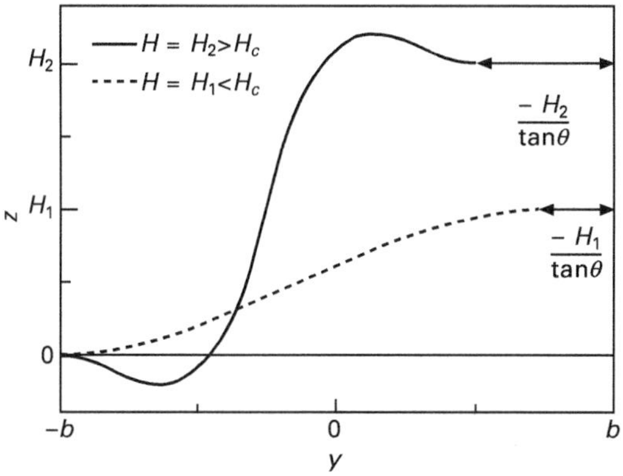

13.9 Theoretical profile associated with the fundamental solution $(y + v_1(y), w_1(y))$ of a film undergoing displacements imposed by deformation of the substrate. When the step height H is higher than a critical value H_c (H_2 in the figure), the profile takes negative values incompatible with the presence of the substrate.

$H > H_c$, preventing thus model from any discontinuity, and is given by Foucher *et al.* (2006):

$$w_2(y) = \frac{H}{2}\left(\frac{1}{\pi}\sin\frac{\pi}{b}y + \frac{y}{b} + 1\right) + \frac{w_m}{2}\left(1 + \cos\frac{\pi}{b}y\right) \qquad [13.20]$$

$$v_2(y) = \frac{1}{32b^2E_f\pi}(a_1 + a_2 + a_3 + a_4) \qquad [13.21]$$

with

$$a_1 = -2E_f\pi(3bH^2 + 3bHw_m + b\pi^2w_m^2 + 3H^2y + \pi^2w_m^2y)$$

$$a_2 = 32b^2\pi(b + y)(1 - v_f^2)(\sigma_0 - \sigma_c)$$

$$a_3 = -2bHE_f\pi w_m\left(4\cos\frac{\pi}{b}y + \cos\frac{2\pi}{b}y\right)$$

$$a_4 = -8bH^2E_f\sin\frac{\pi}{b}y + bE_f(\pi^2w_m^2 - H^2)\sin\frac{2\pi}{b}y$$

and

$$w_m = h\sqrt{\frac{4}{3}\left(\frac{\sigma_0}{\sigma_c} - 1\right) - \frac{3H^2}{\pi^2h^2} + \frac{8Hb}{\pi^2h^2\tan\theta}} \qquad [13.22]$$

The profile of the buckle $(y + v_2(y), w_2(y))$, is plotted in Fig. 13.10.

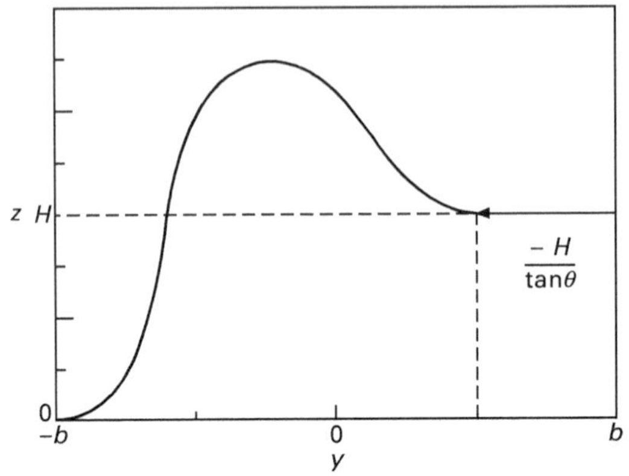

13.10 Theoretical profile of a straight-sided buckle localized above a step of height H given by $(y + v_2(y), w_2(y))$.

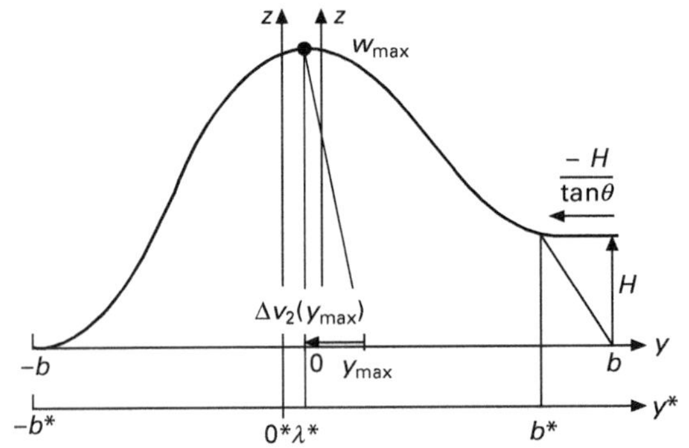

13.11 Theoretical profile of a straight-sided buckle localized above a step of height H given by $(y + v_2(y), w_2(y))$. A new axis is labelled by (0^*y^*), taking into account the in-plane displacement on the right-hand side. The asymmetry of the structure is given by the parameter λ^* corresponding to the abscissa of the maximal deflection in the final state.

This curve is in good agreement with the experimental profiles, a sinusoid with a vertical shift of height H between its edges (Foucher *et al.*, 2006). The buckle profile is asymmetric in such a way that the maximal deflection is not located in the middle of the structure. In the final state, the width of the buckle is given by:

$$2b^* = 2b - \frac{H}{\tan\theta} \qquad [13.23]$$

After buckling, the width that can be experimentally measured is $2b^*$. In order to quantify this asymmetry, it is useful to define a new axis labelled (0^*y^*), y^* varying between $-b^*$ and b^* (see Fig. 13.11). The maximum of the deflection, given by $w_{max} = w_2(y_{max})$, is thus localized in $y_{max} + v_2(y_{max})$. The initial abscissa y_{max} of the point associated with the maximal deflection is given by:

$$y_{max} = \frac{b}{\pi}\arccos\left(\frac{w_m^2\pi^2 - H^2}{w_m^2\pi^2 + H^2}\right) \quad [13.24]$$

The maximal deflexion of the buckle is then given by:

$$w_{max} = w_m + \frac{H}{2}\left(1 + \frac{y_{max}}{b}\right) \quad [13.25]$$

The abscissa of w_{max} in the final benchmark (y^*O^*z) is labelled λ^*:

$$\lambda^* = b - b^* + y_{max} + v_2(y_{max}) \quad [13.26]$$

Experimentally, $H << b$ in such a way that the parameter of asymmetry λ^* is very small (Foucher *et al.*, 2006).

In conclusion, the fundamental solution $w_1(y)$ is only valid for $H \leq H_c$ and the buckled solution $w_2(y)$ is the only solution for $H \geq H_c$. Whereas the fundamental and the buckle solutions coexist in the classical case of buckling on plane substrates, this is not the case for crystalline substrates.

13.4 Discussion

13.4.1 Localization of buckling structures

In order to explain the localization of the buckling structures on the areas of dislocation emergence observed in Figs 13.5 and 13.12, the criteria for buckling on step areas has been compared with those for plane areas. The critical step height H_c is associated to a critical initial stress value, labelled σ_c^{step} and given by:

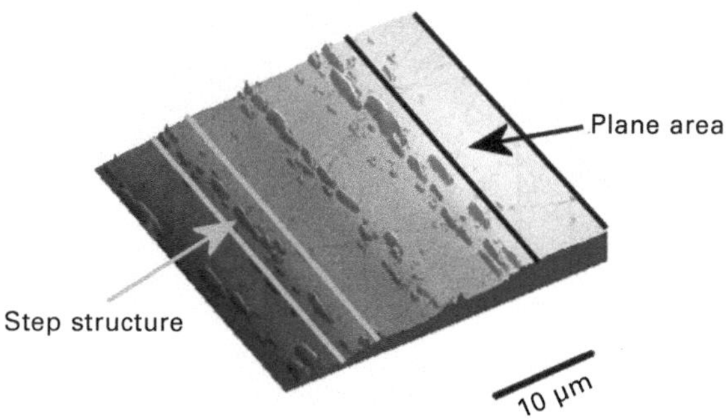

13.12 A 20 nm thick nickel film deposited on a single LiF crystal strained at –2%. AFM image showing the localization of the buckles above the steps formed during the dislocations in the substrate.

$$\sigma_c^{step}(H) = \sigma_c\left[1 - \frac{3}{\pi^2}\left(\frac{2H}{b\tan\theta} - \frac{3H^2}{4b^2}\right)\left(\frac{b}{h}\right)^2\right] \qquad [13.27]$$

If σ_c^{step} is lower than σ_c, the buckling preferentially occurs on the steps. This condition is verified when the angle of emergence θ is lower than a critical angle θ_c given by (Foucher *et al.*, 2006):

$$\theta_c = \arctan\frac{8b}{3H} \qquad [13.28]$$

As previously stated, the shape ratio H/b is small. For instance, for H/b = 1/10 the critical angle is equal to 87.85°. For the film and the substrate used in the experiments shown in Fig. 13.5 (nickel thin film on LiF single crystal), this crystallographic angle is equal to 45°, explaining thus the localization of the buckles above the steps. Moreover, for a parallelepiped sample, the calculation of the Schmidt's factor leads to an angle θ between 30° and 60°. The relation $\sigma_c^{step} < \sigma_c$ is thus always verified whatever the crystalline substrate. However, it is possible to consider particular configurations where the angle of emergence is equal to 90°, as shown in Fig. 13.13. Such a configuration could then be used in order to prevent buckling induced by substrate plasticity.

13.4.2 Slip systems

Several slip systems can be activated and form steps of various orientations during crystalline plasticity. In the case where slip systems are activated beneath the same buckle, the boundary conditions need to be modified as:

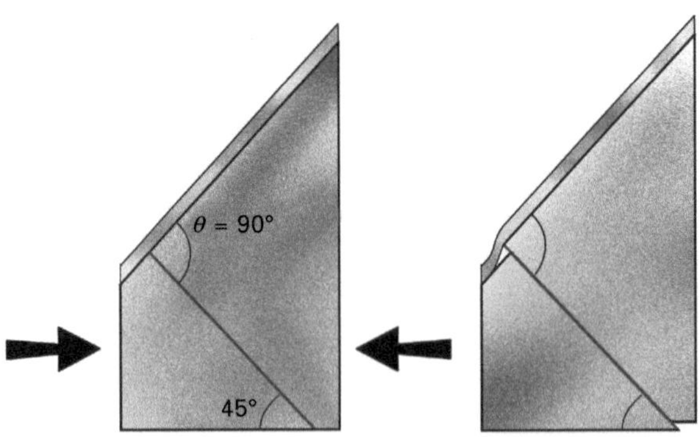

13.13 Example of the effects of uniaxial strain on a thin film deposited on a crystalline substrate where the angle between the active slip planes and the surface is 90°. In this case, $\theta > \theta_c$ in such a way that the film does not buckle on the step.

$$v(b) = -\sum_i \frac{H_i^r}{\tan\theta_i} - \sum_j \frac{H_j^d}{\tan\theta_j} \qquad [13.29]$$

$$w(b) = -\sum_i H_i^r - \sum_j H_j^d \qquad [13.30]$$

with H_i^r the height of the rising step i and H_j^d the height of the descending step j as shown in Fig. 13.14. The shape equation of the buckle can be determined as outlined in Section 13.3.2.

13.4.3 Tensile tests

Complementary to the previous sections, two other solutions can be found, related to the case where the film is unstressed on the step ($\sigma_{yy} = 0$) and the case where the film is in tension ($\sigma_{yy} > 0$). These solutions are obtained for thin films undergoing tensile internal stresses after deposition or when the step height H is negative. Experimentally, $H < 0$ corresponds to a tensile test; the vertical displacement induced by the step in $y = +b$ being thus oriented to the bottom, as shown in Fig. 13.15. The parameter H is thus equal to $N \times h_e$ in compression and to $-N \times h_e$ in tension. Note that the angle θ is set as $0 < \theta \leq 90°$.

The solution for the unstressed film ($\sigma_{yy} = 0$) is given by:

$$w_{nul}(y) = \frac{H}{2}\left(-\frac{y^3}{2b^3} + \frac{3y}{2b} + 1\right) \qquad [13.31]$$

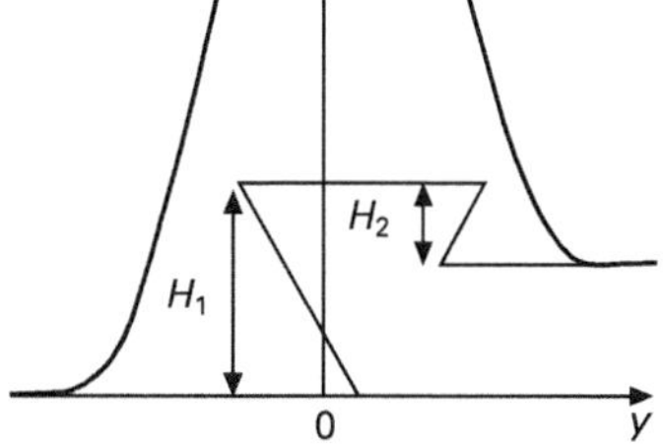

13.14 Sketch showing a straight-sided buckle formed during the activation of two symmetric slip systems.

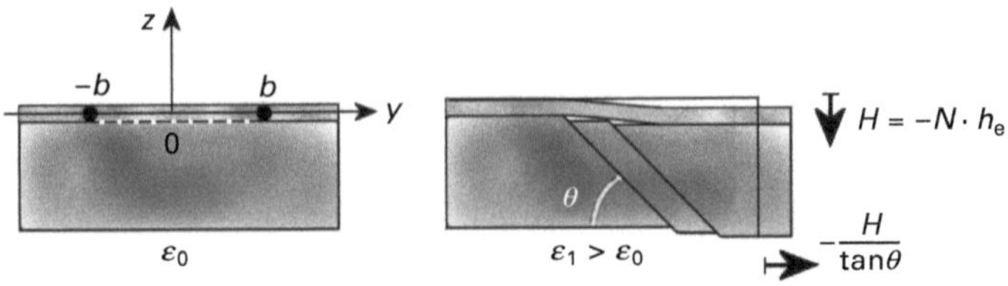

13.15 Evolution of a thin film deposited on a crystalline substrate during a tensile test. The emergence of dislocations leads to the formation of a negative step of height $H = -N{\cdot}h_e$ associated with an elongation of $-H/\tan\theta > 0$ $(0 < \theta < 90°)$.

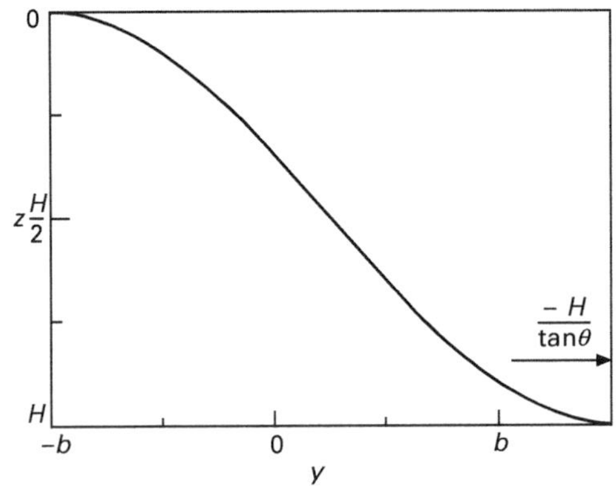

13.16 Theoretical profile of an unstressed thin film ($\sigma_{yy} = 0$) on a step formed during the emergence of dislocations from a crystalline substrate during tensile strain ($y + v_{null}(y)$, $w_{null}(y)$).

$$v_{nul}(y) = -\frac{3H^2}{160b}\left[3\left(\frac{y}{b}\right)^5 - 10\left(\frac{y}{b}\right)^3 + 7\frac{y}{b}\right] - \frac{H}{2\tan\theta}\left(\frac{y}{b} + 1\right) \quad [13.32]$$

The solution ($y + v_{null}(y)$, $w_{null}(y)$) is plotted in Fig. 13.16. It corresponds to a film following the step structure. It is associated with a step height H_0 given by:

$$H_0 = \frac{5b}{3\tan\theta} - \sqrt{\frac{20}{3E_f}(1 - v_f^2)\sigma_0 b^2 + \left(\frac{5b}{3\tan\theta}\right)^2} \quad [13.33]$$

The function of $H = H_0$ can also be written as a condition on the initial internal stress σ_0, as $\sigma_0 = \Sigma_0$ with:

$$\Sigma_0(H) = \frac{E_f}{1 - v_f^2}\left[-\frac{1}{2\tan\theta}\frac{H}{b} + \frac{3}{20}\left(\frac{H}{b}\right)^2\right] \quad [13.34]$$

The solution associated with a thin film in tension is given by:

$$w_{\text{ten}}(y) = \frac{H}{2}\left[\frac{1}{1+\gamma b+(\gamma b-1)e^{2\gamma b}}(e^{\gamma(b-y)} - e^{\gamma(b+y)} + \gamma(1+e^{2\gamma b})y)+1\right] \quad [13.35]$$

$$v_{\text{ten}}(y) = -\frac{A^2H^2}{8}[-\zeta_1(e^{\gamma y} + e^{\gamma b} - e^{-\gamma y} - e^{-\gamma b}) + \zeta_2(e^{2\gamma y} + e^{2\gamma b} - e^{-2\gamma y} - e^{-2\gamma b})] + \zeta_3(y+b) \quad [13.36]$$

Where

$$\gamma^2 = h\frac{(\Delta\sigma_{yy} - \sigma_0)}{D}$$

$$A = \frac{1}{1+\gamma b + (\gamma b - 1)e^{2\gamma b}}$$

$$B = \gamma(1 + e^{2\gamma b})$$

$$\zeta_1 = 2Be^{\gamma b}$$

$$\zeta_2 = \frac{\gamma}{2}e^{2\gamma b}$$

$$\zeta_3 = \frac{1-\nu_f^2}{E_f}\left(\frac{D}{h}\gamma^2 + \sigma_0\right) - \frac{A^2H^2}{8}(B^2 + 2\gamma^2 e^{2\gamma b}) - \frac{11}{12}h^2\gamma^2$$

The solution $(y+v_{ten}(y), w_{ten}(y))$ is plotted in Fig. 13.17. It corresponds to a film stretched between its edges and is associated with a step height $H < H_0$ ($\sigma_0 < \Sigma_0$).

Note that these solutions are based on the FvK theory of thin plates which does not take into account the effect of the substrate and the adhesion. However, Foucher and Coupeau (2007) have shown that the emergence of dislocations is modified by the presence of the film in such a way that the step profile appears to follow the equilibrium shape of the unbuckled film. The interaction between thin films and substrate dislocations was discussed previously by Junqua and Grilhé (1997).

13.4.4 Experiment versus modelling: characterization of the mechanical properties of thin films

The main difference between the modelling and the compression test experiments comes from the initial interfacial delamination hypothesis that does not exist in the experiment. This delamination is obtained experimentally when external

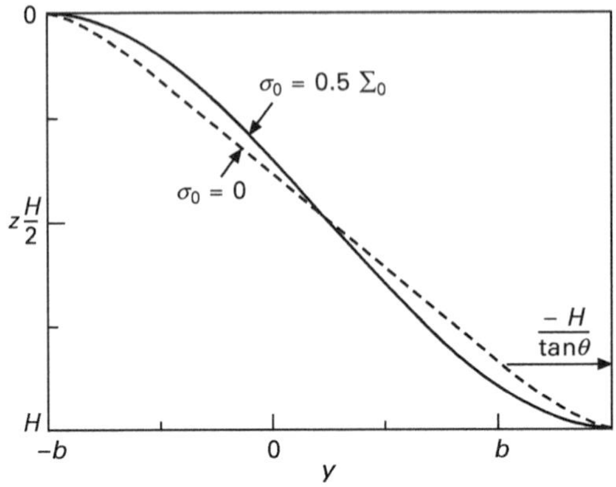

13.17 Theoretical profiles of a thin film strained in tension ($\sigma_{yy} > 0$) on a step formed during the emergence of dislocations from a crystalline substrate under tensile strain ($y + v_{ten}(y)$, $w_{ten}(y)$), for various initial stress intensities σ_0 ($\sigma_0 = 0.5\ \Sigma_0$ and $\sigma_0 = 0$).

compression leads to an increase in the elastic energy stored in the film. When the stored energy is high enough to counterbalance the adhesion energy, the film buckles. The experimentally buckling critical stress σ_c^{exp} is consequently found higher than the theoretical one. This variation in the onset of buckling is shown by a jump from unbuckled to buckled state in the curve $\delta(|\sigma_{yy}^0|)$, as shown in Fig. 13.18. Once the interface is delaminated, the unstuck part of the film can be modelled as a thin plate and the curve $\delta(|\sigma_{yy}^0|)$ follows the theoretical curve. It is noted that if the initial internal stress σ_0 is smaller than the critical stress σ_c, the film can go down onto the substrate during the release of the external applied stress.

This discrepancy has been used in the past to estimate the adhesion energy Γ (Coupeau *et al.*, 1999a) between the film and the substrate using the following equation:

$$\Gamma = \frac{h}{2E_f}(1 - \nu_f^2)(\sigma_c^{exp} - \sigma_c) \qquad [13.37]$$

The model has also been used to estimate the Young's modulus or the initial internal stress in the film (Matuda *et al.*, 1981). Indeed, the internal stress in the film is related to the width and the deflection of the buckle by the equation:

$$\sigma_0 = \frac{\pi^2}{12}\frac{E_f}{1 - \nu_f^2}\left(\frac{3\delta^2}{4h^2} + 1\right)\left(\frac{h}{b}\right)^2 \qquad [13.38]$$

If the internal stress has been determined using another technique, Eq. 13.38 can be used to estimate the Young's modulus of the film. Finally, the

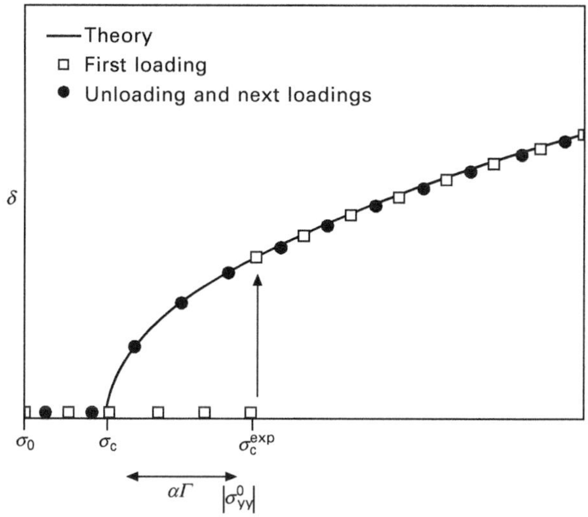

13.18 Schematic diagram showing the effect of adhesion on the evolution of buckle height δ versus the applied stress on its edges. During the first loading, the film adheres to the substrate and the stress must be greater than the theoretical critical stress σ_c for the film to buckle. After delamination, the film buckles suddenly for a critical stress σ_c^{exp}. There is a jump from unbuckled to buckled state (arrow) and the deflexion follows the theoretical curve for thin plates. If the initial internal stress σ_0 is lower than the critical stress σ_c, the film touches the substrate after relaxation of the applied stress and the deflexion follows the theoretical curve for the next loadings.

measurement of the geometric parameters of the buckles appears to be an interesting way to estimate the energy of adhesion, the internal stress or the Young's modulus, parameters that are relatively difficult to determine by other techniques. However, the compliance of the substrate, considered to be infinitely rigid in the model, has been shown to be a limiting process. Indeed, experiments carried out with thin metallic films deposited on polymeric substrates show that the maximal deflection of the buckles was experimentally higher than predicted and that the film sinks into the substrate at the base of the buckles (Parry *et al.*, 2005). It has even been shown that this effect could induce cracking of the film and plastic strain in the substrate (Cordill *et al.*, 2005). The characterization of the mechanical properties of the film appears thus difficult in the case of polymeric substrates.

The critical stress for buckling in the region of steps can be used to estimate the adhesion energy using the formula:

$$\Gamma = \frac{h}{2E_f}(1 - \nu_f^2)(\sigma_c^{exp} - \sigma_c^{step}) \qquad [13.39]$$

By neglecting the asymmetry of the structure, it can be set that $w_{max} \sim w_m + H/2$. The initial internal stress can then be determined using the following formula:

$$\sigma_0 = \frac{\pi^2}{12}\frac{E_f}{1-\nu_f^2}\left[\frac{3}{4h^2}\left(\left(w_{max}-\frac{H}{2}\right)^2+\frac{3H^2}{\pi^2}-\frac{8H}{\pi^2\tan\theta}\left(b^*+\frac{H}{2\tan\theta}\right)\right)+1\right]$$
$$\times\left(\frac{2h\tan\theta}{2b^*\tan\theta+H}\right)^2 \qquad [13.40]$$

Crystalline materials being less compliant than polymers, the mechanical properties of the film can be characterized from the geometric parameters of the buckles with a better precision. Buckling induced by substrate crystalline plasticity is thus a good way to estimate the adhesion energy and the initial internal stress.

13.5 Conclusions

In this chapter dedicated to thin film buckling phenomena, it has been shown that buckling can occur spontaneously or under the action of external forces. Compressive tests carried out to induce buckling show that the nature of the substrate plays a key role. For polymeric substrates, the buckling leads to straight-sided buckles perpendicular to the applied stress. These structures can be modelled analytically using the FvK theory of thin plates. However, it is shown that the experimental and analytical onsets of buckling are different, due to the adhesion energy neglected in the model, and that the experimental shape is slightly modified by the compliance of the substrate.

Experiments carried out on crystalline substrates show the formation of pseudo straight-sided buckles, shifted vertically on their edges, localized on the steps formed by substrate plasticity. This shift modifies the analytical model. It is shown that the critical stress for buckling is decreased on the area of dislocation emergence explaining the localization of the buckles observed experimentally. The model has been extended for all kinds of external and initial internal stresses in the film. All the solutions are displayed in Table 13.1.

Finally, it is shown that compressive tests can be used to estimate the adhesion energy and the internal stress from the difference between the experimental and theoretical onsets of buckling and from the shape of the buckles. Since it is difficult to experimentally determine the mechanical properties of thin films, buckling induced by compression tests appears to be a good way to characterize.

Table 13.1 Vertical displacement *w* of a thin film in the equilibrium state on a crystalline substrate, whatever the initial internal stress in the film and the substrate deformation

	Tensile test				Spontaneous		Compression test			
	$H < H_0$, $\sigma_0 < \Sigma_0$	$H = H_0$, $\sigma_0 = \Sigma_0$	$H_c > H > H_0$, $\sigma_c^{step} \sigma_0 > \Sigma_0$	$H > H_c$, $\sigma_0 > \sigma_c^{step}$	$H = 0$, $\sigma_0 < \sigma_c$	$H = 0$, $\sigma_0 > \sigma_c$	$H < H_0$, $\sigma_0 < \Sigma_0$	$H = H_0$, $\sigma_0 = \Sigma_0$	$H_c > H > H_0$, $\sigma_c^{step} > \sigma_0 > \Sigma_0$	$H > H_c$, $\sigma_0 > \sigma_c^{step}$
$\sigma_0 < 0$	w_{ten}	X	X	X	$w = 0$	X	w_{ten}	w_{null}	w_1	w_2
$\sigma_0 = 0$	w_{ten}	$w = 0$	X	X	$w = 0$	X	X	$w = 0$	w_1	w_2
$\sigma_0 > 0$	w_{ten}	w_{null}	w_1	w_2	$w = 0$	w_b	X	X	w_1	w_2

X = no solution.

13.6 References

Audoly B (2003), 'Self-similar structures near boundaries in strained systems', *Phys. Rev. Lett.*, 91, 086105.

Colin J, Cleymand F, Coupeau C and Grilhé J (2000), 'Worm-like delamination patterns of thin stainless steel films on polycarbonate substrates', *Philos. Mag. A*, 80, 2559–2565.

Cordill M J, Moody N R and Bahr D F (2005), 'The effects of plasticity on adhesion of hard films on ductile interlayers', *Acta Mater.*, 53, 2555–2562.

Coupeau C and Grilhé J (2001), 'Atomic force microscopy observations of *in situ* deformed materials: application to single crystals and thin films on substrates', *J. Micro.*, 203, 99–107.

Coupeau C, Naud J F, Cleymand F, Goudeau P and Grilhé J (1999a), 'Atomic force microscopy of *in situ* deformed nickel thin films', *Thin Solid Films*, 353, 194–200.

Coupeau C, Girard J C and Grilhé J (1999b), 'Atomic force microscopy of single crystal surfaces during plastic deformation', *Probe Mic.*, 1, 313–321.

Coupeau C, Cleymand F and Grilhé J (2000), 'Atomic force microscopy of dislocation locking effects at gold film LiF substrate interface', *Scripta Mater.*, 43, 187–192.

Coupeau C, Girard J C and Rabier J (2004), 'Scanning probe microscopy and dislocations', in Nabarro F R N and Hirth J P (eds), *Dislocations in Solids*, Amsterdam, Elsevier, 275–338.

Crosby K M and Bradley R M (1999), 'Pattern formation during delamination and buckling of thin films', *Phys. Rev. E*, 59, 2542–2545.

Föppl A (1907), *Vorlesungen über Technische Mechanik*, 5, 132.

Foucher F and Coupeau C (2007), 'Effect of the dislocation emergence on the mechanical behavior of coated materials: elastic energy relaxation or adhesion modification', *Surf. Coat. Tech.*, 202, 1094–1097.

Foucher F, Coupeau C, Colin J, Cimetière A and Grilhé J (2006), 'How does crystalline substrate plasticity modify thin film buckling?', *Phys. Rev. Lett.*, 97, 096101.

Hutchinson J W and Suo Z (1992), 'Mixed mode cracking in layered materials', *Adv. Appl. Mech.*, 29, 63–191.

Hutchinson J W, He M Y and Evans A G (2000), 'The influence of imperfections on the nucleation and propagation of buckling driven delaminations', *J. Mech. Phys. Sol.*, 48, 709–734.

Jagla E A (2006), 'Morphologies of expansion ridges of elastic thin films onto a substrate', *Phys. Rev. E*, 74, 036207.

Jagla E A (2007), 'Modeling the buckling and delamination of thin films', *Phys. Rev. B*, 75, 085405.

Junqua N and Grilhé J (1997), 'Surface step-dislocation transition and dislocation nucleation at a solid free surface', *Philos. Mag. Lett.*, 75, 125.

Matuda N, Baba S and Kinbara A (1981), 'Internal stress, Young's modulus and adhesion energy of carbon films on glass substrates', *Thin Solid Films*, 81, 301–305.

Moon M W, Chung J W, Lee K R, Oh K H, Wang R and Evans A G (2002), 'An experimental study of the influence of imperfections on the buckling of compressed thin films', *Acta Mater.*, 50, 1219–1227.

Parry G, Coupeau C, Colin J, Cimetière A and Grilhé J (2004), 'Buckling and post-buckling of stressed straight-sided wrinckles: experimental AFM observations of bubbles formation and finite element simulations', *Acta Mater.*, 52, 3959–3966.

Parry G, Colin J, Coupeau C, Foucher F, Cimetière A and Grilhé J (2005), 'Effect

of substrate compliance on the global unilateral post-buckling of coatings: AFM observations and finite element calculations', *Acta Mater.*, 53, 441–447.
Parry G, Cimetière A, Coupeau C, Colin J and Grilhé J (2006), 'Stability diagram of unilateral buckling patterns of strip-delaminated films', *Phys. Rev. E*, 74, 066601.
Paumier F, Gaboriaud R J and Coupeau C (2003), 'Buckling phenomena in Y_2O_3 thin films on GaAs substrates', *Appl. Phys. Lett.*, 82, 2056–2058.
von Kármán T (1910), 'Festigkeitsprobleme in maschinenbau', *Ency. der. Math. Wissenschaften*, IV/4C, 311–385.

14

Controlled buckling of thin films on compliant substrates for stretchable electronics

J. SONG, University of Miami, USA and J. WU, and Y. HUANG, Northwestern University, USA

Abstract: Electronic systems with large stretchability have many applications. This chapter reviews the mechanics of a non-coplanar mesh design whose stretchability can be up to 100%. Mechanics models for non-coplanar mesh designs are developed and used to predict the maximum strains. The models agree well with experiments and give the stretchability of the system in terms of the geometric parameters and fracture strain of the material.

Key words: buckling, interconnect, island, substrate, stretchable electronics.

14.1 Introduction

Stretchable electronics is electronics with performance equal to established technologies that use rigid semiconductor wafers, but in formats that can be stretched and compressed. It enables many new application possibilities. Examples include flexible displays (Crawford, 2005), electronic eye camera (Jin *et al.*, 2004; Ko *et al.*, 2008; Shin *et al.*, 2010), conformable skin sensors (Lumelsky *et al.*, 2001), smart surgical gloves (Someya *et al.*, 2004), and structural health monitoring devices (Nathan *et al.*, 2000). Circuits that use organic semiconductor materials can sustain large deformations (Garnier *et al.*, 1994; Baldo *et al.*, 2000; Crone *et al.*, 2000; Loo *et al.*, 2002; Facchetti *et al.*, 2005; Sekitani *et al.*, 2008), but their electrical performance is relatively poor (several orders of magnitude lower than that of conventional inorganic material such as silicon). Compatibility with well developed, high performance inorganic electronic materials represents a key advantage in this area. The main challenge here is to design inorganic materials (e.g., silicon) for stretchability, which might initially seem challenging since all known inorganic semiconductor materials are brittle and fracture at strains of the order of 1%.

One of the most intuitive approaches to avoid directly straining these brittle materials exploits wavy shaped structures (Bowden *et al.*, 1998; Khang *et al.*, 2006; Choi *et al.*, 2007) to make these high-quality, single-crystal inorganic semiconductor materials stretchable. Figure 14.1a illustrates one example of this design, as applied to arrays of thin ribbons. The initially flat ribbons are bonded at all points on their bottom surfaces to a prestrained elastomeric substrate. The prestrain can be induced by mechanical (or thermal) stretch along the ribbon directions. When the prestrain is released, the ribbons are subject to a compression that leads to a nonlinear buckling response and forms a wavy profile. The resulting wavy deformations have well-defined wavelengths and amplitudes that depend on the material properties and the level of prestrain. Detailed theoretical analysis reveals all aspects of the mechanics of formation of such structures. For small prestrain ε_{pre}, the wavelength $\lambda_0 = 2\pi h_f[\overline{E}_f/(3\overline{E}_s)]^{1/3}$ is independent of the prestrain. It is proportional to the thickness of thin film h_f, and depends only on the plane-strain moduli $\overline{E}_f = E_f/(1 - \nu_f^2)$ and $\overline{E}_s = E_s/(1 - \nu_s^2)$ of the ribbon and substrate. The amplitude $A_0 = h_f\sqrt{\varepsilon_{pre}/\varepsilon_c - 1}$ increases with the prestrain ε_{pre}, where $\varepsilon_c = (3\overline{E}_s/\overline{E}_f)^{2/3}/4$ is the critical buckling strain (Huang *et al.*, 2005; Khang *et al.*, 2006). For large prestrain ε_{pre}, both wavelength and amplitude depend on the prestrain ε_{pre}. The wavelength given by $\lambda_0/\lfloor(1 + \varepsilon_{pre})(1 + \xi)^{1/3}\rfloor$ decreases with ε_{pre} while the amplitude given by $A_0/\left\lfloor\sqrt{1 + \varepsilon_{pre}}(1 + \xi)^{1/3}\right\rfloor$ increases with ε_{pre} (Song *et al.*, 2008b). The wavelengths and amplitudes of the waves can change to accommodate strains in a way that involves small strain in the ribbons. Following the same procedure to generate stretchable nanoribbons (Khang *et al.*, 2006), Choi *et al.* (2007) generated 2D wavy Si nanomembrane as shown in Fig. 14.1(b). The Si nanomembrane is chemically bonded to a heated poly(dimethylsiloxane) (PDMS) substrate which induces a controlled degree of isotropic thermal expansion. Cooling to room temperature released the thermally induced prestrain in PDMS, and formed the 2D wavy patterns on the surface. Recent work by Kim *et al.* (2008a) demonstrated high-performance, stretchable, and foldable integrated circuits (e.g., silicon complementary logic gates, ring oscillators, and different amplifiers) using this strategy.

In a related strategy, the ribbons can be designed to bond to the elastomeric substrate only at certain locations (Sun *et al.*, 2006). When the prestrain is released, the ribbon on the non-bonded regions delaminates from the substrate and forms the one-dimensional non-coplanar design (Fig. 14.1c). Compared to Figs 14.1(a) and (b), this layout has the advantage that the wavelengths can be defined precisely with a level of engineering control to have a higher stretchability.

The other strategy uses stretchable interconnect bridges such as metal wires to link the rigid device islands (i.e., interconnect-island structure) (Lacour *et al.*, 2005; Kim *et al.*, 2008b, 2008c). Mechanical response to stretching or compression involves, primarily, deformations only in these interconnects, thereby avoiding unwanted strains in the regions of the active devices. Lacour *et al.* (2005) developed coplanar mesh designs by using the wavelike interconnect bridges, which are bonded with the substrate. The deformation of the interconnect bridge is similar to that in Fig. 14.1(a).

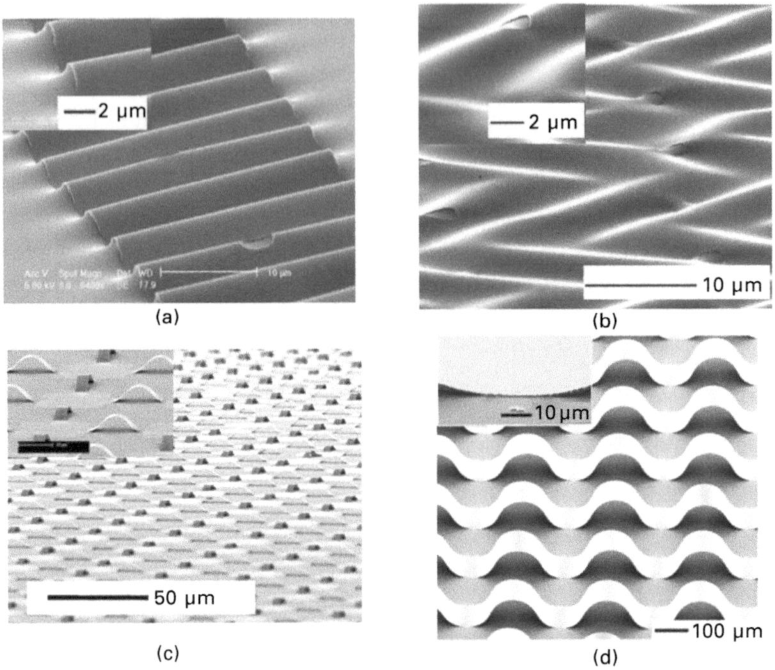

14.1 SEM images of (a) stretchable nanoribbons, (b) stretchable nanomembranes, (c) one-dimensional non-coplanar mesh design, and (d) two-dimensional non-coplanar mesh design. Reprinted with permission from Song *et al.* (2009a), Copyright 2009 American Vacuum Society and Jiang *et al.* (2007), Copyright 2007 American Institute of Physics.

Although such a coplanar mesh design can improve the stretchability to around 40%, the stretchability is still too small for certain applications and large-scale integration can be difficult. Sosin *et al.* (2008) showed that free-standing copper interconnects can be used for stretchable electronics and the stretchability can be a few hundreds percent. By adopting the 'free-standing' idea, Kim *et al.* (2008c) developed a two-dimensional non-coplanar mesh design (Fig. 14.1d) consisting of device islands linked by 'popup' interconnect bridges for stretchable circuits, which can be stretched to rubber-like levels of strain (e.g., up to 100%). Recent work by Ko *et al.* (2008) demonstrated a hemispherical electronic eye camera based on this non-coplanar mesh design.

The fundamental aspects of stretchable coplanar nanoribbons (Fig. 14.1a) and nanomembranes (Fig. 14.1b) such as the edge effect (Koh *et al.*, 2007), finite width (Jiang *et al.*, 2008a), finite deformation (Song *et al.*, 2008a), local versus global buckling modes (Wang *et al.*, 2008), and biaxial stretchability (Song *et al.*, 2008b) have been well studied and reviewed by Jiang *et al.*

(2008b) and Song *et al.* (2009a). This chapter provides a review on the mechanics of non-coplanar mesh designs (Figs 14.1c and d). Section 14.2 reviews the mechanics of one-dimensional non-coplanar mesh design (Fig. 14.1c) and Section 14.3 describes the mechanics of two-dimensional non-coplanar mesh design (Fig. 14.1d).

14.2 Mechanics of one-dimensional non-coplanar mesh design

Figure 14.2 schematically illustrates the fabrication of one-dimensional non-coplanar mesh designs for stretchable electronics on compliant substrates (Sun *et al.*, 2006; Jiang *et al.*, 2007), which combines lithographically patterned surface bonding chemistry and a buckling process. The ribbons are fabricated from SOI wafers, which are patterned by conventional photolithography using AZ 5214 photoresist and etched with SF_6 plasma. After the photoresist is washed away with acetone, the buried oxide layer is then etched in HF (40%). More details can be found in the paper by Sun *et al.* (2006). Figure 14.2(a) shows the photolithography process that defines the bonding chemistry on a pre-stretched PDMS substrate with prestrain $\varepsilon_{pre} = \Delta L/L$ along the ribbon direction to form periodic interfacial patterns. The patterning process creates selected regions with activated sites where chemical bonding occurs between the ribbons (e.g., GaAs or Si) and the PDMS substrate, as well as inactivated sites where there are only weak van der Waals interactions. Let W_{act} and W_{in} denote the widths of activated and inactivated sites, respectively (Fig. 14.2a). Thin ribbons are then attached to the prestrained and patterned PDMS substrate (Fig. 14.2b) with the ribbon direction parallel to the prestreched direction. The relaxation of the prestrain in PDMS leads to the separation of the ribbons from the inactivated sites on the PDMS (Fig. 14.2c). The wavelength of the buckled structures is then given by $2L_1 = W_{in}/(1 + \varepsilon_{pre})$ due to the geometrical constrain, and the amplitude A will depend on the geometries of the interfacial patterns (W_{act} and W_{in}) and the prestrain.

Jiang *et al.* (2007) developed an analytical model to study the buckling behavior of such systems and to predict the maximum strain in the ribbons as a function of interfacial pattern. The energy method is used to determine the buckling geometry. The thin film is modeled as an elastic nonlinear von Kármán beam since the film thickness is much smaller than the other characteristic lengths and the strain in the film is negligibly small (to be shown later). The substrate is modeled as a semi-infinite elastic solid because its thickness (~mm) is much larger than the thickness of the film (~μm). The total energy of the system consists of the bending energy $U_{bending}$ due to thin film buckling, membrane energy $U_{membrane}$ in the thin film and substrate energy $U_{substrate}$.

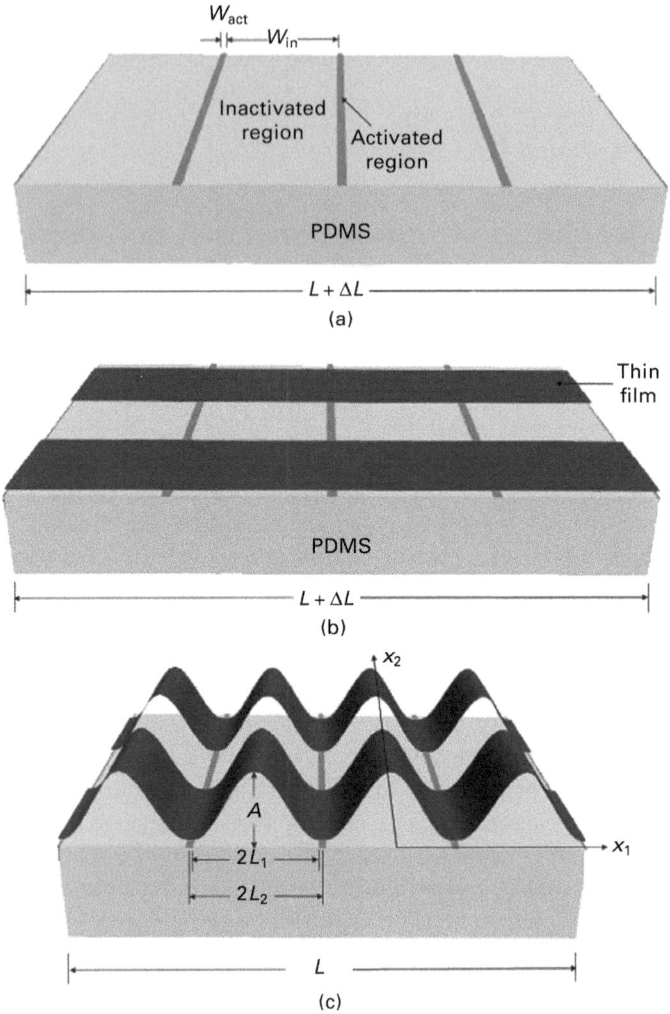

14.2 Processing steps for precisely controlled thin film buckling on elastomeric substrate. (a) Prestrained PDMS with periodic activated and inactivated patterns. L is the original length of PDMS and ΔL is the extension. The widths of activated and inactivated sites are denoted as W_{act} and W_{in}, respectively. (b) A thin film parallel to the prestrain direction is attached to the prestrained and patterned PDMS substrate. (c) The relaxation of the prestrain ε_{pre} in PDMS leads to buckles of thin film. The wavelength of the buckled film is $2L_1$, and its amplitude is A. $2L_2$ is the sum of activated and inactivated regions after relaxation. Reprinted with permission from Jiang *et al.* (2007), Copyright 2007 American Institute of Physics.

The buckling profile of the ribbon can be expressed as

$$w = \begin{cases} w_1 = \dfrac{1}{2}A\left(1 + \cos\dfrac{\pi x_1}{L_1}\right), & -L_1 < x_1 < L_1 \\ w_2 = 0, & L_1 < |x_1| < L_2 \end{cases} \qquad [14.1]$$

where A is the buckling amplitude to be determined, $2L_1 = W_{in}/(1 + \varepsilon_{pre})$ is the buckling wavelength, and $2L_2 = W_{in}/(1 + \varepsilon_{pre}) + W_{act}$ is the sum of activated and inactivated regions after relaxation (Fig. 14.2c). The bending energy in the thin film can be obtained as

$$U_{\text{bending}} = \int_{-L_2}^{L_2} \frac{1}{2}\frac{\bar{E}_f h_f^3}{12}\left(\frac{d^2 w}{dx_1^2}\right)^2 dx_1 = \frac{\pi^4}{96}\frac{\bar{E}_f h_f^3 A^2}{L_1^3} \qquad [14.2]$$

where h_f is the film thickness, and $\bar{E}_f = E_f/(1 - \nu_f^2)$ is the plane-strain modulus of the film.

The membrane strain ε_{11} is related to the out-of-plane displacement w and the in-plane displacement u_1 by $\varepsilon_1 = du_1/dx_1 + (dw/dx)^2/2 - \varepsilon_{pre}$. Huang *et al.* (2005) showed that the shear stress at the film/substrate interface is negligibly small. The force equilibrium gives a constant membrane force, which requires a constant membrane strain ε_{11} given by

$$\varepsilon_{11} = \pi^2 A^2/(16L_1L_2) - \varepsilon_{pre} \qquad [14.3]$$

Then we have the membrane energy in the film as

$$U_{\text{membrane}} = \bar{E}_f h_f\left(\frac{\pi^2 A^2}{16L_1L_2} - \varepsilon_{pre}\right)^2 L_2 \qquad [14.4]$$

The relaxed PDMS has vanishing energy

$$U_{\text{substrate}} = 0 \qquad [14.5]$$

because the substrate has zero displacement at the interface where remains intact and vanishing stress traction at the long and buckled portion.

The buckling amplitude A is determined by energy minimization as

$$A = \frac{4}{\pi}\sqrt{L_1L_2(\varepsilon_{pre} - \varepsilon_c)}, \quad \text{for } \varepsilon_{pre} > \varepsilon_c \qquad [14.6]$$

where $\varepsilon_c = h_f^2\pi^2/(12L_1^2)$ is the critical strain for buckling, which is identical to the Euler buckling strain for a doubly clamped beam with length $2L_1$. When $\varepsilon_{pre} < \varepsilon_c$, the film does not buckle and remains flat. Once the prestrain exceeds the critical strain (i.e., $\varepsilon_{pre} > \varepsilon_c$), the film buckles to form the sinusoidal waves. The critical strain ε_c is very small in most practical applications. For example, for the buckling wavelength $2L_1 \sim 200$ μm and ribbon thickness $h_f \sim 0.1$ μm, ε_c is on the order of 10^{-6}. The buckling amplitude A in Eq. 14.6 then can be approximately by

$$A \approx \frac{4}{\pi}\sqrt{L_1 L_2 \varepsilon_{pre}} = \frac{2}{\pi}\sqrt{W_{in}(W_{in} + W_{act})\varepsilon_{pre}/(1 + \varepsilon_{pre})} \qquad [14.7]$$

which is completely determined by the interfacial patterns (W_{in} and W_{act}) and the prestrain. Figure 14.3 shows the profiles (dashed lines) of buckled GaAs ribbons given by Eqs 14.1 and 14.6 for different prestrain levels for $W_{act} = 10$ μm and $W_{in} = 190$ μm. The experimental images are also shown for comparison. Both wavelength and amplitude agree well with experiments.

The maximum strain in the film is the sum of membrane strain and bending strain. Since the membrane strain ε_{11} is negligible ($\sim 10^{-6}$), the maximum strain in the ribbon can be approximated by the bending strain which is given by

$$\varepsilon_{peak} = \frac{h_f}{2}\max\left(\frac{d^2 w}{dx_1^2}\right) = \frac{h_f \pi}{L_1^2}\sqrt{L_1 L_2 \varepsilon_{pre}} \qquad [14.8]$$

The maximum strain is much smaller than the prestrain. For example, for $h_f = 0.3$ μm, $W_{act} = 10$ μm, $W_{in} = 400$ μm, and $\varepsilon_{pre} = 60\%$, ε_{peak} is only 0.6%,

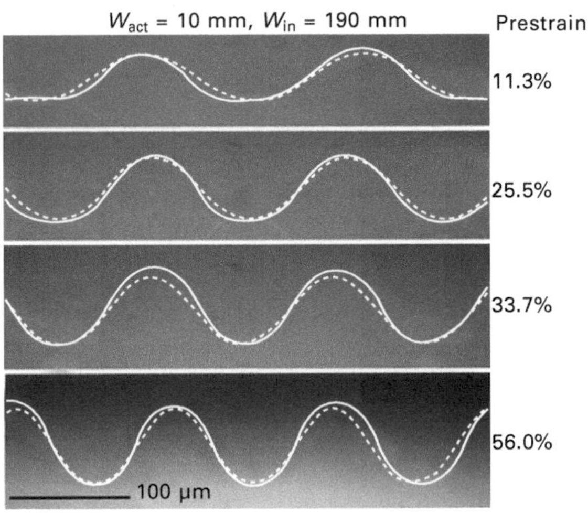

14.3 Buckled GaAs thin films on patterned PDMS substrate with $W_{act} = 10$ μm and $W_{in} = 190$ μm for different prestrain levels, 11.3%, 25.5%, 33.7%, and 56.0% (from top to bottom). The dashed lines are the profiles of the buckled GaAs thin film predicted by the analytical solution. Reprinted with permission from Jiang *et al.* (2007) Copyright 2007 American Institute of Physics.

which is two orders of magnitude smaller than the 60% prestrain. Therefore, the one-dimensional non-coplanar mesh design can significantly reduce the maximum strain in thin ribbon, and improve the system stretchability. Moreover, this small strain ensures that the von Kármán beam assumption for the film is always valid in this study. For much smaller active region (i.e., $W_{act} << W_{in}$), the maximum strain in Eq. (14.8) can be approximated by $\varepsilon_{peak} \approx \frac{h_f \pi}{L_1}\sqrt{\varepsilon_{pre}}$.

It should be pointed out that, for very long inactivated region, buckling profiles different from Eq. 14.1 have been observed in experiments. For example, two or three short waves with wavelength about 100 μm, instead of a long wavelength $2L_1 = W_{in}/(1 + \varepsilon_{pre}) = 586$ μm, have been observed in experiments for $W_{in} = 760$ μm and $\varepsilon_{pre} = 33.7\%$. The thin film and substrate remain bonded between the short waves. The analysis of this different buckling mode can be found in the paper by Wang *et al.* (2008).

14.3 Mechanics of two-dimensional non-coplanar mesh design

Figure 14.4 schematically illustrates the fabrication of a two-dimensional non-coplanar mesh design for stretchable electronics on compliant substrates (Kim *et al.*, 2008c). The silicon (or other semiconductor material) islands, on which the active devices are fabricated, are chemically bonded to a prestrained (e.g., 50%) elastometric substrate poly(dimethylsiloxane) (PDMS). The interconnect bridges between islands are loosely bonded to the substrate. Releasing the prestrain leads to compression, which causes the interconnect bridges to buckle out of the substrate surface and form arc-shaped structures. The out-of-plane deformation localizes only to the interconnect bridges because of their poor adhesion (to PDMS), narrow geometries and low stiffness (compared to device islands). This gives very small strain in islands, which leads to stretchability of the system of device islands and interconnect bridges. The practical use of the non-coplanar mesh design requires an additional encapsulation layer on the top surface to provide mechanical and environmental protection (Kim *et al.*, 2009). A compliant encapsulation layer must be used in order not to restrict free motion of the buckled interconnects, and therefore not to reduce the stretchability.

Song *et al.* (2009b) and Wu *et al.* (2010) established mechanics models to understand the underlying physics and to guide the design of such systems. The analysis for pre-encapsulation is presented in Section 14.3.1 and the analysis for post-encapsulation is discussed in Section 14.3.2.

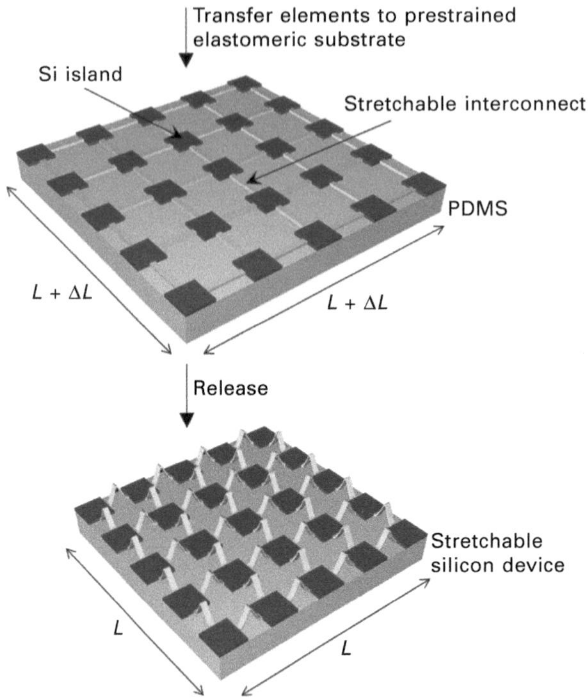

14.4 A schematic illustration of the process for fabricating electronics with two-dimensional non-coplanar mesh designs on a complaint substrate. Reprinted with permission from Song *et al.* (2009b), Copyright 2009 American Institute of Physics.

14.3.1 Analysis for pre-encapsulation

Figure 14.5 shows a schematic diagram of the device islands and interconnect bridges on a prestrained, thick elastomeric substrate prior to encapsulation. The device islands (length L_{island}) and interconnect bridges (length L_{bridge}) are chemically (well) and loosely (poorly) bonded to the substrate, respectively. Release of the prestrain ε_{pre} in the substrate leads to compression, which buckles the interconnect bridges because they are much narrower and have much lower bending stiffness than the device islands.

The width of interconnect bridge is much smaller than the width of device island such that the rotation at the ends of bridges is very small. This is verified by finite element analysis (Kim *et al.*, 2008c). Therefore, the 'bridge'-like interconnect is modeled as a beam with clamped ends since its thickness is much smaller than any other characteristic length. The beam, however, undergoes large rotation once the interconnect bridge buckles. The buckling profile of the interconnect bridge can be expressed as

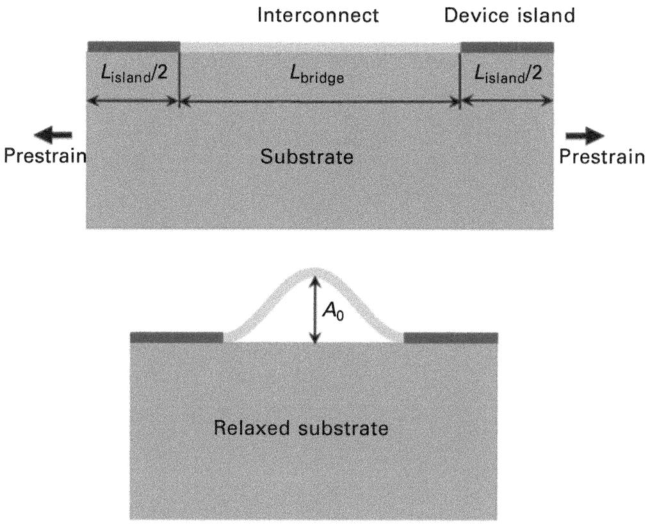

14.5 A schematic diagram of the mechanics model for non-coplanar mesh structure prior to encapsulation. Reprinted with permission from Wu *et al.* (2010), Copyright 2010 Elsevier Ltd.

$$w = \frac{A_0}{2}\left(1 + \cos\frac{2\pi X}{L_{\text{bridge}}}\right) \quad \text{for} \quad |X| \leq L_{\text{bridge}}/2 \qquad [14.9]$$

where A_0 is the buckle amplitude to be determined, and $X = 0$ denotes the center of the bridge. This gives the bending energy in terms of the bending stiffness $\overline{EI}_{\text{bridge}}$ (per unit width) in the interconnect bridge

$$U_{\text{bending}} = \int_{-L_{\text{bridge}}/2}^{L_{\text{bridge}}/2} \frac{\overline{EI}_{\text{bridge}}}{2}\left(\frac{d^2w}{dX^2}\right)^2 dX = \frac{\pi^4 \overline{EI}_{\text{bridge}}}{L_{\text{bridge}}^3} A_0^2 \qquad [14.10]$$

For an interconnect bridge with the Young's modulus E and uniform thickness h, $\overline{EI}_{\text{bridge}} = \overline{E}h^3/12$, where $\overline{E} = E/(1 - v^2)$ is the plane-strain modulus, and v is the Poisson's ratio.

The membrane energy can be obtained through the membrane strain, which is related to the out-of-plane displacement w and the axial displacement u by $\varepsilon_{\text{membrane}} = du/dX + (dw/dX)^2/2$. The force equilibrium, $dN/dX = 0$, requires a constant axial force N in the interconnect bridge, and therefore a constant membrane strain. Releasing the prestrain ε_{pre} in the elastomeric substrate leads to compression in the interconnect bridges. When this compression exceeds the critical value, the interconnect bridges buckle to form the popup structure similar to Fig. 14.1(c). The length of elastomeric substrate underneath the interconnect bridge changes from L_{bridge} before the release of prestrain to

$L_{bridge}/(1+\varepsilon_{pre})$ after the prestrain is completely released. This relates the axial displacement u to the prestrain ε_{pre} by

$$\int_{-L_{bridge}/2}^{L_{bridge}/2} du = \frac{L_{bridge}}{1+\varepsilon_{pre}} - L_{bridge} = -L_{bridge}\frac{\varepsilon_{pre}}{1+\varepsilon_{pre}} \qquad [14.11]$$

It gives the axial displacement $u = \pi A_0^2/(16L_{bridge})\sin(4\pi X/L_{bridge}) - (\varepsilon_{pre}X)/(1+\varepsilon_{pre})$. The membrane strain can be obtained as

$$\varepsilon_{membrane} = \frac{\pi^2}{4L_{bridge}^2}A_0^2 - \frac{\varepsilon_{pre}}{1+\varepsilon_{pre}} \qquad [14.12]$$

The membrane energy in the interconnect bridge is then obtained as

$$U_{membrane} = \int_{-L_{bridge}/2}^{L_{bridge}/2} \frac{\overline{Eh_{bridge}}}{2}\varepsilon_{membrane}^2\, dX$$

$$= \frac{\pi^4\overline{Eh_{bridge}}}{32}\left[\left(\frac{A_0}{L_{bridge}}\right)^2 - \frac{4}{\pi^2}\frac{\varepsilon_{pre}}{1+\varepsilon_{pre}}\right]^2 \qquad [14.13]$$

where $\overline{Eh_{bridge}}$ is the tensile stiffness (per unit width) of interconnect bridges. For an interconnect bridge with the Young's modulus E and uniform thickness h, $\overline{Eh_{bridge}} = \overline{E}h$.

The displacement in Eq. 14.9 induces negligible deformation in the device islands and elastomeric substrate (Song *et al.*, 2009b) such that their strain energy can be neglected. The total energy is the sum of bending and membrane energy in the thin film and the minimization of total energy, $d(U_{bending} + U_{membrane})/dA_0 = 0$, gives the buckle amplitude,

$$A_0 = \frac{2L_{bridge}}{\pi}\sqrt{\frac{\varepsilon_{pre}}{1+\varepsilon_{pre}} - \varepsilon_c} \qquad [14.14]$$

where $\varepsilon_c = \dfrac{4\pi^2\overline{EI_{bridge}}}{\overline{Eh_{bridge}}L_{bridge}^2}$ is the critical buckling strain of a doubly clamped beam. When the prestrain is smaller than $\varepsilon_c/(1-\varepsilon_c)$, the interconnect bridge remains flat and does not buckle. Once the prestrain exceeds $\varepsilon_c/(1-\varepsilon_c)$, the interconnect bridge buckles. The membrane strain in Eq. 14.13 becomes $\varepsilon_{membrane} = -\varepsilon_c$, i.e., the membrane strain is a constant $-\varepsilon_c$ after buckling.

For an interconnect bridge made of n layers of materials, the tensile and bending stiffness can be obtained by $\overline{Eh_{bridge}} = \sum_{i=1}^{n}\overline{E}_i h_i$ and

$\overline{EI}_{bridge} = \sum_{i=1}^{n} \bar{E}_i h_i \left[b - \left(\sum_{j=1}^{i} h_j - \frac{h_i}{2} \right) \right]^2 = \frac{1}{12} \sum_{i=1}^{n} \bar{E}_i h_i^3$, respectively. Here $\bar{E}_i = E_i/(1 - \nu_i^2)$ is the plane-strain modulus, h_i is the thickness of the ith layer from the top surface and $b = \sum_{i=1}^{n} \bar{E}_i h_i \left(\sum_{j=1}^{i} h_j - \frac{1}{2} h_i \right) \Big/ \overline{Eh}_{bridge}$ is the distance between the top surface and neutral mechanical plane. The maximum (compressive) strain in the interconnect bridge is the sum of membrane and bending strains. The bending strain in the interconnect bridge can be obtained by κy, where y is the distance from the neutral mechanical plane, and κ is the maximum curvature given by $\kappa = \max\left(\frac{d^2 w}{dX^2} \right) = \frac{4\pi}{L_{bridge}} \sqrt{\frac{\varepsilon_{pre}}{1 + \varepsilon_{pre}} - \varepsilon_c}$. The membrane strain is shown to be a constant $\varepsilon_{membrane} = -\varepsilon_c$, which is negligible compared to the bending strain.

For the interconnect bridges made of Si (h = 50 nm, L_{bridge} = 20 μm) in experiments, the neutral mechanical plane is $b = h/2$ = 25 nm from the top surface. The critical buckling strain is extremely small, ε_c = 0.0021%. The measured bridge length is 17.5 μm after relaxation, which corresponds to a prestrain ε_{pre} = 14.3%. The amplitude predicted by Eq. 14.10 is A_0 = 4.50 μm, which agrees well with the experimentally measured value 4.76 μm. The maximum strain in the interconnect bridge is 0.56%, which is smaller than the fracture strain of silicon (~1%), and much smaller than the prestrain. The maximum strain versus the prestrain is shown in Fig. 14.6 and for 1% interconnect strain, the prestrain can reach 68%.

For the interconnect bridges made of polyimide ($h_1 = h_4$ = 1.2 μm, $E_1 = E_4$ = 2.5 GPa, $\nu_1 = \nu_4$ = 0.34), Au (h_2 = 0.15 μm, E_2 = 78 GPa, ν_2 = 0.44) and SiO_2 (h_3 = 0.05 μm, E_3 = 70 GPa, ν_3 = 0.17) (Kim *et al.*, 2009), the neutral mechanical plane is b = 1.30 μm from the top surface. The stretchability of the system is determined by the maximum strain in the critical layer, which is made of brittle material. In this case, the critical layer in the interconnect bridge is Au or SiO_2. For an interconnect bridge with L_{bridge} = 460 μm, the critical buckling strain is ε_c = 0.0032%. Figure 14.7 shows the maximum strain $0.0028\sqrt{\varepsilon_{pre}/(1 + \varepsilon_{pre})}$ in Au or SiO_2 versus the prestrain. For a large prestrain 100%, the maximum strain is small, 0.2%.

The finite element method is used to study the silicon island ($L_{island} \times L_{island}$) on PDMS substrate [$(L_{island} + L_{bridge}) \times (L_{island} + L_{bridge})$]. The island is modeled by shell elements since its thickness is much smaller compared with other lengths. The substrate is modeled by 3D solid elements. The island is bonded to the substrate by sharing the same nodes. Periodic conditions are applied on the lateral surfaces (X–Z and Y–Z planes). The buckled interconnect bridge gives the axial force $N = \overline{Eh}_{bridge}\varepsilon_c$ and

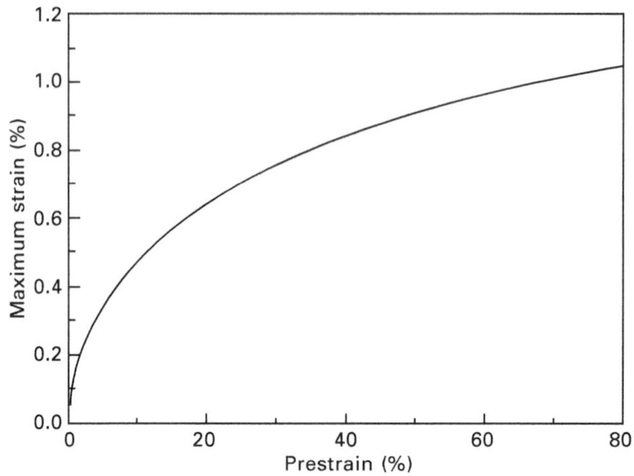

14.6 The maximum strain in the interconnect bridge versus the prestrain. The material properties and geometry of interconnect bridges are given in Section 14.3.1. Reprinted with permission from Song *et al.* (2009b), Copyright 2009 American Institute of Physics.

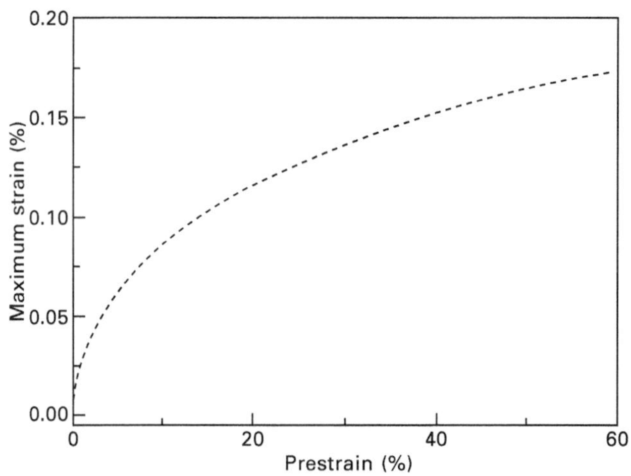

14.7 The maximum strain in the interconnect bridge versus the prestrain. The material properties and geometry of interconnect bridges are given in Section 14.3.1. Reprinted with permission from Wu *et al.* (2010), Copyright 2010 Elsevier Ltd.

bending moment $M = \overline{EI}_{bridge} \frac{2\pi^2 A_0}{(L_{bridge})^2}$, which are applied over the width w_{bridge} on each edge of the island. Since the axial force N scales with the critical buckling strain ε_c, the strain due to the axial force N is negligible compared to the strain due to the bending moment M, i.e., bending dominates.

Figure 14.8 shows the strain distribution ε_{xx} in a Si island (E_{island} = 130 GPa, ν_{island} = 0.27, length L_{island} = 20 μm, and thickness h_{island} = 50 μm) on a PDMS substrate ($E_{substrate}$ = 2 MPa, $\nu_{substrate}$ = 0.48). The axial force and bending moment result from the buckled interconnect bridge(E_{bridge} = 130 GPa, L_{bridge} = 20 μm, h_{bridge} = 50 nm, w_{bridge} = 4 μm, and ε_{pre} = 14.3%). The maximum strain occurs at the interconnect/island boundary.

The maximum strain in the device island is shown (Song *et al.*, 2009b) to be $4\pi y\overline{EI}_{bridge}\sqrt{\varepsilon_{pre}/(1+\varepsilon_{pre})/(L_{bridge}\overline{EI}_{island})}$ where y is the distance from the neutral plane. In practice, the bending stiffness of the device island is much larger than that of the interconnect bridge. Therefore, the strain in the device island is negligible and we only need to focus on the strain in interconnect bridges. Considering the multi-layer non-coplanar mesh design used in the electronic eye camera (Ko *et al.*, 2008), the interconnect bridges consist of a thin layer of patterned metal (360 μm long, 50 μm wide, 150 nm thick gold) on a polyimide (PI) layer and the islands consists of (500 × 500 μm^2 in area, 1.2 μm thick) silicon layer with a capping 1.5 μm thick polyimide. The elastic moduli and Poisson's ratios of metal and PI are E_{metal} = 78 GPa, ν_{metal} = 0.44, E_{PI} = 2.5 GPa and ν_{PI} = 0.34. Figure 14.9 shows the maximum metal strain in interconnects and Si strain in device islands versus the prestrain. The metal strain is much larger than the Si strain and both strains are much smaller than the fracture strain 1% even under 100% prestrain.

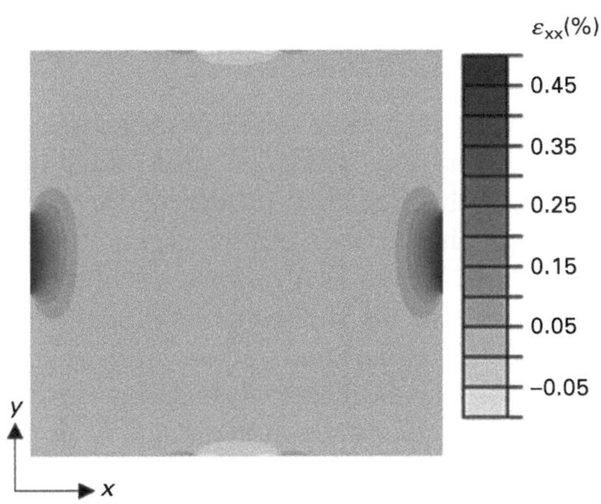

14.8 Distribution of the strain ε_{xx} in islands (20 × 20 μm) when the interconnect bridge relaxes from 20 μm to 17.5 μm. Reprinted with permission from Song *et al.* (2009b), Copyright 2009 American Institute of Physics.

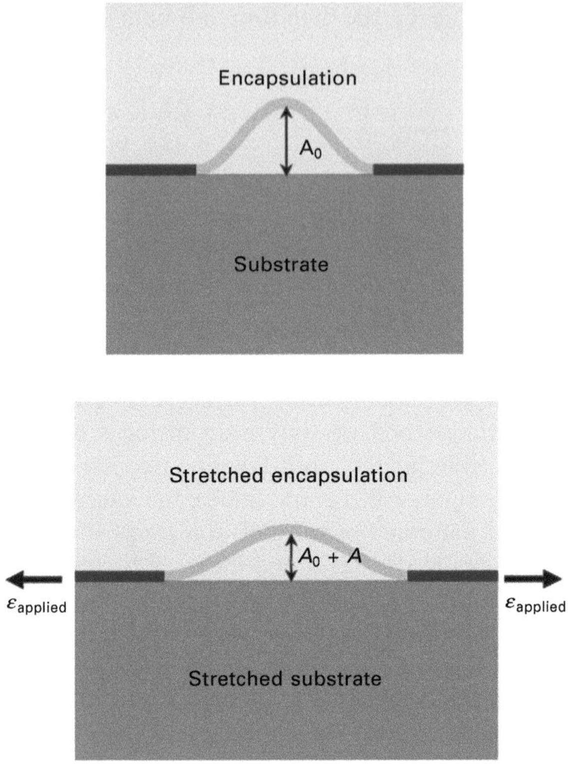

14.9 The maximum metal strain interconnect bridge (metal/PI) and Si strain in islands (PI/Si) versus the prestrain.

14.3.2 Analysis for post-encapsulation

In practice, the non-coplanar mesh designs require top surface encapsulation layers to provide mechanical and environmental protection. In order to provide minimal restriction of the free deformation of the non-coplanar bridges, the encapsulation layer should be an elastomer, with properties not too dissimilar from the substrate. Figure 14.10 shows a schematic diagram of the encapsulated system subject to the applied strain $\varepsilon_{applied}$. A compliant elastomeric material (e.g., PDMS) is cast and cured on top of the buckled interconnect bridges and device islands. The analysis below focuses on the stretchability, i.e., the maximum applied strain that leads to the failure of the system.

Once $\varepsilon_{applied}$ is applied, the amplitude of the interconnect bridge changes from A_0 to $A_0 + A$, where A is the increase of the buckle amplitude due to the applied strain. The buckling profile of the encapsulated interconnect bridges becomes

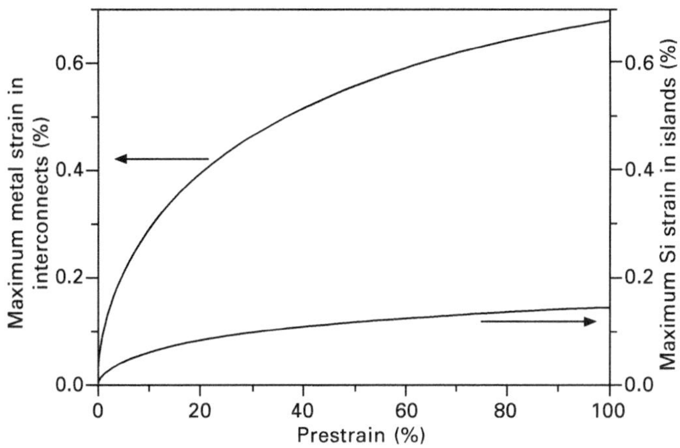

14.10 A schematic diagram of the encapsulated system subject to stretching. Reprinted with permission from Wu *et al.* (2010), Copyright 2010 Elsevier Ltd.

$$w = \frac{A_0 + A}{2}\left(1 + \cos\frac{2\pi X}{L_{\text{bridge}}}\right) \quad \text{for} \quad |X| \le L_{\text{bridge}}/2, \qquad [14.15]$$

where the increase of the amplitude A is to be determined by energy minimization. The total energy of the system consists of the bending energy U_{bending} and membrane energy U_{membrane} in the thin film as well as the substrate energy $U_{\text{substrate}}$ and the energy in the encapsulation layer $U_{\text{encapsulation}}$. The bending and membrane energy can be obtained following the same procedure used in Section 14.3.1. The bending energy in the interconnect bridge is given from Eq. 14.9 by replacing A_0 with $A_0 + A$,

$$U_{\text{bending}} = \frac{\pi^4 \overline{EI}_{\text{bridge}}}{L_{\text{bridge}}^3}(A_0 + A)^2 \qquad [14.16]$$

The axial displacement u is related to the prestrain ε_{pre} and applied strain $\varepsilon_{\text{applied}}$ by

$$\int_{-L_{\text{bridge}}/2}^{L_{\text{bridge}}/2} du = L_{\text{bridge}} \frac{\varepsilon_{\text{applied}} - \varepsilon_{\text{pre}}}{1 + \varepsilon_{\text{pre}}} \qquad [14.17]$$

It gives the membrane strain in the interconect bridge

$$\varepsilon_{\text{membrane}} = \frac{\pi^2}{4L_{\text{bridge}}^2}(A_0 + A)^2 + \frac{\varepsilon_{\text{applied}} - \varepsilon_{\text{pre}}}{1 + \varepsilon_{\text{pre}}} \qquad [14.18]$$

and membrane energy in the interconnect bridge

$$U_{\text{membrane}} = \int_{-L_{\text{bridge}}/2}^{L_{\text{bridge}}/2} \frac{\overline{Eh_{\text{bridge}}}}{2} \varepsilon_{\text{membrane}}^2 \, dX$$

$$= \frac{\pi^4 \overline{Eh_{\text{bridge}}}}{32} \left[\left(\frac{A_0 + A}{L_{\text{bridge}}} \right)^2 + \frac{4}{\pi^2} \frac{\varepsilon_{\text{applied}} - \varepsilon_{\text{pre}}}{1 + \varepsilon_{\text{pre}}} \right]^2 \qquad [14.19]$$

The substrate and encapsulation layer are modeled as semi-infinite solids because they are much thicker than the interconnect bridges and device islands. In order to obtain the energy in the substrate and encapsulation layer, we only need to consider the deformation resulting from the increase of the amplitude because the deformation due to $\frac{A_0}{2}\left(1 + \cos\frac{2\pi X}{L_{\text{bridge}}}\right)$ in Eq. 14.16 is negligible (Song *et al.*, 2009a). The normal displacement on the surface $Y = 0$ is

$$w(X, Y = 0)$$

$$= \begin{cases} \frac{A}{2}\left(1 + \cos\frac{2\pi X}{L_{\text{bridge}}}\right) & \text{for} \quad |X| \le L_{\text{bridge}}/2 \\ 0 & \text{for} \quad L_{\text{bridge}}/2 \le |X| \le (L_{\text{island}} + L_{\text{bridge}})/2 \end{cases}$$

[14.20]

It should be noted that this displacement is periodic over $[-(L_{\text{isaland}} + L_{\text{bridge}})/2, (L_{\text{isaland}} + L_{\text{bridge}})/2]$, and can be expressed in the Fourier series as

$$w(X, Y = 0) = \frac{1}{2} a_0 + \sum_{l=1}^{\infty} a_l \cos\left(\frac{2l\pi X}{L_{\text{island}} + L_{\text{bridge}}}\right) \qquad [14.21]$$

where a_l is obtained by

$$a_0 = A \frac{L_{\text{bridge}}}{L_{\text{island}} + L_{\text{bridge}}} \quad \text{for } l = 0, \qquad [14.22]$$

and

$$a_l = \frac{A}{l\pi} \left[1 - \frac{l^2 L_{\text{bridge}}^2}{(L_{\text{island}} + L_{\text{bridge}})^2} \right]^{-1} \sin\frac{l\pi L_{\text{bridge}}}{L_{\text{island}} + L_{\text{bridge}}} \quad \text{for } l > 0 \qquad [14.23]$$

For the special case of $l = (L_{\text{isaland}} + L_{\text{bridge}})/L_{\text{bridge}}$, $a_l = \frac{A}{2} \frac{L_{\text{bridge}}}{L_{\text{island}} + L_{\text{bridge}}}$.

For a semi-infinite solid subject to normal displacement w(X,Y = 0) = $a_l \cos\left(\frac{2l\pi X}{L_{\text{island}} + L_{\text{bridge}}}\right)$ on its surface, the strain energy stored over the period $L_{\text{isaland}} + L_{\text{bridge}}$ (per unit width) is $\frac{\pi}{4}\bar{E}\sum_{l=1}^{\infty} l a_l^2$ (Huang *et al.*, 2005), where $\bar{E}$ is the plane-strain modulus of the semi-infinite solid. Therefore, the strain energies in the substrate and encapsulation layer (over the period $L_{\text{isaland}} + L_{\text{bridge}}$, per unit width) are

$$U_{\text{substrate}} = \frac{\pi}{4}\bar{E}_{\text{substrate}}\sum_{l=1}^{\infty} l a_l^2 \text{ and } U_{\text{encapsulation}} = \frac{\pi}{4}\bar{E}_{\text{encapsulation}}\sum_{l=1}^{\infty} l a_l^2 \qquad [14.24]$$

where $\bar{E}_{\text{substrate}}$ and $\bar{E}_{\text{encapsulation}}$ are the plane-strain moduli of the substrate and encapsulation layer.

The minimization of total energy, $d(U_{\text{bending}} + U_{\text{membrane}} + U_{\text{substrate}} + U_{\text{encapsulation}})/dA = 0$, gives a cubic equation for the amplitude A,

$$\left(\frac{A_0 + A}{L_{\text{bridge}}}\right)^3 + \frac{4}{\pi^2}\left(\frac{\varepsilon_{\text{applied}} - \varepsilon_{\text{pre}}}{1 + \varepsilon_{\text{pre}}} + \varepsilon_c\right)\frac{A_0 + A}{L_{\text{bridge}}}$$

$$+ \frac{4(\bar{E}_{\text{substrate}} + \bar{E}_{\text{encapsulation}})}{\pi^5 \overline{Eh}_{\text{bridge}}} f\left(\frac{L_{\text{bridge}}}{L_{\text{island}} + L_{\text{bridge}}}\right)A = 0 \qquad [14.25]$$

where $f(x) = \sum_{l=1}^{\infty} \frac{\sin^2(l\pi x)}{l(1 - l^2x^2)^2}$. The numerical method is then used to solve this cubic equation.

The membrane strain in Eq. 14.18 can be rewritten using Eq. 14.25 as

$$\varepsilon_{\text{membrane}} = -\varepsilon_c - \frac{(\bar{E}_{\text{substrate}} + \bar{E}_{\text{encapsulation}})L_{\text{bridge}}}{\pi^3 \overline{Eh}_{\text{bridge}}} f\left(\frac{L_{\text{bridge}}}{L_{\text{island}} + L_{\text{bridge}}}\right)$$

$$\times \frac{A}{A_0 + A} \qquad [14.26]$$

where $-\varepsilon_c$ is the membrane strain from the release of prestrain in Section 14.3.1, and $-\frac{(\bar{E}_{\text{substrate}} + \bar{E}_{\text{encapsulation}})L_{\text{bridge}}}{\pi^3 \overline{Eh}_{\text{bridge}}} f\left(\frac{L_{\text{bridge}}}{L_{\text{island}} + L_{\text{bridge}}}\right)\frac{A}{A_0 + A}$ is the additional membrane strain due to the applied strain. The bending strain is obtained by κy, where the maximum curvature is $\kappa = \max\left(\frac{d^2 w}{dX^2}\right) = \frac{2\pi^2(A_0 + A)}{L_{\text{bridge}}^2}$ and y is the distance from the neutral mechanical plane. Figure 14.11 shows the (absolute value of) membrane strain and maximum bending strain

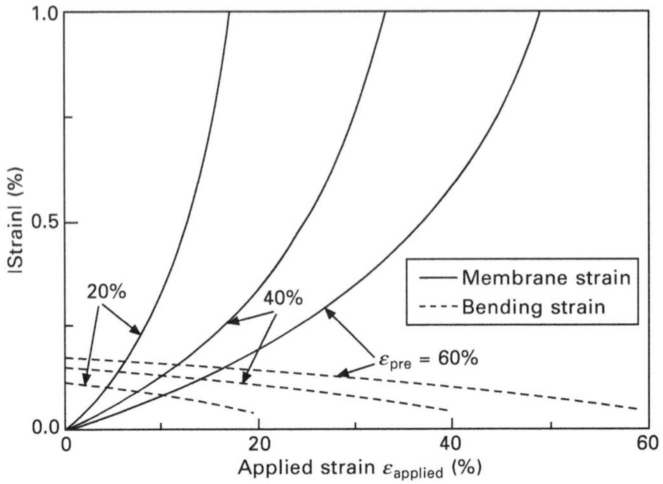

14.11 The maximum bending strain and membrane strain in the interconnect bridge versus the applied strain for the encapsulated system. The prestrain is 20%, 40% and 60%. Reprinted with permission from Wu *et al.* (2010), Copyright 2010 Elsevier Ltd.

in Au or SiO_2 of the interconnect bridge versus the applied strain for ε_{pre} = 20%, 40% and 60%. The interconnect bridge is made of polyimide ($h_1 = h_4 = 1.2$ μm, $E_1 = E_4 = 2.5$ GPa, $\nu_1 = \nu_4 = 0.34$), Au ($h_2 = 0.15$ μm, $E_2 = 78$ GPa, $\nu_2 = 0.44$) and SiO_2 ($h_3 = 0.05$ μm, $E_3 = 70$ GPa, $\nu_3 = 0.17$). The lengths of interconnect bridges and device islands are $L_{bridge} = 460$ μm and $L_{island} = 236$ μm, respectively. The elastic moduli and Poisson's ratios of the substrate and encapsution are $E_{substrate} = E_{encapsulation} = 1.8$ MPa and $\nu_{substrate} = \nu_{encapsulation} = 0.48$. The bending strain decreases as the applied strain increases because the amplitude of the interconnect bridge decreases. However, as the applied strain increases, the membrane strain increases rapidly due to the restriction of the encapsulation layer and dominates the failure of the interconnect bridge.

The stretchability characterizes how much the non-coplanar mesh design can accommodate further stretch. It is defined as the critical applied strain that leads to failure of the interconnect bridge. Because the bending strain is negligible compared to the membrane strain as shown in Fig. 14.11, the stretchability is obtained by equating the membrane strain in Eq. 14.26 to the fracture strain ε_f of the material (e.g., 1%) as

$$(\varepsilon_{applied})_{max} = \left[1 - \left(1 + \frac{\varepsilon_f}{\lambda}\right)^{-2}\right]\varepsilon_{pre} + (1 + \varepsilon_{pre})\varepsilon_f \qquad [14.27]$$

where

$$\lambda = \frac{(\bar{E}_{substrate} + \bar{E}_{encapsulation})L_{bridge}}{\pi^3 \overline{Eh}_{bridge}} f\left(\frac{L_{bridge}}{L_{island} + L_{bridge}}\right) \qquad [14.28]$$

is a non-dimensional parameter that depends on the elastic moduli of substrate and encapsulation layer, tensile stiffness of interconnect bridge, and the lengths of device island and interconnect bridge. When there is no substrate and encapsulation layer (i.e., $\lambda = 0$), the stretchability is given by $\varepsilon_{pre} + (1 + \varepsilon_{pre})\varepsilon_f$.

Figure 14.12 shows the stretchability $(\varepsilon_{applied})_{max}$ versus non-dimensional parameter λ for the failure strain $\varepsilon_f = 1\%$ and prestrain $\varepsilon_{pre} = 60\%$. The geometric and material parameters of the bridges are same for Fig. 14.11 and can also be found in Section 14.3.1. The lengths of interconnect bridges and device islands are $L_{bridge} = 460$ μm and $L_{island} = 236$ μm, respectively. The elastic modulus and Poisson's ratio of the substrate are $E_{substrate} = 1.8$ MPa and $\nu_{substrate} = 0.48$. The Poisson's ratio of the encapsulation layer is $\nu_{encapsulation} = 0.48$. When there is no encapsulation layer, the interconnect bridge can deform freely and the stretchability can reach the prestrain 60%. When the encapsulation layer with $E_{encapsulation} = 1.8$ MPa, which corresponds

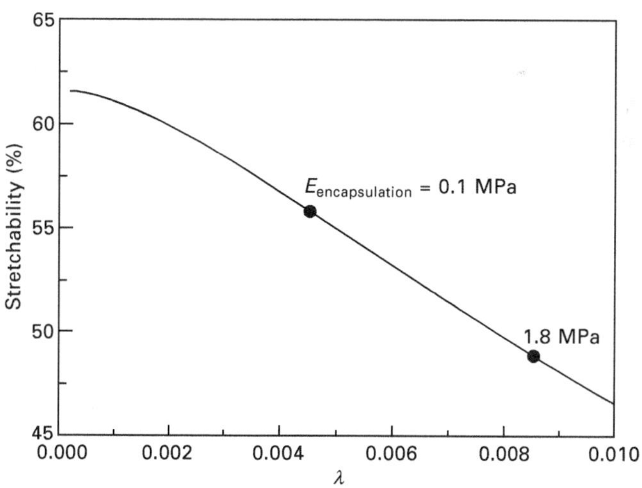

14.12 The stretchability of the encapsulated system versus the non-dimensional parameter, which depends on the material properties and geometry of interconnect bridges, the length of device islands, and material properties of substrate and encapsulation layer. The prestrain is 60%. The points marked on the curves represent the materials properties and geometry used in Section 14.3.1. Reprinted with permission from Wu *et al.* (2010), Copyright 2010 Elsevier Ltd.

to $\lambda = 0.00857$, is used, the stretchability is reduced to 48.8% from Eq. 14.27, which agrees well with the experimental value of 49% (Kim *et al.*, 2009). When a much more compliant encapsulation layer with $E_{\text{encapsulation}} = 0.1$ MPa, which corresponds to $\lambda = 0.00452$, is used, the stretchability becomes 55.8% (closer to the prestrain 60%) from Eq. 14.27, which also agrees well with the experimental measurement of 55% (Kim *et al.*, 2009). Therefore, in order to optimize the stretchability, a compliant encapsulation layer, which helps to retain the stretchability, should be used.

The interconnect bridges described above are straight. To expand the deformability even further, serpentine interconnect bridges may be used (Kim *et al.*, 2009). Figure 14.13 shows an SEM image of such a design after executing the fabrication procedures of Fig. 14.4. Compared to the straight interconnect bridges, the serpentine ones have two major advantages. First, they are much longer than straight interconnect bridges, and therefore can accommodate much larger prestrain. Second, once the applied strain reaches the prestrain, straight interconnect bridges become flat and lose their stretchability, but serpentine interconnect bridges can be stretched much further because large twist will be involved to accommodate larger deformation. Exploring the non-coplanar mesh design with serpentine interconnect bridges represents a fruitful topic for future work.

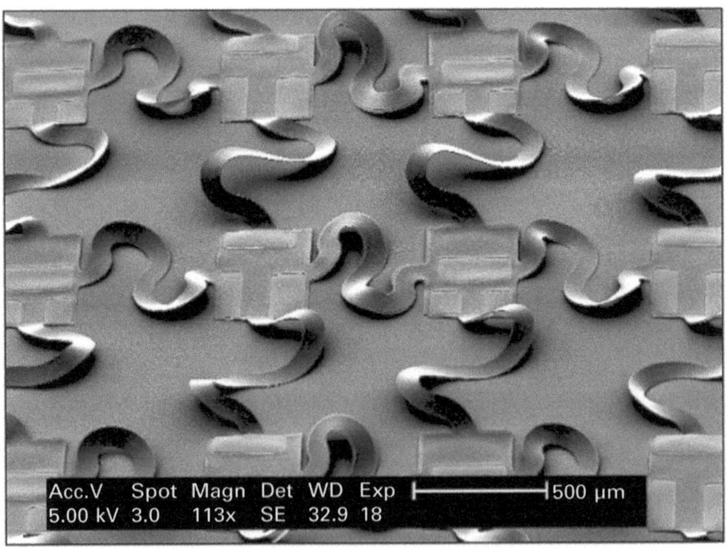

14.13 SEM image of an array of CMOS inverters with non-coplanar mesh design that has serpentine interconnects. Reprinted with permission from Song *et al.* (2009a), Copyright 2009 American Vacuum Society.

14.4 Conclusions

One-dimensional non-coplanar mesh design: A nonlinear buckling model is presented to study the mechanics of this type of film/substrate system. An analytical solution is obtained for the buckling geometry (wavelength and amplitude) and the maximum strain in buckled thin film. The solution agrees very well with the experiments, and shows explicitly how buckling can significantly reduce the thin film strain to achieve the system large stretchability.

Two-dimensional non-coplanar mesh design: Mechanics models are established for the cases of pre-encapsulation and post-encapsulation. The maximum strains in the interconnect bridges are obtained analytically. A simple, analytical expression for the stretchability of the encapsulated system is obtained to help to design the encapsulated electronics via a single, non-dimensional parameter.

14.5 References

Baldo MA, Thompson ME, and Forrest SR, 2000. High-efficiency fluorescent organic light-emitting devices using a phosphorescent sensitizer. *Nature* **403**, 750–753.

Bowden N, Brittain S, Evans AG, Hutchinson JW, and Whitesides GM, 1998. Spontaneous formation ordered structures in thin films of metals supported on an elastomeric polymer. *Nature* **393**, 146–149.

Choi WM, Song J, Khang DY, Jiang H, Huang Y, and Rogers JA, 2007. Biaxially stretchable "wavy" silicon nanomembranes. *Nano Letters* **7**, 1655–1663.

Crawford GP, 2005. *Flexible Flat Panel Display Technology*. John Wiley & Sons, New York.

Crone B, Dodabalapur A, Lin YY, Filas RW, Bao Z, LaDuca A, Sarpeshkar R, Katz HE, and Li W, 2000. Large-scale complementary integrated circuits based on organic transistors. *Nature* **403**, 521–523.

Facchetti A, Yoon MH, and Marks TJ, 2005. Gate dielectrics for organic field-effect transistors: new opportunities for organic electronics. *Advanced Materials* **17**, 1705–1725.

Garnier F, Hajlaoui R, Yassar A, and Srivastava P, 1994. All-polymer field-effect transistor realized by printing techniques. *Science* **265**, 1684–1686.

Huang ZY, Hong W, and Suo Z, 2005. Nonlinear analyses of wrinkles in a film bonded to a compliant substrate. *Journal of Mechanics and Physics of Solids* **53,** 2101–2118.

Jiang H, Sun Y, Rogers JA, and Huang Y, 2007. Mechanics of precisely controlled thin film buckling on elastomeric substrate. *Applied Physics Letters* **90**, 133119.

Jiang H, Khang DY, Fei H, Kim H, Huang Y, Xiao J, and Rogers JA, 2008a. Finite width effect of thin-films buckling on compliant substrate: experimental and theoretical studies. *Journal of Mechanics and Physics of Solids* **56**, 2585–2598.

Jiang H, Song J, Huang Y, and Rogers JA, 2008b. 'Mechanics of stretchable silicon films on elastomeric substrate', in *Unconventional Nanopatterning Techniques and Applications* (eds. Rogers JA and Lee HH), Wiley, Hoboken, NJ, pp. 483–514.

Jin HC, Abelson JR, Erhardt MK, and Nuzzo RG, 2004. Soft lithographic fabrication of an image sensor array on a curved substrate. *Journal of Vacuum Science and Technology* B **22**, 2548–2551.

Khang DY, Jiang HQ, Huang Y, and Rogers JA, 2006. A stretchable form of single-crystal silicon for high-performance electronics on rubber substrate. *Science* **311**, 208–212.

Kim DH, Ahn JH, Choi WM, Kim HS, Kim TH, Song J, Huang Y, Liu ZJ, Lu C, and Rogers JA, 2008a. Stretchable and foldable silicon integrated circuits. *Science* **320**, 507–511.

Kim DH, Choi WM, Ahn JH, Kim HS, Song J, Huang Y, Liu Z, Lu C, Koh CG, and Rogers JA, 2008b. Complementary metal oxide silicon integrated circuits incorporating monolithically integrated stretchable wavy interconnects. *Applied Physics Letters* **93**, 044102.

Kim DH, Song J, Choi WM, Kim HS, Kim RH, Liu Z, Huang Y, Hwang KC, Zhang Y, and Rogers JA, 2008c. Materials and non-coplanar mesh designs for integrated circuits with linear elastic responses to extreme mechanical deformations. *Proceedings of the National Academy of Sciences of the United States of America* **105**, 18675–18680.

Kim DH, Liu Z, Kim YS, Wu J, Song J, Kim HS, Huang Y, Hwang KC, Zhang Y, and Rogers JA, 2009. Optimized structural designs for stretchable silicon integrated circuits. *Small* **5**, 2841–2847.

Ko HC, Stoykovich MP, Song J, Malyarchuk V, Choi WM, Yu CJ, Geddes JB, Xiao J, Wang S, Huang Y, and Rogers JA, 2008. A hemispherical electronic eye camera based on compressible silicon optoelectronics. *Nature* **454**, 748–753.

Koh CT, Liu ZJ, Khang DY, Song J, Lu C, Huang Y, Rogers JA, and Koh CG, 2007. Edge effects in buckled thin films on elastomeric substrates. *Applied Physics Letters* **91,** 133113.

Lacour SP, Jones J, Wanger S, Li T, and Suo Z, 2005. Stretchable interconnects for elastic electronic surfaces. *Proceedings of the IEEE* **93**, 1459–1467.

Loo YL, Someya T, Baldwin KW, Bao Z, Ho P, Dodabalapur A, Katz H, and Rogers JA, 2002. Soft, conformable electrical contacts for organic semiconductors: high-resolution plastic circuits by lamination. *Proceedings of the National Academy of Sciences of the United States of America* **99**, 10252–10256.

Lumelsky VJ, Shur MS, Wagner S, 2001. Sensitive skin. *IEEE Sensors Journal* **1**, 41–51.

Nathan A, Park B, Sazonov A, Tao S, Chan I, Servati P, Karim K, Charania T, Striakhilev D, Ma Q, and Murthy RVR, 2000. Amorphous silicon detector and thin film transistor technology for large-area imaging of X-rays. *Microelectronics Journal* **31**, 883–891.

Sekitani T, Noguchi Y, Hata K, Fukushima T, Aida T, and Someya T, 2008. A rubberlike stretchable active matrix using elastic conductors. *Science* **321**, 1468–1472.

Shin G, Jung I, Malyarchuk V, Song J, Wang S, Ko HC, Huang Y, Ha JS, and Rogers JA, 2010. Micromechanics and advanced designs for curved photodetector arrays in hemispherical electronic eye cameras. *Small* **6**, 851–856.

Someya T, Sekitani T, Iba S, Kato Y, Kawaguchi H, and Sakurai T, 2004. A large-area, flexible pressure sensor matrix with organic field-effect transistors for artificial skin applications. *Proceedings of the National Academy of Sciences of the United States of America* **101**, 9966–9970.

Song J, Jiang H, Liu Z, Khang DY, Huang Y, Rogers JA, Lu C, and Koh CG, 2008a. Buckling of a stiff thin film on a compliant substrate in large deformation. *International Journal of Solids and Structures* **45**, 3107–3121.

Song J, Jiang H, Choi WM, Khang DY, Huang Y, and Rogers JA, 2008b. An analytical study of two-dimensional buckling of thin films on compliant substrates. *Journal of Applied Physics* **103**, 014303.

Song J, Jiang H, Huang Y, and Rogers JA, 2009a. Mechanics of stretchable inorganic electronic materials. *The Journal of Vacuum Science and Technology A* **27**, 1107–1125.

Song J, Huang Y, Xiao J, Wang S, Hwang KC, Ko HC, Kim DH, Stoykovich MP, and Rogers JA, 2009b. Mechanics of non-coplanar mesh design for stretchable electronic circuits. *Journal of Applied Physics* **105**, 123516.

Sosin S, Zoumpoulidis T, Bartek M, Wang L, Dekker R, Jansen KMB, and Ernst LJ, 2008. Free-standing, parylene-sealed copper interconnect for stretchable silicon electronics. *Electronic Components and Technology Conference*, 1339–1345.

Sun Y, Choi WM, Jiang H, Huang Y, Rogers JA, 2006. Controlled buckling of semiconductor nano ribbons for stretchable electronics. *Nature Nanotechnology* **1**, 201–207.

Wang S, Song J, Kim DH, Huang Y, and Rogers JA, 2008. Local versus global buckling of thin films on elastomeric substrates. *Applied Physics Letters* **93**, 023126.

Wu J, Song J, Xiao J, Huang Y, Hwang KC, and Rogers JA, 2010. Mechanics of encapsulated stretchable electronics, *Acta Mechanica Solida Sinica* (in press).

15
The electrocaloric effect (ECE) in ferroelectric polymer films

S.-G. LU and Q. M. ZHANG, The Pennsylvania State University, USA and Z. KUTNJAK, Jozef Stefan Institute, Slovenia

Abstract: The electrocaloric effect (ECE) is the entropy and temperature change of a dielectric when subject to a change of electric field. This chapter will discuss briefly thermodynamic considerations for materials that exhibit large ECE and review previous results of the ECE in polar crystals, ceramics, and thin films. ECE results from direct measurement and deduced from Maxwell's relations will be presented along with the phenomenological modeling for ferroelectric polymers, which have been recently discovered to exhibit a large ECE. Finally the ECE of polar materials are modeled to forecast their potential in future devices.

Key words: electrocaloric effect, ferroelectric polymer, electric displacement–electric field hysteresis loop, isothermal entropy change, adiabatic temperature change.

15.1 Introduction

The electrocaloric effect (ECE) is the temperature change of a material when subject to a change of electric fields (Lines and Glass, 1977; Jona and Shirane, 1993; Lang, 1976). The ECE effect is the inverse of the pyroelectric effect. At constant stress (or strain) and constant electric field, the pyroelectric coefficient ($\mathrm{d}D/\mathrm{d}T$, D = electric displacement, T = temperature) is equal to the electrocaloric coefficient ($\mathrm{d}S/\mathrm{d}E$, S = entropy, E = electric field). This is the Maxwell relation. In addition, secondary pyroelectric effects (Newnham, 2005) caused by change of strain at constant stress may also contribute to the electrocaloric effect.

The ECE may provide an effective means of realizing solid-state cooling devices for a broad range of applications such as on-chip cooling and temperature regulation for sensors or other electronic devices. Refrigeration based on ECE has the potential of reaching high efficiency relative to vapor-compression cycle systems, and with no greenhouse gas emissions.

Solid-state electric-cooling devices based on the thermoelectric effect (Peltier effect) have been used for many decades (Spanner, 1951; Nolas *et al.*, 2001). However, these systems require a large DC current which results in a large amount of waste heat through Joule heating. For example, using the typical coefficient of performance (COP) for these devices, e.g. 0.4 to 0.7, 2.4 to 3.5 W in the form of heat will be generated at the hot end of the system to remove 1 W heat from the cold end. In contrast, no current conducts through the polar materials used in ECE devices, e.g. ferroelectric polymers, thereby eliminating this wasted heat during the refrigeration cycle.

A counterpart of ECE is the magnetocaloric effect (MCE), which has been extensively studied for many years due to the findings of significant MCE in several magnetic materials near room temperature (Gschneidner Jr. *et al.*, 2005; V.K. Pecharsky *et al.*, 2003). Both ECE and MCE devices exploit the change of order parameter brought about by an external electric or magnetic field. However, the difficulty of generating high magnetic field for MCE devices to reach large MCE severely limits their wide application, especially for miniaturized microelectronic devices. In contrast, high electric field can easily be generated and manipulated, which makes it relatively easy to configure ECE cooling devices for a broad range of applications.

This chapter introduces the basic concept of ECE, the thermodynamic considerations for materials with large ECE, and reviews previous investigations of the ECE in polar crystals, ceramics, and thin films. A newly discovered large ECE exhibited by ferroelectric polymers is presented in terms of the direct measurements, and calculations based on Maxwell relations and phenomenological theory. Finally the prospect of ECE devices using polar materials will be discussed.

15.2 Thermodynamic considerations on materials with large electrocaloric effect (ECE)

15.2.1 Maxwell relations

In general the Gibbs free energy *G* for a dielectric material could be expressed as a function of temperature *T*, entropy *S*, stress *X*, strain *x*, electric field *E* and electric displacement *D* in the form

$$G = U - TS - X_i x_i - E_i D_i \qquad [15.1]$$

where *U* is the internal energy of the system, the stress and field terms are written using Einstein notation. The differential form of Eq. 15.1 could be written as

$$dG = -SdT - x_i dX_i - D_i dE_i \qquad [15.2]$$

Entropy *S*, strain x_i and electric displacement D_i can be easily expressed when the other two variables are assumed to be constant,

$$S = -\left(\frac{\partial G_1}{\partial T}\right)_{X,D}, x_i = -\left(\frac{\partial G_1}{\partial X_i}\right)_{T,D}, D_i = -\left(\frac{\partial G_1}{\partial E_i}\right)_{T,X} \qquad [15.3]$$

The Maxwell relation can be derived for (*S*, *T*) and (*D*, *E*) two pairs of parameters (Lines and Glass, 1977),

$$\left(\frac{\partial S}{\partial E_i}\right)_{T,X} = \left(\frac{\partial D_i}{\partial T}\right)_{E,X} \quad [15.4]$$

or

$$\left(\frac{\partial T}{\partial E}\right)_S = \frac{T}{c_E}\left(\frac{\partial D}{\partial T}\right)_E = \frac{Tp_E}{c_E} \quad [15.5]$$

where c_E is the heat capacity, p_E the pyroelectric coefficient. Eqs 15.4 and 15.5 indicate the mutually inverse relationships of ECE and the pyroelectric coefficient. Hence, for the ECE materials with a constant stress *X* imparted, the isothermal entropy change ΔS and adiabatic temperature change ΔT can be expressed as (Lines and Glass, 1977):

$$\Delta S = -\int_{E_1}^{E_2}\left(\frac{\partial D}{\partial T}\right)_E = \mathrm{d}E \quad [15.6]$$

$$\Delta T = -\frac{T}{\rho}\int_{E_1}^{E_2}\frac{1}{c_E}\left(\frac{\partial D}{\partial T}\right)_E \mathrm{d}E \quad [15.7]$$

Equations 15.4–15.7 indicate that in order to achieve large ΔS and ΔT, the dielectric materials should posses a large pyroelectric coefficient over a relatively broad electric field and temperature range. For ferroelectric materials, a large pyroelectric effect exists near the ferroelectric (F)–paraelectric (P) phase transition temperature and this large effect may be shifted to temperatures above the transition temperature when an external electric field is applied. It is also noted that a large ΔT may be achieved even if ΔS is small when the c_E of a dielectric material is small. However, as will be pointed out in the following section, this is not desirable for practical refrigeration applications where a large ΔS is required.

Although a few studies on the ECE were conducted in which direct measurement of ΔT was made (Sinyavsky *et al.*, 1989; Xiao *et al.*, 1998), most experimental studies were based on the Maxwell relations where the electric displacement *D* versus temperature *T* under different electric fields was characterized. ΔS and ΔT were deduced from Eqs 15.6 and 15.7 (see below for details). For dielectric materials with low hysteresis loss and the measurement is in an ideal situation, results obtained from the two methods should be consistent.

In an ideal refrigeration cycle the working material (refrigerant) must absorb entropy (or heat) from the cooling load while in thermal contact with the load (isothermal entropy change ΔS). The material is then isolated from the load while the temperature is increased due to the application of external field (adiabatic temperature change ΔT). The material is then in thermal contact with the heat sink and entropy that was absorbed from the cooling load is rejected to the heat sink. The working material is then isolated from the heat sink and the temperature is reduced back as the field is reduced. The temperature of the refrigerant will be the same as the temperature of the cooling load when they are contacted. The whole process is repeated to further reduce the temperature of the load. Therefore, both the isothermal entropy change ΔS and the adiabatic temperature change ΔT are key parameters for the ECE of a dielectric material for refrigeration (Wood and Potter, 1985).

15.2.2 Phenomenological theory of ECE

Phenomenological theory has been widely utilized to illustrate the macroscopic phenomena that occur in the polar materials, e.g. ferroelectric or ferromagnetic materials near their phase transition temperatures. In principle, the ECE is one of the characteristics of ferroelectric materials that is associated with the phase transition in terms of the order–disorder transition derived entropy change. Therefore, phenomenological theory can be used to estimate the ECE of ferroelectrics. The general form of the Gibbs free energy in terms of the electric displacement can be expressed as (Lines and Glass, 1977):

$$G = \frac{1}{2}\alpha D^2 + \frac{1}{4}\xi D^4 + \frac{1}{6}\zeta D^6 \qquad [15.8]$$

where $\alpha = \beta(T - T_0)$, and β, ξ and ζ are temperature-independent phenomenological coefficients. Since $(\partial G/\partial T)_D = -\Delta S$, one can obtain

$$\Delta S = -\frac{1}{2}\beta D^2 \qquad [15.9]$$

Then the adiabatic temperature change ΔT (= $T\Delta S/c_E$) can be obtained, i.e.

$$\Delta T = -\frac{1}{2c_E}\beta T D^2 \qquad [15.10]$$

Based on Eqs 15.9 and 15.10, the entropy will be reduced when the material changes to a polar state from a non-polar state when an external action, e.g. temperature, electric field or stress, is applied. The entropy change and temperature change are associated with the phenomenological coefficient β and electric displacement D, namely, proportional to β and D^2. Both parameters will affect the ECE values of the materials. A material with large β and large D will generate large ECE entropy change and temperature change near the ferroelectric (F)–paraelectric (P) phase transition temperature.

15.2.3 ECE in several ferroelectric materials

Based on the literature reported values of β and D, the ECE values of various ferroelectric materials could be estimated. For instance, for $BaTiO_3$, $\beta = 6.7 \times 10^5$ ($JmC^{-2}K^{-1}$) and $D = 0.26$ C/m^2 (Jona and Shirane, 1993; Furukawa, 1984), ΔS will be approximately 3 J/(kgK). Using the specific heat $c_E = 4.07 \times 10^2$ J/(kgK) and $T_c = 107$°C (Jona and Shirane, 1993; Akcay *et al.*, 2007), results in a $\Delta T = 2.8$°C. Similarly, for $Pb(Zr_xTi_{1-x})O_3$ ($0.0 < x \leq 0.6$), $\beta = 1.88 \times 10^5$ and $D = 0.39$ C/m^2 (Amin *et al.*, 1981a, 1981b), one will obtain $\Delta S = 1.8$ J/(kgK). Taking $T_c = 250$°C, and $c_E = 3.4 \times 10^2$ J/(kgK) (PI Ceramic, 2009), will result in $\Delta T = 2.7$°C.

For ferroelectric polymers, e.g. P(VDF-TrFE), phenomenological theory predicts large ECE values. For example, P(VDF-TrFE) 65/35 mol% copolymer, with $\beta = 3.5 \times 10^7$ $JmC^{-2}K^{-1}$ and $D = 0.08$ C/m^2 (Furukawa, 1984), will exhibit a $\Delta S = 62$ J/(kgK). Making use of its specific heat capacity $c_E = 1.4 \times 10^3$ J/(kgK) (Furukawa *et al.*, 2006) and Curie temperature $T_c = 102$°C (Furukawa, 1984), yields $\Delta T = 16.6$°C. The large ΔS and ΔT values suggest that a large ECE may be achieved in ferroelectric P(VDF-TrFE) copolymers. Furthermore, relaxor ferroelectric polymers based on P(VDF-TrFE), such as P(VDF-TrFE-CFE) 59.2/33.6/7.2 mol% (CFE-chlorofluoroethylene) relaxor ferroelectric terpolymers also have potential to reach a large ECE because the β and D are still large.

It was found that β of ferroelectric ceramics ($\sim 10^5$) is about two orders of magnitude smaller than that of P(VDF-TrFE) ($\sim 10^7$). D, however, is only several times higher for ceramics, since $\Delta S \sim \beta D^2$, ΔS is still about one order of magnitude smaller than that of the P(VDF-TrFE)-based polymers.

In addition, the heat of phase transition can also be used to estimate the ECE ($Q = T\Delta S$) in the material. For a very strong order–disorder ceramic system (an example of which is the ferroelectric ceramic triglycine sulphate, TGS), the heat of F–P phase transition is 2.0×10^3 J/kg (corresponding to an entropy change of $\Delta S \sim 6.1$ J/(kgK)). For $BaTiO_3$, F–P heat is smaller 9.3×10^2 J/kg ($\Delta S \sim 2.3$ J/(kgK)) (Jona and Shirane, 1993). In other words, although ceramic materials may exhibit a higher adiabatic temperature change, their isothermal entropy change is not very high. In contrast, ferroelectric polymers offer more heat in a phase transition, e.g. P(VDF-TrFE) 68/32 mol% copolymer, the heat of F–P transition is more than 2.1×10^4 J/kg (or $\Delta S \sim 56.0$ J/(kgK)) (Neese *et al.*, 2008). This is approximately 10 times larger than its inorganic counterparts.

15.3 Previous investigations on electrocaloric effect (ECE) in polar materials

15.3.1 ECE studies in ferroelectric ceramics and single crystals

The history of ECE study may be traced back to as early as the 1930s. In 1930, Kobeko and Kurtschatov did a first investigation on ECE in Rochelle salt (Kobeko and Kurtschatov, 1930); however, they did not report any numerical values. In 1963, Wiseman and Kuebler redid their measurements (Wiseman and Kuebler, 1963), obtaining $\Delta T = 0.0036°C$ in an electric field of 1.4 kV/cm at 22.2°C. In their study, the Maxwell relation was used to derive ΔT:

$$\Delta T = -\frac{T}{c_E}\frac{\partial \alpha}{\partial T} D\Delta D$$

where $\alpha = 1/\varepsilon$ as defined in Eq. 15.8 and ε is the permittivity). The isothermal entropy change was 28.0 J/m^{3}K (1.56×10^{-2} J/(kgK)).

Other studies on inorganic materials used KH_2PO_4 crystal, and $SrTiO_3$, $Pb(Sc_{0.5}Ta_{0.5})O_3$, and $Pb_{0.98}Nb_{0.02}(Zr_{0.75}Sn_{0.20}Ti_{0.05})_{0.98}O_3$ ceramics. For KH_2PO_4 crystal, the Maxwell relation was used in the form of

$$\Delta T = -(T/c_E)\left(\frac{\partial D/\partial T}{\partial D/\partial E}\right)dD$$

to obtain $\Delta T = 1°C$ for a 11 kV/cm electric field and an entropy change of 2.31×10^3 J/(m^{3}K) (or 0.99 J/(kgK)) (Baumgartner, 1950). For $SrTiO_3$, $\Delta T = 1°C$ and $\Delta S = 34.63$ J/(m^{3}K) (6.75×10^{-3} J/(kgK)) under 5.42 kV/cm electric field at 4 K from Eq. 15.7 (Lawless and Morrow, 1977). For $Pb(Sc_{0.5}Ta_{0.5})O_3$, a $\Delta T = 1.5°C$ and ΔS of 1.55×10^4 J/m^{3}K (1.76 J/(kgK)) were measured directly for a sample under 25 kV/cm field at 25°C (Sinyavsky and Brodyansky, 1992). For $Pb_{0.98}Nb_{0.02}(Zr_{0.75}Sn_{0.20}Ti_{0.05})_{0.98}O_3$, $\Delta T = 2.5°C$ and $\Delta S = 1.73 \times 10^4$ J/(m^{3}K) (2.88 J/(kgK)) at 30 kV/cm and 161°C deduced from Eq. 15.7 (Tuttle and Payne, 1981).

A direct ECE measurement was carried out for $(1\text{-}x)Pb(Mg_{1/3}Nb_{2/3})O_3\text{-}xPbTiO_3$ ($x = 0.08, 0.10, 0.25$) ceramics near room temperature using a thermocouple when a DC electric field was applied (Xiao *et al.*, 1998). A temperature change of 1.4°C was observed for $x = 0.08$ although at high temperatures (as x increased), this change was reduced. This high ECE can be accounted for by considering the electric field-induced first-order phase transition from the mean cubic phase to 3 m phase.

These results indicate that the ECE in ceramic and single crystal materials is relatively small, i.e., $\Delta T < 2.5°C$, and $\Delta S < 2.9$ J/(kgK), mainly because the breakdown field is low, using applied electric fields that are less than

3 MV/m. Defects existing in bulk materials cause early breakdown and empirically the breakdown electric field was inversely proportional to the material's thickness. For piezoelectric ceramics, the breakdown field (in kV/cm) is related to the thickness (in cm) via the relationship, $E_b = 27.2t^{-0.39}$, indicating that thin films are more appropriate for an ECE study. Additionally, the breakdown field of dielectric polymers can be several orders of magnitude higher than ceramics, suggesting polar-polymers are good candidates for ECE investigations.

15.3.2 ECE in ferroelectric and antiferroelectric thin films

In 2006, Mischenko *et al.* investigated ECE in sol-gel derived antiferroelectric $PbZr_{0.95}Ti_{0.05}O_3$ thin films near the F–P transition temperature. In their study, films with 350 nm thickness were used to allow for electric fields as high as 48 MV/m. An adiabatic temperature change of 12°C was obtained (as deduced from Eq. 15.7 at 226°C, slightly above the phase transition temperature (222°C)) (Mischenko *et al.*, 2006). Both the high electric field and the high operating temperature near phase transition contribute to the large ΔT (= $T\Delta S/c_E$). On the other hand, its isothermal entropy change is estimated to be 8 J/(kgK), which is fairly low compared with magnetic alloys exhibiting a large magnetocaloric effect (MCE) near room temperature, where $\Delta S \geq 30$ J/(kgK) was observed (Provenzano *et al.*, 2004). As stated previously, for high performance refrigerants, a large ΔS is necessary (Wood and Potter, 1985).

To reduce the operational temperature for large ECE in ceramic thin films, Correia *et al.* successfully fabricated $PbMg_{2/3}Nb_{1/3}O_3$-$PbTiO_3$ thin films with perovskite structure using $PbZr_{0.8}Ti_{0.2}O_3$ seed layer on Pt/Ti/TiO_2/SiO_2/Si substrates (Correia *et al.*, 2009). A temperature change of $\Delta T = 9$ K was achieved at 25°C. An entropy change of 9.7 J/(kgK) can be deduced. A significant difference for ferroelectric thin films is that the largest ΔT occurs at 25°C, near the depolarization temperature (18°C), not above the permittivity peak temperature. The large ECE only happens at field heating. Transitions for stable and metastable polar nanoregions (PNR) to nonpolar regions are accounted for by observed phenomena. Interactions of PNRs are similar to that between the dipoles in a glass. The field-induced phase transition has been observed in PMN-PT single crystals (Lu *et al.*, 2005; Ye and Schmid, 1993). Thermal history has a critical impact on the field-induced phase transition. Relaxor ferroelectrics are of great interest due to their phase transition temperatures being near or at room temperature. The field induced phase transition may produce larger polarization, e.g. induced polarization, $\langle P_d \rangle$, which can lead to larger dP/dT as well as large ΔS and ΔT.

For thin film, the substrate must be taken into account as it may exert compressive and tensile stresses on the thin film due to the misfit of the lattices. The free energy of thin film is subject to lateral clamping and may be expressed as (Akcay *et al.*, 2007):

$$G_{\text{film}} = G_0 + \tilde{\alpha}D^2 + \tilde{\xi}D^4 + \tilde{\zeta}P^6 - EP + G_{\text{strain}} \qquad [15.11]$$

where

$$\tilde{\alpha} = \alpha - 2u_{\text{m}}Q_{12}\tilde{C} \qquad [15.12]$$

and

$$\tilde{\beta} = \beta + Q_{12}^2\tilde{C} \qquad [15.13]$$

are the modified phenomenological coefficients, $G_{\text{strain}} = u_{\text{m}}^2\tilde{C}$ is the polarization-free strain energy, $\tilde{C} = C_{11} + C_{12} - 2C_{12}^2/C_{11}$, C_{ij} are the elastic constants at constant polarization, Q_{ij} are the cubic electrostrictive coefficients, and u_{m} is the in-plane misfit strain. The phase transition temperature varies linearly with the lattice misfit strain via Eq. 15.12 while the two-dimensional clamping is illustrated by Eq. 15.13. The excess entropy S_{E}^{XS} and specific heat ΔC_{E} of the ferroelectric phase transition follow the form (Akcay *et al.*, 2007):

$$S_{\text{E}}^{\text{XS}}(T,E) = -T\left(\frac{\partial G(D)}{\partial T}\right)_{\text{E}} \qquad [15.14]$$

$$\Delta C_{\text{E}}(T,E) = -T\left(\frac{\partial^2 G(D)}{\partial T^2}\right)_{\text{E}} \qquad [15.15]$$

It was found that for $BaTiO_3$ (BTO) thin film deposited on substrate, perfect lateral clamping of BTO will transform the discontinuous phase transition (first-order phase transition) into a continuous one. Accordingly the polarization and the specific heat capacity will be reduced near the phase transition temperature. On the other hand, based on Eqs 15.12 and 15.13, adjustment of misfit strain in epitaxial ferroelectric thin films may vary the magnitude and temperature dependencies of their ECE properties.

15.4 Large electrocaloric effect (ECE) in ferroelectric polymer films

15.4.1 ECE obtained from Maxwell relations

As estimated in Section 15.2, the ferroelectric copolymer may produce large ECE near its phase transition temperature. P(VDF-TrFE) 55/45 mol% was chosen because its F–P phase transition is of second-order (continuous), thus avoiding the thermal hysteresis effect associated with the first-order phase transition. In addition, among all available P(VDF-TrFE) copolymers, this composition exhibits the lowest F–P phase transition temperature (~ 70°C), which is favorable for refrigeration near room temperature.

Polymer films used for ECE measurement were prepared using a spin-casting method. Powders were dissolved in a dimethyl formamide (DMF) solvent at a concentration of 12 wt%. Thin films were obtained by filtering the solutions through a 1 μm pore size polytetrafluoroethylene (PTFE) filter onto an Al/Cr coated glass substrate and were spin coated at about 2000 rpm for 2 minutes. The resulting film thickness was in the range of 0.4 μm to 2 μm. Samples were subsequently annealed in a vacuum oven at 140°C for 2 hours for copolymer to further remove the solvent and improve the crystallinity. An aluminum coating was then evaporated on the polymer film surface to serve as a top electrode (1 × 1 cm^2). The dielectric properties as a function of temperature were characterized using a multi-frequency LCR Meter (HP 4284A) equipped with a temperature chamber. The electric displacement–electric field (D–E) loops at different temperatures were measured using a Sawyer–Tower circuit with a temperature chamber. The differential scanning calorimetry (DSC) data were taken using a TA Instrument (TA Q100) which was also used in modulated mode to measure the specific heat capacity.

Figure 15.1 shows the permittivity as a function of temperature for P(VDF-TrFE) 55/45 mol% copolymers measured at 1 kHz. It can be seen that hysteresis is pretty small (~1°C). The remanent polarization as a function of temperature shown in Fig. 15.2 further indicates a second-order phase transition occurred in the material. The phase transition temperature is about 70°C, and the glass transition temperature is about –20°C. At temperatures higher than 100°C, the loss tangent rises sharply, which is associated with

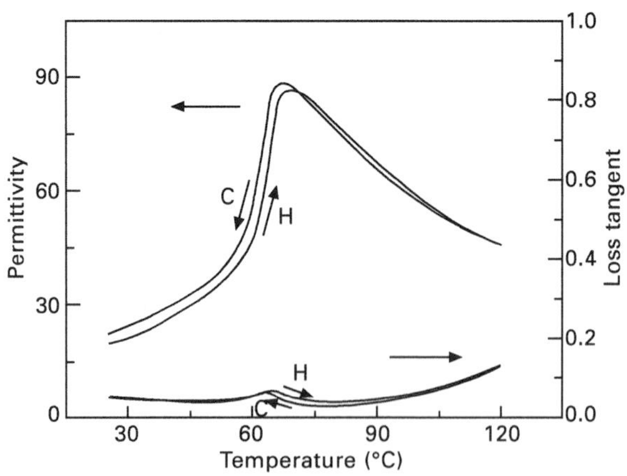

15.1 Permittivity as a function of temperature for P(VDf-TrFE) 55/45 mol% copolymers.

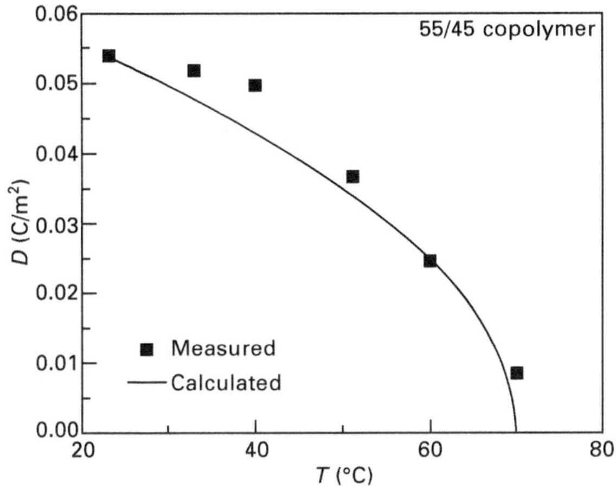

15.2 Remanent polarization as a function of temperature for P(VDf-TrFE) 55/45 mol% copolymers.

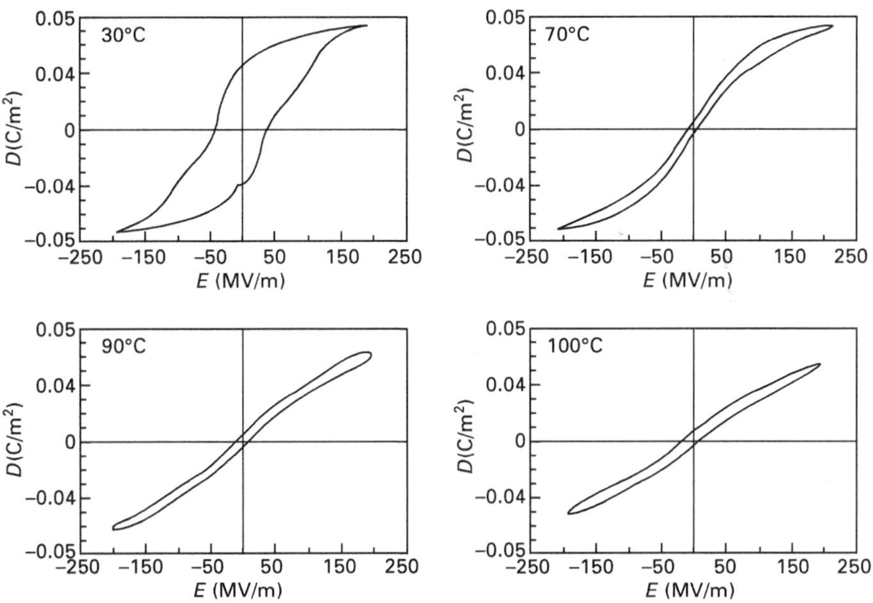

15.3 Electric displacement–electric field hysteresis loops at temperatures below and above the phase transition.

the thermally activated ionic conduction. Hence the electric displacement as a function of electric field at different temperatures can be procured, which is sketched in Fig. 15.3 (Neese, 2009).

At temperatures below the transition temperature, the polymer film is in a ferroelectric state, the normal hysteresis loop is observed while at higher temperatures, the loop becomes slimed, remanent polarization diminishes, and saturation polarization still exists, which is indicative of polarization changing with temperature. The measured data are shown in Fig. 15.4. One can see that the electric displacement monotonically decreases with temperature above the phase transition. The Maxwell relations could be used to calculate the isothermal entropy change and adiabatic temperature change as a function of ambient temperature. The results are demonstrated in Figs 15.5 and 15.6.

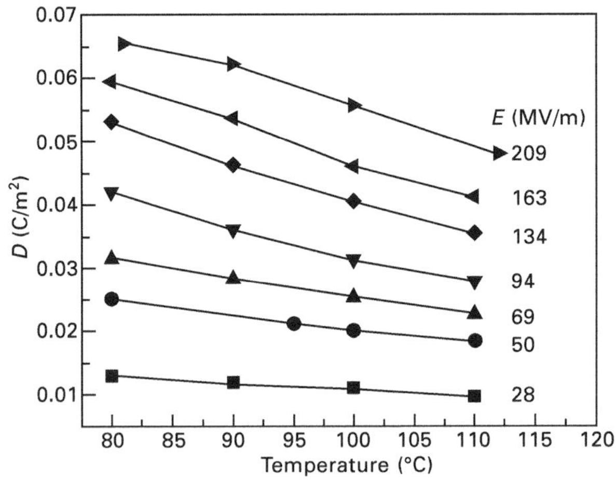

15.4 Electric displacement as a function of temperature at different electric fields for P(VDF-TrFE) 55/45 mol% copolymers.

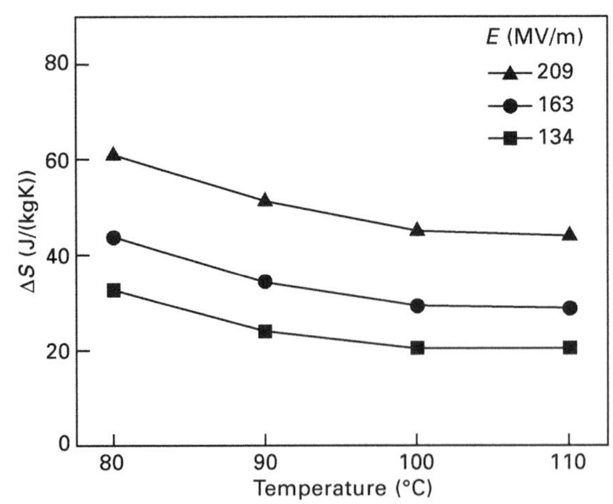

15.5 Isothermal entropy changes as a function of ambient temperature at different electric fields.

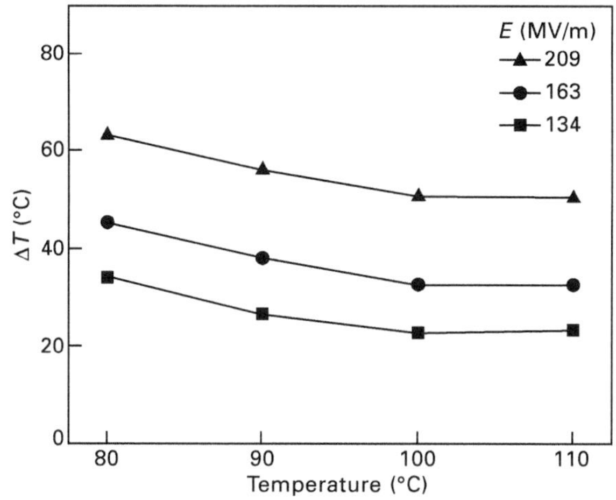

15.6 Adiabatic temperature changes as a function of ambient temperature at different electric fields.

15.4.2 Phenomenological calculations on ECE

It has been proven by many scholars including ourselves (see Fig. 15.2) that the F–P phase transition of P(VDF-TrFE) 55/45 copolymer is of second order. For the copolymer with second-order phase transition, free energy can be written as

$$G = G_0 + \frac{1}{2}\beta(T - T_c)P^2 + \frac{1}{4}\xi P^4 - EP \tag{15.16}$$

where G_0 is the free energy of paraelectric phase, β and ξ are phenomenological coefficients, that are assumed temperature independent, T_c is the Curie temperature, E the electric field, and P the polarization.

Differentiating G with respect to P provides a relationship between the electric field and the polarization,

$$E = \beta(T - T_c)P + \xi P^3 \tag{15.17}$$

When $E = 0$, one may obtain

$$P^2 = \beta(T - T_c)/\xi \tag{15.18}$$

Further differentiating Eq. 15.17 yields the reciprocal permittivity,

$$\frac{1}{\varepsilon} = \beta(T - T_c) + 3\xi P^2 (T < T_c) \tag{15.19}$$

$$\frac{1}{\varepsilon} = \beta(T - T_c)\ (T \geq T_c) \tag{15.20}$$

Using Eqs 15.18 and 15.20, the permittivity versus temperature (Fig. 15.1), and the polarization versus temperature relationships (Fig. 15.2), β and ξ can be obtained. Their values are $\beta = 2.4 \times 10^7$ (Jm $C^{-2}K^{-1}$), and $\xi = 3.9 \times 10^{11}$ (Jm^5 C^{-4}).

Now we may use Eq. 15.17 to calculate the polarization as a function of temperature under various DC bias fields. Before doing the calculation, it should be noted that the F–P transition temperature is a function of DC bias field. This relationship was obtained by directly measuring the permittivity as a function of temperature in different DC bias fields. The results are shown in Fig. 15.7.

However, for $E_{DC} > 100$ MV/m, it is hard to pursue the dielectric measurement. The relationship of $\Delta T_c - E^{2/3}$ (Lines and Glass, 1977) was fitted and extrapolated to get the T_c at $E > 100$ MV/m.

The calculated electric displacement versus temperature relationships with various DC biases are shown in Fig. 15.8. Based on the D-T data, the ΔS and ΔT can be calculated via Eqs 15.6 and 15.7. The results are shown in Figs 15.9 and 15.10.

Phenomenological calculation indicates that there is a large ECE exhibited by P(VDF-TrFE) 55/45 copolymers. The ΔS and ΔT can reach 70 J/(kgK) and 15°C respectively near the phase transition temperature of 68°C. It can also be seen that ΔS has a linear relationship with D^2 (or P^2), the slope being $1/2\beta$.

15.4.3 ECE procured from direct measurements

Direct measurement of ECE is imperative for characterizing ECE materials as refrigerants to be used in solid state cooling devices. There are several

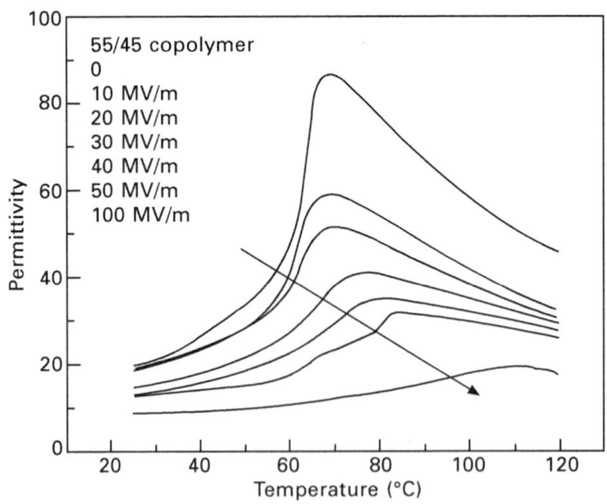

15.7 Permittivity as a function of temperature at 1 kHz in various DC bias fields for 55/45 copolymer.

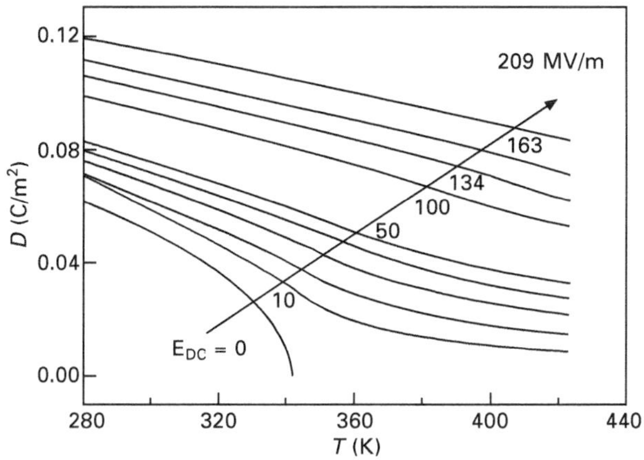

15.8 Polarization versus temperature relationships with various DC biases for 55/45 copolymer.

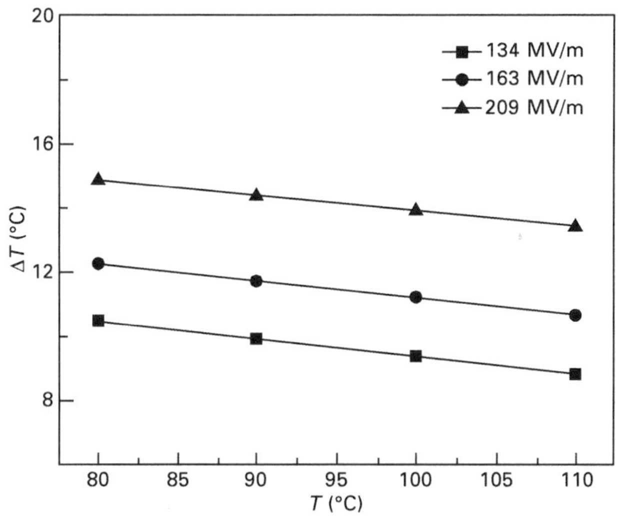

15.9 ECE temperature changes versus temperature for 55/45 copolymer.

methods that have been used to measure the magnetocaloric effect (MCE) in terms of measuring the isothermal entropy change and adiabatic temperature change, such as thermocouple (Dinesen *et al.*, 2002; Lin *et al.*, 2004; Spichkin *et al.*, 2007), thermometer (Gopal *et al.*, 1997), and calorimeter (Tocado *et al.*, 2005; Pecharsky *et al.*, 1997). Here we introduce a method that has been used to directly measure the ECE in P(VDF-TrFE) 55/45 mol% copolymers.

A high resolution calorimeter measured the sample temperature variation

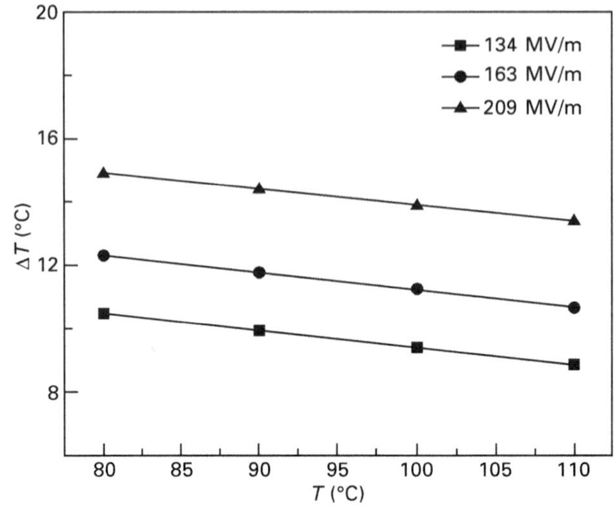

15.10 ECE entropy changes versus temperature for 55/45 copolymer.

due to ECE when an external electric field was applied (Yao *et al.*, 1998). The temperature of the bath in which the calorimeter was located was precisely controlled within 0.1 mK. The temperature signal was measured by a small bead thermistor. A step-like pulse was generated by the functional generator to heat the sample in a quantified amount, and the width of the pulse was chosen so that the sample can reach thermal equilibrium with surrounding bath. Due to the fast electric as well as thermal response (ECE) of the polymer films (in the order of tens of milliseconds; Furukawa, 1989), a simple zero-dimensional model to describe the thermal process can be applied with sufficient accuracy. In a relaxation mode, the temperature $T(t)$ of the whole sample system can be measured, which has an exponential relationship with time, i.e.

$$T(t) = T_{\text{bath}} + \Delta T e^{-t/\tau} \tag{15.21}$$

where T_{bath} is the temperature of the bath, ΔT the temperature change of the polymer film. The total temperature change ΔT_{EC} of the whole sample system was measured, which can be expressed as $\Delta T_{\text{EC}} = \Delta T \sum C_{\text{p}}^{\text{i}} / C_{\text{p}}^{\text{EC}}$. Here, C_{p}^{i} represents the heat capacity of each subsystem, C_{p}^{EC} is the heat capacity of the polymer film covered with electrode. ΔT_{EC} was measured as a function of temperature at constant electric field and as a function of electric field at constant temperature. Typical results for P(VDF-TrFE) 55/45 mol% copolymers are presented in Figs 15.11 and 15.12. The direct measured ECE results are consistent with the results derived from Maxwell relations and the phenomenological calculations.

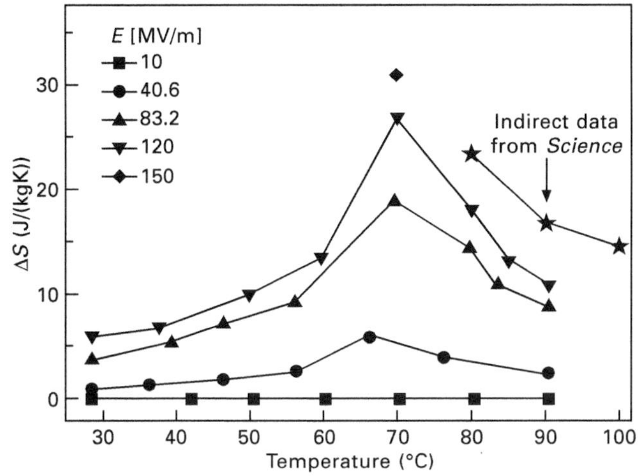

15.11 ECE entropy change as a function of temperature for P(VDF-TrFE) 55/45 mol% copolymers.

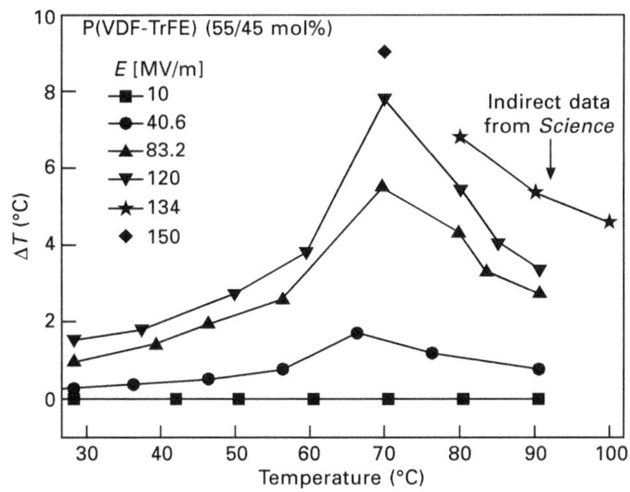

15.12 ECE temperature change as a function of temperature for P(VDF-TrFE) 55/45 mol% copolymers.

15.5 Future trends

Recently reported ECE results for inorganic ferroelectric PZT thin films, PMN-PT thin films, and organic P(VDF-TrFE) based copolymer films indicate that the adiabatic temperature change and isothermal entropy change are over 10°C and 30 J/(kgK), respectively, which demonstrate intriguing prospects for solid state refrigeration devices. On the one hand, high quality inorganic ferroelectric thin films possess large polarization, especially induced polarization coming from fluctuating dipoles in relaxor ferroelectrics, and field-induced phase transition. On the other hand, high quality organic films have a high breakdown field and a moderate polarization of the dipoles which make them generate large ECE. These two categories of materials can be tailored using doping for inorganic materials, or doping or electron irradiation for organic materials so that the phase transition occurs at room temperature. Direct ECE measurements indicated that the isothermal entropy change and adiabatic temperature change are comparable with those in very large magnetocaloric effect materials, e.g. $Gd_5Si_2Ge_2$ (ΔT = 21 K, ΔS = 24 J/(kgK) @ 270 K (A.O. Pecharsky *et al.*, 2003)). Compared with MCE devices, ECE devices will have smaller size and more flexible designs for a broader range of uses.

Prototype refrigerators using ceramic ECE materials have been tested by a few research groups starting from the 1980s (Sinyavsky and Brodyansky, 1992; Kucherov, 1997; Mathur and Mischenko, 2006). The low breakdown electric field of the materials used resulted in low ECE preventing them from realizing practical devices. Recently discovered polymer materials and inorganic ferroelectrics may be coupled with thin film heat switches (e.g. liquid crystal; Epstein and Malloy, 2009), and work as refrigerators. Conventional solid-state cooling devices based on the thermoelectric effect (Peltier effect) possess very low efficiency, as measured by large energy loss in moving the heat from the cold end to the hot end. For this reason, the solid-state cooling devices based on the very large ECE materials offer the potential to achieve a much higher efficiency.

In addition, the order-parameter induced large ECE and MCE in ferroelectrics and ferromagnetics also suggest the possibility of achieving large entropy change associated with temperature change induced by elastic stress (elastocaloric effect) in polymers. This approach can be exploited in cooling devices.

15.6 Conclusion

The electrocaloric effect was introduced, and the history of investigations on the ECE in inorganic materials was reviewed. General considerations for polar materials to achieve larger ECE were presented in terms of phenomenological theory analysis. It was found that the ferroelectric polymers may generate large entropy changes as well as large temperature change based on the phenomenological estimation. Ferroelectric P(VDF-TrFE) 55/45 mol% copolymers films were prepared and the ECE was measured in terms of Maxwell relations and direct measurements. An adiabatic temperature change over 12°C and an isothermal entropy change over 30 J/(kgK) were obtained, consistent with the phenomenological estimation. The prospect of using large ECE materials to drive solid-state refrigeration is believed to be imperative for obtaining high efficiency devices.

15.7 Acknowledgements

The work at Penn State University was supported by the US Department of Energy, Division of Materials Sciences, under Grant No. DE-FG02-07ER46410. The work at Jozef Stefan Institute was supported by the Slovenian Research Agency. The authors thank B. Neese, B. Chu, Y. Wang, and E. Furman for their contributions to ECE work, and Lee J. Gorny for revising the manuscript.

15.8 References

Akcay G, Alpay S P, Mantese J V, and Rossetti Jr. G A (2007), 'Magnitude of the intrinsic electrocaloric effect in ferroelectric perovskite thin films at high electric fields', *Appl Phys Lett*, 90, 252909/–252909/3.

Amin A, Cross L E, and Newnham R E (1981a), 'Calorimetric and phenomenological studies of the $PbZrO_3$-$PbTiO_3$ system', *Ferroelectrics*, 37, 647–650.

Amin A, Newnham R E, Cross L E, and Cox D E (1981b), 'Phenomenological and structural study of a low-temperature phase-transition in the $PbZrO_3$-$PbTiO_3$ system', *J Solid State Chem*, 37, 248–255.

Baumgartner H (1950), 'Elektrische Sättigungserscheinungen und elektrokalorischer Effect von Kaliumphosphat KH_2PO_4', *Helv Phys Acta*, 23, 651–696.

Correia T M, Young J S, Whatmore R W, Scott J F, Mathur N D, and Zhang Q (2009), 'Investigation of the electrocaloric effect in a $PbMg_{1/3}Nb_{2/3}O_3$-$PbTiO_3$ relaxor thin film', *Appl Phys Lett*, 95, 182904/1–182904/3.

Dinesen A R, Linderoth S, and Mørop S (2002), 'Direct and indirect measurement of the magnetocaloric effect in a $La_{0.6}Ca_{0.4}MnO_3$ ceramic perovskite', *J Magn Magn Mater*, 253, 28–34.

Epstein R and Malloy K J (2009), 'Electrocaloric devices based on thin-film heat switches', *J Appl Phys*, 106, 064509/1–064509/7.

Furukawa, T (1984), 'Phenomenological aspect of a ferroelectric vinylidene fluoride/trifluoroethylene copolymer', *Ferroelectrics*, 57, 63–72.

Furukawa T (1989), 'Piezoelectricity and pyroelectricity in polymers', *IEEE Trans Electr Ins*, 24, 375–394.

Furukawa T, Nakajima T, and Takahashi Y (2006), 'Factors governing ferroelectric switching characteristics of thin VDF/TrFE copolymer films', *IEEE Trans Diel Electr Ins*, 13, 1120–1131.

Gopal B R, Chahine R, and Bose T K (1997), 'A sample translatory type insert for automated magnetocaloric effect measurements', *Rev Sci Instrum*, 68, 1818–1822.

Gschneidner Jr K A, Pecharsky V K, and Tsokol A O (2005), 'Recent developments in magnetocaloric materials', *Rep Prog Phys*, 68, 1479–1539.

Jona F and Shirane G (1993), *Ferroelectric Crystals*, Dover Publications, New York.

Kobeko Von P and Kurtschatov J (1930), 'Dielektrische Eigenschaften der Seignettesalzkristalle', *Z Phys*, 66, 192–205.

Kucherov Y R (Thermodyne, Inc.) 1997, *Piezo-pyroelectric energy converter and method*, US Patent, US 5644184, 1 July 1997.

Lang S B (1976), 'Cryogenic refrigeration utilizing electrocaloric effect in pyroelectric lithium sulfate monohydrate', *Ferroelectrics*, 11, 519–523.

Lawless W N and Morrow A J (1977), 'Specific heat and electrocaloric properties of a $SrTiO_3$ ceramic at low temperatures', *Ferroelectrics*, 15, 159–165.

Lin G C, Xu C D, and Zhang J X (2004), 'Magnetocaloric effect in $La_{0.80}$-$x$$Ca_{0.20}Sr_xMnO_3$ (x = 0.05, 0.08, 0.10)', *J Magn Magn Mater*, 283, 375–379.

Lines M and Glass A (1977), *Principles and Applications of Ferroelectrics and Related Materials*, Clarendon Press, Oxford.

Lu S G, Xu Z K, and Chen H (2005), 'Field-induced dielectric singularity, critical exponents, and high-dielectric tunability in [111]-oriented $(1-x)Pb(Mg_{1/3}Nb_{2/3})O_3$-$xPbTiO_3$ (x = 0.24)', *Phys Rev B* 72, 054120/1–054120/4.

Mathur N and Mischenko A (Cambridge University Technical Services Limited) 2006, *Solid state electrocaloric cooling devices and methods*, World Patent, WO 2006/056809 A1, 1 June 2006.

Mischenko A S, Zhang Q, Scott J S, Whatmore R W, and Mathur N D (2006), 'Giant electrocaloric effect in thin-film $PbZr_{0.95}Ti_{0.05}O_3$', *Science*, 311, 1270–1271.

Neese B (2009), PhD dissertation, The Pennsylvania State University.

Neese B, Chu B J, Lu S G, Wang Y, Furman E and Zhang Q M (2008), 'Large electrocaloric effect in ferroelectric polymers near room temperature', *Science*, 321, 821–823.

Newnham R E (2005), *Properties of Materials: Anisotropy, Symmetry, Structure*, Oxford University Press, Oxford.

Nolas G, Sharp J, and Goldsmid H (2001), *Thermoelectrics*, Springer-Verlag, Berlin.

Pecharsky A O, Gschneidner Jr K A, and Pecharsky V K (2003), 'The giant magnetocaloric effect of optimally prepared $Gd_5Si_2Ge_2$', *J Appl Phys*, 93, 4722–4728.

Pecharsky V K, Moorman J O, and Gschneidner Jr K A (1997), 'A 3–350 K fast automatic small sample calorimeter', *Rev Sci Instrum*, 68, 4196–4207.

Pecharsky V K, Holm A P, Gschneidner Jr K A, and Rink R (2003), 'Massive magnetic-field-induced structural transformation in Gd_5Ge_4 and the nature of the giant magnetocaloric effect', *Phys Rev Lett*, 91, 197204/1–197204/4.

PI Ceramic (2009), www.piceramic.de/site/piezo_002.html (accessed January 2009).

Provenzano V, Shapiro A J, and Shull R D (2004), 'Reduction of hysteresis losses in the magnetic refrigerant $Gd_5Ge_2Si_2$ by the addition of iron', *Nature*, 429, 853–857.

Sinyavsky Y V and Brodyansky V (1992), 'Experimental testing of electrocaloric cooling with transparent ferroelectric ceramic as a working body', *Ferroelectrics*, 131, 321–325.

Sinyavsky Y V, Pashkov N D, Gorovoy Y M, Lugansky G E, and Shebanov L (1989), 'The optical ferroelectric ceramic as working body for electrocaloric refrigeration', *Ferroelectrics*, 90, 213–217.

Spanner D C (1951), 'The Peltier effect and its use in the measurement of suction pressure', *J Experm Botany*, 2, 145–168.

Spichkin Y I, Derkach A V, Tishin A M, Kuz'min M D, Chernyshov A S, Gschneidner Jr K A, and Pecharsky V K (2007), 'Thermodynamic features of magnetization and magnetocaloric effect near the magnetic ordering temperature of Gd', *J Magn Magn Mater*, 316, e555–e557.

Tocado L, Palacios E, and Burriel R (2005), 'Direct measurement of the magnetocaloric effect in $Tb_5Si_2Ge_2$', *J Magn Magn Mater*, 290–291, 719–722.

Tuttle B A, and Payne D A (1981), 'The effect of microstructure on the electrocaloric properties of $Pb(Zr,Sn,Ti)O_3$ ceramics', *Ferroelectrics*, 37, 603–606.

Wiseman G G, and Kuebler J K (1963), 'Electrocaloric effect in ferroelectric Rochelle salt', *Phys Rev*, 131, 2023–2027.

Wood M E and Potter W H (1985), 'General analysis of magnetic refrigeration and its optimization using a new concept: maximization of refrigerant capacity', *Cryogenics*, 25, 667–683.

Xiao D Q, Wang, Y C, Zhang R L, Peng S Q, Zhu J G, and Yang B (1998), 'Electrocaloric properties of (1-x)$Pb(Mg_{1/3}Nb_{2/3})O_3$-$x PbTiO_3$ ferroelectric ceramics near room temperature', *Mater Chem Phys*, 57, 182–185.

Yao H, Ema K, and Garland C W (1998), 'Nonadiabatic scanning calorimeter', *Rev Sci Instrum*, 69, 172–178.

Ye Z G, and Schmid H (1993), 'Optical dielectric and polarization studies of the electric field-induced phase transition in $Pb(Mg_{1/3}Ng_{2/3})O_3$ [PMN]', *Ferroelectrics*, 145, 83–108.

16
Network behavior in thin films and nanostructure growth dynamics

H. GUCLU, University of Pittsburgh, USA,
T. KARABACAK, University of Arkansas at Little Rock, USA
and M. YUKSEL, University of Nevada – Reno, USA

Abstract: We present a new network modeling approach for various thin film growth techniques that incorporates re-emitted particles due to the non-unity sticking coefficients. We model re-emission of a particle from one surface site to another as a network link, and generate a network model corresponding to the thin film growth. Monte Carlo simulations are used to grow films and dynamically track the trajectories of re-emitted particles. We performed simulations for normal incidence, oblique angle, and chemical vapor deposition (CVD) techniques. Each deposition method leads to a different dynamic evolution of surface morphology due to different sticking coefficients involved and different strength of shadowing effect originating from the obliquely incident particles. Traditional dynamic scaling analysis on surface morphology cannot point to any universal behavior. On the other hand, our detailed network analysis reveals that there exist universal behaviors in degree distributions, weighted average degree versus degree, and distance distributions independent of the sticking coefficient used and sometimes even independent of the growth technique. We also observe that network traffic during high sticking coefficient CVD and oblique angle deposition occurs mainly among edges of the columnar structures formed, while it is more uniform and short-range among hills and valleys of small sticking coefficient CVD and normal angle depositions that produce smoother surfaces.

Key words: nanostructures, thin-film growth, network modeling, scale-free networks, sputter deposition, oblique-angle deposition, chemical vapor deposition.

16.1 Introduction

Thin film coatings have been the essential components of various devices in industries including microelectronics, optoelectronics, detectors, sensors, micro-electro-mechanical systems (MEMS), and more recently nano-electro-mechanical systems (NEMS). Commonly employed thin film deposition

An earlier version was published by the American Physical Society ©: Karabacak T, Guclu H, and Yuksel M, (2009), 'Network behavior in thin film growth dynamics', *Physical Review B*, 79, 195418. http://prb.aps.org/abstract/PRB/v79/i19/e195418.

[1–3] techniques are thermal evaporation, sputter deposition, chemical vapor deposition (CVD), and oblique angle deposition. Unlike the others, oblique angle deposition [4–11] is typically used for the growth of nanostructured arrays of rods and springs through a physical self-assembly process. In many applications, it is often desired to have atomically flat thin film surfaces. However, in almost all of the deposition techniques mentioned above, the surface morphology generates a growth front roughness. The formation of growth front is a complex phenomenon and very often occurs far from equilibrium. When atoms are deposited on a surface, atoms do not arrive at the surface at the same time uniformly across the surface. This random fluctuation, or noise, which is inherent in the process, may create the surface roughness. The noise competes with surface smoothing processes, such as surface diffusion (hopping), to form a rough morphology if the experiment is performed at either a sufficiently low temperature or at a high growth rate.

Due to its intractability, a conventional statistical mechanics treatment cannot be used to describe the complex phenomenon of surface morphology formation in thin film growth. About two decades ago, a dynamic scaling approach [12, 13] was proposed to describe the morphological evolution of a growth front. Since then, numerous modeling and experimental works have been reported based on this dynamic scaling analysis [2, 3, 14]. On the other hand, there has been a significant discrepancy among the predictions of these growth models and the experimental results published [15–17]. For example, various growth models have predictions on the dynamic evolution of the root-mean-square roughness (RMS) [18], which is defined as $\omega(t) = \sqrt{[h(r,t) - \langle h \rangle]^2}$, where $h(r, t)$ is the height of the surface at a position r and time t, and $\langle h \rangle$ is the average height at the surface. In most of the growth phenomena, the RMS grows as a function of time in a power law form [2, 3, 14, 19], $\omega \sim t^{\beta}$, where β is the 'growth exponent' ranging between 0 and 1. $\beta = 0$ for a smooth growth front and $\beta = 1$ for a very rough growth front (the RMS could be as large as the film thickness). Figure 16.1 shows a collection of experimental β values reported in the literature* and compares them with the predictions of growth models. Theoretical predictions of growth models in dynamic scaling theory basically fall into two categories. One involves various surface smoothing effects, such as surface diffusion, which lead to $\beta \leq 0.25$ [2, 3, 14, 19]. The other category involves the shadowing effect (which originates from the preferential deposition of obliquely incident atoms on higher surface points and always occurs in sputtering, CVD, and oblique angle deposition) during growth which would lead to $\beta = 1$ [20]. However, it can be clearly seen in Fig. 16.1 that experimentally reported

* Survey of experimental β values is obtained by updating the data presented in Ref. 15.

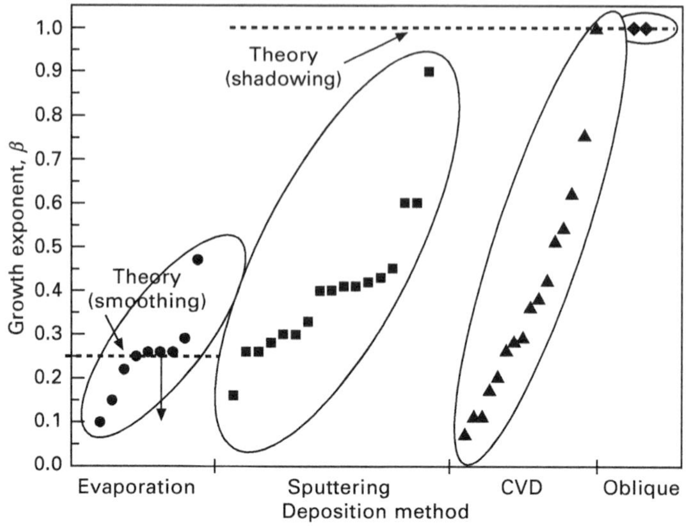

16.1 A survey of experimentally obtained values of growth exponent β reported in the literature for different deposition techniques is compared to the predictions of common thin film growth models in dynamic scaling theory.

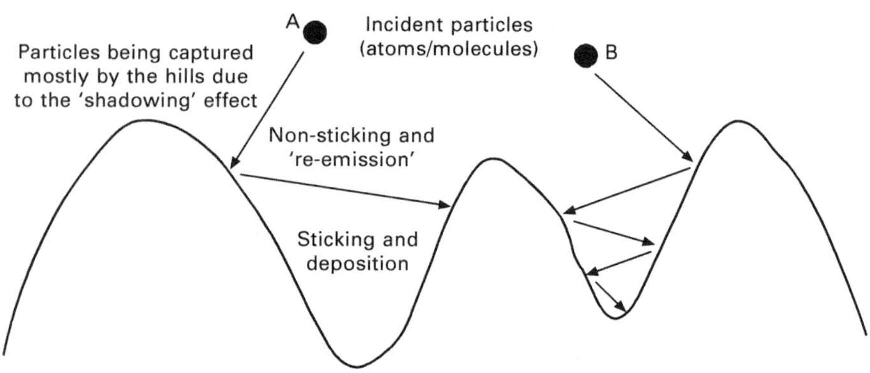

16.2 Surface of a growing thin film under shadowing and re-emission effects.

values of growth exponent β are far from agreement with the predictions of these growth models* In particular, sputtering and CVD techniques are observed to produce morphologies ranging from very small to very large β values indicating a 'non-universal' behavior.

Only recently, it has been recognized that in order to better explain the dynamics of surface growth one should take into account the effects of both

* Survey of experimental β values is obtained by updating the data presented in Ref. 15.

'shadowing' and 're-emission' processes [14, 16, 17, 21, 22]. As illustrated in Fig. 16.2, during deposition, particles can approach the surface at oblique angles and be captured by higher surface points (hills) due to the shadowing effect. This leads to the formation of rougher surfaces with columnar structures that can also be engineered to form 'nanostructures' under extreme shadowing conditions, as in the case of oblique angle deposition that can produce arrays of nanorods and nanosprings [6–11]. In addition, depending on the detailed deposition process, particles can either stick to or bounce off from their impact points, which is determined by a sticking probability, also named 'sticking coefficient' (s). Non-sticking particles are re-emitted and can arrive at other surface points including shadowed valleys. In other words, re-emission has a smoothing effect while shadowing tries to roughen the surface. Both the shadowing and re-emission effects have been proven to be dominant over the surface diffusion and noise, and act as the main drivers of the dynamical surface growth front [9, 10]. The prevailing effects of shadowing and re-emission rely on their 'non-local' character: The growth of a given surface point depends on the heights of near and far-away surface locations due to shadowing and the existence of re-emitted particles that can travel over long distances.

Figure 16.3 summarizes some of the experimentally measured sticking coefficient values reported in the literature during evaporation [23], sputtering [24–31], and CVD [32–38] growth of various thin film materials. Names of incident atoms/molecules on the growing film are also labeled. It can be clearly seen from Fig. 16.3 that incident particles can have sticking probabilities much less than unity in many commonly used deposition systems, which further indicates that re-emission effects should be taken into account in attempts for a realistic thin film growth modeling.

Due to the complexity of the shadowing and re-emission effects, no growth model has been developed yet within the framework of dynamical scaling theory that takes both these effects into account and that still can be analytically solved to predict the morphological evolution of thin film or nanostructure deposition [39]. Only recently, shadowing and re-emission effects could be fully incorporated into the Monte Carlo lattice simulation approaches [10, 14–17, 21, 22, 39–42]. These simulations successfully predicted the experimental results including the β values reported in the literature (see Fig. 16.1). However, like in experiments, β values from simulations ranged all the way from 0 to 1 depending on the sticking coefficients used. For example, Fig. 16.4 shows β values for a Monte Carlo simulated CVD growth obtained for various sticking coefficient and surface diffusion values [15, 17, 21, 22]. It has been observed that re-emission and shadowing effects dominated over the surface diffusion processes due to their long-range non-local character. At small sticking coefficients (e.g. $s < 0.5$) re-emission was stronger than the roughening effects of shadowing and Monte Carlo simulations produced

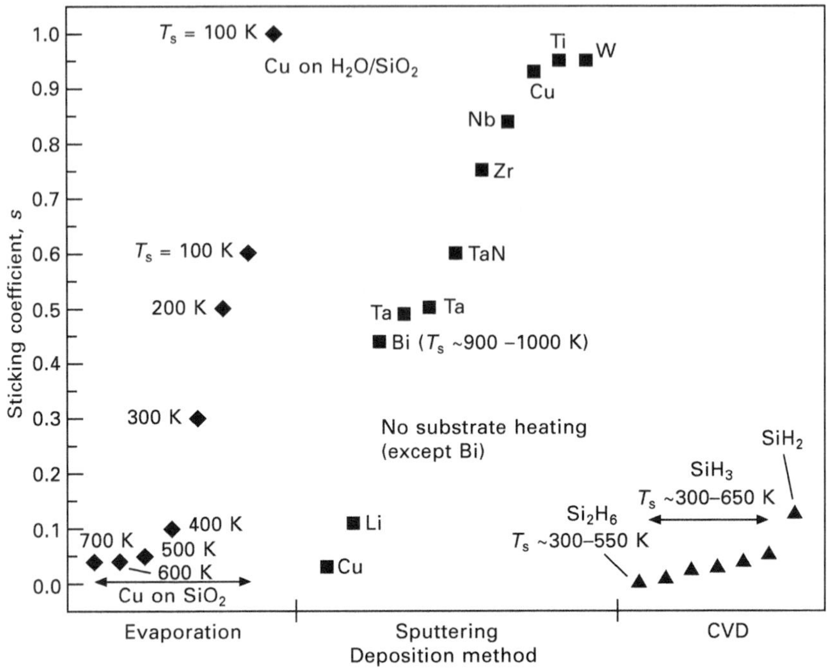

16.3 Some of the experimentally measured sticking coefficient values reported in the literature during evaporation [23], sputtering [24–31], and CVD [32–38] growth. Names of incident atoms/molecules on the growing film are also labeled. In same cases, depositions were done with substrate heating at temperatures denoted as T_s in the figure.

smooth surfaces with small β values. At higher sticking coefficient values, the shadowing effect becomes the dominant process and columnar rough morphologies start to form. On the other hand, like in experiments, it was not possible to capture a 'universal' growth behavior using Monte Carlo simulation approaches, which would lead to dynamically common aspects of various thin film growth processes.

Moreover, it has very recently been revealed that shadowing effect can lead to the breakdown of dynamical scaling theory due to the formation of a mounded surface morphology [16, 17]. In these studies, using Monte Carlo simulations it has been shown that for common thin film deposition techniques, such as sputter deposition and CVD, a 'mound' structure can be formed with a characteristic length scale that describes the separation of the mounds, or 'wavelength' λ. It has been found that the temporal evolution of λ is distinctly different from that of the mound size, or the lateral correlation length, ξ. The formation of the mound structure is due to non-local growth effects, such as shadowing, that lead to the breakdown of the self-affinity of the morphology described by the dynamic scaling theory. The wavelength

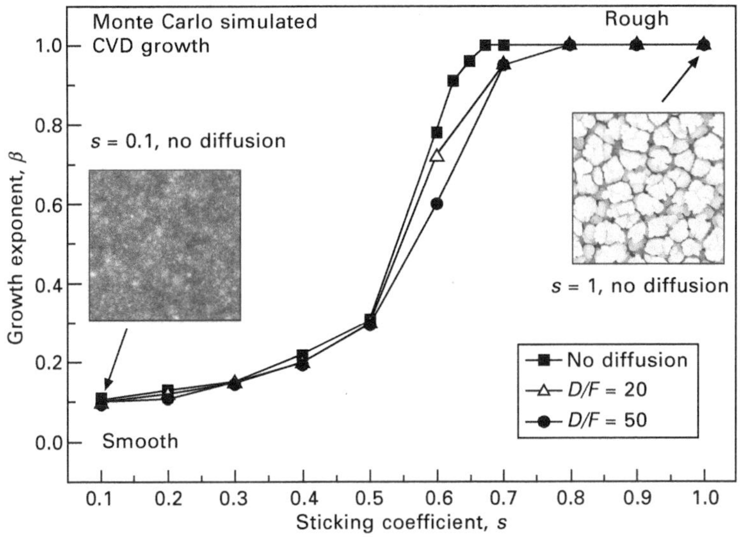

16.4 Growth exponent β values for a Monte Carlo simulated chemical vapor deposition (CVD) growth obtained for various first-impact sticking coefficient (s) and surface diffusion (D/F) values. The sticking coefficient at the second impact after re-emission was set to 1. Two sample surface morphologies are also included for a small $s = 0.1$ (left) and high $s = 1$ (right) sticking coefficient value, which leads to a smooth and rough surface topography, respectively.

grows as a function of time in a power law form, $\lambda \sim t^p$, where $p \approx 0.5$ for a wide range of growth conditions, while the mound size grows as $\xi \sim t^{1/z}$, where $1/z$ depends on the growth conditions.

In brief, conventional growth models in dynamic scaling theory cannot explain most of the experimental results reported for dynamic thin film growth; and dynamic scaling theory itself often suffers from a breakdown if shadowing effect is present, which is the case for most of the commonly used deposition techniques. On the other hand, simulation techniques were not successful in revealing the possible universal behavior in various growth processes. Furthermore, simulations that can successfully predict the experimental results cannot always be easily implemented by other researchers. Therefore, there is an immense need for alternative and robust modeling approaches for the dynamical growth of thin film surfaces that incorporates easy-to-implement analytical and/or empirical relations which in turn can lead to universal growth behavior aspects of thin films. In this work, we explore a radically new 'network' modeling approach for the dynamical growth of a large variety of thin film growth systems that can potentially capture universal properties of film growth processes and at the same time not suffer from the shortcomings of dynamic scaling theory and Monte Carlo simulation approaches mentioned above.

Network modeling pervades various areas of science ranging from sociology to statistical physics or computer science, see Ref. [43] and the references therein. A network in terms of modeling can be defined as a set of items, referred to as nodes with links connecting them. Examples of real-life complex networks include the Internet, the World Wide Web, metabolic networks, transportation networks, social networks, etc. Recent studies show that many of these networks share common properties such as having a low degree of separation among the nodes (modeled as small-world networks [44]), having a power-law degree/connectivity distributions (modeled as scale-free networks [45]), etc.

16.2 Origins of network behavior during thin film growth

Interestingly, non-local interactions among the surface points of a growing thin film that originate from shadowing and re-emission effects (Fig. 16.2) can lead to non-random preferred trajectories of atoms/molecules before they finally stick and get deposited. For example, during re-emission, the path between two surface points where a particle bounces off from the first and heads on to the second can define a 'network link' between the two. If the sticking coefficient is small, then the particle can go through multiple re-emissions that form links among many more other surface points. In addition, due to the shadowing effect, higher surface points act as the locations of first-capture and centers for re-emitting the particles to other places. In this manner, hills on a growing film resemble the network 'nodes' of heavy traffic, where the traffic is composed of the amount of particles re-emitting from the nodes. In terms of network traffic, nodes can be classified as: *source*, *sink*, or *router*. So, the initial point/hill where an atom re-emits can correspond to a 'source' in a network, and the final point where the atom sticks/settles can be thought as a 'sink' in the network. Similarly, the intermediate re-emission points/hills can be thought as the 'routers'. Therefore, a 'traffic model' for thin film growth can then be constructed by counting the number of atoms starting from a point on the film and ending at another point on the film.

16.3 Monte Carlo simulations

Development of network models by our approach requires the track record of the trajectories of re-emitted and deposited atoms/particles. Since it is not possible to *experimentally* track the trajectories of re-emitted and deposited atoms during dynamic thin film growth, we used 3D Monte Carlo simulation approaches instead, which were already shown to efficiently mimic the experimental processes and predict the correct dynamical growth morphology [9, 10, 16, 17, 21, 22, 40, 41]. In these simulations, each incident particle (atom/molecule) is represented with the dimension of one lattice point. As substrate, we used a $N \times N = 512 \times 512$ size lattice with continuous boundary conditions. A specific angular distribution for the incident flux of particles is chosen depending on the deposition technique being simulated. During normal angle deposition, all the particles are sent from the top along the substrate normal (polar angle $\theta = 0°$), while during oblique angle deposition simulations we used a grazing incidence flux where all particles are emitted at an angle of $\theta = 85°$ from the substrate normal. For CVD, the incident flux had an angular spread according to the distribution function $dP(\theta, \phi)/d\Omega = \cos\theta/\pi$, where ϕ is the azimuthal angle [39].

At each simulation step, a particle is sent toward a randomly chosen lattice point on the substrate surface. Depending on the value of sticking coefficient (s), the particle can bounce off and re-emit to other surface points. Re-emission direction is chosen according to a cosine distribution centered around the local surface normal [39]. At each impact, sticking coefficient can have different values represented as s_n, where n is the order of re-emission ($n = 0$ being for the first impact). In this study, we use a constant sticking coefficient value for all impacts (i.e. $s_n = s$ for all n) during a given simulation, which is a process also called 'all-order re-emission' [39]. In all the emission and re-emission processes, shadowing effect is included, where the particle's trajectory can be cut off by long surface features on its way to other surface points. After the incident particle is deposited onto the surface, it becomes a so-called 'adatom'. Adatoms can hop on the surface according to some rules of energy, which is a process mimicking the surface diffusion. However, as noted before, non-local processes of re-emission and shadowing are generally dominant over local surface diffusion effects. Therefore, in this work we did not include surface diffusion in order to better distinguish the effects of re-emission and shadowing effects. After this deposition process, another particle is sent, and the re-emission and deposition are repeated in a similar way.

In our simulations, deposition time t is represented by number of particles sent to the surface. Final simulation time (total number of particles sent) for all the simulations was $t_{final} = 25 \times 10^7$. Because of re-emission, deposition rate and therefore average film thickness (d) depended on the sticking coefficient s used, and changed with simulation time t approximately according to $d \approx t \times s/(N \times N)$, where lattice size N was 512.

Furthermore, in our simulations, trajectories of particles during each re-emission process can be tracked in order to reveal the dynamic network behavior in detail. When the simulation time reaches a pre-set value that we called the 'snapshot state', we label each particle sent to the surface and start recording the coordinates of lattice point where the particle impacts and also the lattice point where it is re-emitted and makes another impact. Therefore, especially a small sticking coefficient particle can potentially make multiple re-emissions among the surface points and have multiple trajectory data. In order to increase the number of trajectory data for a better statistical analysis while keeping the surface morphology unchanged, we cancelled the final deposition of the particle sent during the trajectory data collection process. In other words, when the simulation time reached the pre-set value, particles were still being sent for re-emission and collection of trajectory data; however, they were not depositing to the surface, therefore not changing the surface morphology. We collected the trajectory data of about 10^6 re-emitted particles for each snapshot state. We did not include the trajectory data of particles as they re-emitted into the space or if they crossed the lattice boundaries, since cross-boundary particles can lead to artificially long trajectories due to the continuous boundary conditions used. All the simulation results are an average of 10 runs (realizations), each time using a different seed number for the random number generator.

16.4 Results and discussion

Figure 16.5 shows the snapshot top view images of two surfaces simulated for a CVD type of deposition, at two different sticking coefficients. Figure 16.5 also displays their corresponding particle trajectories projected on the lateral plane. Qualitative network behavior can easily be realized in these simulated morphologies as the trajectories of re-emitted atoms 'link' various surface points. It can also be seen that larger sticking coefficients (Fig. 16.5(b) and (d)) lead to fewer but longer range re-emissions, which are mainly among the peaks of columnar structures. Therefore, these higher surface points act as the 'nodes' of the system. This is due to the shadowing effect where initial particles preferentially head to hills. They also have less chance to arrive down to valleys because of the high sticking probabilities (see for example particle A illustrated in Fig. 16.2). On the other hand, at lower sticking coefficients (Fig. 16.5(a) and (c)), particles now go through multiple re-emissions and can link many more surface points including the valleys that are normally shadowed by higher surface points (particle B in Fig. 16.2). This behavior is better realized in 'surface-degree' and their corresponding height matrix plots of Fig. 16.6 measured for CVD grown films at two different sticking coefficients $s = 0.1$ and $s = 0.9$. The high values (darker shades) in surface-degree plots correspond to the highly connected surface sites where these sites get or re-distribute most of the re-emitted particles. At smaller sticking coefficients (Fig. 16.6(a)), which lead to a smoother morphology, surface-degree values are quite uniform indicating a uniform re-emission process among hill-to-hills and hill-to-valleys. On the other hand, at high sticking coefficients (Fig. 16.6(b)), the high degree nodes are mainly located around the column borders suggesting a dominant column-to-column re-emission. This is consistent with the shadowing effect

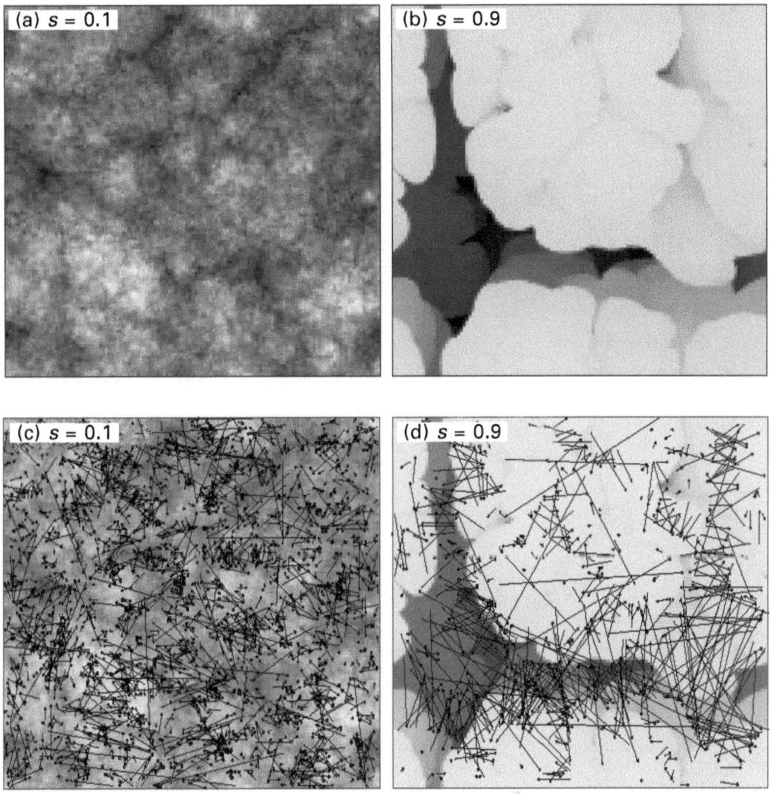

16.5 Top view images from Monte Carlo simulated thin film surfaces grown under shadowing, re-emission, and noise effects (no surface diffusion is included in these simulations) for sticking coefficients (a) $s = 0.1$ and (b) $s = 0.9$ and with unity sticking coefficient at the second impacts. Each image corresponds to a 128×128 portion of the total lattice. The incident flux of particles has an angular distribution designed for chemical vapor deposition (CVD). Corresponding projected trajectories of the re-emitted particles are also mapped on the top view morphologies for (c) $s = 0.1$ and (d) $s = 0.9$. Qualitative network behavior can be seen among surface points linked by the re-emission trajectories.

where columns capture most of the incident particles because of their larger heights, and also their borders are more likely to re-distribute the particles towards the neighboring column sides because of the re-emission process used (i.e. cosine distribution centered along the local surface normal).

Another interesting observation revealed in our Monte Carlo simulations was the dynamic change of network behavior on the trajectories of re-emitted particles. Figure 16.7 shows top view images and their corresponding particle trajectories obtained from the CVD simulations for a sticking coefficient of

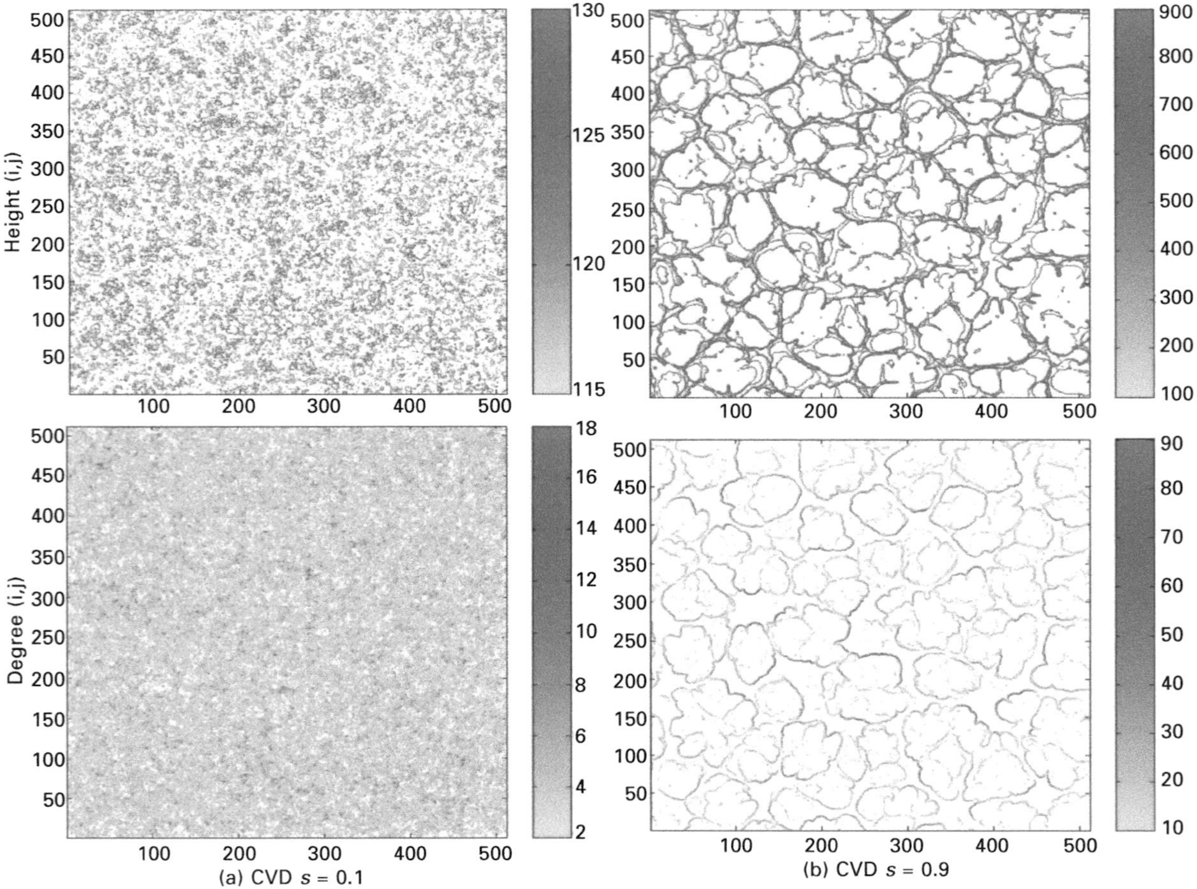

16.6 Height matrix and corresponding surface-degree values are plotted for CVD grown films with sticking coefficients (a) s = 0.1 and (b) s = 0.9. Total lattice size is 512 × 512 and simulation time for these snapshot states was t = 23.75 × 10^7 particles.

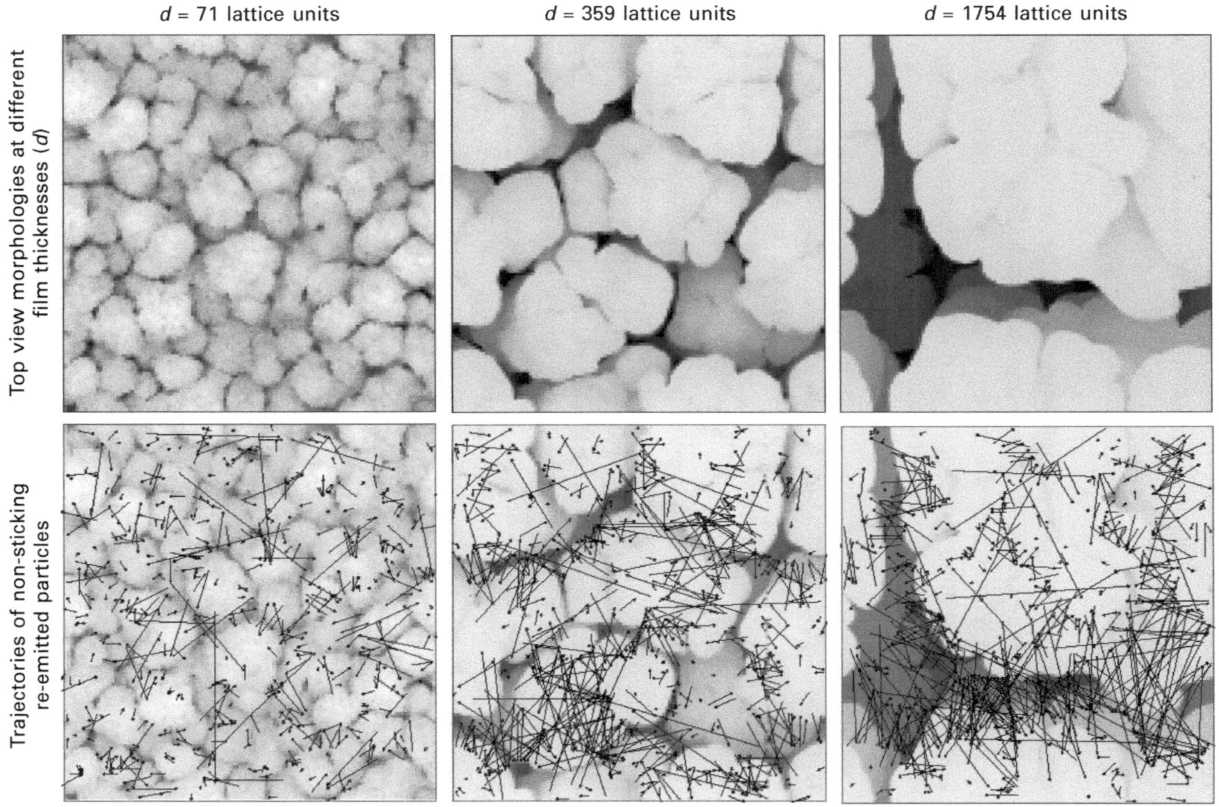

16.7 First row: Top view images from Monte Carlo simulated thin film surfaces for a CVD growth with $s = 0.9$ at different film thicknesses d, which is proportional to growth time. Bottom row: Corresponding projected trajectories of the re-emitted particles qualitatively show the dynamic change in the network topography.

$s = 0.9$, but this time at different film thicknesses proportional to the growth time. The dynamic change in the network topography can be clearly seen: at initial times, when the hills are smaller and more closely spaced, the re-emitted particles travel from one hill to another one or to a valley. However, as the film gets thicker, and some hills become higher than the ossthers and get more separated, particles travel longer ranges typically among these growing hills. The shorter hills that get shadowed become the valleys of the system.

It is expected that this dynamic behavior should be strongly dependent on the values of sticking coefficients and angular distribution of the incident flux of particles, which determine the strength of re-emission and shadowing effects, respectively. In other words, each deposition technique and material system can have different dynamic network behavior that can lead to various kinds of network systems. For example, as we will show later, the dynamic network among the surface points of a mounded CVD grown film can be quite different from the one among the nanorod and nanospring structures formed in an oblique angle deposition system, where the shadowing effect is most dominant, and also the one during normal angle evaporation, where the shadowing effect is almost absent (re-emitted particles during normal angle deposition can still lead to a minimal short-range shadowing effect).

In order to make a more quantitative analysis of the network characteristics of thin film growth dynamics, in Fig. 16.8, we plotted the degree distributions $P(k)$ (i.e. proportional to the percentage of surface points having 'degree (k)' number of links through incoming or outgoing re-emitted particles), average distance $<l_k>$ versus degree k (i.e. the average 'lateral' distance particles travel that are re-emitted from/to surface sites having k number of links), and distance distributions $P(l)$ (i.e. proportional to the probability of a re-emitted particle traveling lateral distance l) for Monte Carlo simulated normal incidence evaporation, oblique angle deposition, and CVD thin film growth for various sticking coefficients. The left and right columns in Fig. 16.8 correspond to the initial (thinner films) and later (thick films) stages of the growth times, respectively. First, the comparison of degree distributions (Fig. 16.8(a) and (b)) of normal incidence and oblique angle growth reveals that independent of most sticking coefficients used and also their growth time, universal behavior exists for both deposition techniques. There is an exponential degree distribution for normal angle evaporation (confirmed in the semi-log plots, not shown here), while this behavior is mainly power-law for oblique angle deposition with an exponential tail. Interestingly, quantitative values of degree distributions for both normal and oblique angle depositions also seem to be independent of the sticking coefficient used, which becomes clearer at later stages of the growth (Fig. 16.8(b)), leading to two distinct distributions for each deposition. The power-law observed in degree distribution of oblique angle deposition has a $P(k) \sim k^{-2}$ behavior apparent

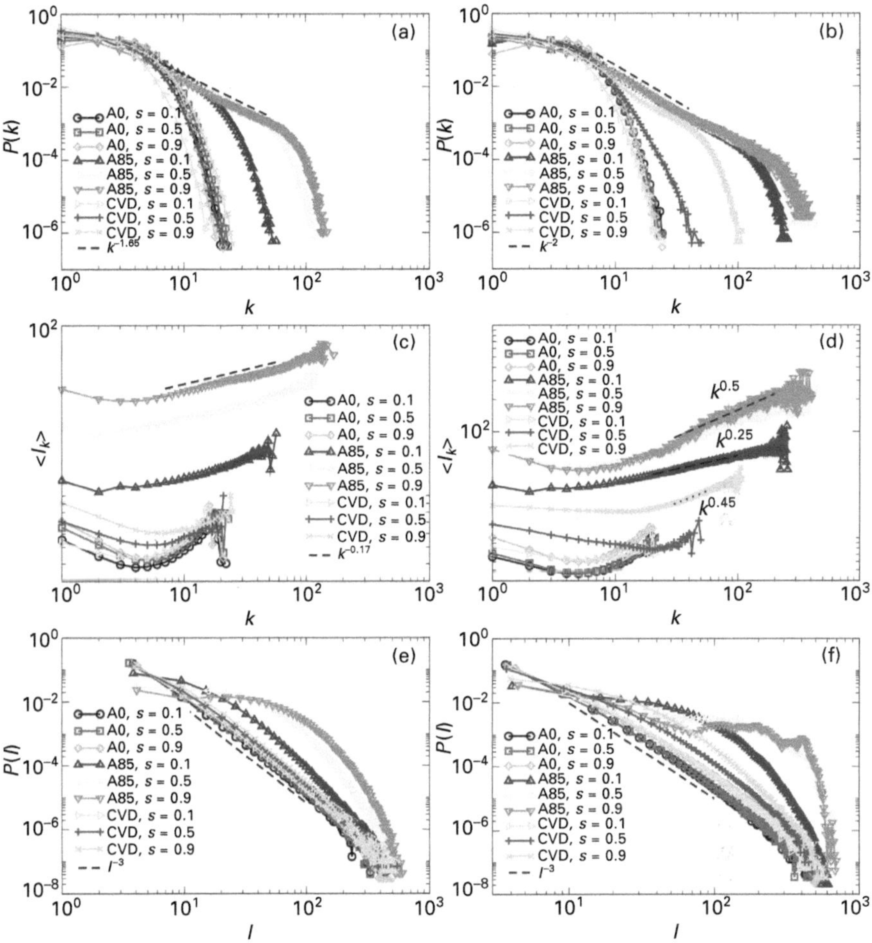

16.8 Behavior of degree distributions $P(k)$ (top row), average distance $<l_k>$ versus degree (middle row), and distance distributions $P(l)$ (bottom row) for network models of a Monte Carlo simulated normal incidence evaporation (A0), oblique angle deposition (A85), and CVD thin film growth for various sticking coefficients s and for two different deposition times t (left column: $t = 1.25 \times 10^7$ particles, and right column: $t = 23.75 \times 10^7$ particles).

at later stages. All these suggest the possibility of a universal behavior in normal and oblique angle growth independent of the sticking coefficient. This is quite striking since each different sticking coefficient corresponds to a different type of morphological growth (i.e., smoother surfaces for smaller sticking coefficients and rougher surfaces for higher sticking coefficients), yet the degree distribution in network traffic of re-emitted particles seems to reach a unique universal state.

As can be seen in Fig. 16.8(a) and (b), the re-emission process which is the dominant process in normal angle growth promotes an exponential degree distribution; while shadowing which is the governing effect during oblique angle deposition leads to a power-law distribution. On the other hand, CVD shows an exponential degree distribution at initial times of growth, while it becomes closer to power-law type for higher sticking coefficients, $s > 0.5$. This is believed to be competing forces of re-emission and shadowing effects, where the re-emission is more dominant for smaller sticking coefficients and at initial times of the growth when the film is smoother, leading to an exponential degree distribution. However, the shadowing effect originating from the obliquely incident particles within the angular distribution of CVD flux can lead to a power-law behavior at higher sticking coefficients especially when the film gets rougher at later stages of growth. A power-law degree distribution corresponds to a more correlated network that is consistent with the long-range, column-to-column traffic observed in surface-degree plots of high sticking coefficient CVD above (Fig. 16.6(b)). It is also realized that especially for high sticking coefficients, there exist high degree nodes represented with data points at the tails of the degree distributions. These relatively small percentage but highly connected nodes are mainly located at the column edges as seen in the surface degree plot of Fig. 16.6(b) and are likely to be the 'hubs' of the network. Therefore, briefly, degree distribution during CVD growth can be similar to the universal line of normal incidence growth for smaller sticking coefficients ($s < 0.5$) showing an exponential behavior with a short range network traffic; or it can converge to the universal power-law degree distribution of oblique angle deposition for higher sticking coefficients ($s > 0.5$) leading to a highly correlated network driven mostly at column edges.

In addition, it is revealed from average distance versus degree plots of Fig. 16.8(c) and (d) that nodes with high degree are mainly linked with long-distance surface points. Independent of the deposition method, the average distance changes with degree k according to a power-law behavior, where the value of the exponent increases as the flux becomes more oblique (i.e., A0 $\rightarrow$ A85 $\rightarrow$ CVD in Fig. 16.8(c) and (d)), sticking coefficient increases, and the film gets thicker (i.e., Fig. 16.8(c) $\rightarrow$ (d)). In other words, when the shadowing effect becomes more dominant and film morphology gets more columnar, high degree nodes can exchange atoms with longer distance surface points. This also implies that high degree nodes placed on column edges (Fig. 16.6(b)) of high sticking coefficient growth are more likely to transfer particles with other more distant column edges as well. This process is further supported by the distance distribution plots of Fig. 16.8(e) and (f) where as the sticking coefficient is increased (i.e. less re-emission) and more obliquely deposited particles are introduced (more shadowing effect), a higher percentage of particles start to travel longer distances. On the other hand, for smaller sticking coefficients and normal angle deposition, where the re-emission effect is more dominant, average distances particles travel from high degree nodes become significantly less compared to high sticking coefficient and oblique angle depositions. This suggests networking during re-emission dominated growth occurs mainly among smaller size hills and valleys, consistent with the surface-degree plot of Fig. 16.6(a).

A more interesting universal behavior is observed in a 'weighted and scaled average distance' versus degree plots of Fig. 16.9(a) and (b). Here we re-scale average distance for nodes with degree k, $\langle l_k \rangle$, first with degree k, then with the average distance value of all nodes $\langle l \rangle$ (average distance of all links), and plot $\langle l_k \rangle/(k\langle l \rangle)$ versus k. After re-scaling, independent of the deposition technique used, sticking coefficients, and the growth time, all curves fall on a similar line obeying a power-law behavior with $\langle l_k \rangle/(k\langle l \rangle) \sim k^{-1.2}$. The origin of the −1.2 value of the exponent is not clear and is under investigation.

Another universal behavior is observed in distance distribution plots. Independent of sticking coefficients, normal incidence growth shows a power-law behavior with $P(l) \sim l^{3}$. A similar power-law behavior with an exponent of −2.75 has been observed in the distance distribution plots during a normal incidence growth simulation with re-emission (p. 83 of Ref. [14]). The authors of that work did not use a snapshot state approach, surface morphology changed continuously, and therefore they measured a kind of average distance distribution of the whole growth simulation. However, their exponent value is still close to our results and agrees with our findings that dynamic network behavior during normal incidence deposition does not change significantly due to the relatively smooth morphology throughout the growth. On the other hand, the behavior in distance distribution plots is exponential for oblique angle deposition (confirmed in the semi-log plots, not shown here). CVD has a power-law behavior similar to that of normal incidence growth with $P(l) \sim l^{-3}$ at smaller coefficients and at initial times of growth, and becomes exponential similar to oblique angle deposition at higher sticking coefficients apparent especially in later stages of growth.

16.5 Conclusions

In conclusion, we presented a new network modeling approach for various thin film growth techniques that incorporates re-emitted particles due to the non-unity sticking coefficients. We define a network link when a particle is re-emitted from one surface site to another. Monte Carlo simulations are used to grow films and dynamically track the trajectories of re-emitted particles. We performed simulations for normal incidence, oblique angle, and CVD techniques. Each deposition method leads to a different dynamic evolution of surface morphology due to different sticking coefficients involved and

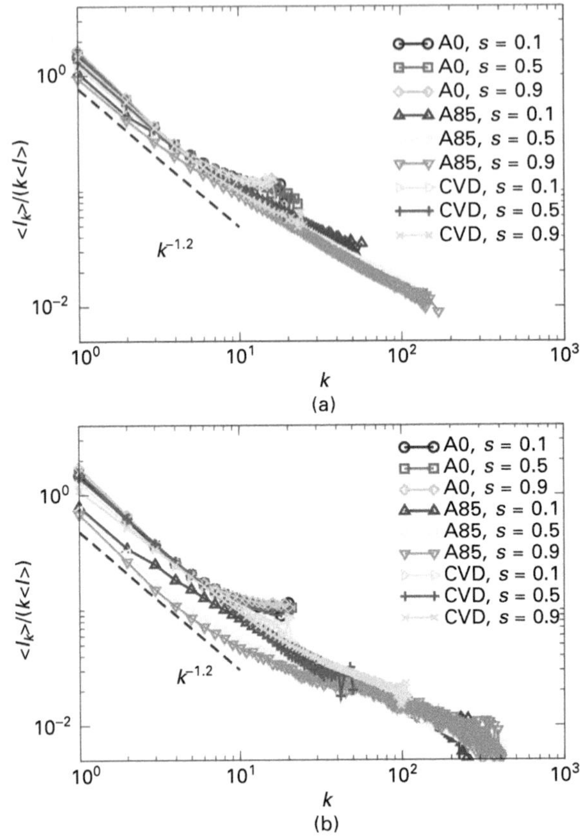

16.9 Weighted average distance $<l_k>/(k<l>)$ versus degree k for network models of a Monte Carlo simulated normal incidence evaporation (A0), oblique angle deposition (A85), and CVD thin film growth for various sticking coefficients s and for two different deposition times t (a: $t = 1.25 \times 10^7$ particles, and b: $t = 23.75 \times 10^7$ particles).

different strength of shadowing effect originating from the obliquely incident particles. Traditional dynamic scaling analysis on surface morphology cannot point to any universal behavior. On the other hand, our detailed network analysis reveals that there exist universal behaviors in degree distributions, weighted average degree versus degree, and distance distributions independent of the sticking coefficient used and sometimes even independent of the growth technique. We also observe that network traffic during high sticking coefficient CVD and oblique angle deposition occurs mainly among edges of the columnar structures formed, while it is more uniform and short-range among hills and valleys of small sticking coefficient CVD and normal angle depositions that produce smoother surfaces.

16.6 References

[1] Smith D L (1995), *Thin-Film Deposition: Principles and Practice*, New York, McGraw-Hill.

[2] Meakin P (1998), *Fractals, Scaling, and Growth far from Equilibrium*, Cambridge, Cambridge University Press.

[3] Barabasi A L and Stanley H E (1995), *Fractal Concepts in Surface Growth*, Cambridge, Cambridge University Press.

[4] Young N O and Kowal J (1959), 'Optically active fluorite films', *Nature*, 183, 104–105.

[5] Motohiro T and Taga Y (1989), 'The film retardation plate by oblique deposition', *Appl. Opt.*, 28, 2466–2482.

[6] Azzam R M (1992), 'Chiral thin solid films: method of deposition and applications', *Appl. Phys. Lett.*, 61, 3118–3120.

[7] Robbie K, Brett M J and Lakhtakia A (1996), 'Chiral sculptured thin films', *Nature*, 384, 616.

[8] Robbie K, Beydaghyan G, Brown T, Dean C, Adams J and Buzea C (2004), 'Ultrahigh vacuum glancing angle deposition system for thin films with controlled three-dimensional nanoscale structure', *Rev. Sci. Instrum.* 75, 1089–1097.

[9] Karabacak T and Lu T M (2005), 'Shadowing growth and physical self-assembly of 3D columnar structures', in Rieth M and Schommers W, *Handbook of Theoretical and Computational Nanotechnology*, Stevenson Ranch, CA, American Scientific Publishers, 729.

[10] Karabacak T, Wang G C and Lu T M (2004), 'Physical self-assembly and the nucleation of three-dimensional nanostructures by oblique angle deposition', *J. Vac. Sci. Technol. A*, 22, 1778–1784.

[11] Lakhtakia A and Messier R (2005), *Sculptured Thin Films: Nanoengineered Morphology and Optics*, Bellingham, WA, SPIE Press.

[12] Family F and Vicsek T (1985), 'Scaling of the active zone in the Eden process on percolation networks and the ballistic deposition model', *J. Phys. A*, 18, L75–L81.

[13] Family F (1986), 'Scaling of rough surfaces: effects of surface diffusion', *J. Phys. A*, 19, L441–L446.

[14] Pelliccione M and Lu T M (2007), *Evolution of Thin Film Morphology: Modeling and Simulations*, New York, Springer.

[15] Lu T M, Zhao Y P, Drotar J T, Karabacak T and Wang G C (2003), 'Novel mechanisms on the growth morphology of films', *Mat. Res. Soc. Symp. Proc.*, 749, 3.

[16] Pelliccione M, Karabacak T, and Lu T M (2006), 'Breakdown of dynamic scaling in surface growth under shadowing', *Phys. Rev. Lett.*, 96, 146105.

[17] Pelliccione M, Karabacak T, Gaire C, Wang G C and Lu T M (2006), 'Mound formation in surface growth under shadowing', *Phys. Rev. B*, 74, 125420.

[18] Zhao Y P, Wang G C and Lu T M (2001), *Characterization of Amorphous and Crystalline Rough Surfaces: Principles and Applications*, San Diego, CA, Academic Press.

[19] Family F and Viscek T (1991), *Dynamics of Fractal Surfaces*, Singapore, World Scientific.

[20] Karunasiri R P U, Bruinsma R, and Rudnick J (1989), 'Thin-film growth and the shadow instability', *Phys. Rev. Lett.*, 62, 788–791.

[21] Karabacak T, Zhao Y P, Wang G C and Lu T M (2001), 'Growth-front roughening in amorphous silicon films by sputtering', *Phys. Rev. B*, 64, 085323.

[22] Karabacak T, Zhao Y P, Wang G C and Lu T M (2002), 'Growth front roughening in silicon nitride films by plasma-enhanced chemical vapor deposition', *Phys. Rev. B*, 66, 075329.

[23] Xu X and Goodman D W (1992), 'Metal deposition onto oxides: an unusual low initial sticking probability for copper on SiO_2', *Appl. Phys. Lett.*, 61, 1799–1801.

[24] van Veldhuizen E M and de Hoog F J (1984), 'Analysis of a Cu-Ne hollow cathode glow discharge at intermediate currents', *J. Phys. D: Appl. Phys.*, 17, 953–968.

[25] Bogaerts A, Naylor J, Hatcher M, Jones W J and Mason R (1998), 'Influence of sticking coefficients on the behavior of sputtered atoms in an argon glow discharge: modeling and comparison with experiment', *J. Vac. Sci. Technol. A*, 16, 2400–2410.

[26] Obara K, Fu Z, Arima M, Yamada T, Fujikawa T, Imamura N and Terada N (2002), 'Collision processes between sputtered particles on high speed rotating substrate and atomic mass dependence of sticking coefficient', *J. Crystal Growth*, 237–239, 2041–2045.

[27] Migita S, Sakai K, Ota H, Mori Z and Aoki R (1996), 'The influence of Bi-sticking coefficient in the growth of Bi(2212) thin film by ion beam sputtering', *Thin Solid Films*, 281–282, 510–512.

[28] Bogaerts A, Wagner E, Smith B W, Winefordner J D, Pollmann D, Harrison W W and Gijbels R (1997), 'Three-dimensional density profiles of sputtered atoms and ions in a direct current glow discharge: experimental study and comparison with calculations', *Spectrochimica Acta Part B*, 52, 205–218.

[29] Toprac A J, Jones B P, Schlueter J and Cale T S (1995), 'Modeling of collimated titanium nitride physical vapor deposition using a combined specular-diffuse formulation', *Mat. Res. Soc. Symp. Proc.*, 355, 575.

[30] Yamazaki O, Iyanagi K, Takagi S and Nanbu K (2002), 'Modeling of Cu transport in sputtering using a Monte Carlo simulation', *Jpn. J. Appl. Phys.*, 41, 1230–1234.

[31] Liu D, Dew S K, Brett M J, Smy T and Tsai W (1994), 'Compositional variations in Ti-W films sputtered over topographical features', *J. Appl. Phys.*, 75, 8114–8120.

[32] Buss R J, Ho P, Breiland W G and Coltrin M E (1988), in Rubloff G W, *Deposition and Growth: Limits for Microelectronics*, AIP Conf. Proc. 167, 34.

[33] Buss R J, Ho P, Breiland W G and Coltrin M E (1988), 'Reactive sticking coefficients for silane and disilane on polycrystalline silicon', *J. Appl. Phys.*, 63, 2808–2819.

[34] Tsai C C, Shaw J G, Wacker B and Knights J C (1987), 'Film growth mechanisms of amorphous silicon in diode and triode glow discharge systems', *Mat. Res. Soc. Symp. Proc.*, 95, 219.

[35] Perrin J and Broekhuizen T (1987), 'Modeling of $Hg(^3P_1)$ photosensitization of SiH_4 and surface reactions of the SiH_3 radical', *Mat. Res. Soc. Symp. Proc.*, 75, 201–208.

[36] Perrin J and Broekhuizen T (1987), 'Surface reaction and recombination of the SiH_3 radical on hydrogenated amorphous silicon', *Appl. Phys. Lett.*, 50, 433–435.

[37] Robertson R and Gallagher A (1986), 'Mono- and disilicon radicals in silane and silane-argon dc discharges', *J. Appl. Phys.*, 59, 3402–3411.

[38] Robertson J (2000), 'Growth mechanism of hydrogenated amorphous silicon', *J. Non-Cryst. Solids*, 266–269, 79–83.

[39] Drotar J T, Zhao Y P, Lu T M and Wang G C (2000), 'Mechanisms for plasma and reactive ion etch-front roughening', *Phys. Rev. B*, 61, 3012–3021.

[40] Karabacak T, Singh J P, Zhao Y P, Wang G C and Lu T M (2003), 'Scaling during shadowing growth of isolated nanocolumns', *Phys. Rev. B*, 68, 125408.

[41] Karabacak T, Wang G C and Lu T M (2003), 'Quasi-periodic nanostructures grown by oblique angle deposition', *J. Appl. Phys.*, 94, 7723–7728.

[42] Smy T, Vick D, Brett M J, Dew S K, Wu A T, Sit J C, and Harris K D (2005), 'Three-dimensional simulation of film microstructure produced by glancing angle deposition', *J. Vac. Sci. Technol. A*, 18, 2507–2512.

[43] Boccaletti S, Latora V, Moreno Y, Chavez M and Hwang D U (2006), 'Complex networks: structure and dynamics', *Physics Reports*, 424, 175–308.

[44] Watts D J and Strogatz S H (1998), 'Collective dynamics of small-world networks', *Nature*, 393, 440–442.

[45] Barabasi A L and Albert R (1999), 'Emergence of scaling in random networks', *Science*, 286, 509–512.

Index